Sensor networks in theory and practice

Ansgar Meroth • Petre Sora

Sensor networks in theory and practice

Successfully realize embedded systems projects

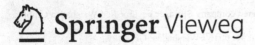

Ansgar Meroth
Fakultät für Mechanik und Elektronik
Hochschule Heilbronn
Heilbronn, Germany

Petre Sora
Fakultät für Mechanik und Elektronik
Hochschule Heilbronn
Heilbronn, Germany

ISBN 978-3-658-39575-9   ISBN 978-3-658-39576-6   (eBook)
https://doi.org/10.1007/978-3-658-39576-6

This book is a translation of the original German edition "Sensornetzwerke in Theorie und Praxis" by Meroth, Ansgar, published by Springer Fachmedien Wiesbaden GmbH in 2021. The translation was done with the help of artificial intelligence (machine translation by the service DeepL.com). A subsequent human revision was done primarily in terms of content, so that the book will read stylistically differently from a conventional translation. Springer Nature works continuously to further the development of tools for the production of books and on the related technologies to support the authors.

This Springer Vieweg imprint is published by the registered company Springer Fachmedien Wiesbaden GmbH, part of Springer Nature.
The registered company address is: Abraham-Lincoln-Str. 46, 65189 Wiesbaden, Germany

We dedicate this book to our families

Preface

The "Internet of things" is on everyone's lips. An important prerequisite for the digital revolution in everyday life is the availability of the smallest microcontrollers, which are connected and networked with sensors and actuators. Here, computing power plays a lesser role than power consumption and installation space. With the availability of miniaturized, highly precise and inexpensive sensors, the number of applications is growing rapidly, opening up previously unimagined possibilities for smaller entrepreneurs, students and ambitious hobbyists. At the same time, understanding how sensors work and how to control them is becoming increasingly difficult, especially since many sensors are designed by manufacturers to be integrated into networks. The high dynamics on the component market and the documentation, which is often insufficient for beginners to understand, require a great deal of training. In many cases, only the basic functions of the components are required, so this book contains very concrete instructions and explanations.

A well thought-out, modular control architecture enables a hardware-independent structure of the control hardware and software that can be easily adapted to the specific environment. Most examples are formulated in such a way that they are also suitable for other platforms and only the direct hardware connection needs to be replaced.

However, the book not only contains concrete programming instructions for the components used, but also many tips and tricks for the efficient use of the C programming language in microcontrollers. It is intended to stimulate further thinking and development.

The authors therefore welcome feedback, ideas and suggestions for future editions!

Heilbronn, Germany
November 2017

Ansgar Meroth
Petre Sora

Preface to the Second Edition

The development of the IoT is dynamic and causes content to age very quickly. In addition, we received some valuable tips for the further development of the book. The time until the first edition went to press was also limited, and many ideas could no longer be implemented. Therefore, we are happy to comply with the request for a second, considerably expanded and revised edition. In this process, the whole book has been restructured, and quite a few building blocks and also some algorithms have been added. Following the request of the readers, we have added notes on the functionality of the blocks. At the time of going to press, all blocks are available in stores.

We have found that readers are overwhelmed with the introductory chapter on the C programming language. This book is no substitute for a beginner's book in programming. Rather, a good knowledge of C is a prerequisite for understanding; Chapter 2 is only for quick reference. We refer the reader to the introduction, which further elaborates on the prerequisites for understanding and what to expect from reading this book.

Heilbronn, Germany
April 2021

Ansgar Meroth
Petre Sora

Contents

Introduction

Abstract

A brief overview of what to find in this book and how to use it.

1.1   What Do You Find in This Book?

The present work starts at the point where books about the functioning of sensors usually end. In addition to a basic section, in which the general programming in C is compactly presented, there is a concrete manual on the control of selected sensors. This is based on a standard architecture and is described down to source code level using the example of control with an AVR microcontroller, as used in many embedded systems.[1] The selection of sensors was made in such a way that they are as representative as possible of a wide range of components available on the market. Thus, the examples are suitable not only for reprogramming but also for step-by-step building of understanding and for transfer to your own applications with your own components. Because of the hardware abstraction used, the vast majority of examples in this book can be transferred to other processor families without much effort. With these considerations in mind, the book is divided into the following parts:

- A brief introduction to the C language, written more as a reference for quick reference. People who have never programmed in C should use the numerous tutorials and beginner courses available on the Internet to first practice and understand C

---

[1] This is also the case in the Arduino® family.

A. Meroth, P. Sora, *Sensor networks in theory and practice*,
https://doi.org/10.1007/978-3-658-39576-6_1

- A short introduction to the microprocessor. The examples in this book are without exception tested on processors of the ATmega8x series and generally also run on the other 8-bit AVR processors, in particular the authors use the AT90CAN128 in everyday life. The company Atmel has published an excellent, instructive manual, which is maintained after the takeover by Microchip, in addition, numerous authors describe this family, especially on the page www.mikrocontroller.net are detailed instructions and answers, with which the readers can deepen what is written here.
- A description of important software mechanisms that can be used in an embedded program – independent of the processor used. These are described independently of the hardware.
- Chapter on bus systems and networks, in which both the communication interfaces UART, SPI and I^2C (TWI)[2] of the processor and, more abstractly, the communication via selected bus systems (CAN, Modbus, LIN, OneWire, PSI5) and radio interfaces (ISM bands, Bluetooth, ZigBee, LoRa) are described. This allows a wide range of sensor networks to be set up.

The core of the book consists of various chapters describing selected sensors, displays and other networkable devices, for example flash data memories, port expanders, AD converters, variable resistors and radio controllers. The examples are introduced in a hardware-abstract manner with the described bus drivers and can therefore be ported to other processor families. For the vast majority of sensor types that can be used in everyday life, readers will find practical examples that reduce the often complex components to a few typical applications.

At the end there is a chapter in which a typical example project and its realization as a whole are shown. Table 1.1 shows the basic structure of the book for quick location of the individual chapters.

1.2   For Whom Is This Book Written?

Accordingly, the book is aimed at students in higher semesters who, for example, are looking for concrete solutions for networked sensor circuits as part of their project and research work, as well as at electronics developers with measurement tasks. However, the circuit examples and programming instructions are also suitable for ambitious amateurs with the appropriate basic skills.

---

[2] In the context of the AVR family, the term two-wire interface (TWI) is used. In the context described, TWI is equivalent to I^2C.

Table 1.1 Structure of the book

Part	Detail
1: Introduction and introduction to C	1 Introduction
	2 Short introduction to C
2: Programming in the AVR family	3 Structure and programming of AVR controllers
3: Hardware independent mechanisms	4 Software framework
	5 Memory concepts and algorithms
4: Communication	6 Communication – theory
4.1: Microcontroller interfaces	7 Asynchronous serial interfaces
	8 The SPI interface
	9 The I^2C interface
4.2: Bus systems	10 The CAN bus
	11 The Modbus
	12 Single-wire bus systems
	13 Wireless networks
5: Sensors	14 Sensor technology – system considerations
	15 Environmental sensors
	16 Acceleration sensors
	17 Rate of rotation sensors
	18 Magnetic field sensors
	19 Proximity sensors
6: Network devices in sensor networks	20 DA and AD converters
	21 Serial EEPROMS
	22 Serial flash memory
	23 Building blocks for audio technology
	24 Further networkable modules
7: Further considerations	25 Display devices
	26 Example projects

1.3   What Knowledge Does the Book Presuppose?

Although the book contains an overview of the C programming language and the physical measurement principles of the sensors used in short summaries, it is primarily intended for readers who already have programming knowledge in C and have mastered the basics of measurement technology. The books [1, 2], for example, are recommended for familiarization. English books can be found accordingly. At the end of each chapter there are also extensive references to literature for further study.

1.4   Why Does the Book Not Describe an Arduino?

Arduino is a fine and practical thing to quickly get to programming small embedded systems. However, some of the examples described in this book are very complex and are not supported by the Arduino library. In industrially deployed circuits, one will tend to save the memory required by the Arduino bootloader, and the serial ports. Of course, you will be able to run the examples with Arduino as well, and in this case you should replace the functions for running the serial ports with the Arduino functions. For this purpose, the Arduino library provides a complete hardware abstraction of the used interfaces UART, SPI, I^2C (TWI) and the port pins [3]. However, since this book is also intended to help understand the rudimentary operations in a microcontroller, it would not make sense to hide these through the Arduino library.

1.5   Additional Materials

The authors are constantly working on the programs described in the book. The source codes are therefore subject to a revision process. On the website http://www.springer.com/ de/ selected sources and schematics are also offered for download under the detailed description of the book.

1.6   Disclaimer

All circuit and code examples in this book have been created to the best of our knowledge and belief. Nevertheless, for the sake of readability and comprehensibility of the examples, they have only been constructed as prototypes. They are in no way suitable for safe and reliable code, since the authors have generally refrained from error catching or plausibility checks in order to avoid distracting attention from the actual function. Therefore, the circuits and code examples shown here may not be used in military or commercial products. Operation of the circuits and software described in the book can be potentially dangerous. The authors and the publisher disclaim any liability should circuit or code examples fail to function in operation or cause damage to property or persons.

This book was written by two people and cross-read by several others. Nevertheless, given the complexity of the subject matter, errors can creep in that no one notices. We ask that these be reported to the authors so that they can be corrected in the next edition. A post-print erratum will be made available on the publisher's website.

1.7   Thanks

This book has a longer prehistory and we would like to thank the colleagues in our research group and in the Faculty of Mechanics and Electronics, who accompanied us with ideas, suggestions or tips up to this book. In particular, we thank Petre Sora jr. and Nico Sußmann, as well as our esteemed colleague em. Prof. Dr.-Ing. Wolfgang Wehl, who read the book with great care and gave valuable hints. Our wives are both engineers themselves and are very familiar with the subject matter. We thank them for many years of companionship, support and understanding.

References

1. Hering, E., & Schönfelder, G. (Hrsg.). (2018). *Sensoren in Wissenschaft und Technik* (2. Aufl.). Vieweg & Teubner.
2. Küveler, G., & Schwoch, D. (2006). *Informatik für Ingenieure und Naturwissenschaftler 1 – Grundlagen Programmierung mit C/C++ – Großes C/C++ Praktikum* (5. vollständig überarbeitete und aktualisierte Aufl.). Vieweg.
3. Arduino: Sprachreferenz. https://www.arduino.cc/en/Reference/HomePage. Accessed 19 Aug 2020.

Introduction to the Programming Language C

Abstract

In this chapter a short overview of the programming language C is given, as far as it is necessary for the understanding of the rest of the book and especially the programming examples.

This chapter is deliberately kept in key words and does not replace the study of a C course for programming beginners. Rather, it serves as a reference aid. First, key words and symbols in C are listed with the respective reference to the corresponding explanations, then the basic concepts: Variables and constants, loops, branches and functions, complex data types and pointers are explained, and at the end the basic structure of a C program is described. This introduction deliberately avoids a complete presentation of the language; for example, you will not find any syntax diagrams here. For that, please refer to the relevant literature. Instead, you will find tips for the correct use of the C language in the target system AVR microcontroller.

2.1   The C Language: Background and Structure

The programming language C was developed in 1971 in the Bell Laboratories of the company AT&T by Dennis M. Ritchie as a revision of the language B with the goal of rewriting the operating system UNIX in a machine-independent way. Ken Thompson had developed this in 1968 in beginnings in assembly language. Then in 1973 Thompson and Ritchie rewrote UNIX in C. AT&T made the source codes publicly available and thus achieved a rapid spread of UNIX and C. C, which was initially specified on only a few

A. Meroth, P. Sora, *Sensor networks in theory and practice*,
https://doi.org/10.1007/978-3-658-39576-6_2

pages, is now an international standard (ANSI-C) that has evolved in several steps (ISO/IEC 9899) until currently 2018 (C18) [1]. Bjarne Stroustrup further developed C into an object-oriented programming language C++, which has been codified in an ISO standard since 1998 (ISO/IEC 14.882:2014) [2]. UNIX has been increasingly commercialized over the years, and through the open-source UNIX derivative LINUX by then-student Linus Thorvalds, it has gained new popularity since 1991, especially on ×86 platforms. C is currently the world's most widely used programming language for embedded systems, not least because it is designed to be very close to the hardware and therefore leads to efficient machine code. Since work is being done in many places on the revision of LINUX for functionally secure real-time systems, it may be expected that the operating system will also become even more widespread.

The development of software for embedded systems, however, is nowadays driven in model-based form. Modeling languages such as Matlab/Simulink from Mathworks generate C code that is so efficient that it can be run directly in embedded environments [3].

C is an imperative programming language. Programs are processed from a starting point command by command, so-called control structures (branches, loops) are used to control the sequence, absolute jump instructions (goto Sect. 2.7.4) are permitted in C in principle, but should be avoided if possible. The structured/procedural programming style (paradigm) of the C language does without these jumps.

To make C programs clearer, they are structured in functions and modules. More about this later.

One of the strengths of C is its direct access to the hardware. Thus, the common processors have an instruction set that allows efficient translation of C code into machine code. This work is done by the compiler, which is usually called from an integrated development environment, abbreviated IDE (Integrated Development Environment). The following example is intended to illustrate this:

Example

The operation a = c + 7; assigns to the variable a the value of the sum of the value of the variable c and the constant 7. Let each of the variables used be of type integer (16 bits = 2 bytes). After translation, the machine code is:

```
ldd    r24, Y + 1    ;Load from memory the low byte of c
ldd    r25, Y + 2    ;Load from memory the higher byte of c
adiw   r24, 0x07     ;Add to the lower byte of c the number 7
std    Y + 6, r25    ;Store result (high byte) to the memory location of a
std    Y + 5, r24    ;Store result (low byte) to the memory location of a ◄
```

Many elementary operations in C can thus be translated directly into machine code. For more complex data types and operations that are not directly supported by the processor, it gets more complicated. Examples of this will be given later.

The whole process of translating C sources can be found in Fig. 2.1.

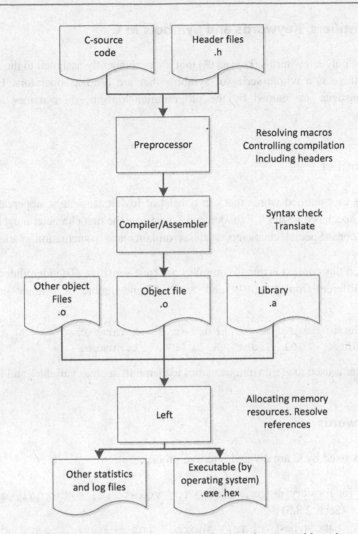

Fig. 2.1 Sequence of the translation of a C source code up to the executable code

Header and C files (modules) form the sources (.c .cpp .h). Modules that have already been translated can be subsequently included as libraries (.a) or binary files (.o). Individual steps of the translation process will be discussed later.

The following description of the C language is a very abbreviated summary of all the possibilities that can be used in C. For an intensive study of the language, one of the numerous available online tutorials (for example [4]) or books [5–10] should be used.

2.2   Identifiers, Keywords and Symbols in C

In C, there are only a few terms (keywords) that are permanently assigned to the language. In addition, there is a whole series of symbols that are used as operators. Functions, variables, constants are named by the programmer himself, their names are called *identifiers*.

2.2.1 Identifier

Identifiers are user-defined words that can consist of lowercase letters, uppercase letters, digits and the special character "_" (underscore), whereby the first character must be a letter or an underscore. Special characters such as umlauts and punctuation marks are not allowed.

Important in this context is that C compilers are case-sensitive. The identifier "value" is therefore different from "VALUE" and "value" and denotes a different variable or function.

Valid identifiers: umu_foo   Number_5   n_number
Invalid identifier: 1001_Night   B_or_C?!   1.number

An identifier is used to assign unique names to elements such as variables and functions.

2.2.2 Keywords

The keywords fixed by C are divided into five groups:

1. Keywords for memory usage: automatic, volatile, external, static, register (Sect. 2.8.5)
2. Elementary data types: char, short, int, long, float, double, unsigned, (const) (Sect. 2.3)
3. Control structures if, else, switch, case, break, continue, default, for, do, while, return, goto (Sect. 2.6)
4. Complex data types: enum, struct, union (Sect. 2.9)
5. Other keywords: void (Sect. 2.8), typedef (Sect. 2.9.2), sizeof (Sect. 2.9)

2.2.3 Symbols

C uses a whole set of fixed symbols. These are for example:

1. Arithmetic/logic operators: + − * / % < > ! ~ ^ & | = ( ) and combinations of these
2. Symbols for structuring and controlling the program: { } ; // /* */ : ?
3. Symbols for accessing complex data structures and pointers: & * [ ] . − >

Obviously, symbols occur twice here, that is, their meaning depends on the linguistic context in which they are used.

2.2.4 Instructions

The C language consists of a sequence of *statements*, which we can interpret as actions to be executed. Statements can be declarations, expression statements, jump statements, branches, loops, and so on. Sequences of statements are also statements in C (block statement). These are grouped together by curly brackets.

Example

The statement.

```
c = a * (b + c + 7);
```

assigns the value of the calculation to the variable c. The assignment forms an *expression*, the right side, the calculation forms an expression, which takes a value depending on the value of the variables a, b and c and in turn consists of expressions. So, abstractly, expressions can be said to consist of expressions and/or variables composed using operators.

Expression statements always end with a semicolon (;). ◀

2.3   Comments

To improve the understanding of a program listing, it is recommended to insert comments directly into the source code. These comments should describe the functionality in more detail and thus facilitate troubleshooting or modification (inline documentation).

2.3.1   Single-Line Comment

```
//this is a comment
```

The one-line comment starts with the double slash "//" and ends automatically at the end of the line.

2.3.2 Multiline Comment

```
/* This comment can
be as long as he wants */
```

In a multiline comment, the beginning is marked by "/*" and the end by "*/".

2.3.3 Doxygen Comments

Special keywords in comments can be used by the free software tool *Doxygen* to create an automatic documentation of the source code.

A typical Doxygen file comment (at the beginning of a module file) might look like this:

```
/*!
    \file       ad_basic.c
    \brief      functions for reading the ADC converter.\n
    \see        Defines for the hardware abstraction in hardware.h
    \author     Ansgar Meroth
    \version    V1.0 - 1.10.2020 Ansgar Meroth - HHN\n
*/
```

Each function can then still be described, for example, as follows:

```
/*!
    \brief
    Returns the temperature of the selected sensor
    \return
    8 bit value of the temperature (uint8_t) between -40 and 87.5°C.
    The temperature is calculated from (value - 40)/2
    \par input parameter: uint8_t number of the temperature sensor
    \par example:
    (to demonstrate the parameters/return values)
    \code
#define TEMPSENS0 0
#define TEMPSENS1 1
uint8_t result;
result = ucTemperatureGet(TEMPSENS0);
    \end code
**********************************************************************/
```

Doxygen generates very readable files in Rich Text Format (RTF – Word), HTML, LaTex, XML, UNIX manpages etc. from these comments and from the analysis of the code and thus facilitates the documentation of the entire code. Thereby a disciplined commenting is enforced, which also increases the readability in the code. However, Doxygen can already generate documentation from uncommented code. The following Fig. 2.2 shows how Doxygen generates a rich-text file (MS Word) from the comment examples shown above.

With Doxygen, which is available from the developer [11] and the also free GraphViz library with the tool dot [12], the software also generates call diagrams and include hierarchies, so that the created software is elegantly and completely documented. In C++, further UML diagrams, such as class diagrams, are also generated.

2.4   Types, Variables and Constants

The concept of variables is fundamental to any programming language. Variables are symbolic placeholders for data. In their use, constants (immutable) and variables (mutable) differ only in that a value can be assigned to a variable by placing it to the left before an equal sign (equivalent to the assignment operator, Sect. 2.5). Variables and constants need a type so that the compiler can organize the appropriate memory space.

2.4.1   Fundamental Data Types

Each variable has a *data type*. This determines the amount of memory and the way in which operators must be applied.

The strength of C is that new data types can be generated in various ways. A few types are already predefined, they are called *fundamental*. We distinguish among the fundamental data types the *integer types* and the *floating-point types*. A logical data type has existed since the ANSI standard C99 as `_Bool`. In this book, however, we always use `unsigned char` for logical expressions for backward compatibility.

For integer types, the prefix *unsigned* can be used. This shifts the value range into the number range ≥ 0 (see Table 2.1). The basis for this is the two's complement representation, which ensures that any mathematical operation can be mapped within the value range. The negative number is formed from the bitwise negation of the positive number plus a binary 1. That is, in 8-bit representation, the -1 would be represented by 0000 0001 complementary which is 1111 1110 + 1 = 1111 1111. The -128 is 1000 0000 complementary 0111 1111 plus 1 = 1000 0000. The zero thus remains a zero and $-1 + 1$ also results in a zero, since there is no overflow across the 8-bit boundary.

The word length of `int` is not defined. If defined word lengths are important and the code should be portable, `int` should be avoided as a data type.

Floating point numbers consist of a mantissa m and an exponent e (see Table 2.2). In the case of the number $1.345 \cdot 10,^3$ 1.345 is the mantissa and 3 is the exponent. 10 is the radix r, so $r = m * b^e$ is a floating point number. Note the international notation using the dot instead of the German comma as the decimal point! In the C standard, the internal use of floating-

File documentation

ad_basic.c file reference

Functions for reading the ADC converter.

```
#include <avr/io.h>
```

Functions

- unsigned char **temp** (unsigned char sens_no)
  Returns the temperature of the selected sensor.
- int **main** (void)

Detailed description

Functions for reading the ADC converter.

See also:

Defines for the hardware abstraction in hardware. h

Author:

Ansgar Meroth

Version:

V1.0 - 1.10.2015 Ansgar Meroth - HHN

Documentation of the functions

unsigned char temp (unsigned char *sens_no*)

Returns the temperature of the selected sensor.

Return:

8 bit value of the temperature (unsigned cliar) between -40 and 87.5°C. The temperature is calculated from (value - 40)/2

Input parameter: unsigned char Number of the temperature sensor

Example:

(to demonstrate the parameters/return values)
```
1 #define TEMPSENS0 0
2 #define TEMPSENS1 1;
3 unsigned char result;
4 result = ucTemperatureGet(TEMPSENS0);
```

3

Fig. 2.2 Example of a Doxygen output as Word (RTF) file

Table 2.1 Integer data types

Name	Size	Value range (signed)	Value range (unsigned)
char	1 byte (8 bit)	$-128 \ldots +127$	$0 \ldots 255$
short	2 byte (16 bit)	$-32{,}768 \ldots +32{,}767$	$0 \ldots 65{,}535$
int	Word length of the system with ATMega: 2 bytes, 16 bits	$-32{,}768 \ldots +32{,}767$	$0 \ldots 65{,}535$
long	4 byte (32 Bit)	$-2{,}147{,}483{,}648$ $\ldots +2{,}147{,}483{,}647$	$0 \ldots 4{,}294{,}967{,}295$
long long	8 byte (64 bit)	$-2^{63} \ldots +2^{63}-1$ $-9{,}223{,}372{,}036{,}854{,}775{,}808$ $\ldots +9{,}223{,}372{,}036{,}854{,}775{,}807$	$0 \ldots 2^{64}-1$

Table 2.2 Floating point types

Name	Size	Typical value ranges
float	4 byte (32 bit)	$\pm 1.175494351 * 10^{-38} \ldots 3.402823466 * 10^{38}$
double	8 byte (64 bit)	$\pm 2.2250738585072014 * 10^{-308} \ldots$ $1.7976931348623158 * 10^{308}$
long double	$8 \ldots 16$ byte (128 bit)	$\pm 3.3621031431120935062677817321752602598 * 10^{-4932}$ $\ldots$ $1.18973149535723176502126385303097 0205169 * 10^{4932}$

point numbers is largely kept free. For example, for float, one bit is reserved for the sign, eight bits for the exponent to base 2 in *signed notation* (two's complement), and 23 bits for the mantissa as a binary number. So in a very low-precision format, 1 would be $1.000 \cdot 2^{0}$, 2 would be $1.000 \cdot 2^{1}$, and 0.375 would be $1.100 \cdot 2^{-2}$, where the first 1 after the decimal point counts as 1/2, the second as 1/4, and so on. For a properly scaled (or normalized) floating-point number on base 2, the digit before the decimal point is always 1. For this reason, it is usually omitted (although this requires a special representation for the 0). This then results in somewhat "crooked" ranges of values.

In general, the floating point types are implementation-dependent in their size; the compilers provide macros through which the limits of the value ranges can be retrieved (e.g. LLONG_MIN, LLONG_MAX or DBL_MIN, DBL_MAX). These can be found in the compiler documentation. The software presented in this book does not use floating-point numbers at all. They are generally not recommended in small system environments because many embedded processors, including the 8-bit AVR family, do not have a floating point unit and arithmetic expressions lead to significant code overhead. For example, the

following operation leads to highly different assembler code with a, b, c of type
unsigned char.

```
      c=a*b;
17c:  90 91 1a 01   lds   r25, 0x011A
180:  80 91 0a 01   lds   r24, 0x010A
184:  98 9f         mul   r25, r24
186:  80 2d         mov   r24, r0
```

Whereas with x, y, z of type float....

```
      z=x*y;
136:  80 91 0b 01   lds    r24, 0x010B
13a:  90 91 0c 01   lds    r25, 0x010C
13e:  a0 91 0d 01   lds    r26, 0x010D
142:  b0 91 0e 01   lds    r27, 0x010E
146:  20 91 16 01   lds    r18, 0x0116
14a:  30 91 17 01   lds    r19, 0x0117
14e:  40 91 18 01   lds    r20, 0x0118
152:  50 91 19 01   lds    r21, 0x0119
156:  bc 01         movw   r22, r24
158:  cd 01         movw   r24, r26
15a:  57 d3         rcall  .+1710        ; 0x80a <__mulsf3>
15c:  dc 01         movw   r26, r24
15e:  cb 01         movw   r24, r22
160:  80 93 0f 01   sts    0x010F, r24
164:  90 93 10 01   sts    0x0110, r25
168:  a0 93 11 01   sts    0x0111, r26
16c:  b0 93 12 01   sts    0x0112, r27
```

... much more code is produced and in adress 0x15A a function is called, which is also
copied into the program memory and swallows another 488 bytes there.

▶   In the embedded-C environment on small microcontrollers, floating point types
      should not be used. Further considerations are given in Sect. 6.1.6.

In C, there are no separate data types for printable characters or character strings.
Whether a character is printable or not depends on the interpretation of the printing function
or the output unit. If characters are represented in an 8-bit code (for example, an ASCII
code),[1] character variables are declared as char. Character strings are arrays of char
(Sect. 2.9.1).

---

[1] The ASCII code is a 7-bit code, but various conventions extend this to an 8-bit code.

At this point, it should be noted that many C compilers use synonyms for the fundamental data types; these are catchier to remember and have become widely accepted. The Gnu-C compiler uses the following naming conventions, among others:

```
typedef unsigned char uint8_t;
typedef unsigned short uint16_t;
typedef unsigned int uint16_t;
```

2.4.2 Declaration of Variables

In C, the principle of "Declare before Use" applies. All elements must be declared, i.e. introduced, before they can be used. The data type and the name of a variable must be declared by entering the data type and then the variable name. This defines the value ranges and the memory usage.

Examples

```
int iNumber;
```

Declares a variable of the data type int and the name iNumber. The variable iNumber can therefore be assigned an integer value from this declaration. The prefix i can be used to remember the value range.

```
double rNumber;
```

Declares a variable of the data type double and the name rNumber. ◄

Each declaration line is terminated with a semicolon. If several variables have the same data type, they can simply be written as a list with a separating comma.

The following line declares two variables of the data type int and the names number1 and number2.

```
int number1, number2;
```

No values are assigned at the declaration itself, so it is not clear what value the variable contains at that time.

If a value is already assigned to the variable when it is declared, this is referred to as initialization.

Examples:

```
unsigned int number = 47;
```

declares a variable of the data type `unsigned int` and the name `number`. The variable is immediately assigned the value 47. ◄

2.4.3   Constants

Constants are divided into *literal constants* (literals) and *symbolic constants*. Literals are the representation of basic types, for example:

- Decimal integer constants e.g. 123 or −14
- Hexadecimal integer constants: These are prefixed with 0x. The number 0x10 thus corresponds to the decimal 16. The number 0xF corresponds to the decimal 15.
- Binary integer constant 0b01101010 (equivalent to decimal 106 or hexadecimal 0x6A)
- Floating point numbers e.g. 1.234 or 14.3e10 or − 12e-3 (the e indicates the exponent to the respective radix)
- Character constants e.g. 'a' (in single quotation marks, are interpreted as `char`)
- String constants e.g. "This is a text" (in double inverted commas, automatically terminated by the compiler with a binary 0)
- Special case of character constants are control sequences that are headed with a '\' (backslash), like '\n' for line break, '\t' for a tab jump or '\' for a backslash. Note that these are not strings, but single characters!

Literal constants can be assigned types by a postfix, for example:
```
129u (unsigned)
123987L (long)
987ul (unsigned long)
-24.8E10 (double)
3.14159f (float)
1.414213562373095L (long double)
```

Literal constants that represent a number are also called numeric constants.

Symbolic constants are named by identifiers. At compile time or runtime, depending on the language and usage, these identifiers are replaced by their value. In earlier C versions, there was no separate construct for defining symbolic constants. Instead, the so-called *precompiler statement* #define was used. Before the compiler executes, the *precompiler* loads and evaluates the statements marked with a # gate. By

```
#define NAME expression
```

NAME is assigned to an expression and each occurrence of NAME is replaced by the expression before the compiler run. NAME is also called a *macro,* and you can build powerful constructs with it. Examples:

```
#define PI 3.1415926535
#define length 32
#define ENDLESS while(1)
```

The replacement text can extend over several lines, the line break must then be indicated with a \ (backslash) to include the contents of the subsequent line in the statement, since precompiler statements do not know a terminating character other than the line break. For more on macros, see Sect. 2.11.

Nowadays, compilers use the keyword (type qualifier) const to indicate to the compiler that a variable may no longer be assigned a value at runtime.

```
const int Length = 32;
const double PI = 3.1415926535;
```

When a constant is declared with const, initialization must occur immediately. It is important to understand that although a variable declared with const can no longer be assigned a value during the course of the program, it can still change its value in some cases, for example if it is a register or the memory location is overwritten by reference.

► In general, the use of numeric constants should be avoided in the source code and instead symbolic constants should be used instead, this applies in particular to field lengths, physical constants and other constant expressions.

2.5 Operators

Operators are used to link one or two *operands* with each other. Operators with one operand are called *unary operators,* those with two operands are called *binary operators* (this has nothing to do with the binary system!). C uses the infix notation, i.e. binary operators are placed between the two operands (e.g. a + b). The left operand is called *lvalue*, the right operand *rvalue*. Operands can be complicated, i.e. they can be composed. In general, we speak of an *expression* as a linguistic construct that yields a value as a result after its evaluation.

An important operator is the *assignment operator* (Table 2.3):

Important operators for various tasks are described below:

Table 2.3 Assignment operator

Operator	Description	Note
a = b	The assignment operator assigns the value of the rvalue to the lvalue, the latter can consist of a complex expression, the former must be a variable	

Table 2.4 Arithmetic operators

Operator	Description	Note
i++	The post-increment operator (++) increments a variable by the value 1 **after** the entire statement has been processed.	Similar to i = i + 1
i−−	The post-decrement operator (−−) decrements a variable by the value 1 **after** the entire statement has been processed.	Similar to i = i − 1
++i	The pre-increment operator (++) increments a variable by the value 1 **before** the entire statement is processed.	
−−i	The pre-decrement operator (−−) decrements a variable by the value 1 **before** the entire statement is processed.	
a + b	Addition (+)	
a − b	Subtraction (−)	
a*b	Multiplication (*)	
a/b	Division (/)	
a%b	Modulo division (%)	Remainder from an integer division

2.5.1 Arithmetic Operators

Arithmetic operators are applied to numbers. Table 2.4 gives an overview.

2.5.2 Logical Operators

Logical operators are applied to boolean expressions. Since in C originally no logical data type exists which could take the values true or false, the following applies: false corresponds to the value 0 (zero), true corresponds to any other value not equal to 0. Since the ANSI standard C99 a data type _Bool exists. However, in the standard library *stdbool.h,* a macro bool is defined, as well as the macros true and false. The logical operators are listed in Table 2.5.

2.5.3 Bit Operators

Bit operators (Table 2.6) in C allow the manipulation of individual data bits within an expression. To understand these operators, knowledge of Boolean algebra is required, a

Table 2.5 Logical operators

Operator	Description	Note
! a	Logical not	returns 1^a if a $= 0$ or 0 if a $\neq 0$
a && b	AND operation logical	returns 1 if a and b are different from 0, otherwise 0
a || b	OR operation logical	returns 1 if a or b is different from 0, otherwise 0
a < b	Small	returns 1 if a is less than b otherwise 0
a > b	Larger	returns 1 if a is greater than b, otherwise 0
a <= b	Smaller or equal	returns 1 if a is less than or equal to b otherwise 0
a >= b	Greater or equal	returns 1 if a is greater than or equal to b otherwise 0
a == b	Equal (comparison operator, attention! Two equal signs)	returns 1 if a equals b, otherwise 0
a != b	Unbalanced	returns 1 if a is not equal to b, otherwise 0

[a]In the following, the one is to be understood in the sense of "not equal to zero"

Table 2.6 Bit operators

Operator	Description	Note
a >> n	Shift right	a is shifted to the right by n bits (corresponds to a division by 2^n)
a << n	Shift left	a is shifted n bits to the left (corresponds to a multiplication by 2^n)
~a	Bitwise negation	Bitwise negation of a
a & b	Bitwise AND operation	Bitwise AND of a and b
a | b	Bitwise OR operation	Bitwise OR of a and b
a ^ b	Bitwise XOR (Exclusive OR)	Bitwise XOR of a and b

look in the corresponding search engines can help those who have not yet dealt with it. In the chapter about AVR programming some examples can be found.

2.5.4   Operators for Memory Accesses

Table 2.7 lists the operators for memory accesses. These are used for pointers and pointer arithmetic. See Sect. 2.9.5 for more details.

Table 2.7 Operators for memory accesses

Operator	Description	Note
&a	Address operator: returns the memory address of a	Sect. 2.9.5
*a	Derefencing operator: If a is the address of a variable, *a returns the value	Sect. 2.9.5
sizeof (a)	Returns the size in bytes that a occupies in memory.	No function! An exception in C
.	Access operator. This is used to access individual elements in structures and unions.	Sect. 2.9.5
->	Access operator, if access is to be made via a pointer	Sect. 2.9.5
[]	Indexing operator. Given an array a, a[i] returns the object at position i	Sect. 2.9.1

Table 2.8 Further operators

Operator	Description	Note
()	Type conversion operator: (typ)a returns the value of a in the data type specified by typ	Sect. 2.3
()	Function call: fname() calls the function named fname	Sect. 2.8
()	Parenthesis: Expressions in parentheses are evaluated before the rest of the expression is evaluated.	
,	Comma operator: Combines two independent expressions in one statement	
?:	Condition operator: condition ? expression1 : Expression2 If the condition is fulfilled, expression 1 is evaluated, otherwise expression 2 – the only ternary operator in C.	Sect. 2.6.1

2.5.5 Other Operators

The operators listed in Table 2.8 are further described in the sections listed in the table.

2.5.6 Associativity and Priority of Operators

According to the well-known "dot-before-dash rule" from mathematics, there are also priorities in C for operators. The higher an operator is in Table 2.9, the higher its priority. It is recommended to bracket expressions if possible.

Table 2.9 Associativity of operators

Art	Associativity	Operators
	Left to right	() [] –>
Unär	Right to left	~ ! + + – – sizeof (type)
Unär		– * (Reference) &(Address)
Binary	Left to right	* (mult)/(div) % (modulo)
Binary	Left to right	+ (plus) – (minus)
Binary	Left to right	<< >> (Shift)
Binary	Left to right	< <= > >= (logical)
Binary	Left to right	== != (Logical)
Binary	Left to right	& (bitwise)
Binary	Left to right	^ (bitwise)
Binary	Left to right	| (bitwise)
Binary	Left to right	&& (logical)
Binary	Left to right	|| (logical)
Binary	Right to left	?:
Binary	Right to left	= += –= *= /= %= etc.
Binary	Left to right	, (comma operator)

2.5.7 Type Conversion

C requires the knowledge of a data type to be able to perform an operation meaningfully. Therefore, variables are always declared. However, sometimes it is necessary to mix between different data types, for example by multiplying an integer (char, integer, short, long) by a fixed-point number (double, float). The following code example shows this:

```
int a = 5;
double x = 1.5;
double result;
result = a* x;
```

In order to execute this instruction, the compiler must first cast all variables involved in the operation to a common data type. This is done by the so-called *implicit cast*. The rule is that the "weaker" type is converted to the stronger one. In this case, a is first interpreted as double and then the operation * is executed. The result is then of type double.

Implicit type conversion can also involve pitfalls. The following example makes it clear:

```
int a = 1;
int b = 3;
double result;
result = a/ b;
```

The result of this operation, which is stored in result, is 0.0! The reason is simple. The two integer operands are divided integer, result is 0, because decimal places are truncated in integer operations. Then, only during assignment, the result is implicitly converted to double. This is not really a surprise.

To protect against implicit type conversion, C has *explicit type conversion*. Here, the target type is specified in round brackets, for example:

```
int a = 2;
int b = 3;
double result;
result = (double)a / (double)b;
```

Here a and b are explicitly converted to double and then correctly divided. In principle, it would have been sufficient to convert only a or b, the other operand would then have been converted implicitly. Explicit type conversions can be performed in an extending way (e.g. int → double) or in a restricting way (e.g. double → int), in which case information is lost.

2.6   Control Structures

Control structures are used to control the program flow beyond the simple sequence. C knows the branches and the loops as control structures.

2.6.1   Branches (Selection)

The branch is also called a conditional statement, *if* statement or selection. A **condition** represents an expression of the data type integer. The expression is calculated and interpreted as a truth expression, i.e. if it supplies a value other than 0, the statement is executed, otherwise it is skipped and, if necessary, the alternative branch is passed through.

For example, a branch looks like this:

```
if (condition)
{
    Statement_1;
    ...
    Statement_n;
}
else
{
    Statement_1;
    ...
    Statement_n;
}
```

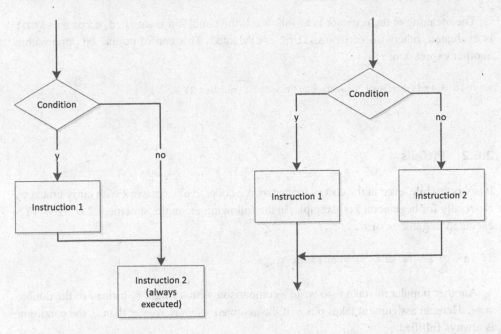

Fig. 2.3 Program flow chart for a branch without (left) and with else (right)

The first statement block is only processed if the condition is fulfilled. The lower block is only processed if the condition is NOT fulfilled. This second block, which opens with *else*, can also be omitted. Any expressions that return 0 or not 0 can be used as conditions, for example.

```
(a > 0)
(value == 128)
(4 * a != b)
```

If only single statements are provided in the alternatives, the brackets can be omitted:

```
if (condition) statement_1;
else statement_2;
```

Figure 2.3 shows the branching as a program flow chart according to DIN 66001 once with and once without alternative.

An elegant alternative to simple branching is the *condition operator*. It consists of only one line (condition ? expression1: expression2), as shown in the following example:

```
return (a < b) ? a : b;
```

The meaning of the operator is as follows: If the condition is satisfied, `expression1` is evaluated, otherwise `expression2` is evaluated. This can of course be done within another expression:

```
result = 4*((number1 > number2)?number1:number2) + 9);
```

2.6.2 Pitfalls

It is worthwhile, even at the cost of longer source code, to always work with curly brackets, especially for beginners! For example, in the following example, statement 2 is ALWAYS executed, regardless of a.

```
if (a < 0) statement_1; statement_2;
```

Another popular mistake is to write a comparison with a single = instead of the double ==. Here an assignment takes place, if the assigned value is greater than 0, the condition is always fulfilled:

```
if (a = 10)
{
    Statement_1;
}
```

results in statement 1 always being executed!

It is better (and mandatory in safe code) to write the constant on the left when comparing, this way the compiler would detect an erroneous notation in any case and abort:

```
if (10 == a) statement_1;
```

2.6.3 Multiple Branching

Branches can also be combined with each other in such a way that, in effect, a multiple selection results. Example:

```
if (a < 0)
{
    Statement_block_1;
}
else if (a < 5)
```

```
{
    Statement_block_2;
}
else if (a < 10)
{
    Statement_block_3;
}
else
{
    Statement_block_4;
}
```

Statement block 1 is only executed if a- < 0. If a ≥ 0 (and only then!) the next condition a < 5 is queried. Only if no condition is met, in this case a ≥ 10, statement block 4 is queried. For such multiple branches, you should always use an unconditional *else* to avoid creating uncovered cases.

As an alternative to the examples shown above, there is the *switch statement*. The expression specified in the brackets, which is used to select the statement block in the *switch statement*, must be of the integer data type. If the expression matches the value 1, statement block 1 is executed. If the expression takes the value 2, then statement block 2 is processed, and so on. If the expression does not match any value, then the *default statement block* is executed. If the default *statement block* is missing, then the entire block is skipped if there is no match.

```
switch (expression)
{
    case value_1: statement_1; break;
    case value_2: statement_2; break;
    ...
    case value_n: statement_n; break;
    default: statement_n+1;
}
```

The keyword *case* serves here as a jump label, in other words: After evaluating the expression, processing continues at the point where there is a corresponding value after the *case*. However, there are pitfalls to be considered. The following example illustrates this:

```
switch(a)
{
    case 1: statement_1; break;
    case 5: statement_2;
    case 10: statement_3;
    default: statement_4;
}
```

If a = 1 before the switch statement, statement_1 is executed, then the statement block is exited. If, on the other hand, a = 5, statement_2, statement_3 and statement_4 are executed one after the other; if a = 10, statement_3 and statement_4 are executed. If a were equal to any other number, only statement_4 would be executed (*default*). The reason for this behavior is the missing *break* after the jump marks at 5 and 10. Basically, the switch statement should be used with care!

2.7   Loops

Loops are used in programming to execute a program section several times as long as a condition is fulfilled. A distinction is made between loops that are executed at least once and thus check the condition only at the end of the loop body, and those that decide right at the beginning of the loop body whether the loop is executed at all.

2.7.1   Head Controlled Loops

The statement block (loop body) of the head-controlled loop or while loop is executed as long as the condition in the loop head is fulfilled. However, this loop is not executed at all if the loop header condition is not fulfilled at the beginning.

```
while(condition)
{
    Statement_1;
    ...
    Statement_n;
}
```

The condition expression can be any expression that can take the value 0 (not fulfilled) or not 0 (fulfilled), for example a comparison, or an expression that is composed of Boolean combinations of these expressions. Head controlled loops are suitable whenever failure to satisfy the condition would produce a failure in the loop body. A typical application is to prevent a memory object accessed in the loop body from not existing and to query this in the condition. More on this later. It is important to note that the condition must change as the loop progresses, otherwise it is an endless loop:

```
while(1)
{
    Instruction;
}
```

In this case, the loop is never terminated. Thus, in a non-endless loop, the loop body always contains an iteration statement or the determination of a value that is used in the condition expression.

In principle, any loop can also be terminated from within the loop body using the *break* statement, but this should only be used with extreme caution in the interests of program clarity.

2.7.2 Foot-Controlled Loops

Foot-controlled loops or *do-while loops* are used whenever the statement block (loop body) must be passed at least once, for example, to query a condition. The instructions are repeated after execution once as long as the condition queried in the foot of the loop is fulfilled.

```
do
{
    Statement_1;
    ...
    Statement_n;
} while(condition);
```

The flowcharts for header and footer controlled loops are shown in Fig. 2.4.

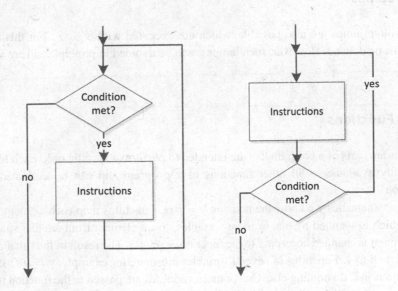

Fig. 2.4 Head-controlled (left) and foot-controlled (right) loops in the flowchart

2.7.3 Counting Loops

A counting loop or *for-loop* is used if the number of loop passes is known at the beginning of the program design. In principle, this is a head-controlled loop whose iteration expression is in the loop head.

```
for(start condition; run condition; iteration statement)
{
    Statement_1;
    ...
    Statement_n;
}
```

The counting loop can also be executed endlessly, as shown in the following code example:

```
for( ; ; )
{
    Statement_1;
    ...
    Statement_n;
}
```

2.7.4 Jumps

In C absolute jumps are also possible, which are executed with a *goto*. For this a jump mark must be defined. However, such jumps are to be avoided in principle and are also not required.

2.8 Functions

Functions are parts of a program that are intended to perform a specific task. Each function is basically available to all other functions of a program and can be evaluated in an expression.

From mathematics we know the notation $y = f(x)$. Functions map each element from a space, which is spanned by one or more variables, to an element from another space, the input element is changed according to the function, e.g. (x^2). The result of the calculation is then assigned to y. Functions of several variables are then, for example, $y = f(a,b)$;

Functions in C do nothing else. One or more variables are passed to the function (further called parameter(s)), handled there according to the statement, and then the result value (return parameter) is evaluated so that it can be assigned to another variable, for example. In contrast to mathematics, not only numbers can be processed in C, but also other objects. Therefore, it is important to define the data type of the pass and return parameters precisely. Of course, you can also use functions in expressions without assignment, for example, let the function sqrt (x) be of type double and calculate the square root of a number x, then the function can be used in an expression.

```
y = a * sqrt (x) + 3;
```

or even directly as a passing parameter for another function:

```
printf (sqrt (x * x + y * y));
```

Basically, C works according to the principle of **call by value,** which means that not the variable itself is passed, but only its value. A called function can therefore usually not change the original variable.

2.8.1 int main()

main () is the most important function of a C program. It is the starting and ending point of the execution. It also does not have to be declared, its name and function are fixed from the beginning. main () is of type *int*, so it returns a value.

2.8.2 Definition and Declaration

When developing your own functions, the following scheme should be observed;
The function prototype is used to *declare* each function. In doing so, the function is formally introduced. The principle of **Declare before Use** also applies to functions. Therefore, the function prototype should also be at the beginning of the source code.
The form of a functional prototype can be seen in the following example:

```
double length (double x, double y);
void wait (int t);
```

The length () function has two input parameters x and y. Its type is double, in other words it takes a floating point value after it has been executed. wait (), on the other hand, is of type *void*, and does not return any value, so it cannot be used in an expression, but is used to bundle program parts that are needed again and again.
The function can only be used after the declaration. It does not have to be defined yet. The actual function is not assigned to the call until it is linked; it can then come from

another source, usually a library. The definition is the actual programming of the function as a sequence of statements and control structures. These are grouped together by curly brackets { } (block brackets). At the end, the *return* statement determines what value is then returned. The definition of the function can be anywhere in the source code.

So with the sqrt () function already described above, length () could be executed as follows:

```
/**************************************************/
/*         Returns the length of a vector        */
/**************************************************/
double length(double x, double y)
{
    double result;
    result = sqrt(x * x + y * y);
    return result ;
}
```

The keyword *return* cancels the function and returns the expression that follows *return*. This can be of a complex nature. Functions of type *void* (empty) do not need a *return*.

Generally speaking, then:

```
Memory class Data type Function name (parameter list)
{
    Declarations of the local variables
    Statements
}
```

The memory class is explained in Sect. 2.8.5.

C generally makes it possible to declare and define a function at the same time by using the function only after it has been defined. However, a definition always declares the function at the same time. However, you should get into the habit of separating the declaration and the definition right from the start, since the concepts used in the following only make sense then.

For very short functions, since C99 it has been possible to recommend to the compiler with the keyword *inline* that the function call be replaced by the function body. This saves the saving of the call context, i.e. the registers and the program counter, if short functions are to be called very often, which otherwise has to be carried out by the processor in the case of jumps in functions, in order to be able to return to the call location after the function has been processed. An example should clarify this:

```
static inline int max(int a, int b)
{
    return (a < b) ?b:a;
}
```

The call is made in the same way as for a normal function.

Instead of the keyword *static*, the function can also be declared as *external* in a header file (see Sect. 2.8.4), in which case it looks like this in the header:

```
external inline int max(int a, int b);
```

And in the source code:

```
inline int max(int a, int b)
{
    return (a < b)?b:a;
}
```

Another alternative to the inline function is to define macros with parameters. Macros have already been introduced in Sect. 2.4. Macros with parameters are resolved by the precompiler before the translation process, for example:

```
#define MAX(a,b) (a < b)?b:a
```

However, the use of such constructs is subject to great caution and should be avoided by beginners. See also Sect. 2.11.1.

2.8.3 Visibility and Lifetime of Variables in Functions

In general, in C, a variable is only valid where it is declared. A variable that is declared within a function or another statement block is also called a *local* variable, i.e. it is only visible and valid within the function. It does not matter whether the name of the variable has ever been declared elsewhere. As soon as the statement block is finished, the variable is destroyed and its contents are deleted. The following example illustrates this without claiming to be complete:

```
/* Module Example.c */
int i,j,k; //i,j and k are global, i.e. visible everywhere
void function_1(int j, int m) //j and m are local parameters,
                        // the global j is overridden
{
    int k, l; //k and l are local, the global k is overridden
    i = 5; //i is the global variable
    j = 9;
}
void function_2 (int m) //m is local parameter here
{
```

```
   int l; //l is a different l here than in function_1
   int n; //n is only locally valid here
   m = 18;
}
int main()
{
   int m = 1; //this m is valid inside main(), but not in the Functions
   int o = 8;
   function_2(o); //content of o is copied into function_2 (call by value)
}
```

Variables declared outside of each statement block are globally valid as long as they are not covered. Outside modules, you can declare global variables (in header files) that are then valid across multiple modules. See Sect. 2.8.5.

In earlier C versions, variables always had to be declared at the beginning of a statement block; since the C99 standard, this is no longer necessary. Nevertheless, for reasons of compatibility and readability, it is advisable to avoid declaring variables "wildly" if possible.

2.8.4   Header

Library functions, which are loaded from libraries on demand, extend the functionality of a program. These are loaded via library files in the binding process (see later) while the function declaration is included via header files (file extension .h).

C compilers usually provide some standard libraries with basic functions that are needed again and again. An example: The macro sei(), which enables the handling of interrupts, is made available by including the header file interrupt.h. But also mathematical functions like sine and cosine become available through library functions, or the output of data to streams. In general, headers are used to make module interfaces (i.e., their function calls and macros) available to other modules without having to expose the module's source code. So for each module (.c file), an identically named .h file is created with the declarations to be published.

Header files are included as follows:

```
#include <avr/io.h>
#include "Module1.h"
```

The name of the file to be included is in the angle brackets if it is stored in certain directories known to the preprocessor (include path). If the name is in " " , the path name of the project is used. With the compiler instruction −I an own search path can be entered, this is done user-friendly in the properties dialog of the project (include paths). The #include instruction is a "preprocessor instruction", i.e. it is processed before the actual translation process.

In general, the #include statement works like a copy/paste: the text from the header file is inserted into the source code at this point and thus becomes an integral part of the program. With this it is of course possible that there are further #include statements in a header file, but it becomes dangerous when a file is recursively included by itself via detours. This can be prevented quite simply by the following construction:

```
/*** driver.h ***/
#ifndef DRIVER_H
#define DRIVER_H
//the contents of the header file are in here
#endif //DRIVER_H
```

The keywords #ifndef (if not defined) and the concluding #endif form a meaningful parenthesis in a header file to prevent circular references from occurring in the event of accidental double inclusion, because everything in the parenthesis is only read in if the macro (here DRIVER_H) is not yet defined. DRIVER_H is defined when it is included for the first time. See also Sect. 2.11.1.

▶ Note that functions may not be defined in the header itself, but only declared! Syntactically it is possible, but it can go wrong if the header is included several times within a program. Then the linker reports an error.

2.8.5 The Keywords External, Volatile and Static

The keyword external can be used to declare a variable in a header file without reserving memory for it. The following example is intended to illustrate this:

Example

In a driver file a status is to be passed as a global variable. This could be realized in the file Driver.c as follows:

```
/*** driver.c ***/
#include driver.h
uint8_t Status;
void setStatus(uint8_t s)
{
    Status = s;
}
```

The file driver.h now contains an external declaration of status.

```
/*** driver.h ***/
#ifndef DRIVER_H
#define DRIVER_H
external uint8_t Status;
#endif //DRIVER_H ◄
```

As soon as another function, which uses this driver, binds the file driver.h, it is ensured by the keyword extern that no new variable Status is to be created. Only when binding with the library or the object file containing driver.c, the linker allocates the corresponding memory location to the now open reference of Status.

The use of external global variables should only be done with extreme caution. In the example mentioned, a so-called "getter" function could do a better job:

```
/*** driver.c ***/
#include driver.h
uint8_t Status;
void setStatus(uint8_t s)
{
    Status = s;
}
uint8_t getStatus(void)
{
    return status;
}
```

The header file then looks like this:

```
/*** driver.h ***/
#ifndef DRIVER_H
#define DRIVER_H
uint8_t getStatus();
void setStatus(uint8_t s);
#endif //DRIVER_H
```

This procedure is explained in more detail in Sect. 4.3.

Volatile can be used to prevent an optimizing compiler from optimizing away a variable that can be directly manipulated by hardware or another process. For example, the variable testvar would.

```
int testvar;
void function(void) {
    testvar = 0;
```

```
while (testvar != 255)
/*loop body without manipulation of testvar : infinite loop*/ ;
}
```

be set to `true` (1) by an optimizing compiler, since it does not change in the loop. If this variable were changed by the hardware or by a process that has external access to the memory area, this change would be ineffective.

With the volatile you can prevent this:

```
volatile int testvar;
void function(void) {
    testvar = 0;
    while (testvar != 255)
    /*loop body without manipulation from testvar until testvar = 255*/ ;
}
```

A handy keyword is also `static`. It specifies that a *local* variable (see Sect. 2.8.3) "survives" after exiting the function, i.e. it is not visible outside its local context, but it remains in memory and retains its value outside. Example:

```
#include <stdio.h>
void function() {
    static int x = 0;
    //x is only initialized at the first call of function and will only
    //be incremented on the next calls. At the end it has the value 5.
    x++;
    printf("%d\n", x); //print the value of x
}
int main()
{
    function(); //prints the number 1
    function(); //prints the number 2
    function(); //prints the number 3
    function(); //prints the number 4
    function(); //prints the number 5
    return 0;
}
```

For example, you can use this technique to count function calls or split functions without losing context.

2.9   Complex Data Types

2.9.1   Arrays, Fields and Strings

In the software, there is usually a need to process several data structured in the same way. For example, strings consist of a sequence of individual characters, vectors consist of several floating point numbers or time series consist of a sequence of measured values of the same data type. These sequences are called *fields* or *arrays*. In C, arrays are defined by storing the start address (pointer) of the field in a variable and accessing the individual element via an *index*.

```
uint8_t ucMyField[10];
```

defines an array with the name ucMyField and the length 10. This means that ten variables of the type unsigned char are stored here. Since the field size is unchangeable, the field is also called *static*. We will not say anything about dynamic fields and data structures in this brief summary; the reader is referred to the relevant C literature. The variable ucMyField itself does not contain any data but the address of the field. To access a variable within the field, it is selected by an index using the square bracket. The first variable has the index 0, the last (nth) the index n-1.

```
ucMyField[0] = 15;
ucMyField[9] = 99;
```

An access to an index outside the allocated (assigned) memory is usually not recognized by the compiler, in a runtime environment (Windows) the process is then aborted during execution. In an embedded system such an access leads to an unpredictable behavior, the cause of which is also difficult to find.

It is often necessary to traverse a field completely once, for example to initialize or copy it. If fields are used as vectors in the mathematical sense, vectorial operations (for example the scalar product) are also implemented by iteration. Counting loops are suitable for this purpose:

```
#define UC_MY_FIELD_LEN 10
uint8_t ucMyField[UC_MY_FIELD_LEN];
int i;
...
for (i = 0; i < UC_MY_FIELD_LEN; i++)
{
    ucMyField[i]....
}
```

Multidimensional fields are also possible:

```
#define MAT_COL 10
#define MAT_ROW 8
uint8_t ucMat [MAT_ROW] [MAT_COL];
int i,j;
...
for (i = 0; i < MAT_ROW; i++) //throughs first columns and then rows
{
    for (j = 0; i < MAT_COL; j++)
    {
        ucMat [i] [j] ....
    }
}
```

Another special array is the string. There is no separate data type for this in C, but it is formed from an array of char:

```
char name [20];
char ort [] = "Berlin";
```

In an embedded environment, strings are often used when data is to be output via a serial interface or when the line of a display is to be written. As early as 1963, the American Standards Association standardized the American Standard Code for Information Interchange (ASCII, alternatively US-ASCII) as a 7-bit code. It encodes the letters, digits, the space character, a number of special characters and various control characters (printed in italics in Table 2.10) that are not displayed but are used as protocol characters, for example, for the remote control of teleprinters. Table 2.10 shows the ASCII code in hexadecimal form, first the corresponding line is searched and then the column. The character "F" is thus hexadecimal 0x46, the character "Z" is 0x5A.

In the array, the characters are interpreted as ASCII characters by the respective output function. It is interesting that many strings, especially those formed by constants with double apostrophes, are *zero-terminated*. This means that next to the visible characters there is another one, the 0x00, since this is not printable in the ASCII code – an ASCII zero, i.e. the printable digit 0, is assigned to the number 0x30 (decimal 48) in the ASCII code.

Of the non-printable control characters, particular mention should be made of:

- DLE: The usual escape character for bytestuffing in the link layer.
- CR: The character for carriage return (\r in C)
- LF: The character for a new line (\n in C)
- HT: The horizontal tab (\t in C)

So in the above program example, the string "Berlin" would consist of the following characters.

Table 2.10 ASCII table – the italic characters are not printable

0x.	..0	..1	..2	..3	..4	..5	..6	..7	..8	..9	..A	..B	..C	..D	..E	..F
0...	*NUL*	*SOH*	*STX*	*ETX*	*EOT*	*ENQ*	*ACK*	*BEL*	*BS*	*HT*	*LF*	*VT*	*FF*	*CR*	*SO*	*SI*
1...	*DLE*	*DC1*	*DC2*	*DC3*	*DC4*	*NAC*	*SYN*	*ETB*	*CAN*	*EM*	*SUB*	*ESC*	*FS*	*GS*	*RS*	*US*
2...	Space	!	"	#	$	%	&	'	(	)	*	+	,	-	.	/
3...	0	1	2	3	4	5	6	7	8	9	:	;	<	=	>	?
4...	@	A	B	C	D	E	F	G	H	I	J	K	L	M	N	O
5...	P	Q	R	S	T	U	V	W	X	Y	Z	[	\	]	^	_
6...	`	a	b	c	d	e	f	g	h	i	j	k	l	m	n	o
7...	p	q	r	s	t	u	v	w	x	y	z	{	|	}	~	*DEL*

0x42	0x65	0x72	0x6c	0x69	0x6e	0x00
B	e	r	l	i	n	NUL

Many functions use null termination to detect the end of the string, including the well-known output function `printf()` and its derivatives, which are not discussed here in the embedded environment. You can acquire the necessary knowledge by a simple internet search.

It is often necessary to convert a number into a string in order to output it via a serial interface (for example to a terminal). Alternatively, the number can of course be transmitted in binary form, but this is usually not useful for displays and terminals. For this purpose there are different variants of the functions `itoa()` or `ftoa()`, which can be found in *stdlib.h*. The usage can be seen in the following example:

```
char text[4];
itoa(523,text,10);
```

The number 523 is output as ASCII in a zero-terminated string, i.e. 0x35 0x32 0x33 0x00, so the array `text` is also longer than the three digits of the number. In general.

```
char* itoa (int number, char* reserved_string, int number_base)
```

`itoa()` can also be implemented manually by a simple loop:

```
do
{
    text[1 - i] = value % 10 + 48;
    value = value / 10;
}while(value > 0);
```

Here the digits are isolated by a modulo-10 operation and then shifted to the right by an integer division.

Note: Arrays can consist not only of elementary types but also of structure types, more on this in Sect. 2.9.2.

2.9.2   Structure

In contrast to the array, you can combine different types of variables in a structure variable and access them through so-called selectors [13]. The variables are called members. The following example illustrates this (note the semicolon at the end of the definition):

```c
struct date
{
    uint8_t day;
    char month[10];
    uint8_t year;
};
```

With this structure definition you can now declare structure variables, for example.

```c
struct date birthday, school_enrollment;
```

The components can then be accessed during the `birthday` as follows:

```c
birthday.day = 15;
birthday.year = 1980;
```

An assignment between two structure variables causes the contents to be copied.

```c
school_enrollment = birthday;
```

You can initialize a structure with a curly bracket:

```c
struct date birthday = {15, "August", 1980};
```

You can also declare structure variables directly with the structure definition, but this is not recommended:

```c
struct test
{
    int a;
    char b;
} variable_a, variable_b;
```

It makes more sense to create a new data type with the structure definition and to store this in the .h file.

Structures can also be nested:

```c
struct time {
    uint16_t hr;
    uint16_t min;
    uint16_t sec;
};
struct date {
    uint16_t day;
```

```
    uint16_t month;
    int year;
};
struct appointment {
    struct date d;
    struct time t;
};
struct appointment a = { {19, 12, 2017}, {20, 15, 0} };
```

Access is then accordingly:

```
int hour = a.time.hr;
```

The keyword typedef can be used to define your own complex data type, so that struct can be omitted for further use.

```
typedef struct sMeasurement {
    uint16_t uiTemp;
    uint16_t uiBrightness;
} tMeasurement;
```

From this moment on, variables with the type tMeasurement can be declared:

```
tMeasurement measurement_series [10];
```

defines an array measurement_series, in which ten structure variables of the type tMeasurement are stored. Typedef is also valid for other things. Typedef unsigned int u16; means, for example, that a new data type u16 is introduced, which corresponds to an unsigned int.

Generally speaking:

```
typedef struct
{
    //structure components
} Structure type;
```

To determine the size of a structure required in memory, you need the C keyword sizeof.

```
sizeof(struct date)
sizeof(tMeasurement)
```

provide the sizes of the structures, which do not necessarily have to be identical to the sum of the individual components, since the compilers have the option of optimizing access. It should be noted that computers whose processing width is greater than one byte (e.g. 16, 32 or 64 bytes) usually also inflate the structures accordingly. For example, if a component on a 64-bit compiler is of type char, a 64-bit field (= 8 bytes) is usually created for it. You can optimize this behavior in some compilers with the #pragma pack (1) option.

In the following definition:

```
struct store {
    char x;
    int z;
};
#pragma pack(1)
struct store2 {
    char x;
    int z;
};
```

the commands sizeof (store) and sizeof (store2) output different values, namely 8 and 5, in a 32-bit GCC compiler. The unpacked structure consists of two times four bytes. This behavior is called alignment. This makes the access faster, but the memory becomes fuller. In general, the latter plays a minor role, which is why many companies forbid packing structures in their code. In the pointer arithmetic Sect. 2.9.5.2, these alignment ratios are already taken into account. For more on the #pragma keyword, see Sect. 2.11.2.

2.9.3 Unions

Unions are complex data types where several structures or data types can be superimposed in memory. They can then be accessed in different ways depending on how they are used. In the following example:

```
union vector3d {
    struct { float x, y, z; } vkart;
    struct { float r, phi, theta; } vpolar;
    struct { float r, z, phi; } vzyl;
    float vec3 [3];
};
```

can be used to access a vector with the selectors x, y and z as Cartesian coordinates or as polar coordinates with the selectors r, phi and theta or as cylindrical coordinates r, z, phi or as an array:

Table 2.11 Example for the interpretation of a serial message

SID	Byte 6	Byte 5	Byte 4	Byte 3	Byte 2	Byte 1	Byte 0
0x01	char temp	short pressure		char ax	char ay	short velocity	
0x02	char[7] text						
0xFF	char errCode	char status	long timestamp				char flag

```
#include <math.h>
vector3d pa, pb;
pa.vpolar.r = 1.0;
pa.vpolar.phi = M_PI;
pa.vpolar.theta = M_PI/2;
pb.vkart.x = pa.vpolar.r * sin(pa.vpolar.theta) * cos(pa.vpolar.phi);
pb.vkart.y = pa.vpolar.r * sin(pa.vpolar.theta) * sin(pa.vpolar.phi);
pb.vkart.z = pa.vpolar.r * cos(pa.vpolar.theta;
```

However, unions are particularly useful when it comes to allowing different interpretations of the stored content, as the following example shows.

```
union ui32
{
    struct {char i,j,k,l ;} bytes;
    long lNumber;
    float fNumber;
    char data[4];
} z;
```

z can therefore be viewed as float, long or in four individual bytes. A common application is when data is read in from a serial port and is then present as a byte array. Depending on what is in the first byte, the rest of the message is interpreted differently. For example, let there be a message consisting of eight bytes. The first byte is interpreted as the Service ID (SID), which defines the rest of the message as follows (Table 2.11):

As a union, this is how it would be implemented:

```
union umsg {
    char raw_msg[8];
    struct {
        uint8_t sid;
        char temp;
        short pres;
        char ax;
        char ay;
        short v;} proc_msg;
    struct { uint8_t sid; char text[7];} text_msg;
```

```
struct {
    uint8_t sid;
    char errCode;
    char status;
    long timestamp;
    char flag; } err_msg;
};
```

With a function `ReadDataFromSerial(char* data)`[2] a corresponding inter-preter could look like this:

```
umsg msg;
ReadDataFromSerial(&msg.raw_msg);
switch (msg.proc_msg.sid)
{
    case 0x01: ProcessProcMsg(msg); break;
    case 0x02: ProcessTextMsg(msg); break;
    case 0xFF: ProcessErrMsg(msg); break;
}
```

2.9.4 Enumeration Types

Enumeration types facilitate access to fixed values via symbolic constants, thus replacing elaborate #defines or const variables.

For example, replaces.

```
enum color {
    Blue, Yellow, Red, Green, Black };
```

the following code:

```
#define Blue 0
#define Yellow 1
#define Red 2
#define Green 3
#define Black 4
```

You can then use this as follows:

---

[2]To use the char* pointer, see Sect. 2.9.5.

```
typedef struct {
    double x, y;
} Point;
struct {
    Color color;
    Point x,y;
} line;
line.color = Green;
```

Accessing the line color would then return the result 3. Enumerations are always numbered incrementally in ascending order, but you can break this by explicitly defining the number directly:

```
enum prime {
    Two = 2, Three, Five = 5, Seven = 7
};
```

However, this should be used sparingly!

The keyword enum together with `typedef` can be used elegantly as a type with a controlled range of values:

```
typedef enum { monday=1, tuesday, wednesday, thursday, friday, saturday,
sunday} tDay;
typedef enum {jan=1, feb, mar, apr, may, jun, jul, aug, sep, oct, nov, dec}
tMonth;
void printDate(tDay day, tMonth month);
```

The call is then structured like this:

```
printDate(monday, apr);
```

and is preferable to the pure use of constants for the sake of readability and type safety.

2.9.5 Pointer

In Sect. 2.9.1 it was already mentioned that the field variable contains the address of the field, this is called a *pointer* to the field. Pointers can also be used to do some very elegant things. Describing the full power of pointers here is beyond the scope of this book. Instead, we refer you to the C literature. Nevertheless, a few examples should demonstrate the possibilities of pointers. For basic usage: A declaration with * first returns a pointer variable similar to the declaration with square brackets [] in Sect. 2.9.1:

```
char *text;
int *array;
double *vector;
```

However, without reserving memory space for this purpose. A field access with index would lead to a memory protection violation in a complex microprocessor, but in a small embedded system without operating system to an uncontrolled overwriting of values from other variables.

This is where the library function `malloc(size_t size)`, provided in many standard libraries, comes in handy. It reserves *size* memory (in bytes) in the so-called *heap* and returns a pointer to this memory, which is stored in a pointer variable. `void* calloc (size_t num, size_t size)` however returns such a pointer to a field containing num objects of size `size`. If the variable is used elsewhere during the course of the program, the memory must be freed again with `free()`. For the moment, we refrain from a further explanation.

To work with a pointer variable, you have the following options:

```
int a; //normal integer variable
int *b; //pointer variable (Attention! Without memory)
&a //delivers the address (pointer) of the variable a
*b //content of the variable to which b points
b = &a //sets the pointer variable b to the address of the variable a
a = *b //copies the contents of the variable pointed to by b into a
*b = a; //assigns the value of a to the variable pointed to by b
```

In the second to last line b = &a of the above example, the behavior becomes clear. The variable b refers to the memory location of a, i.e. when *b is changed:

```
a = 3;
b = &a;
*b = 9;
```

the content of the variable a would also change! This can be used to some interesting applications, as shown in the following.

2.9.5.1 Use in Function Calls

The use of pointers can be explained clearly when you want to manipulate fields of unknown size in functions. As an example, consider the following function ToCapital (), which converts the letters of a text to uppercase. Of course, for a null-terminated string, Sect. 2.9.1 could also detect the terminating null as the end-of-text character, but the same technique can be used in many other examples, which will also be described throughout the book.

The function looks like this:

```
int ToCapital(char *text, int len)
{
    int i, j = 0;
    for (i = 0; i < len; i++) //over the entire string
    {
        //WHEN letter between a and z
        if (text[i] >= 0x61 && text[i] <= 0x7A)
        {
            text[i] -= 0x20; //reduce value to A..Z
            j++;
        }
    }
    return j; //number of hits
}
```

A call.

```
char text[] = "Hello World!";
int y;
y = ToCapital(text,11);
```

returns the string "HELLO WORLD" in text and sets y to 7, since a total of seven characters were converted.

Another application of pointers in function calls is given by functions which are to have several return values. The "classic" is swapping two variables. The function swap(double *a, double *b) does this as follows:

```
void swap(double *x, double *y)
{
    double tmp;
    tmp = *x;
    *x = *y;
    *y = tmp;
}
```

Without pointers, the call-by-value principle would make this function meaningless.

2.9.5.2 Pointer Arithmetic
Basically, you can do math with pointers. An example:

```
int x[10];
int *y;
y = &x[5];
```

At the end of these lines, y points to the sixth element in the field x. With:

```
y++;
```

y now points to the seventh element of the field x. The compiler automatically recognizes from the type (pointer to integer) how large the elements are and calculates the next position. This is a fundamental difference between pointer arithmetic and working with adresses in variables. If you want to output the value of the next element immediately after incrementing the memory, you have to consider the binding priority of the dereferencing operator (Sect. 2.5.6):

```
*y++; //returns the value of the next element
(*y)++ ; //increases the value of the current element by 1
```

2.9.5.3 Pointers to Functions

You can also set pointers to functions, since they also represent objects in memory. The declaration:

```
int (*fun) (int);
```

Returns in the variable fun the reference to a function that is of type int and has an input parameter of type int . If we now had two functions to choose from:

```
int Increment (int a) {return ++a;}
int decrement (int a) {return --a;}
```

of which one function is to increment the number a and the other is to decrement the number a, we could now write:

```
typedef enum {large, small} sizes;
int SmallOrLarge(int number, sizes direction)
{
    int (*fun) (int);
    if (direction == small)
    {
        fun = &decrement;
    }
    else fun = &increment;
    return fun(number); //executes the function pointed to by the pointer
}
```

Depending on whether the passed parameter `direction` is "small" or "large", the function returns the increment or decrement of the passed number.

```
y=SmallOrLarge(5,small);
```

so set y to 4;
We will see examples with more sense in the next chapters.

2.9.5.4 Const Pointer

You often see the keyword `const` in the context of pointers. Here are three examples:

```
const char * myPtr = &char_A;
char * const myPtr = &char_A;
const char * const myPtr = &char_A;
```

The first line of `const` declares a pointer to a constant character. You cannot use this pointer to change the value it points to. Of course, another pointer pointing to the same memory location could still change the character.

The second line with `const` declares a constant pointer to a character. The position stored in the pointer cannot be changed. So you cannot change the address to which this pointer points.

The third line with `const` declares a pointer to a character, neither the pointer value (i.e. the address) nor the value referenced by this pointer can be changed.

2.10   Structure of an Embedded C Program

For the following considerations it is useful to define a basic program structure. This will be expanded in the following chapters.

We start from the following principles:

- The main function is the initial function that controls the general flow.
- The main function is to be kept as short as possible
- Each program consists of an initialization part and a working part or process part (main loop), which is constantly executed and does the actual work.
- All functions that access hardware (registers) are additionally described in separate modules to make the "core" program hardware independent.
- For each source file there is a header file in which the (public) functions are declared! Public means that these should also be accessed from other modules. First of all, we assume that all functions are public.

So we need two files `programname.c` and `programname.h`. In the interest of readability, we have omitted elaborate commenting here. The extensively commented sources can be found in the accompanying sources in the download area of the book.

A completely empty program structure then looks like this:

```
#include "programname.h" //headerfile of the main module
/***********************************************/
/* Declaration of the module-global variables */
/***********************************************/
// none available yet
/***********************************************/
/* Main program */
/***********************************************/
int main()
{
    Init(); //initialization part
    while(1) //endless loop
    {
        //here the process functions are called
    }
    return 0;
}
/***********************************************/
/* Initialization */
/***********************************************/
void Init(void)
{
    //here the initializations are called
}
```

The declaration of the Init() function is done in the header file Programname.h.

```
#ifndef PROGRAMNAME_H
#define PROGRAMNAME_H
/***********************************************/
/*      Declaration of the functions      */
/***********************************************/
void Init(void);
#endif
```

By the construct #ifndef PROGRAMNAME_H, as already mentioned, accidental double inclusion is prevented by searching for the macro PROGRAMNAME_H. If it is not present, the header file was not yet included, the macro is then defined and the header information can be read. However, if the macro was defined, the rest is skipped (Sect. 2.8.4). To achieve a unique designation of the macros, the name of the header file with the suffix _H is used.

In the following Chaps. 3 and 4, this framework is further extended to a small real-time operating system.

2.11 Working with the Precompiler

The precompiler or preprocessor is a powerful tool to control the compilation process. It can be used to switch compiler options on and off or to modify or switch on and off entire code blocks in order to generate more efficient, more flexible and more readable code at runtime. However, the commands often hide pitfalls. For reasons of space, only the most important features of the precompiler commands are mentioned here. In the bibliography you will find books that devote much more space to this topic.

2.11.1 #define and Working with Macros

In Sect. 2.4.3, the precompiler command #define was introduced as a method for defining macros. In Sects. 2.8.2 and 2.8.3, a macro with a parameter list was introduced. At this point, other features of macros will be discussed.

(a) Since macros can also be passed to the compiler via the command line, the macro is a popular way to compile code for different target hardware:

```
#ifdef TARGET_1
    #include <target1.h>
#elif TARGET_2
    #include <target2.h>
#endif
#define INCFILE "global_include.h"
#include INCFILE
```

If the macro TARGET_1 is defined, then another header file is inserted, as if TARGET_2 is defined. This is controlled by #ifdef (= if defined), #elif (=else if defined) and #endif (as delimiter). Code can also be inserted at this point, which then contains different hardware abstractions, for example (Chap. 4).

```
#ifdef _DEBUG_
//paste debugging/simulation code here
    Get_simulated_event();
#elif _TARGET_
//insert target code here
    Get_event();
#else
    #error No target defined
#endif
```

The macro #error aborts the process and draws attention to itself with an error message from the compiler.

(b) Usually header files are enclosed with #ifdef ... #endif to prevent erroneous multiple declarations, many IDEs do this automatically when a new header file is created (this is also described in Sect. 2.8.4):

```
#ifndef PROGRAMNAME_H
    #define PROGRAMNAME_H
    /* Here come the function declarations or struct and constants */
    void Init(void);
#endif
```

Alternatively, with GCC you can replace the #include directive with an #import, which has the advantage of checking itself for multiple uses and preventing them. However, we do not recommend this for compatibility reasons.

(c) There are a number of standard macros that the compiler replaces with code, examples can be seen in Table 2.12. The list is deliberately not complete in order not to confuse you. Many compilers come with dozens of their own standard macros, including AVR GCC, which we use here.

(d) As mentioned in Sect. 2.8.2, you can also add parameters to macros:

```
#define square(x) ((x) * (x))
#define MAX(x,y) ((x) > (y) ? (x) : (y))
```

This tool is very powerful and we would like to point out at this point that beginners should handle it very carefully! Since the resolution of the macros is done in the source code, nasty errors can creep in when using semicolons, for example. Further interesting features can be

Table 2.12 Standard macros

__DATE__	The current compile date in "MMM DD YYYY" format.
__TIME__	The compile time in "HH:MM:SS" format
__FILE__	The name of the current file as a literal string constant
__LINE__	The line number in the current file as a numeric constant
__func__	The name of the function in which the macro is located
ZERO	The null pointer
offsetof(type, member-designator)	Returns the offset, i.e. the position of a component in a structure.

found in the online documentation of the GCC compiler (CPP Manual) [14] in the respective version in Chap. 3, for example the use of macros with a variable number of parameters or the chaining of parameters.

2.11.2 #pragma

In the context of GCC it is sometimes necessary to switch individual compiler options in places. A very important example is the use of the optimizer. This generates considerably more efficient code according to semantic rules than a pure one-to-one translation. However, often deliberately introduced commands are lost, which the optimizer does not attach any importance to. In other words, sometimes you have to turn off the optimizer in order for the code to still run. Usually this is done in the compiler's option settings (in the IDE or via command line parameters). On the other hand, you often reach the limits of flash memory. The optimum would be to turn off the optimizer only at the crucial points. This is where #pragma comes into play. It is a powerful tool that has dozens of setting options and can practically control the entire compiler from within the coder (see [14] GCC Manual Chap. 6). We use at times:

```
#pragma GCC optimize (string, ...)
```

and set string to the respective optimizer option, for example.

```
#pragma GCC push_options
#pragma GCC optimize("O0")
```

Turns off the optimizer. At the end of the code block for which the options should apply, we then write:

```
#pragma GCC pop_options
```

#pragma GCC push_options stores the existing compiler options and #pragma GCC pop_options retrieves them from memory, restoring the old state.

2.11.3 Summary of the Precompiler Commands

Table 2.13 summarizes the common precompiler commands again.

Table 2.13 Precompiler commands

Keyword	Meaning
#define	Defines a macro
#include	Copies a header file to the current location in source code
#undef	Deletes an existing macro
#ifdef	Returns true if the named macro exists, the code in the following conditional block will then be compiled by the compiler (otherwise skipped)
#ifndef	The opposite: returns "true" if the macro does not exist
#if	Checks whether a compiler condition is fulfilled
#else	This branch will be translated if the previous #if or #ifdef condition was not met (alternative)
#elif	#else and #if in one statement, so an alternative with an additional condition
#endif	End of a condition block
#terror	Stops the compilation process and outputs an error message on the channel stderr
#pragma	Instructs the compiler to execute the following compiler command

2.12   Translating and Binding

Usually, an integrated development environment (IDE) is used to create a program. This controls the entire development process from the creation of the source code to the upload to the target system. Well-known IDEs are Eclipse, Visual Studio or the environments of Greenhill and Keil. We used Microchip Studio, which Microchip distributes for free at www.microchip.com and which is based on the well-known Visual Studio, when writing this book. In principle, however, you can also use command-line oriented tools, for example the well-known GNU tool chain, which also provides the C compiler for Microchip Studio.

The process is the same in each case and can be found in Fig. 2.1: .c and .h files form the source code, which is created in a suitable editor. Common editors offer syntax highlighting (color-coding of language elements) and auto-completion of known function and variable names. The precompiler controls the inclusion of .h-files and the solving of macros. This ultimately results in the final source code, which is then first checked by the compiler for its syntax. Afterwards a so-called object code is generated. This is not executable, because the references to functions and external variables are not yet resolved. This can only be done when all source files have been compiled individually.

The linker links the object files together and resolves the external references to them, i.e. it allocates a memory location to each function and each variable. Precompiled object code collections (libraries, usually named .a) can also be included here. These have the advantage that the user does not see the source code, but only the interface, i.e. the declared functions, macros and external variables, via the .h file. In this way a commercial use of libraries is possible. The executable code is finally transferred to the target system in a suitable form using a tool for flashing, which is usually also integrated into the development environment. For this purpose, a programming device is required, which the processor manufacturers usually offer at low cost. For Atmel processors with programming interface according to the JTAG standard (IEEE standard 1149.1), for example, the JTAG-ICE can be used for upload and on-chip debugging. The STK500 programmer is suitable for processors used for so-called In-System-Programming (ISP) or other programming modes of the AVR family via serial interfaces. For this system very inexpensive replicas or building instructions exist on the net.

Furthermore, many IDEs contain a debugger. This links the lines in the source code with the executable target code and monitors the state of the variables and registers. The program can be executed step-by-step or continuously, individual functions can be called specifically and the sequence can be stopped at any time at a breakpoint or due to a condition. Microchip Studio has an integrated simulator that can be used to debug the code without a target system. When using a programming device with JTAG, the process can take place directly on the processor. So-called emulators are also common, which emulate the target system in terms of hardware and have extended diagnostic options.

References

1. ISO/IEC 9899:2011. *Information technology – Programming languages – C*. www.iso.org/iso/home/store/catalogue_ics/catalogue_detail_ics.htm?csnumber=57853. Accessed 20 Aug 2020.
2. ISO/IEC 14882:2014. *Standard for programming language C++*. https://isocpp.org/std/the-standard. Accessed 20 Aug 2020.
3. Mathworks. *Embedded coder*. http://de.mathworks.com/products/embedded-coder/features.html. Accessed 20 Aug 2020.
4. Wikibooks. *C-Programmierung*. https://de.wikibooks.org/wiki/C-Programmierung. Accessed 20 Aug 2020.
5. Erlenkötter, H. (1999). *C: Programmieren von Anfang an* (23. Aufl.). Rowohlt Taschenbuch.
6. Kernighan, B. W., & Ritchie, D. M. (1990). *Programmieren in C: Mit dem C-reference manual in deutscher Sprache*. Hanser.
7. Kernighan, B. W., & Ritchie, D. M. (2010). *The C programming language*, 2. Auflage, Prentice Hall (englisch).
8. Gookin, D. (2010). *C für Dummies*, 2. Wiley-VCH.
9. Goll, J. (2014). *C als erste Programmiersprache- Mit den Konzepten von C11* (8. Aufl.). Springer.
10. Wiegelmann, J. (2017). *Softwareentwicklung in C für Mikroprozessoren und Mikrocontroller* (7., neu bearbeitete und erweiterte Aufl.). VDE.
11. Dimitri van Heesch. *Doxygen*. https://www.doxygen.nl/download.html. Accessed 11 März 2021.
12. Emden Gansner GraphVIz dot (online). https://graphviz.org. Accessed 11 März 2021.
13. C-Programmierung: Komplexe Datentypen. (2015, 27 Juli). *Wikibooks*, Die freie Bibliothek. https://de.wikibooks.org/w/index.php?title=C-Programmierung:_Komplexe_Datentypen&oldid=764916. Accessed 11 März 2021.
14. GCC Online Dokumentation. (2020). *GCC10.2*. http://gcc.gnu.org/onlinedocs/. Accessed 20 Aug 2020.

Programming of AVR Microcontrollers

Abstract

This chapter describes the most important peripheral elements of the AVR family, i.e. the basic functions with which a microcontroller of the AVR family communicates with the outside world. The structure of peripheral circuits is described, which can of course also be applied to other representatives of the AVR family or other microcontroller families. Interrupts, digital inputs and outputs, AD converters, timers, PWM, the EEPROM memory and methods of power management are the main topics of the chapter. The ideas underlying them are also transferable to other processors.

At the end of this chapter you will be able to read and switch simple digital I/O connections, read analog sensors, control a motor with a full bridge, write to and read from the integrated EEPROM and generate time-dependent sequences and signals from the timer. Power management is also briefly discussed. All examples in this and the following chapters have been tested and compiled with MicrochipStudio[1] (version 6 and higher).

This chapter does not claim to be complete, but explains the relationships to the extent that they are necessary for further understanding of the examples in the rest of the book. In the literature appendix, a number of books are identified in which individual functions are discussed in detail. Advanced readers are recommended to read the very detailed and well-written data sheet and the extensive documentation [1] afterwards. The data sheet alone has over 660 pages and is constantly being updated! Therefore, it should always be consulted

---

[1] Previously: AtmelStudio.

as a reference in a specific development, which this book cannot do. Besides that, the authors recommend to consult the excellent website mikrocontroller.net [2].

3.1   Architecture of the AVR Family

Every microcomputer, including the processors of the AVR family (Fig. 3.1), initially consists of a CPU (Central Processing Unit). This contains at least two components: a control unit which is responsible for the processing and interpretation of instructions and controls the other components, and an operation unit, which consists of the work or data registers and the arithmetic unit, the ALU (Arithmetic Logic Unit). It is responsible for arithmetical and logical operations. The control unit and the arithmetic logic unit can be designed in two ways:

- For very simple CPUs, they consist of hardwired logic
- In the case of more complex CPUs, they are themselves microprogrammed, i.e. an instruction starts its own program stored in the controller.

With the processor clock, which is generated by an internal oscillator or an external quartz oscillator, a sequence of states is now processed according to (roughly simplified) the following scheme:

- A program counter (PC program counter) is incremented by one and addresses an instruction in a program memory. This is usually stored temporarily in an instruction register (instruction fetch cycle).

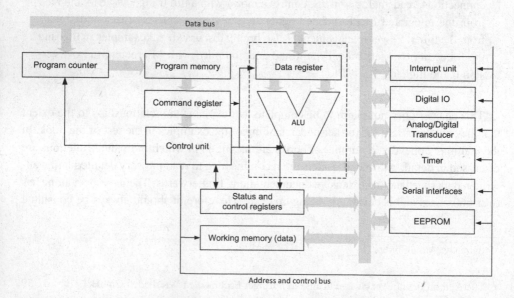

Fig. 3.1 Basic architecture of AVR microcontrollers

- The instruction is decoded and, if necessary, the operands belonging to the instruction are loaded into the working register. For example, an addition requires two operands, namely the two numbers to be added. This state is called Instruction Decode/Register Fetch Cycle. In a routing unit implemented with hardwired logic, the two cycles coincide. In a microprogrammed control unit, the instruction execution is prepared here over several clock cycles.
- In the third step the ALU calculates the result of the operation (execute cycle), if necessary memory accesses are also executed here. With a hardwired ALU, the execution coincides with the last cycle; with a microprogrammed ALU, a program is also started here over several clock cycles. In the status register certain values are set, for example whether an operation ends with the result 0 or generates an overflow (carry), which can be used for following operations.
- In the fourth step, the calculation results are written back to the working registers (Write Back Cycle).

More complex microcomputers cycle through more than these four states, while the simplest architectures make do with two (Fetch/Decode/Execute and Write Back). To speed things up, so-called pipelines can be used, which can execute cycles on successive instructions in parallel. However, this is beyond the scope of this introduction. Microprogrammed pipelines and operations allow complex instructions, but their execution takes a long time (CISC: Complex Instruction Set Computer). The Intel and AMD processors for PCs are such CISC processors. In contrast, the RISC (Reduced Instruction Set Computer) concept is based on very simple instructions that can be processed in a single clock cycle. The AVR controller described in this book is such a RISC computer. It is very fast at processing simple instructions and then simply needs more instructions for more complex tasks.

Basically, a distinction is made between the following types of instructions that a CPU can execute:

- Arithmetic/logic commands, e.g., addition, subtraction, multiplication, Boolean functions, negation, etc.
- Data transfer instructions, i.e. copying data from registers to working or program memory or vice versa from memory to registers, between registers or between memory cells.
- Jump commands: The program counter reading is not simply incremented, but loaded with a value that may, for example, depend on the previous calculation result. This allows the program sequence to be changed depending on the calculation (conditional jumps).

Many commands (operations) require one or more operands. These can be fixed or variable. Variable operands and the calculation results are stored in the working registers or in the working memory. Control commands, addresses and data are distributed in the system via central lines (buses). Basically, a distinction is made between two types of architecture, which are characterized by the bus connection of working and program memory.

In the "von Neumann" architecture, program memory and main memory are connected to a bus, access to programs and data is therefore sequential ("von Neumann bottleneck"), but access is more flexible. In a PC, program memory and main memory are even identical; the programs are copied from the hard disk to the main memory as required. In a memory cell, it is no longer possible to distinguish between data and instruction.

Embedded controllers, on the other hand, are often designed in Harvard architecture, as are the processors of the AVR family. This architecture, developed at Harvard University in Boston, requires strict separation between program, data, arithmetic unit and peripherals. The advantage of this concept is that the software is compactly stored in a (normally non-volatile) memory, the data is located in the main memory, or in further non-volatile memories (EEPROM). Since the program is not changed during runtime anyway, the flexibility of the von Neumann architecture is not necessary. Access to program and data is much faster and more efficient than with the von Neumann architecture. Furthermore, program and data memory can be optimized separately in terms of data and instruction word width, which can lead to better memory utilization. In addition, program code is not overwritten in the event of software errors, which is possible with the von Neumann architecture. With the AVR family, the separation is not quite as strict. There is a direct access from the program memory to the calculation register, so constant operations, which are part of the program, can be included in the calculation without detour over the working memory.

3.2   Packaging and Pin Assignments

The AVR family contains hundreds of different processors. In this book, we refer to the ATmega48/88/168 in each of the examples, especially because they come in convenient 28-pin PDIP packages that are also suitable for quick board assembly. For product applications, the 28- and 32-pin MLF or the TQFP package in SMD solder technology are suitable. For very small applications, for example in dimmers, fire alarms, model making etc. the ATtiny is suitable in different variants, these have less memory and less periphery than the ATmega, but they are already available in the 8-pin PDIP or in the 20-pin QFN/MLF package and are therefore very compact. For applications with CAN networking, the AT90CANxx with significantly more features in the 64-pin TQFP package is suitable in the ATmega family.

Figure 3.2 shows the pinout of the ATmega48/88/168. The individual pins will be described in detail later.

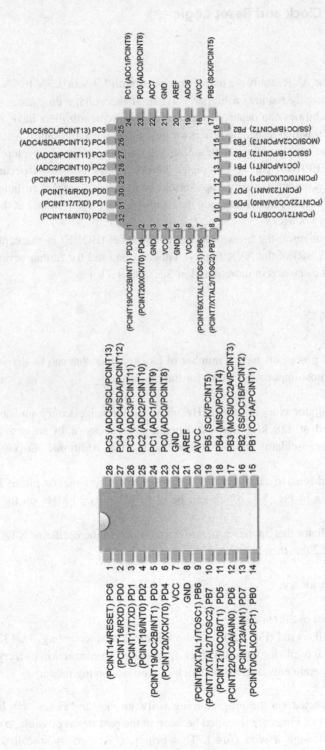

Fig. 3.2 Pin assignment of the ATMegax8 in the PDIP package (left) and in the TQFP package (right). (According to [1])

3.3   Supply, Clock and Reset Logic

3.3.1   Supply

The processors of the AVR family are usually supplied with 1.8 V or 3.3 V to 5 V. A higher crystal frequency usually requires a higher voltage. If the voltage drops, e.g. due to an empty battery, instabilities can occur. Therefore, many microcontrollers have a so-called brown-out detection, which resets the processor below an adjustable supply voltage level.

By measuring the supply voltage with an internal, supply-independent reference, write operations in the EEPROM can be completed before a brown-out reset occurs. In real circuits, for example, batteries or supercaps should be used to ensure that sufficient energy is still available to process these emergency measures (usually a few 10 ms) in the event of a drop in the supply voltage.

After a brown-out reset, the brown-out reset flag is set (BORF) in the central MCU status register (MCUSR) of the AVR family. This can be used for further actions after a reset. The supply is discussed in more detail in Sect. 3.8.1.

3.3.2   Clock

The AVR family of processors have a number of clock signals that can be derived from a central clock. The most important sources for the clock are:

- An internal oscillator clocked at 8 MHz without external circuitry and an internal oscillator clocked at 128 kHz that can operate the processor in an extremely low power mode. The oscillator inputs can then be used as additional IO pins on most processors
- An oscillator connected to an external resonator (quartz or ceramic) on pins XTAL1 and XTAL2 as shown in Fig. 3.3, which can be clocked up to 20 MHz on the ATmega family.
- An external oscillator that applies a square wave signal to the oscillator XTAL1 of the processor, XTAL2 can then be used as an IO pin.

Typical usage scenarios are:

- Use of the internal oscillator only
- Use of an external crystal (Fig. 3.3) with the full clock frequency (e.g. 18.432 MHz)
- Use of the internal oscillator with full clock frequency and connection of a crystal with 32.768 kHz to operate only a real time clock in a power saving mode.

The oscillator is selected via the programming software via the "Fuses" tab by setting the fuse: SUT_CKSEL. Fuses are so called because in the past it was possible to configure the hardware by blowing a wire (fuse). This principle is used in so-called PROMs

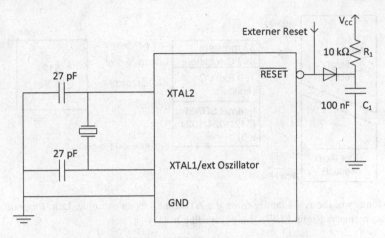

Fig. 3.3 Wiring of the oscillator and the RESET input

(Programmable Read Only Memory), in which each memory cell could be written once and irreversibly by a current pulse. Today, the fuses in the microcontroller are no longer irreversible. However, they should be operated with care, especially if you are using an SPI programmer and the processor is soldered in. A wrong setting can completely paralyze the system (for example by disabling the programming interface or wrong choice of the oscillator).

3.3.3 Reset Logic

Resetting the processor (RESET) causes the program counter to point to address 0 in the program memory and, among other things, the outputs to be switched to high impedance. At address 0 there is a jump instruction to the memory location of the main program main(), i.e. the program starts there.

The processor can be reset externally via various mechanisms:

- During power-on reset, the uncharged capacitor at the RESET pin forms a short circuit to ground. This causes reset pin 1 to be briefly set to 0 until C1 has charged via R1.
- A RESET button on the RESET pin causes a manual reset.
- A so-called watchdog is connected to the RESET pin. This regularly tries to reset the processor after a certain time of some ms to seconds. To prevent it from doing so, the watchdog must be reset before its "time-out" time expires. This is done via a digital IO line, which is controlled from the program. If the program "hangs" for any reason, the watchdog time expires and the processor is reset.
- Programming of the ATmega88 can only be done when the processor is in RESET mode, therefore the RESET line from the programmer is pulled to ground

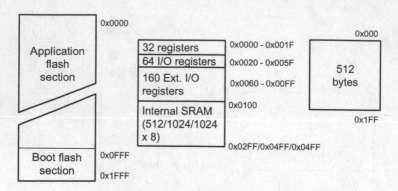

Fig. 3.4 Memory of the AVR family using the ATmega88 as an example. Left: Program memory, Middle: Data memory, Right: EEPROM. (According to [1])

Note for advanced users: Internally the processor is reset by brown-out (undervoltage) or by the watchdog, which is also present internally, if it is active. This can be a nasty source of errors. Therefore Microchip recommends to always clear the Watchdog System Reset Flag (WSRF) in the MCU Control Register (MCUCR) and the Watchdog System Reset Enable (WDE) Flag in the Watchdog Timer Control Register, even if the watchdog is not used.

3.3.4   Memory

Figure 3.4 shows the three memory types of the AVR family.

Figure 3.4a shows the flash memory, which is 8 kByte in the ATmega88. In the notion of the Harvard architecture it represents the program memory (firmware). On the right side of the picture you can see the EEPROM memory, which holds persistent data (i.e. data that is valid beyond power-down), e.g. error memory entries or parameterizations that you want to set independently of the source code. For example, the behavior of the software can be adjusted or calibrated at runtime.

The Flash and EEPROM are programmed in two ways:

- As long as the RESET input is held at 0 (low), Flash and EEPROM are programmable via the SPI interface. This is done by many programming devices (e.g. ISP programmer).
- Both memories are also programmable from within the program, so a program can "reprogram itself", usually this is done from a bootloader routine or from a routine that reads program code from the serial port.
- Some representatives of the AVR family have their own boot flash section. This holds the bootloader routine and can be varied in size and optionally locked, so that it can only be overwritten if it has been unlocked before. Optionally, this section can be executed

after a RESET or jumped to as a subroutine, for example by a corresponding message via a serial interface. This makes it possible to reprogram the firmware of the system via a diagnostic access, for example via a serial interface.

In the center of Fig. 3.4 the registers and the main memory are indicated. In the case of the ATmega88 this has 1 kByte free memory space[2] and is used to store variables and fields temporarily. Since the compiler prefers to keep local variables in working registers rather than in working memory, it is usually only needed for global variables. The figure outlines the 32 working registers, which can be addressed directly by the computational unit and therefore allow correspondingly fast instructions. Data in the working memory must first be copied to a working register before processing, which takes a corresponding amount of time.

Furthermore, you can see the so-called IO registers in the figure. These control the functions of the microcontroller. The following sections describe the most important of them. A detailed reference can be found in [1].

3.4   Handling Registers

To understand the operation of all I/O ports and modules, one must first know something about the memory organization of the AVR. The access to the periphery is done by *registers,* which are located in the address area of the working memory at the data bus (Fig. 3.4 middle).

Using the ATmega88 as an example, this is explained in more detail: The first 32 bytes (0x0000 to 0x001F) are for the working registers, which provide fast access and hold the operands during operations. The next 64 bytes (0x0020 to 0x005F) are for the I/O registers, another 160 bytes are for the extended I/O registers, and then (starting at address 0x0100) the ATmega88 has another 1024 bytes of working memory where the variables can be stored. The compiler automatically determines where variables are stored, local variables are usually located in registers, while global and static variables are stored in main memory, as well as fields.

The access to the periphery is done in the same way as to a variable. The registers are one byte in size according to the data type `unsigned char`.

Direct access to an address is error-prone and also not very intuitive, so the registers have names that are usually defined in the include file `<avr/io.h>`.[3]

---

[2] The picture also shows the memory expansions of the ATmega48 and the ATmega168 with 512 bytes and 1 Kbyte respectively.

[3] To be precise, io.h includes the file corresponding to the processor type, e.g. iomx8.h for the ATmegax8 series.

For those who want to know exactly: The register macros represent a reference to the port address, e.g.:

```
#define PORTB _SFR_IO8 (0x05)
```

The macro `_SFR_IO8(io_addr)` also included by io.h:

```
#define _SFR_IO8(io_addr) ((io_addr) + ___SFR_OFSET)
with __SFR_OFFSET = 0x020.
```

With this definition, it is possible to access the register with the name PORTB instead of the address (0x25 in this case), using the assignment operator "=". Example:

```
PORTB = 0x17;
uint8_t value = PORTB;
```

Often, however, you only want to change or query a single bit of a register. Since C does not know any bit manipulation instructions, one helps oneself with the bit operators:

&	Bitwise AND
|	Bitwise OR
~	Bitwise negation, one's complement
<<	Bitwise shifting left
>>	Bitwise shift to the right

Thus, setting a single bit is done by "ooding" the register with a single 1, i.e.
0000 0001 corresponds to hexadecimal 0x01 (0th bit)
0000 0010 corresponds to hexadecimal 0x02 (1st bit)
0000 0100 corresponds to hexadecimal 0x04 (2nd bit)
0000 1000 corresponds to hexadecimal 0x08 (3rd bit)
0001 0000 corresponds to hexadecimal 0x10 (4th bit)
0010 0000 corresponds to hexadecimal 0x20 (5th bit)
0100 0000 corresponds to hexadecimal 0x40 (6th bit)
1000 0000 corresponds to hexadecimal 0x80 (7th bit)
So if you want to set the fourth bit in port B (ATTENTION: counting starts from 0!), you OR port B with a 0x10. All other bits remain unchanged according to the definition of the OR operator.

```
PORTB = PORTB | 0x10;
```

or in shorthand:

```
PORTB |= 0x10;
```

Another variant is to put a 1 in the right place by left shifting, for this the necessary definitions are already available in the include files, in the case of port B there are PB0 (=0), PB1 (=1), PB2 (=2), PB3 (=3), PB4 (=4) etc. up to PB7 (=7).

The access is therefore called.

```
PORTB |= (1 << PB4);
```

In other words, the 1 is shifted four places to the left (result binary: 0001 0000) and then ORed with the contents of the PORTB register. Since the zero is the neutral element of the OR operation, only the fourth bit of PORTB is set to 1, the others remain unchanged.

If you want to set a bit to 0 in the opposite case, you have to AND the register with the one's complement, in the case of the fourth bit with 1110 1111. According to the definition of the AND-operator all other bits remain unaffected. To avoid complicated conversions, the complement is calculated with the negation operator:

```
PORTB &= ~0x10 ; //or in shift notation
PORTB &= ~ (1 << PB4);
```

sets the fourth bit to 0, all others remain as they were before the operation.

Another case where this technique is usefully applied is to query a single bit, for example to check the state of a port. In this case, one can mask out of the register all but the bit of interest. The expression.

```
(PINB & (1 << PB5))
```

is 0 if the fifth bit in the PINB register is 0, otherwise it is not equal to 0. Here we are helped by the definition of logical expressions in C, which states that an expression is "true" if it is not equal to 0. If it is equal to 0, it is "false". Thus, the condition.

```
if (PINB & (1 << PB7))
{
}
```

into the parenthesis if the seventh (counted from 0!) bit in PINB is set to 1(true).

Of course, all bit masks can be cascaded:

```
DDRC &= ~ (1 << DDC0) & ~ (1 << DDC1) & ~ (1 << DDC2) & ~ (1 << DDC3);
```

alternatively:

```
DDRC &= ~ ((1 << DDC0) | (1 << DDC1) | (1 << DDC2) | (1 << DDC3));
```

masks the register DDRC with 1111 0000, so sets the four lowest bits of the register to 0. So if you want to set single register bits to 0 or 1, you need two lines, one for the ones and one for the zeros.

```
PORTB |= (1 << PB3) | (1 << PB2) | (1 << PB0);
```

sets bits 0, 2 and 3 of register PORTB to 1.

The setting of a single bit can of course be written more nicely by defining two macros.

```
#define setbit(reg, bit) reg |= (1 << bit)
#define clrbit(reg, bit) reg &=~(1 << bit)
//call:
setbit(PORTB, PB0);
clrbit(PORTB, PB0);
```

and then set or reset the bits one at a time.

At this point it should be pointed out that an extensive function library AVR Libc exists, which for example comes along with the Microchip Studio or the WinAVR compiler. In this library there is a rich number of functions, which are easier to use than register programming would allow. We also use these functions in a few places in the rest of the book. A detailed description can be found in [3, 4].

3.5   Digital Input/Output

3.5.1   Basic Structure

Almost every pin of an AVR processor, with the exception of the power supplies for processor and A/D converter and the reference voltage inputs, can be switched as a digital input or output. Figure 3.5 shows roughly the structure.

Figure 3.6 shows possible input circuits. Instead of a pushbutton, the output stage of a digital circuit or the phototransistor of an optocoupler can of course also be assumed.

The data direction register (DDRxn)[4] now determines whether the pin represents an input or an output. If it is set to 0 (automatically on reset), the output switch is set to "tristate", i.e. switched off with high impedance, and the operating mode is "input". In this case the pull-up resistor can be switched on via a 1 in the PORT register. All pull-up resistors can be switched off independently by setting bit 4 (PUD = Pull-up Disable) in the MCU Control Register (MCUCR):

---

[4] n stands for the corresponding bit, x stands for "B", "C" or "D" in the case of the ATmega88, i.e. for one of the three port registers.

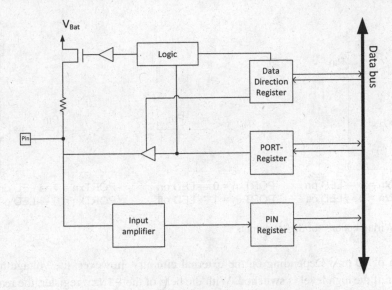

Fig. 3.5 Simplified schematic diagram of the digital inputs and outputs. (According to [1])

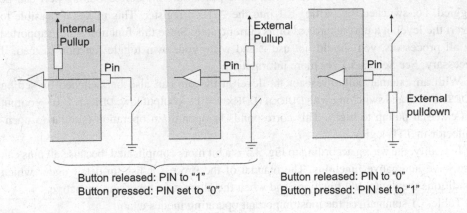

Fig. 3.6 Wiring options for digital inputs

```
MCUCR |= (1 << PUD);
```

Note: For all inputs, care should be taken to maintain a defined level by either activating the internal pull-up resistor or setting a level through external circuitry.

The level at the input can be read via the PINxn register. This takes the input level after the next clock edge due to the synchronization circuit.

Figure 3.7 shows possible wiring for digital outputs. If the DDRxn is set to 1, the output driver is active, the port is switched as output port in the so-called. "Push-pull operation". Depending on PORTxn, the pin is either high, i.e. low impedance to the supply voltage (PORTxn = 1) or low, low impedance to ground (PORTxn = 0). The device can drive a

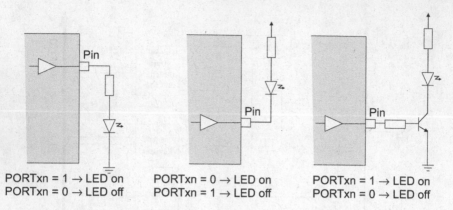

PORTxn = 1 → LED on   PORTxn = 0 → LED on   PORTxn = 1 → LED on
PORTxn = 0 → LED off   PORTxn = 1 → LED off   PORTxn = 0 → LED off

Fig. 3.7 Wiring options for digital outputs

maximum of 40 mA. Depending on the external circuitry, however, the voltage may be lower even if the high level is switched. With the help of the PINxn register, the real level at the output can be read out with a clock delay.

Note: Independently of DDRxn, with some processors the corresponding port can be toggled, i.e. switched, by writing a 1 into the PINxn register. This makes it possible to invert the level of a port regardless of its current state. Since this function is not supported by all processors, you should not use it and write your own toggle function instead, if necessary. See Sect. 3.5.2 for more information.

With an external pullup resistor the level at the pin can also be switched by setting PORTxn to 0 and switching with DDRxn (DDRxn = 1 → output low, DDRxn = 0 → output via external pullup to high). This corresponds to open-drain operation (similar to open-collector in TTL logic).

In reality, the wiring according to Fig. 3.5 is a bit more complicated, because all pins can also have alternative functions. The manual of the processor lists in detail under which conditions the digital I/O functions and when the alternative functions are active.

Table 3.1 summarizes the most important operating modes again:

3.5.2 Programming

The digital inputs/outputs are programmed by register access, as already discussed above. In the following example, an LED is connected to PORTB at output PB2 according to the pattern shown in Fig. 3.7 on the left.

In order to get a hardware abstraction for application programming, it is advisable to write a module LED.c (corresponding of course with a header file LED.h) the functions LED_Init(), LED_On() and LED_Off(). Alternatively, one can of course also imagine the function LED(char state), via the input parameter state, which can take the values ON and OFF, it is controlled whether the LED should be on or off.

Table 3.1 Operating modes of the digital inputs/outputs

Operating mode	DDRx	PORTx	PINx	Electrical connection
Input (tristate)	0	0	Open	Tristate
Input (high)	0	1	1 (0 if connected to ground)	High above approx. 30 kΩ
Output	1	0	0	Low
Output	1	1	1	High

The file LED.h looks like this:

```
#ifndef LED_H_
#define LED_H_
#define OFF 0
#define ON 1
/*Declaration of the functions*/
void LED_Init(void);
void LED_On();
void LED_Off();
void LED(uint8_t state);
#endif /* LED_H_ */
```

In the file LED.c the actual routines are executed:

```
#include <avr/io.h>
#include "LED.h"
void LED_Init()
{
    DDRB |= (1 << DDB2); //set data direction register to 1: Output
    PORTB &= ~(1 << PB2); //LED initially off
}
void LED_On()
{
    PORTB |= (1 << PB2);
}
void LED_Off()
{
    PORTB &= ~(1 << PB2);
}
void LED(uint8_t state)
{
    if (ON==state) { LED_On(); }
    else { LED_Off(); }
}
```

Some readers will wonder in the comparison if (ON == state) why the constant ON is on the left and not the variable state. The explanation can be found in Sect. 2.6.2: When programming, even experienced programmers confuse the assignment operator (=) with the comparison operator (==). The compiler does not notice this, because in both cases it is a syntactically correct statement. If one gets used to the fact that the constant to be compared is on the left *(lvalue)*, the compiler would react with an error message in case of an accidental assignment, since a variable must always stand as *lvalue* in an assignment.

Another function for toggling the LED, i.e. for switching at each activation, can look as follows:

```
void LED_toggle()
{
    if (PORTB & (1 << PB2)) PORTB &= ~(1 << PB2);
    else PORTB |= (1 << PB2);
}
```

The conditional expression PORTB & (1 < <PB2) takes the value 0 for every bit except bit 2. If this is set in PORTB, the whole expression is not equal to 0 and thus logically "true". If it is not set, the expression is 0 and thus "false". This executes the else branch.

Alternatively, the function can be written to set the PORT register bitwise XOR 1:

```
void LED_toggle()
{
    PORTB ^= (1 << PB2);
}
```

The two files can now be included in the basic structure from Sect. 2.10. Of course, LED.h must be included in the main program using the appropriate #include statement. LED.c can also be part of a library. The initialisation functions (including the following ones) are called one after the other in the Init() function in Sect. 2.10. Sometimes it is necessary to repeat an initialisation while the program is running.

In the second application example, let four pushbuttons or other switching devices be installed in ports D2, D3, D4, and D5 as shown in Fig. 3.6 left or center. The initialization function in a file key.c looks as follows with external pullup:

```
void key_Init(void)
{
    DDRD &= ~(1 << DDD2) &~(1 << DDD3) &~(1 <<DDD4) &~(1 << DDD5);
}
```

If you want to use the internal pullup instead, you can write the function like this:

```
void key_Init(void)
{
    DDRD &= ~(1 << DDD2) & ~(1 << DDD3) & ~(1 <<DDD4) & ~(1 << DDD5);
    #ifdef INTERNAL_PULLUP
        MCUCR &= ~(1 << PUD); //Pullup disable to 0
        //Pullup switch on
        PORTD |= (1 << PD2) | (1 << PD3) |(1 << PD4) | (1 << PD5);
    #endif
}
```

In this case the internal pullup can be switched on by a #define INTERNAL_PULLUP
in a separate configuration file that should generally be included. This technique of
compiler control is very popular, especially since the defines can also be executed from
the IDE or from the compiler call line with the -D parameter. Thus one can adapt the code
effectively and leanly to the hardware.

The keys are queried by masking:

```
char key_get(char key)
{
    return !(PIND & (1 << key));
}
```

The key_get() function returns a value other than 0 when the key is pressed and
0 when the key is released. In the key.h file handy names for the keys can be defined, for
example:

```
#define keyS1 PIND2
#define keyS2 PIND3
#define keyS3 PIND4
#define keyS4 PIND5
```

Through the query.

```
if (key_get(keyS1)) LED_On();
else LED_Off();
```

the LED from the above example is switched on when switch S1 is pressed. Note: This
takes advantage of the fact that in C a value other than 0 is interpreted as "true" in a
conditional expression.

In Chap. 4 we will deal with the query and especially the debouncing of pushbuttons in
the context of the software framework.

3.6   Interrupts

An interrupt is an interruption of the normal program flow.[5] As soon as an interrupt is requested (the triggering event is called interrupt request, IRQ), the normal program pauses and an interrupt service routine (ISR) is executed. Once this has been processed, the normal program is continued.

3.6.1   Getting Started with Interrupts

The ATmega88 knows 26 different interrupts, which can be triggered by the periphery and are listed in Table 3.2. In the following these are explained in more detail.

This means that there are basically two operating options for responding to peripheral interfaces: Sequential polling or interrupt operation. With polling all peripheral blocks are checked one after the other in an endless loop. In interrupt mode, the interrupt source must first be activated and, of course, an ISR must be provided. The memory contains a jump table with the memory addresses from Table 3.2, there is the entry address of the respective ISR. This is determined by the compiler via a macro. This is called the entry table.

As soon as the interrupt event occurs, the processor stores both the value of the program counter and the registers used in a stack and processes the ISR. The registers and the program counter are then reloaded from the stack and the program is thus continued at the original location. Three conditions must therefore be fulfilled for an interrupt to be triggered:

1. Enabling of the interrupts in the status register by $I = 1$
2. Enabling of the relevant interrupt in a mask register
3. Occurrence of the interrupt event

The enabling of all interrupts is done in many processors, so also generally in the AVR family by the macro sei(), which in the case of the AVR family is defined in the file avr/interrupt.h. This .h file must be included where either ISR are defined or the macros sei() and cli() are used. Cli() is used for basic disabling of interrupts. Since the AVR family does not know interrupt priority, this macro can be used to prevent another interrupt from arriving while an ISR is being processed.

An interrupt service routine (ISR) is assigned to each of the interrupts listed above. The instructions that are to be executed in the event of an interrupt are stored in this ISR. The macro ISR (..._vect), which is also defined in avr/interrupt.h, helps to define the ISR. The names of the macros are listed in Table 3.3 as examples.

For the RESET no ISR is stored, because this is as well known the main() function.

---

[5]From Latin: interrumpere.

Table 3.2 Interrupts of the ATmega88 [1]

Vector number	Program address	Triggered by	Interrupt definition
1	0x000	RESET	External pin, power-on reset, Brown-out reset and watchdog system reset
2	0x001	INT0	External interrupt request 0
3	0x002	INT1	External interrupt request 1
4	0x003	PCINT0	Pin change interrupt request 0
5	0x004	PCINT1	Pin change interrupt request 1
6	0x005	PCINT2	Pin change interrupt request 2
7	0x006	WDT	Watchdog time-out interrupt
8	0x007	TIMER2 COMPA	Timer/Counter2 compare match A
9	0x008	TIMER2 COMPB	Timer/Counter2 compare match B
10	0x009	TIMER2 OVF	Timer/Counter2 overflow
11	0x00A	TIMER1 CAPT	Timer/Counter1 capture event
12	0x00B	TIMER1 COMPA	Timer/Counter1 compare match A
13	0x00C	TIMER1 COMPB	Timer/Counter1 compare match B
14	0x00D	TIMER1 OVF	Timer/Counter1 overflow
15	0x00E	TIMER0 COMPA	Timer/Counter0 compare match A
16	0x00F	TIMER0 COMPB	Timer/Counter0 compare match B
17	0x010	TIMER0 OVF	Timer/Counter0 overflow
18	0x011	SPI, STC	SPI serial transfer complete
19	0x012	USART, RX	USART Rx complete
20	0x013	USART, UDRE	USART, data register empty
21	0x014	USART, TX	USART, Tx complete
22	0x015	ADC	ADC conversion complete
23	0x016	EE	EEPROM ready
24	0x017	ANALOG COMP	Analog comparator
25	0x018	TWI	Two-wire serial Interface
26	0x019	SPM READY	Store program memory ready

Table 3.3 Names of the ISR for the ATmega88

Vector number	Triggered by	Interrupt service routine ISR (..._vect) {...}
1	RESET	
2	INT0	INT0_vect
3	INT1	INT1_vect
4	PCINT0	PCINT0_vect
5	PCINT1	PCINT1_vect
6	PCINT2	PCINT2_vect
7	WDT	WDT_vect
8	TIMER2 COMPA	TIMER2_COMPA_vect
9	TIMER2 COMPB	TIMER2_COMPB_vect
10	TIMER2 OVF	TIMER2_OVF_vect
11	TIMER1 CAPT	TIMER1_CAPT_vect
12	TIMER1 COMPA	TIMER1_COMPA_vect
13	TIMER1 COMPB	TIMER1_COMPB_vect
14	TIMER1 OVF	TIMER1_OVF_vect
15	TIMER0 COMPA	TIMER0_COMPA_vect
16	TIMER0 COMPB	TIMER0_COMPB_vect
17	TIMER0 OVF	TIMER0_OVF_vect
18	SPI, STC	SPI_STC_vect
19	USART, RX	USART_RX_vect
20	USART, UDRE	USART_UDRE_vect
21	USART, TX	USART_TX_vect
22	ADC	ADC_vect
23	EE	EE_READY_vect
24	ANALOG COMP	ANALOG_COMP_vect
25	TWI	TWI_vect
26	SPM READY	SPM_READY_vect

3.6.2 Interrupt Programming Using the Example of the Pin Change Interrupt

A pin change interrupt (PCI) is triggered when at least one bit of a port changes, e.g. when a key is pressed. The AVR family knows the simple PCI, which can be triggered by any port and the so called External Interrupts, which can only be triggered by the pins INT0 and INT1 of the ATmega88. To trigger an interrupt, different registers have to be set. In the case of the simple PCI it is the PCICR (Pin Change Interrupt Control Register) and one of the PCMSKn (Pin Change Mask Register). Each pin of one of the three ports can trigger an interrupt as soon as its value changes (positive and negative edge). Each port forms its own

Table 3.4 Port B – PCMSK0 (Pin Change Mask Register 0)

Bit 7	Bit 6	Bit 5	Bit 4	Bit 3	Bit 2	Bit 1	Bit 0
PCINT7	PCINT6	PCINT5	PCINT4	PCINT3	PCINT2	PCINT1	PCINT0

Table 3.5 Port C – PCMSK1 (Pin Change Mask Register 1)

Bit 7	Bit 6	Bit 5	Bit 4	Bit 3	Bit 2	Bit 1	Bit 0
–	PCINT14	PCINT13	PCINT12	PCINT11	PCINT10	PCINT9	PCINT8

Table 3.6 Port D – PCMSK2 (Pin Change Mask Register 2)

Bit 7	Bit 6	Bit 5	Bit 4	Bit 3	Bit 2	Bit 1	Bit 0
PCINT23	PCINT22	PCINT21	PCINT20	PCINT19	PCINT18	PCINT17	PCINT16

Table 3.7 Pin Change Interrupt Control Register – PCICR

Bit 7	Bit 6	Bit 5	Bit 4	Bit 3	Bit 2	Bit 1	Bit 0
–	–	–	–	–	PCIE2	PCIE1	PCIE0

group of interrupts (PORTB: PCINT 0 ... 7, PORTC: PCINT 8 ... 14,[6] PORTD: PCINT 15 ... 23. The assignment given in Tables 3.4, 3.5 and 3.6 applies:

If the button on PIND4 from Sect. 3.5.2 is to trigger an interrupt, the corresponding bit PCINT20 in register PCMSK2 must be set.

Subsequently, the entire PCI group must be enabled; this is done in the PCICR register (Table 3.7).

Accordingly, the code for the initialization routine looks like this:

```
void PCI_Init(void)
{
    PCICR |= (1 << PCIE2); //PCI group 2
    PCMSK2 |= (1 << PCINT20); //PCI at PORTD PD4 free
    sei();
}
```

The interrupt service routine for the PCI can then look like this:

```
ISR(PCINT1_vect)
{
    cli(); //disable other interrupts if desired
    uiPCIcnt++;
```

---

[6] Port C has only seven inputs.

```
      sei (); //enable other interrupts again
}
```

The global variable uiPCIcnt counts only the edges in this case. It is defined in the range of module global variables from Sect. 2.10.

```
uint16_t uiPCIcnt = 0;
```

In general, interrupt service routines should be kept as short as possible, especially if further interrupts are to be permitted while the ISR is being processed (these are referred to as "nested interrupts"). Further examples of interrupt service routines follow in later chapters.

3.6.3   The External Interrupts INTx

The processors of the AVR family have in addition to the pin change interrupts much more powerful external interrupts, the ATmega88 for example the interrupts INT0 and INT1, which are located at pins D2 (pin 4 in the DIP package) and D3 (pin 5) in the connection diagram.

They are controlled via the registers EICRA – External interrupt control register A and EIMSK – External interrupt mask register (Table 3.8).

The bits ISC00 to ISC11 control the behavior according to Table 3.9, whereby the table is to be read in such a way that the "x" is to be replaced by the number of the interrupt. Setting all four bits to 1 would therefore trigger an interrupt on a rising edge on both terminals.

Before an interrupt request can be generated, it must be enabled accordingly. This is done in register EIMSK, as described in Table 3.10.

Where INT0 turns on external interrupt 0 and INT1 turns on external interrupt 1.

In the following code example, external interrupt 0 is triggered on a rising edge.

Table 3.8 EICRA – External interrupt control register A

Bit 7	Bit 6	Bit 5	Bit 4	Bit 3	Bit 2	Bit 1	Bit 0
–	–	–	–	ISC11	ISC10	ISC01	ISC00

Table 3.9 Control of the interrupt behaviour with external interrupt

ISCx1	ISCx0	Description
0	0	"Low" level (0) at INTx (0 or 1) generates an interrupt request
0	1	Any level change at INTx generates an interrupt request
1	0	A falling edge at INTx generates an interrupt request
1	1	A rising edge at INTx generates an interrupt request

Table 3.10 EIMSK – External interrupt mask register

Bit 7	Bit 6	Bit 5	Bit 4	Bit 3	Bit 2	Bit 1	Bit 0
–	–	–	–	–	–	INT1	INT0

```
void INT0_Init()
{
    EICRA |= (1 << ISC00) | (1 << ISC01);
    EIMSK |= (1 << INT0);
}
```

In the second example, it is triggered to a 0 level:

```
void INT0_Init()
{
    EICRA &= ~(1 << ISC00) & ~(1 << ISC01);
    EIMSK |= (1 << INT0);
}
```

The appropriate ISR toggles the LED as described above:

```
ISR (INT0_vect)
{
    LED_toggle();
}
```

3.7   Timer

A heart of the periphery of every microcontroller and at the same time the most versatile components are the timers. They generate defined time intervals, count events and can be used for PWM control of actuators or for time measurement.

3.7.1   Timer Basics

The ATMega88 contains three timer/counter TCNT. TCNT0 and TCNT2 are 8 bit, TCNT1 16 bit. They are initialized by registers and run independent of the program. When a certain state is reached, they trigger an interrupt or set an output port. In PWM mode, the setting and resetting of the outputs is controlled to achieve a variable duty cycle at a constant frequency.

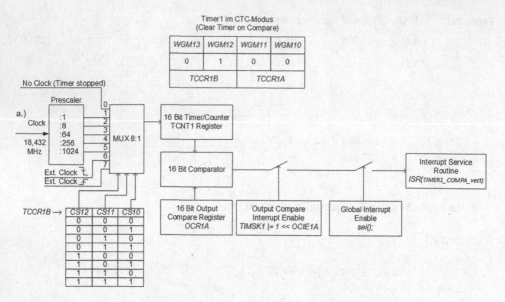

Fig. 3.8 Basic structure of the 16-bit timer 1

The heart of the timer/counter is the counter. It counts with 8 or 16 bits the events, which come either from a specially designated pin or from the system clock. Because this is usually much too fast to realize a reasonable time measurement, a so-called prescaler is integrated in each timer module. This prescaler then divides the system clock by an adjustable factor. A typical system clock is 18.432 MHz. With a factor of 1024 you get exactly 18 kHz as timer clock.

Figure 3.8 on the left shows the input of the "prescaler". This can be set to fCLK_I/O, fCLK_I/O/8, fCLK_I/O/64, fCLK_I/O/256 or fCLK_I/O/1024 with the bits CS10, CS11 and CS12 in the timer/counter control registers TCCR0B, TCCR1B and TCCR2B (depending on the timer number) as shown in Table 3.11. The choice is realized by a multiplexer.

Timers 0 and 1 can also be controlled via a pin instead of the prescaler, namely pins T0 and T1, which are assigned to port PD4 and PD5. Thus the device is able to count events.

Timer 2 does not know any external clock sources, but the prescaler can be adjusted in finer steps, according to Table 3.12.

To make a defined time interval out of the preset input clock, the value of the counter is compared with a preset value in the Output Compare Register (OCR 0/1/2). If this value is reached, an interrupt is triggered (Compare Mode) and/or an output bit is set (thus a hardware clock can be derived directly without loading the μC). Two different values can be used for comparison (OCR A and B). It should be noted that the data width of the timer is limited in this case (i.e. numbers of any size cannot be compared). Once this value has been reached, the value of the counter can be set to 0 (Clear Timer on Compare CTC). This ensures that the next interrupt also occurs after the same period of time.

Table 3.11 Selection of the clock source in register TCCR1B (Timer 1, analog to Timer 0)

CS12	CS11	CS10	Description
0	0	0	No clock source (stop timer)
0	0	1	fCLK_I/O (from prescaler)
0	1	0	fCLK_I/O/8 (from prescaler)
0	1	1	fCLK_I/O/64 (from prescaler)
1	0	0	fCLK_I/O/256 (from prescaler)
1	0	1	fCLK_I/O/1024 (from prescaler)
1	1	0	External clock source (pin T0 or T1) clocking on falling edge
1	1	1	External clock source (pin T0 or T1) clocking on rising edge

Note: The underlined number is the timer number, so it can be 0 or 1

Table 3.12 Selection of the clock source in register TCCR2B (Timer 2)

CS12	CS11	CS10	Description
0	0	0	No clock source (stop timer)
0	0	1	fCLK_I/O (from prescaler)
0	1	0	fCLK_I/O/8 (from prescaler)
0	1	1	fCLK_I/O/32 (from prescaler)
1	0	0	fCLK_I/O/64 (from prescaler)
1	0	1	fCLK_I/O/128 (from prescaler)
1	1	0	fCLK_I/O/256 (from prescaler)
1	1	1	fCLK_I/O/1024 (from prescaler)

Alternatively, one can trigger the interrupt when the timer overflows (i.e. from 0xFF to 0x00 or 0xFFFF to 0x0000). One could initialize the timer to a specific value from the ISR and achieve the same effect as with the Output Compare Register. Usually, however, the CTC mode is selected for timing purposes.

3.7.2 Programming the Timer/Counter

In the following, three use cases of timers are considered in more detail:

Case 1:	Generating a second cycle
Case 2:	Generating a fixed frequency at an output pin
Case 3:	Generating a PWM signal at an output pin

First the registers for timer 0 are considered. The other two timers of the ATmega88 work similarly, but the count and compare registers of the 16-bit timer 1 are two bytes long, furthermore it handles twice as many modes and therefore has more mode bits. Timer 2 also has no external clock sources. Therefore, the registers are assigned a little differently. In this book only a few modes are deliberately discussed, reference is made to the very

Table 3.13 Timer/Counter Control Register TCCR0A

Bit 7	Bit 6	Bit 5	Bit 4	Bit 3	Bit 2	Bit 1	Bit 0
COM0A1	COM0A0	COM0B1	COM0B2	–	–	WGM01	WGM00

Table 3.14 Timer/Counter Control Register TCCR0B

Bit 7	Bit 6	Bit 5	Bit 4	Bit 3	Bit 2	Bit 1	Bit 0
FOC0A	FOC0B	–	–	WGM02	CS02	CS01	CS00

Table 3.15 Timer Interrupt Mask Register TIMSKn

Bit 7	Bit 6	Bit 5	Bit 4	Bit 3	Bit 2	Bit 1	Bit 0
–	–	–	–	–	OCIEnB	OCIEnA	TOIEn

extensive and well described manual [1]. Tables 3.13 and 3.14 describe the registers TCCR0A and TCCR0B of Timer 0. They are used to control the modes, the prescaler and the behaviour of the timers at the outputs.

The meaning of each bit is explained in more detail in the respective examples. The other necessary registers are:

- The actual timer/counter register TCNT0
- The two Output Compare Registers OCR0A and OCR0B

and furthermore the Timer Interrupt Mask Register (Table 3.15).

This register is used to select whether a "compare event" of Output Compare Register A or B or an overflow of the timer should trigger an interrupt:

OCIEnB:	Output Compare Interrupt Enable Output Compare Match B
OCIEnA:	Output compare interrupt enable output compare match A
TOIEn:	Timer counter overflow interrupt enable

3.7.2.1 Generation of a One Second Cycle

A recommended oscillating crystal for the ATmega88 generates a frequency of 18.432 MHz as system clock. If this frequency is divided by the prescaler by 1024, the result is a frequency f of 18 kHz. By the formula

$$T = \frac{1}{f} = \frac{1}{18\ \text{kHz}} = 55.\overline{5}\ \mu s \tag{3.1}$$

this results in a period T of 55.555... μs. This means that the interrupt is triggered every approx. 55.5 μs. This is of course too short for our application, so this system clock must be divided further. This is the task of the Output Compare Register. In the above example, the

Table 3.16 Some waveform generation modes for timer 0

WGM02	WGM01	WGM00	Description
...	...	...	
0	1	0	Delete timer after comparison (CTC)
0	1	1	Fast PWM
...	...	...	

OCR0A value of 179 causes the interrupt service routine to be called only every 180th time (Please note: 0–179 is 180 passes). This causes the ISR to be called every $180 \cdot 55.\overline{5}$ μs = 10 ms. However, this still does not give us a second clock. Therefore the ISR divides the clock by a factor of 100. This results in a clock of exactly one second. With the settings described above, we select the initialization routine as follows.

```
void Timer0_Init()
{
    //initialization Timer 0
    //Clear Timer on Compare (CTC)
    TCCR0A |= (1 << WGM01);
    //prescaler set to divide by 1024
    TCCR0B |= ((1 << CS02) | (1 << CS00));
    //Output Compare Register A
    OCR0A = 179;
    //interrupt Timer0 Output Compare Match A
    TIMSK0 |= (1 << OCIE0A);
    sei();
}
```

Setting WGM01 sets the CTC (Clear Timer on Compare Match) mode (Table 3.16), so the timer counts up to 179 and then starts again from 0. In the following this mode will be considered first, later (Sect. 3.7.2.5) also some of the PWM modes.

By setting OCIE0A in the TIMSK0 register, the interrupt is triggered and we need a corresponding ISR:

```
ISR (TIMER0_COMPA_vect)
{
    ucFlag10ms = 1;
    ucCount10ms++;
    if (ucCount10ms == 100)
    {
        ucCount10ms = 0;
        ucFlag1s = 1;
    }
}
```

Note the three global variables used here, which of course must be declared outside of a function:

```
volatile uint8_t ucFlag10ms = 0;
volatile uint8_t ucFlag1s = 0;
volatile uint8_t ucCount10ms = 0;
```

After each 10 ms the variable ucFlag10ms is set, after 1 s also the ucFlag1s, this can be queried in an endless loop, where the variables must then also be set to 0 again immediately, so that they are back to 1 after the corresponding times. The keyword volatile here prevents the optimizer in the following code from optimizing away the branch, since the variable ucFlag1s is not set anywhere in the program flow, but in an interrupt service routine (unknown at compile time).

```
while(1)
{
    if (ucFlag1s)
    {
        ucFlag1s = 0;
        //here are the things that
        //should be executed every second
    }
}
```

If the timers are programmed in an external module Timer.c, then the two global flags must of course be declared as external in the Timer.h file.

```
external volatile uint8_t ucFlag10ms;
external volatile uint8_t Flag1s;
```

Alternatively, and usually the better way, is to define query routines in the Timer.c file instead of the cross-module flags, which query the flags and also delete them again. This is explained in more detail in Chap. 4.

3.7.2.2 Generation of a Fixed Frequency at an Output
In this example, a signal of a certain frequency is to be applied to output PORTB1, for example to output a sound via a loudspeaker. The frequency is to be adjustable. The choice of output is due to the fact that six pins named OCxy (Fig. 3.2) are directly connected to the timers. The numbers x stand for the timer number, the letters y for the Output Compare Registers (A and B), for example OC1A, which is identical to PORTB1.

For this example we use the 16-bit timer 1. It is initialized exactly as before, but this time without a prescaler, i.e. with the quartz clock f_{OSC} at the input of the counter register.

```
void Generator_Init(void)
{
    DDRB |= 1 << DDB1; //pin 1 of port B is set to output
    TCCR1A = (1 << COM1A0); //timer 1 toggles PB1 when counter
                          //TCNT1 = OCR1A
    TCCR1B = (1 << CS10); //system clock without prescaler
    TCCR1B |= 1 << WGM12; //CTC mode selected
}
```

The CTC mode ensures that the timer is reset to 0 after each Compare Match. Setting bit COM1A0 causes pin 1 to be toggled for each Compare Match. This means that the frequency at pin 1

$$f = \frac{f_{clk\_IO}}{2 \cdot n \cdot (OCR1A + 1)} \tag{3.2}$$

The number n here stands for the prescaler ratio, which is 1 in the above example. The frequency f_{clk_IO} corresponds to the crystal frequency without further settings. The OCR1A register is now set in another function, the variable uiFreq is of course not the frequency, this is derived from Eq. 3.2.

```
void Generator_ChangeFreq(uint16_t uiFreq)
{
    OCR1A = uiFreq;
}
```

Note that the 16-bit registers of timer 1 are of type unsigned int. To turn the signal on and off, we set or clear the COM1A0 bit that connects the output pin to the timer.

```
void Generator_Set_On(void)
{
    TCCR1A |= (1 << COM1A0);
}
void Generator_Set_Off(void)
{
    TCCR1A &= ~(1 << COM1A0);
}
```

Alternatively, the clock source could be switched off by clearing bit CS10 in register TCCR1B or switched on by setting it.

In general, the meaning of bits COM1A0, COM1A1, COM1B0 and COM1B1 in register TCCR1A can be taken from the following Table 3.17. However, this only applies if no

Table 3.17 Registers for controlling the output behaviour at pins OC1A and OC1B

COM1A1/COM1B1	COM1A0/COM1B0	Behavior
0	0	Pins not connected to timer (normal port outputs)
0	1	Pins are toggled during the output compare match
1	0	Pins are set to 0 at output compare match
1	1	Pins are set to 1 during output compare match

PWM mode is selected. Bits COM1Ax control output OC1A, bits COM1Bx control output OC1B.

3.7.2.3 Output of a PWM Signal

In this example, a PWM signal (pulse width modulation) is to be output from port B1 in order to control a motor. The idea of pulse width modulation is to output a square wave pulse with a changed duty cycle[7] D (with $0 < D < 1$). The average value of the signal amplitude of such a signal is:

$$\bar{y} = \frac{1}{T}\int_0^T y(t)dt = \frac{1}{T}\left(\int_0^{D \cdot T} y_{max}\,dt + \int_{D \cdot T}^T y_{min}\,dt\right) \qquad (3.3)$$

And with that

$$\bar{y} = \frac{1}{T}(D \cdot T \cdot y_{max} + T(1-D)y_{min}) = D \cdot y_{max} + (1-D)y_{min} \qquad (3.4)$$

If $y_{min} = 0$, as in Fig. 3.9, then

$$\bar{y} = D \cdot y_{max} \qquad (3.5)$$

and thus proportional to the duty cycle.

When outputting a signal on a pin as in Sect. 3.7.2.2, the duty cycle can be set by running the timer to the maximum value and switching the pin when a threshold is reached. There are twelve different modes of PWM for timer 1. In the following two basic modes are described.

In the left picture of Fig. 3.10 you can see the so-called fast PWM . Here the timer runs from 0 to the maximum value. This can be either 0x00FF (8 bits or 255), 0x01FF (9 bits or 511), 0x03FF (10 bits or 1023) or 0xFFFF (16 bits or 65,535) for timer 1 or can be defined

---

[7]Ratio between on-time and total time of a square wave signal.

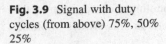

Fig. 3.9 Signal with duty cycles (from above) 75%, 50% 25%

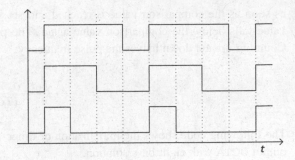

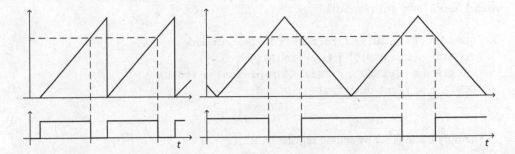

Fig. 3.10 Pulse width modulation (left (**a**) fast, right (**b**) phase and frequency correct)

by the value in register OCR1A or in other ways.[8] The timer is then reset and starts counting at 0 (BOTTOM). When the timer reaches the value set in the OCRnX register (X stands for A or B), the corresponding bit is set or cleared at the output. For example, COM1A1 = 1 clears input OC1A (= 0) when the comparison value in register OCR1A is reached. On timer overflow and reset to 0, the output is set again (non-inverting mode).

The frequency of the PWM signal results from the setting of the maximum value (also called TOP) and the value of the prescaler. In the case of fast PWM, this is – again with n as the prescaler ratio -

$$f = \frac{f_{clk\_IO}}{n \cdot (TOP + 1)} \tag{3.6}$$

The figure shows that in this mode the rising edge of the output signal, which results from the periodic overflow, is repeated isochronously, while the falling edge "wanders" with the state of the comparison register. The phase position of the pulse center thus changes. This is contrasted with the phase correct mode (Fig. 3.10 right). Here the timer counts up to the maximum value and then counts back to zero. The output pin is cleared when counting up

---

[8] For the sake of readability, a complete description is omitted here. In [1, 7] are detailed descriptions.

as soon as the comparison value is reached and set when counting down as soon as the value falls below the comparison value again. The pulse center is thus always in phase. Counting up and down halves the pulse frequency:

$$f = \frac{f_{clk\_IO}}{2 \cdot n \cdot (TOP)} \tag{3.7}$$

The following code shows the initialization of timer 1 for a phase correct PWM signal at output OC1A with eight bit resolution.

```
void Timer1_PWM_Init(void)
{
    DDRB |= (1 << DDB1); //port B bit 1 - PWM output
    TCCR1A = (1 << WGM10) | (1 << COM1A1);
    TCCR1B = ( 1<< CS10); //CS10 = 1 no prescaler (fOSC)
    OCR1A = 0; //set duty cycle to 0
}
```

The duty cycle is set by setting register OCR1A:

```
void Motor_Set_Speed(uint16_t uiSpeed)
{
    OCR1A = uiSpeed;
}
```

The difference between fast PWM and phase corrected PWM is usually marginal, with phase corrected PWM the pulses are symmetrical and in phase for each duty cycle. This may be necessary in some bridge circuits to limit the commutation currents.

3.7.2.4 Debouncing the Keyboard with the Timer

Buttons can assume two states "closed" and "open". In between are the events "key was pressed" and "key was released".

These events cannot simply be detected by a "pin change", because keys "bounce". This means that after mechanical triggering, the potential at the connected pin may fluctuate several times around the triggering threshold, causing the binary value of the pin to oscillate back and forth between the values "0" and "1" until the value then comes to rest after a few milliseconds (Fig. 3.11). Similarly, short-term interference pulses could fool the input into thinking that a switching operation is taking place.

In order to reliably detect the events "pressed" and "released", no pin change interrupt can be used. Instead, the key states must be checked at intervals of approx. 5 to 10 ms and then it must be determined on the basis of the change whether a key was currently pressed or not. In this way, line- or field-related interference pulses can also be suppressed.

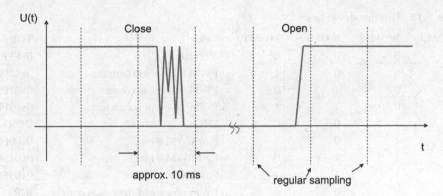

Fig. 3.11 Bouncing a button

```
if (ucFlag10ms)
{
    ucFlag10ms = 0;
    event = keyCheckEvent();
}
```

The appropriate "event generator" then looks as follows (in the key.h file Sect. 3.5.2).

```
#define EVENT_void 0
#define EVENT_S1_PRESSED 1
#define EVENT_S1_RELEASED 10
```

The corresponding code is then:

```
uint8_t S1Old;
uint8_t S1State;
char keyCheckEvent()
{
    S1Old = S1State; //how was the key 10 ms ago?
    S1State = key_get(keyS1); //what is the key now?
    if ((S1State) && (S1Old != S1State)) return EVENT_S1_PRESSED;
    else if ((!S1State) && (S1Old != S1State)) return EVENT_S1_RELEASED;
    else return EVENT_void;
}
```

The function returns whether the key was pressed (EVENT_S1_PRESSED) or released (EVENT_S1_RELEASED) since the last call of the function. If nothing has happened, EVENT_void is returned.

Table 3.18 Timer modes of timer 1

WGM13	WGM12	WGM11	WGM10	Mode	TOP
0	0	0	0	Normal	0xFFFF
0	0	0	1	PWM 8 bit phase correct	0x00FF
0	0	1	0	PWM 9 bit phase correct	0x01FF
0	0	1	1	PWM 10 bit phase correct	0x03FF
0	1	0	0	CTC	OCR1A
0	1	0	1	Fast PWM 8 bit	0x00FF
0	1	1	0	Fast PWM 9 bit	0x01FF
0	1	1	1	Fast PWM 10 bit	0x03FF
1	0	0	0	PWM phase and frequency correct	ICR1
1	0	0	1	PWM phase and frequency correct	OCR1A
1	0	1	0	PWM phase correct	ICR1
1	0	1	1	PWM phase correct	OCR1A
1	1	0	0	CTC	ICR1
1	1	0	1	Reserved	
1	1	1	0	Fast PWM	ICR1
1	1	1	1	Fast PWM	OCR1A

3.7.2.5 PWM for Advanced Users

The following Table 3.18 summarizes the timer modes depending on the bits WGM13 and WGM12 in register TCCR1B and WGM11 or WGM10 in register TCCR1A.

The PWM signals in phase-correct and phase/frequency-correct mode do not differ as long as the maximum value (Top), which defines the frequency, and the comparison value (OCRnx) (for the duty cycle) do not change. The OCRnx registers are double buffered and are updated in the two modes once on top and once on bottom as can be seen in Fig. 3.12. For a detailed description please refer to the data sheet [1].

A typical application for a continuous change of the OCRx registers is the output of audio data or a sine signal. For this you need two timers. One is responsible for the sampling (here: timer 0). With each sampling pulse a new value is generated at the PWM output by rewriting the OCRxn register of the second timer (Timer 2 in the following example). This second timer is then responsible for the output of the PWM signal whose duty cycle is present until the following sampling pulse.

The values of the OCRxn register are stored in a table, which represents the course of the (integrated) output signal as an integer type. For example, a quadrant of a sinusoidal signal $(0 \ldots 90°$ or $0 \ldots \pi/2)$ with a resolution of 64 time steps and 128 value steps looks like this:

```
uint8_t pucSintab[] = {0,3,6,9,12,16,19,22,25,28,31,34,37,40,43,
                46,49,51,54,57,60,63,65,68,71,73,76,78,81,83,85,
                88,90,92,94,96,98,100,102,104,106,107,109,111,
                112,113,115,116,117,118,120,121,122,122, 123,124,
                125,125,126,126,126,127,127,127};
```

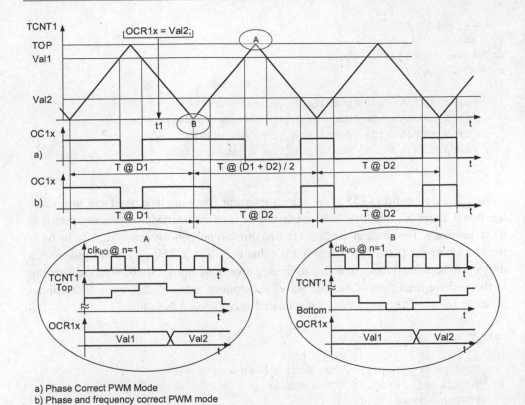

a) Phase Correct PWM Mode
b) Phase and frequency correct PWM mode

Fig. 3.12 Difference between phase-correct and phase- and frequency-correct PWM for AVR

In order to obtain a complete sinusoidal signal after integration, for the first quadrant, with each expiry of timer 0, the value at the index of the table corresponding to the respective time is added up to 128 (corresponding to 50% duty cycle). In the second quadrant, count down the table. In the third quadrant one subtracts the values counted forward again from 128, in the fourth quadrant one runs through the list backwards again. This is how a complete sine curve is assembled. With a time resolution of 64 steps per quarter wave, a full sine curve with the frequency results after 256 steps:

$$f_{sin} = \frac{f_{timer0}}{256} \tag{3.8}$$

The interrupt service routine of timer 0 (the sampling timer) contains the following code[9]:

---

[9]We first assume that the variable `ucMult = 1`.

```
ucCnt1 += ucMult;
ucQuad = ucCnt1 / 64;
ucCnt = ucCnt1 % 64;
switch(ucQuad)
{
    case 0: OCR2A = 128 + pucSintab[ucCnt++]; break;
    case 1: OCR2A = 128 + pucSintab[63 - ucCnt++];break;
    case 2: OCR2A = 128 - pucSintab[ucCnt++];break;
    case 3: OCR2A = 128 - pucSintab[63 - ucCnt++];break;
}
```

ucCnt1 counts up to 255 at each timer interrupt (the sampling time) and then starts again at 0. From this, the table index ucCnt is derived by the modulo operation from 0 to 63 respectively. The quadrant ucQuad results from an integer division of ucCnt by 64 in the range between 0 and 255. The result is either 0, 1, 2 or 3. Thus, a PWM signal with a duty cycle between 0 (corresponding to the negative minimum) and 100% (corresponding to the positive maximum of the sine wave) is output at output OC2A, i.e. a sine with an offset of $U_{Bat}/2$. The prerequisite is that timer 2 is configured as follows:

```
void Timer2_PWM_Init(void)
{
    DDRB |= (1 << DDB3); //port B bit 3 - PWM output
    TCCR2A = (1 << WGM20) | (1 << COM2A1);
    TCCR2B = (1 << CS20);
}
```

Figure 3.13 illustrates this situation.

By choosing 64 sampling steps (an integer divisor of 256) it is possible to generate a whole series of sine signals (namely 1,2,3,4,5 ... times the frequency of the basic signal described above) by increasing the counter not by 1 but by 2,3,4,5 ... and so on. However, the discrete-time and discrete-value resolution of the sinusoidal signal will be worse as a result. The frequency of the output sine signal is then:

$$f_{sin} = \frac{f_{timer0}}{256} \cdot ucMult \tag{3.9}$$

In order for the PWM signal to be completely generated, timer 2 must theoretically output at least one complete PWM pulse per sampling step. Practically, it should be about three to five pulses so that the signal is integrated correctly. With a resolution of n bits, Timer 2 must thus be clocked at least with $3 \cdot 2^n f_{timer0}$.

Of course, you can also play entire audio streams (as far as the memory allows) with it, whose sampling frequency does not need to be made nearly as fine, because the Nyquist-Shannon sampling theorem applies, according to which a signal band-limited to f_{max} must

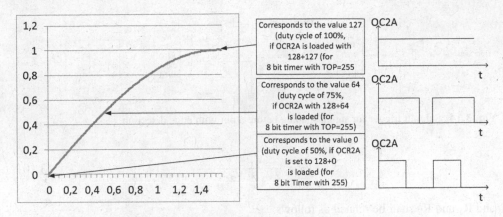

Fig. 3.13 Quarter sine wave from a PWM signal

be sampled with at least $2 \cdot f_{max}$ in order to reconstruct it completely from the time-discrete signal. Setting timer 0 to 125 μs, for example, results in a representable bandwidth of 4 kHz, which is quite sufficient for speech output. Timer 2 would then have to be clocked at $128 \cdot 3 \cdot 4$ kHz = 1.536 MHz with a completely sufficient resolution of seven bits. Discretization with 7 bits results in a theoretical signal-to-noise ratio of $7 \cdot 6$ dB = 42 dB (i.e. a ratio of signal to discretization noise of $1{:}2^{14}$). This quickly reaches the limits of the method due to the limited processor clock, so in practice other types of modulation (delta-sigma modulation) are used, which require only one bit of resolution per sampling step. Details about the sampling of analog signals and further literature recommendations can be found for example in [5].

With an external flash memory (Chap. 22), for example, an answering machine or voice output can be implemented.

3.7.2.6 Continuous Timing Between Input Pulses

Another interesting application is the continuous time measurement between two input pulses. This can be used, for example, to realize a tachometer, to measure a frequency or also the phase between two voltage curves. Basically, the signal should be shaped, i.e. a Schmitt trigger should be provided in the hardware if the sensor does not output a clean level. An example of an inverting Schmitt trigger is shown in Fig. 3.14. The threshold voltage V_T is set by a potentiometer, for example, or is fixed at half the supply voltage. The switching has a comparatively high input resistance, but requires an additional inverter. If the input voltage V_{In} approaches the upper hysteresis value V_2 from below, the output voltage V_{Out} jumps from the supply voltage V_S, i.e. 5 V or 3.3 V, to 0 V. If it approaches the lower hysteresis value V_1 from above, V_{Out} jumps back to the supply voltage. V_T is interpreted as follows:

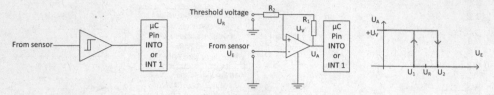

Fig. 3.14 Schmitt trigger for signal shaping before a pin with edge detection

$$V_T = \frac{V_1 + V_2}{2V_S + V_2 - V_1} \tag{3.10}$$

and R_1 and R_2 shall be related as follows:

$$R_1 = \frac{V_S - V_1}{V_1 - V_R} R_2 \tag{3.11}$$

If you do not want to set the hysteresis yourself, a Schmitt trigger can also be implemented with an IC from the 74×14 family, which contains six of these circuits.

The idea of the following program module is to start the time measurement with the falling (or rising, as you like) edge of an event and to read out and restart it with the next corresponding edge. The time measurement itself is implemented by a timer. Here a careful planning is necessary in advance. For example, one can set a suitable frequency by the prescaler and read the elapsed time directly from the counter register TCNTx. Depending on the used quartz you have to convert the time afterwards. For example, the time in μs can be set with an 8 MHz crystal and a prescaler ratio of 8. For longer times it is recommended to measure the time with a variable via the timer interrupt. In the following two examples are realized. The measurement of the phase position of two 50 Hz sine signals and the measurement between two pulses in milliseconds.

3.7.2.6.1 Time Measurement in the Millisecond Range

For this we need one timer interrupt per millisecond. In the interrupt a global variable is incremented (soft timer). With each external interrupt (pinchange) the value of this variable is re-stored and it is reset.

Timer 1 is used here and set so that one interrupt occurs per millisecond:

```
void Timer1_Init (void)
{
    TCCR1B |= 1<<CS10 | 1<<CS12 | 1<<WGM12; //prescaler 1024 -> 18 kHz
                                            //free running
    OCR1A = 17;
    TIMSK1 |= 1<<OCIE1A;
}
```

In the timer interrupt a variable is then simply incremented.

```
ISR(TIMER1_COMPA_vect)
{
    cntval++;
}
```

In our case, the signal to be measured is applied to the pin of INT0 and an interrupt is triggered:

```
ISR (INT0_vect)
{
    Timer1_Reset();
    SlopeDetected = 1;
}
```

This interrupt resets the count variable cntval to 0 and at the same time sets a flag which can be read in the main program.

```
void Timer1_Reset(void)
{
    TCNT1 = 0;
    counter = cntval;
    Cntval = 0;
}
```

On the flag SlopeDetected the variable counter can be read in the program in the main loop. The timer register TCNT1 is set to 0 at this point, since the interrupt is triggered asynchronously and the register in between can assume a value between 0 and 17, so that a small measurement error could result.

3.7.2.6.2 Measurement of the Phase of Two Sinusoidal Signals (Active and Reactive Power Measurement)

The phase position of two 50 Hz sinusoidal signals is of interest if the active and reactive power of inductive loads is to be separated, for example in order to bill only the active power. If current and voltage are out of phase, the apparent power S measured with the RMS values is composed of the active power P and the reactive power Q which merely oscillates back and forth between the load and the generator, whereby the following applies:

$$P = U \cdot I \cos \varphi \tag{3.12}$$

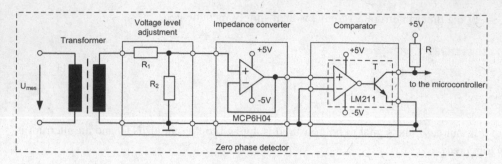

Fig. 3.15 Zero phase detector

where φ is the phase angle. So all you have to do is transform the voltage to a normalised value and convert the current to a voltage, ideally using a transformer coil, convert both to RMS values electronically and measure them in analog (see Sect. 3.8.5). In this section we discuss phase measurement, which is obtained by using the zero crossing of the voltage to start a timer and the zero crossing of the current to stop the time and hence the phase angle.

At 50 Hz (T = 20 ms), 1° of the phase angle corresponds exactly to 55.$\overline{5}$ μs, i.e. 18 kHz. With little effort, it is therefore possible to measure the phase angle to an accuracy of 1°, which corresponds to an accuracy at the cosine of less than 0.1% to a maximum of 1.7% for small phase angles.

If you use a 18.432 MHz crystal, you don't need a software counter at all, but configure the prescaler of the timer to 1024, so that the timer is driven with 18 kHz. With one pin the zero crossing of the voltage triggers an interrupt which sets the TCNTx register to zero, with a second pin the zero crossing of the current triggers an interrupt which reads the TCNTx register. This value is then exactly the phase angle in degrees (DEG).

To detect the zero crossing of a sinusoidal signal, a circuit as shown in Fig. 3.15 can be used.

The voltage to be measured U_{mess}, which usually has larger amplitudes than the supply voltage of the circuit, is transformed down with a transformer; if required, the voltage level is additionally adjusted. The transformer also ensures galvanic isolation between the measurement voltage and the microcontroller circuit. The voltage level adjustment must not change the phase of the voltage, so a voltage divider consisting of R_1 and R_2 must be used to set the desired ratio between the original voltage and the value measured by the microcontroller.

The following impedance converter, implemented with a rail-to-rail operational amplifier, is designed to prevent high current through the voltage divider and further voltage drops.

The zero crossing of the voltage is detected by a fast comparator whose inverted output controls an open-collector transistor. Access to the emitter and collector of the output transistor T enables the generation of TTL-compatible voltage levels via the presented circuitry, which can be switched directly to the microcontroller input. For example, the signal is applied to the INT0 input. The current is handled in the same way: Practice has shown that an inductive current transformer is best suited for a measurement circuit of this type; the measuring coil supplies a small voltage that is proportional (i.e., also sinusoidal) to the current to be measured. One can up-transform this voltage, which additionally ensures galvanic isolation. After that, it must be treated completely analogous to the voltage measurement circuit. The phase detection is then applied to the INT1 input.

On the software side, the timer that counts the time between zero crossings is set to 18 kHz by prescaler, reset at ISR of INT0, and read at ISR of INT1.

```
ISR (INT0_vect)
{
    TCNT1 = 0;
}
ISR (INT1_vect)
{
    counter = TCNT1;
    MeasurementComplete = 1;
}
```

In this way, the phase angle in degrees is obtained directly in the variable counter at 50 Hz. This only has to be evaluated on the MeasurementComplete flag, which can be done in the main loop. The most effective way to determine the cosine is to use a table (see Sects. 3.7.2.5 and 3.8.4).

3.7.2.7 Hardware Control of Motors with PWM Signals

Pulse width modulation is often used in conjunction with the so-called H-bridge, also called four-quadrant controller. The principle is shown in Fig. 3.16. Of the four transistors, 2 are controlled at a time. If T1 and T4 switch, the motor turns clockwise, if T2 and T3 switch, it turns anticlockwise. If T2 and T4 switch, the motor is braked. The diodes (free-wheeling diodes) pick up the current which flows on in the switching gap due to the magnetic and mechanical energy stored in the motor.

The circuit is made in such a way that T2 or T4 (low-side drivers) are switched through, while the PWM signals are applied to the so-called high-side drivers T1 and T3. Of course you have to be careful that the transistors of one string (T1 – T2 and T3 – T4) are not switched on at the same time. This would lead to a short circuit and to the destruction of the bridge. Therefore, a logic built in hardware usually protects the circuit from this condition and only provides the user with a PWM input and a direction input, which then has to be controlled by an additional IO pin of the microcontroller. Furthermore, the two high-side driver transistors must be driven with special charge pumps to ensure that the gate voltage is above the source voltage and thus higher than the supply voltage. In integrated H-bridges (e.g. ATA6823 from Atmel/Microchip without switching transistors or L6205 from ST Microelectronics with switching transistors), charge pump, logic and power driver as well

Fig. 3.16 Four-quadrant
controller

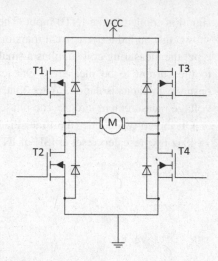

as protection against overheating and overcurrent are available in one package or even on
one chip.

An integrated bridge with selectable direction by an input pin has the special feature that
you can also apply the PWM signal to the direction input. As soon as the PWM input is
high, the motor switches on. At a duty cycle of 50% it stops, because it is reversed with the
PWM frequency, the resulting speed is therefore 0 but the motor generates a counter torque
at standstill. Duty cycles less than 50% then lead to counter-clockwise rotation, duty cycles
greater than 50% lead to clockwise rotation (if the direction input defines clockwise
rotation at 1). Figure 3.17 shows a simple circuit with bipolar transistors. Again, the fly
back diodes are used to protect the transistors from overvoltages when switching motors or
inductive loads. Transistors of type BC327 (PNP) and BC337 (NPN) are suitable for
simple modeling mini-motors, in which case base resistors around 1 KΩ and fly back
diodes of type 1 N4148 can be used. The 2–4 decoder 74HC139 is used to protect the
bridge against short circuits. The PWM signal of the microcontroller and a digital output
which sets the direction are connected to the inputs.

3.8   Analog Interface

The ATMega88 has two analog input interfaces. The *analog* comparator compares two
analog input voltages. The *analog-to-digital converter* (AD converter or ADC for Analog
Digital Converter) measures the voltage at one of the analog multiplex inputs with a
resolution of eight or ten bits and outputs it relative to a comparison voltage. The reference
voltage can be applied externally, generated from an internal reference (bandgap) with
1.1 V or simply be the supply voltage.

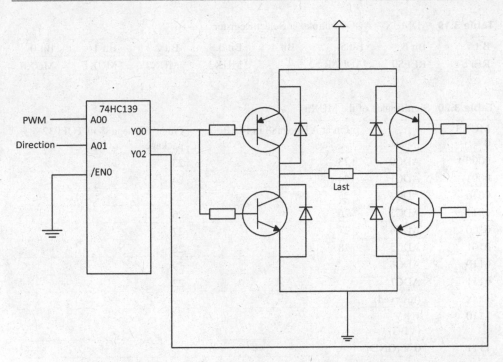

Fig. 3.17 H-bridge with bipolar transistors, for inductive loads with free-wheeling diodes

3.8.1 Analog Multiplexer

Several ports of the AVR family processors can be used as analog inputs, but only one at a time. The inputs ADC0 to ADC5 of the ATmegax8 correspond to the pins PC0 to PC5, depending on the housing ADC6 and ADC7 are led out separately or omitted. The respective input, if it is also a digital input, must of course be configured as such (Sect. 3.5), the pull-up resistor must be switched off. The ADMUX register (ADC Multiplexer Select Register) can be used to set which input is to be used (Table 3.19):

MUX3, MUX2, MUX1 and MUX0 binary code the input, whereas MUX3 is kept free in the described ATmegax8. Only in the combination 1110 and 1111 the internal reference voltage of 1.1 V or ground is applied to the (inverting) input of the measuring amplifier. The corresponding inputs are listed in Table 3.20.

The REFS1, REFS0 and ADLAR bits are described in Sect. 3.8.3.

3.8.2 Analog Comparator

The analog comparator has an internal differential amplifier that compares the voltages at the inputs AIN0 and AIN1 or ADC0 ... 7. If the voltage at the positive input (AIN0) is higher than the voltage at the negative input (AIN1 or ADC0 ... 7), the output ACO

Table 3.19 ADMUX – ADC Multiplexer Select Register

Bit 7	Bit 6	Bit 5	Bit 4	Bit 3	Bit 2	Bit 1	Bit 0
REFS1	REFS0	ADLAR	–	MUX3	MUX2	MUX1	MUX0

Table 3.20 Significance of the MUXn

MUX3 ... 0	Input	Pin at ATmega88 in PDIP package	Pin at ATmega88 in TQFP32 package
0000	ADC0	23	23
0001	ADC1	24	24
0010	ADC2	25	25
0011	ADC3	26	26
0100	ADC4	27	27
0101	ADC5	28	28
0110	ADC6	–	19
0111	ADC7	–	22
1xxx	(reserved)		
1110	1.1 V (VBG)		
1111	0 V (GND)		

Table 3.21 ACSR – Analog Comparator Status Register

Bit 7	Bit 6	Bit 5	Bit 4	Bit 3	Bit 2	Bit 1	Bit 0
ACD	ACBG	ACO	ACI	ACIE	ACIC	ACIS1	ACIS0

(Analog Comparator Output) is 1, otherwise 0. The analog comparator is controlled by register ACSR (see Table 3.21).

Bit ACBG controls whether the internal reference voltage source should be used instead of the comparison input AIN0 (ACBG = 0). Bit ACD = 1 switches the analog comparator off; this should always be set if it is not used and the processor is to save energy. ACO provides the output value (truth value AIN0 > AIN1) and if ACIE is set (Analog Comparator Interrupt Enable) the analog comparator interrupt service routine ISR (ANALOG_COMP_vect) is executed, this will cause the ACI bit to go to 1 until the interrupt service routine is started. The mode of execution is controlled by ACIS1 and ACIS0 according to Table 3.22.

Finally, ACIC can be used to set the analog comparator to trigger timer 1. Timer 1 can thus be configured to start running or counting when something changes at ACO. More about this in the datasheet of the respective processor.

Table 3.22 Interrupt execution of the analog comparator

ACIS1	ACIS0	Interrupt in progress ...
0	0	... when ACO changes
0	1	... (reserved)
1	0	... when ACO becomes 0 (falling edge)
1	1	... when ACO becomes 1 (rising edge)

3.8.3 AD Converter (ADC)

3.8.3.1 Functionality

The AD converter allows a simple measurement of analog sensor values with a resolution of 10 bits, whereby the manufacturer specifies the accuracy to ± 2 LSB,[10] an 8-bit measurement thus already sufficiently exhausts the accuracy of the analog input.

A simplified overview of the structure of the AD converter can be found in Fig. 3.18.

The measurement technique is based on the principle of successive approximation, in which half the reference voltage[11] $V_c = V_{REF}/2$ is first applied and compared with the input voltage V_{in}. If this is larger, the comparison voltage is increased by a quarter of the reference voltage, otherwise it is decreased by a quarter. In the next step, the change is one eighth of the reference voltage until the reference voltage and input voltage finally match.[12]

```
If Vin < Vc → Vc =Vc - VREF /4
otherwise →Vc =Vc + VREF /4
If Vin < Vc → Vc =Vc - VREF /8
otherwise → Vc =Vc + VREF /8
etc.
```

This is a step-by-step process that requires **clocking** and thus also **time.** The clock is generated from an own prescaler (see below) and should be between about 50 kHz and 200 kHz according to the manufacturer. To estimate the measurement time, it is important to know the timing. Ten clocks are necessary for the approximation procedure, three more for the preparation and post-processing of the measurement. For example, at 18.432 MHz we set the prescaler to $128 \rightarrow f_{ADC} = 144$ kHz.

Thus, if the measurement requires 13 clocks @f_{ADC} holds:

---

[10] LSB means "least significant bit", i.e. the least significant bits.

[11] Here, for compatibility with the manual, the letter V is used for voltage.

[12] See also Sect. 20.2.3

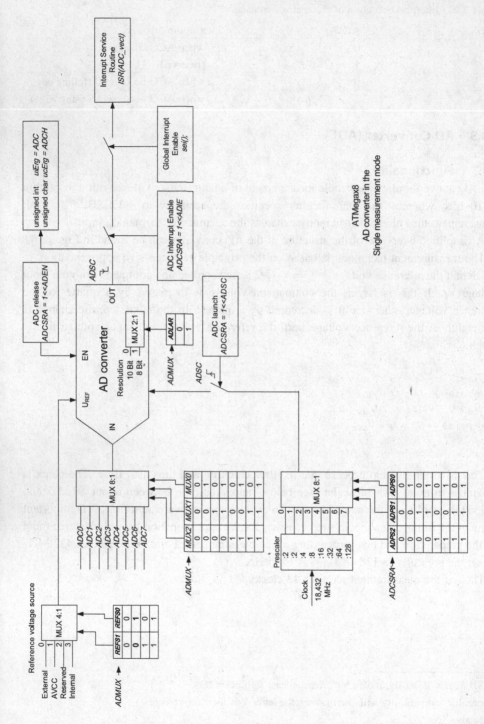

Fig. 3.18 Simplified structure of the AD converter

$$t_{Sample} = \frac{13}{144 \cdot 10^3}s = 9.027 \cdot 10^{-5}s \approx 90 \ \mu s \ \text{respectively} \tag{3.13}$$

$$f_{Sample} = \frac{144 \cdot 10^3}{13}Hz \approx 11 \ kHz \tag{3.14}$$

The ADC has a total resolution of 10 bits (1024), so that the value read out is the ratio:

$$ADC = 1024\frac{V_{in}}{V_{ref}} \tag{3.15}$$

represents. To avoid disturbances during the measurement, a *sample-and-hold* circuit decouples the input voltage to be measured from the ADC during the measurement.

The reference voltage can be applied externally and may then be at most as high as the supply voltage of the ADC, which is isolated from the rest of the system. Alternatively, the supply voltage of the ADC coupled via filter is used as reference voltage. This must be taken into account during initialization. Furthermore, the internal reference voltage source with 1.1 V can also be selected.

3.8.3.2 Triggering the Measurement

The measurement can be started in different ways. In the simplest case, the ADC Start Conversion bit (ADSC) is set in ADC control and status register A (ADCSRA). A measurement is then started immediately. The bit is automatically reset when the measurement is finished. By setting the ADATE bit (ADC auto trigger enable), automatic triggering of the measurement by various sources can be achieved. In the so-called "free running" mode, a new measurement is triggered by the ADC interrupt flag, which is set by the hardware when a measurement is completed ("conversion complete"). This ensures that a new measurement is started immediately after completion. Further sources are the external interrupt input 0, the analog comparator and various events of timers 0 and 1. These can be found in Table 3.25.

3.8.3.3 Registers for AD Programming

The results of the AD conversion can be found in the two 8-bit registers ADCH (high byte) and ADCL (low byte). At 10-bit resolution, only the two lowest bits in ADCH are relevant (right alignment).

Nice of the C compiler to allow direct access to both registers simultaneously with the 16-bit variable ADC:

```
uint16_t uiData;
uiData = ADC;
```

Table 3.23 Selection of the reference voltage source

REFS1	REFS0	Reference voltage source
0	0	AREF, internal reference voltage source is off
0	1	AVCC with external capacitor connected to AREF pin
1	0	Reserved
1	1	Internal 1.1 V reference voltage source

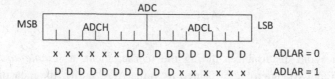

Fig. 3.19 Meaning of the left alignment in the AD converter

In addition to the MUX2 ... 0 bits mentioned above, the ADMUX register also contains the bits for selecting the reference voltage REFS1 and REFS0. This is set according to the following Table 3.23:

An interesting variant is offered by ADLAR (ADC Left Adjust Result) in the same register (Table 3.19): If the bit is set to 1, the result (10 bits) in the 16-bit register is left-aligned, i.e. the MSB[13] of the result is bit 15 of the result register. In order to read out the ADC with only eight bits, which is usually quite sufficient and does not involve any real loss of accuracy, it is therefore sufficient to read out the high-order byte of the result register ADCH with 8 bits of resolution when ADLAR is set. Figure 3.19 illustrates this situation. The bits marked with D are the data bits containing information.

The ADCSRA and ADCSRB registers (ADC Control and Status Register A and B) control the conversion process with the following bits:

ADCSRA (see Table 3.24):

- ADEN (ADC enable, bit 7): Switching on (1) or off (0) the ADC
- ADSC (ADC start conversion, bit 6): Start of measurement (1) or end of measurement (0). The bit is set to 0 by the controller in the "Single measurement" operating mode as soon as the measurement is completed.
- ADATE (ADC auto trigger enable select, bit 5): Here, it is possible to switch between single measurement mode (0) and auto trigger mode (1). In auto-trigger mode, the ADCSRB register can then be used to select whether the measurement is performed in free-running mode or in response to a trigger signal.
- ADIF (ADC interrupt flag, bit 4): The bit is set to 1 by the controller as soon as the measurement is completed. If an interrupt is triggered by the ADC, the bit is set to

---

[13] MSB: Most significant bit, i.e. the most significant bits.

Table 3.24 ADCSRA – AD control and status register A

Bit 7	Bit 6	Bit 5	Bit 4	Bit 3	Bit 2	Bit 1	Bit 0
ADEN	ADSC	ADATE	ADIF	ADIE	ADPS2	ADPS1	ADPS0

Table 3.25 ADCSRB – AD control and status register B

Bit 7	Bit 6	Bit 5	Bit 4	Bit 3	Bit 2	Bit 1	Bit 0
	ACME				ADTS2	ADTS1	ADTS0

0 again. Without interrupt control, the bit can be queried in a wait loop and cleared by writing a 1 (actually! A one) when the measurement data has been read out.

- ADIE (ADC Interrupt Enable, bit 3): If this bit is set, an end of measurement triggers an interrupt. This is a recommended method. With a global variable unsigned int uiResult the interrupt service routine looks like this:

```
ISR (ADC_vect)
{
    uiResult = ADC;
}
```

- ADPS2, ADPS1, ADPS0 (ADC Prescaler Select Bits): Set the divider ratio of the prescaler and thus the measurement speed: 000 is half-quartz clock, 010 is quarter-quartz, 011 is eighth-quartz, 100 is sixteenth-quartz, 101 is thirty-second-quartz, 110 is sixty-fourth-quartz, and 111 is 1/128-quartz. A measuring frequency of 50. 200 kHz is recommended by the manufacturer for maximum accuracy.

The ADCSRB register (Table 3.25) has the following bits:

- ACME (Analog Comparator Multiplex Enable, bit 6): Switches the analog multiplex inputs between ADC (0) and Analog Comparator (1). We leave it at 0.
- ADTS2, ADTS1, ADTS0 (ADC Auto Trigger Source): These bits code the source for the automatic triggering of the measurement if ADATE in ADCSRA is set to 1 (Table 3.26).

To save power, the corresponding digital inputs can be switched off during the measurement by setting the respective bit ADC0D ... ADC5D to 1 in register DIDR0 (Digital Input Disable Register).

Another power saving option is to switch the ADC from "free running" mode to single measurement by setting the PRADC value in the Power Reduction Register (PRR) to 0.

If a line is required exclusively as an analog input line and the digital input can be dispensed with, the digital input should be switched off to save power. This is done in register DIDR0 (Digital Input Disable Register 0, see Table 3.27). The corresponding digital input is switched off (disabled) by setting a 1.

Table 3.26 Trigger sources for starting an analog measurement

ADTS2	ADTS1	ADTS0	Trigger source
0	0	0	Free running mode
0	0	1	Analog comparator
0	1	0	External interrupt request 0
0	1	1	Timer/Counter0 compare match A
1	0	0	Timer/Counter0 overflow
1	0	1	Timer/Counter1 compare match B
1	1	0	Timer/Counter1 overflow
1	1	1	Timer/Counter1 capture event

Table 3.27 DIDR0 – Digital Input Disable Register 0

Bit 7	Bit 6	Bit 5	Bit 4	Bit 3	Bit 2	Bit 1	Bit 0
		ADC5D	ADC4D	ADC3D	ADC2D	ADC1D	ADC0D

3.8.3.4 ADC Initialization

To initialize the ADC of the µECU, we first assume the scenario of starting a free running measurement at input ADC4.

In addition, the supply voltage according to Fig. 3.20 is to serve as an external reference voltage, thus the values of the AD converter refer to the supply voltage.

We choose ADC4 as multiplexer input and a measurement frequency of 144 kHz, i.e., a prescaler ratio of 1/128 (144,000 * 128 = 18,432,000). This completes the initialization:

```
void ADC_Init(void)
{
    ADMUX = (1 << REFS0) | (1 << MUX2); //Vcc reference + channel 4 selected
    ADCSRA |= (1 << ADPS2) | (1 << ADPS1) | (1 << ADPS0); //prescaler :128
    DIDR0 |= (1 << ADC4D); //disable digital input
    ADCSRA |= (1 << ADEN); //AD converter is enabled
    ADCSRA |= (1 << ADSC); //AD converter is started for the 1st time
    ADCSRA |= (1 << ADIE); //AD interrupt is enabled
}
```

Note: In the second last line the first measurement is started. For various reasons, this measurement returns a wrong value (1024) and must be discarded. In this code the interrupt is "armed" immediately after the start of the measurement. So of course the first call of the interrupt service routine returns a wrong measurement value. Normally this is not of further importance, but if nevertheless, the ADC interrupt should only be enabled when the first measurement is safely finished. This can happen in the initialization phase for example by a delay loop.

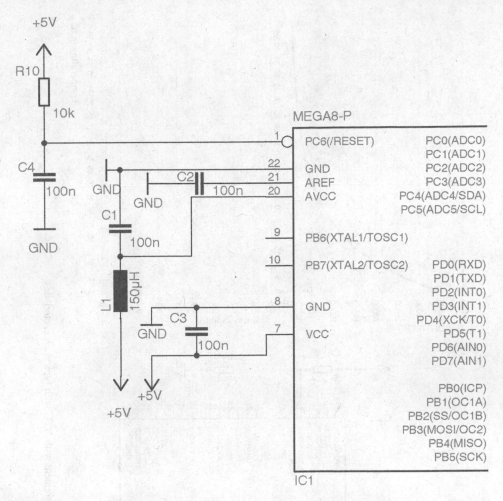

Fig. 3.20 Example of using the supply voltage as an external reference voltage

A second hint refers to possible noise of the system: In "Noise Reduction Mode" the processor is stopped during the measuring process to avoid disturbances by the CPU. More details can be found in the data sheet [1].

3.8.4 Example: Thermometer

A simple analog sensor circuit is a thermometer. Using a measuring resistor with a negative temperature coefficient (NTC or thermistor), the simple circuit shown in Fig. 3.21 can be designed. Even the crystal is omitted, via fuses, the internal oscillator with 8 MHz is

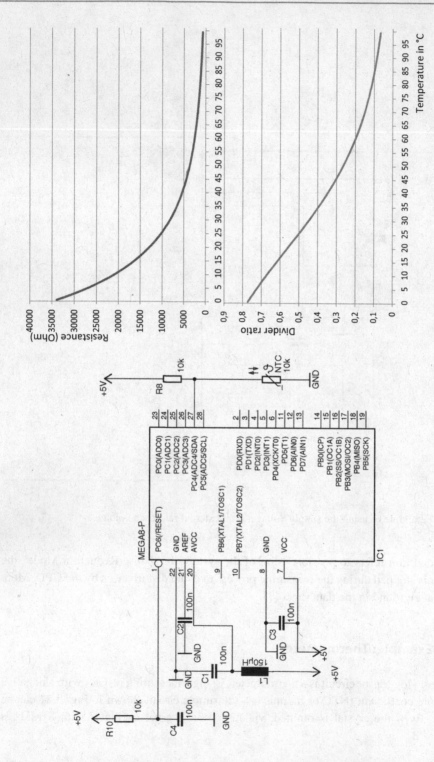

Fig. 3.21 Circuit of a simple temperature measurement with NTC with typical characteristics

selected (Sect. 3.3.2). The circuit omits how the measured temperature is evaluated (e.g. via a display, a serial interface or a digital output).

In this case the prescaler is set to 64 so that the measuring frequency is between 50 kHz and 200 kHz, according to Eqs. 3.13 and 3.14 it is then 125 kHz.

Hot conductors have a characteristic that satisfies the following function:

$$R_T = R_R \cdot e^{B \cdot \left(\frac{1}{T} - \frac{1}{T_R}\right)} \tag{3.16}$$

Thereby is:

- R_T the resistance at temperature T
- R_R the resistance at the reference temperature T_R in Kelvin,
- for example, $R_{25} = 10$ kΩ at $T_R = 25$ ° C $= 298.15$ K .
- B the thermistor constant in Kelvin, which is specified in the data sheet or determined experimentally.

B can be measured by determining the resistance value at two temperatures $R_R = R(T_R)$ and: $R_T = R(T)$

$$B = \frac{T \cdot T_R}{T_R - T} \ln \frac{R_T}{R_R} \tag{3.17}$$

In the circuit shown, the NTC is used in a voltage divider, which has a divider ratio of 0.5 at 25 °C, i.e. the voltage is 2.5 V. The upper diagram shows the resistance value of the NTC 10 k, the lower diagram shows the voltage at the analog input in each case as a function of temperature, so that the value of ADC is output according to Eq. 3.15.

To infer the temperature from the applied voltage, the following procedure can be used. Define a table (lookup table) containing the temperature values with the desired accuracy, in the following example from 0 to 40 °C in 1 K steps.

```
uint16_t uiAnaTemp[40] = {791,781,771,761,751,740,730,719,708,697,
                          686,674,663,652,640,628,617,605,593,582,
                          570,558,547,535,523,512,501,489,478,467,
                          456,445,434,424,413,403,393,383,373,363};
```

The value of the ADC is located in the variable `uiData` and was previously read from the ADC register. In a loop, the value is now successively compared with those in the table. The table index at the position that corresponds to the measured value or is just smaller than the measured value is output as the temperature.

```
for (j = 1; j < 40; j++)
{
    if ((uiAnaTemp[j-1] >= uiData) && (uiAnaTemp[j] < uiData))
    {
        temp = j - 1;
        break;
    }
}
```

3.8.5   Example: RMS Value Measurement on a Sinusoidal Voltage

To measure an AC voltage with the ADC, e.g. for a power measurement according to Sect. 3.7.2.6), you could sample it with high frequency and mathematically infer the parameters peak value and RMS value with a known curve (e.g. sine). According to the Nyquist criterion, the signal would have to be sampled with at least twice the fundamental frequency to obtain these values, but this does not work in practice, since the sampling time would have to be exactly at the phase angles 90° and 270° (see also Sect. 14.1). In practice, therefore, the signal would have to be oversampled considerably in order to obtain the mathematical curve and thus the RMS value correctly.

RMS is important because it denotes a constant voltage that produces the same power conversion at a resistive load as the measured AC voltage. The power at a resistive load:

$$P = UI = U\frac{U}{R} = \frac{U^2}{R} \sim U^2 \tag{3.18}$$

is proportional to the square of the voltage U, the power of an alternating voltage is the mean value of the power from the time-varying curve:

$$P = \overline{u(t) \cdot i(t)} = \frac{1}{T} \int_{t_0}^{t_0+T} \frac{u(t)^2}{R} \, dt \tag{3.19}$$

It follows that the RMS value of a voltage can be obtained from the square root of the root mean square of the voltage:

$$U_{eff} = \sqrt{\overline{u(t) \cdot i(t)}} = \sqrt{\frac{1}{T} \int_{t_0}^{t_0+T} \frac{u(t)^2}{R} \, dt} = \sqrt{\overline{u(t)^2}} \tag{3.20}$$

With a little calculation, if the curve is exactly sinusoidal, the RMS value is

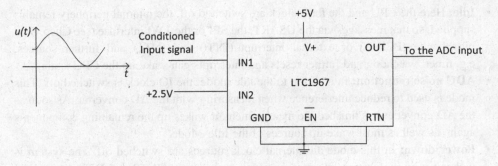

Fig. 3.22 Wiring of the LTC1967. (According to [6])

$$u(t) = \widehat{U} \sin \omega t \qquad U_{eff} = \frac{\widehat{U}}{\sqrt{2}} \qquad (3.21)$$

Technically, an RMS measurement is best handled with an external circuit, for example the IC LTC1967 from Linear Technology [6]. The chip calculates the true RMS value (RMS = root mean square) of a voltage of arbitrary shape. Figure 3.22 shows the basic circuitry of the device. As a rule, the signal is to be conditioned linearly, e.g. to turn the bipolar signal ($\pm \widehat{U}$) into a unipolar one ($0 \ldots 2\widehat{U}$). This offset can be compensated again via input IN2. No further software precautions are necessary except for the calibration of the measured value.

3.9   Power Management

Embedded systems often have to manage with very low energy reserves.[14] Like most microcontrollers, the AVR family has several mechanisms to adapt power consumption to computing requirements. Especially for battery-powered circuits that cannot or should not be explicitly switched off, it makes sense to specifically "put to sleep" the processor and peripheral components and only wake them up again when needed. In motor vehicles, many control units even switch the power supply itself to emergency mode, so that power supply losses can also be reduced. The main reason for power consumption in microcontrollers is gate reloading. So switching off or reducing clock is the common power saving method. This can be done individually per component (see below) or globally, depending on the power management concept.

The AVR family has a multi-level power management concept and knows the five sleep modes:

---

[14] See also Sect. 6.2

- **Idle:** Here the CPU and the flash clock are switched off, the internal periphery remains supplied, so that messages via the USART, the SPI or the TWI interface (see Chap. 5), a pin change (PCINT) or external interrupt (INT0 and INT1) and further sources, e.g. timer, watchdog and further resets and interrupts can wake up the CPU again.
- **ADC noise reduction:** In addition to the idle mode, the IO clock is switched off. This mode is used to reduce interference when measuring with the AD converter. As soon as the AD-converter has finished the measurement, it wakes up the remaining components again, as well as most wake-up sources of the Idle-Mode.
- **Power down:** In this mode all internal clock sources are switched off. The system is woken up by pin change interrupt, level interrupt on pins INT0 and INT1, a received TWI message or a reset.
- **Power safe:** This mode is almost identical to Power down, except that Timer 2 also generates a wake-up reason. You can use this mode for example to put the system into sleep mode and to wake it up periodically, for example to keep a clock running. You can operate Timer 2 asynchronously with a 32.768 kHz crystal and generate the remaining system clock from the internal 8 MHz oscillator. This mode is suitable for battery powered devices with clocks.
- **Standby:** Also almost identical to Power down, except that the external oscillator continues to run. This costs power, but has the advantage that the CPU is available again after waking up after only six clocks. Crystal oscillators have a long settling time of 10 ... 100 ms due to their very high Q (very narrowband oscillator circuit). The standby mode is ideal when a crystal is used and shortest response times after wake-up are required. Since all clocks are stopped, only the asynchronous modules still function.

The power consumption in power-down mode drops dramatically. Processors of type ATmega88V consume about 250 µA in normal operation at 1.8 V operating voltage and 1 MHz system clock, while they need only 0.1 µA in power-down mode. At 8 MHz and 5 V operating voltage the current consumption increases to about 12 mA while the power down mode still remains in the small single digit microamp range.

The sleep modes can be selected by writing to the SMCR (Sleep mode control register) and then started with the assembler command SLEEP. However, it is recommended to use the macros from avr/sleep.h provided by the manufacturer. A program for falling asleep and waking up by an interrupt looks as follows:

```
if (key_get(PIND4))
{
    set_sleep_mode(SLEEP_MODE_PWR_DOWN);
    sleep_mode();
}
```

As soon as PIND4 goes to zero, the microcontroller switches to sleep mode. Of course, an external interrupt must be set up to wake it up again (see Sect. 3.6.3). It is important that only the level interrupt is available in power down mode, i.e. only the configuration:

Table 3.28 Power Reduction Register (PRR)

Bit 7	Bit 6	Bit 5	Bit 4	Bit 3	Bit 2	Bit 1	Bit 0
PRTWI	PRTIM2	PRTIM0	–	PRTIM1	PRSPI	PRUSART0	PRADC

```
EICRA &= ~(1 << ISC00) & ~(1 << SC01);
```

Further energy savings can be achieved by specifically switching off the peripheral units that are not required. The Power Reduction Register is available for this purpose (Table 3.28).

Setting a 1 in any of the fields disables the two-wire interface (TWI) interface (bit7), timer 2 (bit 6), timer 0 (bit 5), timer 1 (bit 3), SPI interface (bit 2), USART (bit 1), and ADC (bit 0). Again, the libc in avr/power.h supports this. For example, `power_adc_disable ()` and `power_adc_enable()` turn the AD converter off and on. The same is true for `power_twi_enable()` and `power_twi_disable()` and others. Here it is recommended to read [3].

3.10   Internal EEPROM

The internal EEPROM serves the purpose of storing data persistently, i.e. permanently and independently of the processor's power supply. These then also survive a reset or an interruption of the power supply. EEPROMs need some time until the data is stored, therefore they are not suitable as working memory. Typical applications are error memory entries and parameterizations or adaptations of functions that should not be in the source code, such as controller parameters, serial numbers, teach-in parameters, characteristic diagrams. Since the number of EEPROM write accesses is limited, it should be used sparingly. The internal EEPROM can be written to and read from via the programming interface. In the first case, it can be used to parameterize a module (e.g. with an individual serial number or with a value that defines a SW variant). In the second case, the EEPROM can be used as an error memory or to "save" variable values via restarts. The EEPROM is not particularly suitable as a data logger because of its small size, here the use of an external flash memory is recommended, for this you will find notes in Chap. 22.

To access the internal EEPROM, it is recommended to use the macros from the include file avr/eeprom.h.

3.10.1 Declaration of a Variable in the EEPROM

The macro `EEMEM` instructs the compiler to move a variable into the EEPROM, for example.

```
uint8_t ucEPByte EEMEM; //one byte
uint16_t uiEPWord EEMEM; //one word - i.e. a 16 bit integer
uint8_t pucEPByte[10] EEMEM; //pointer to a field
```

The address of the variable is then used for the following operations. It should be mentioned that this definition also applies to float and to dword (corresponds to long).

3.10.2 Reading from the EEPROM

Reading from the EEPROM is a simple way to read in external parameterizations at runtime. For this purpose, the EEPROM is written via the programming interface and the data is read and used in the code. For example, functions can be enabled at runtime, time constants can be adjusted or the behavior of controllers can be influenced.

The reading is done by eeprom_read_byte():

```
uint8_t ucValue;
uint8_t pucField[10];
ucValue = eeprom_read_byte(&ucEPByte);
eeprom_read_block(pucField, pucEPByte,10);
```

The same applies to float, word and dword. Note the order of block transfer. The first parameter is the destination, i.e. the pointer to a field in the working memory, the second parameter is the address in the EEPROM generated with the EEMEM macro, the third parameter is the length.

3.10.3 Writing to the EEPROM

Writing into the EEPROM is done by.

```
uint8_t ucValue = 0x1A;
uint16_t uiValue = 0x2345;
uint8_t pucField[] = "0123456789";
cli();
eeprom_write_byte(&ucEPByte, ucValue);
eeprom_write_word(&uiEPWord, uiValue);
eeprom_write_block(pucField, pucEPByte, 10);
sei();
```

The same applies to float and dword. With block transfer, first the address of the field in the working memory is transferred, then the address in the EEPROM and finally the length of the field.

Instead of `eeprom_write_xxx()` also `eeprom_update_xxx()` can be used. These functions check before, if the value of the variable has changed compared to the EEPROM and write only the changes into the EEPROM. This can save considerable write cycles under certain circumstances and thus increase the durability of the EEPROM.

Interrupts must not occur during writing, as the timing must be strictly adhered to. The write times are relatively long (3.4 ms per byte for erase and reset) and the microcontroller blocks during these macros. They are therefore not recommended for fast persistent memory tasks. Chapter 7 presents more suitable networkable memory devices for logging tasks. These are usually written via SPI or TWI interface and buffer their data themselves, which leads to considerable time advantages.

To be sure that the EEPROM is ready to read and write, you can query the EEPROM Control Register (EECR). This is also done by a macro from the eeprom.h file called `eeprom_is_ready()`. You can also signal the completion of a read or write operation by an interrupt if larger block transfers are pending. This at least mitigates the blocking of the system.

In the following example, the number of resets of the microcontroller is stored by writing the following lines in the initialization part and erasing the EEPROM at the corresponding position before the first call by the programmer.

```
if (eeprom_is_ready())
        ucValue = eeprom_read_byte(&ucEPByte);
    eeprom_write_byte(&ucEPByte, ucValue + 1);
```

3.11   Dynamic Memory Use

The standard lib (<avr/stdlib.h>) allows dynamic memory usage for the AVR family as well. Although one should be extremely careful with this because of the limited size of the internal RAM, the functionality will be briefly described here. The ratios are shown in Fig. 3.23.

The AVR-C compiler knows four different working memories [3, 4]:

- The range of initialized data, i.e. variables that have been preassigned in the code, with the label.data

```
char err_str[] = "Fatal error occurred in code";
```

- The area of uninitialized data, which are global variables or variables marked as `static`. The area is called .bss
- The stack (or basement storage)
- The heap ("pile")

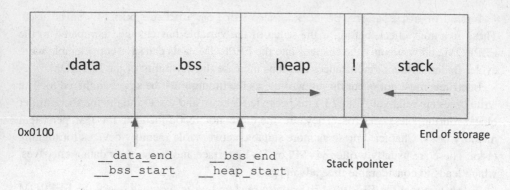

Fig. 3.23 Structure of the memory [3, 4]

The stack starts at the end of the memory area and is used for calling functions and interrupt service routines. As soon as a call is made, local variables, register contents and the program counter are pushed onto the stack and "saved" again after processing. This is done automatically by the processor during a function call (rcall) or an ISR. With the call depth (i.e. the number of functions called from functions or with the number of nested interrupts) the stack grows forward. Especially if you call functions recursively or use many local variables, the stack can grow quickly.

The heap contains the so-called *dynamic* memory. It starts directly after the .bss section and grows with the demand up to an adjustable limit between stack and heap.

To use the heap, you need two functions: malloc() and free(), both of which are available in the standard lib.

Example:

```
struct sListItem {char i; struct sListItem *next;};
typedef struct sListItem tListItem;
```

creates a data type tListItem. This consists of three bytes, namely the character i and a pointer to a structure next. The C keyword sizeof can be used to determine the size of the data type (Sect. 2.9).

To allocate (create) dynamic memory in the heap,[15] the malloc(size) function is used.

```
tListItem *first;
...
first = (tListItem*) malloc (sizeof(tListItem));
```

---

[15] Allocate literally means "to assign a place".

The call to `malloc(size)` returns a valid address in the heap or a `null pointer` if no more heap is available. Thus the heap grows slowly like a pile until it "collides" with the stack, which is quite fast in the case of an ATmega88. In addition to the heap, `malloc (size)` also creates a so-called "freelist", which stores the address and size of the allocated memory. With the call `free(address)` the memory is freed as soon as it is no longer needed.

```
free(first);
```

Forgetting to do this can result in a memory overflow.

For an example of how dynamic memory management is used, see Sect. 4.4.

3.12 Moving Data to the Program Memory

Larger tables take up a lot of memory. If these tables are constant (e.g. for the conversion of a sensor curve as in Sect. 3.8.4), the PROGMEM attribute can be used to prompt the compiler to write this data to program memory. In the case of the thermometer, this looks like this:

```
#include <avr/pgmspace.h>
(...)
const uint16_t uiAnaTemp[40] PROGMEM = {791,781,771,761,751,740,730,719,
                                        708,697,686,674,663,652,640,628,
                                        617,605,593,582,570,558,547,535,
                                        523,512,501,489,478,467,456,445,
                                        434,424,413,403,393,383,373,363};
```

Contrary to the recommendations from [3] to [4], the Gnu-C compiler 4.8.1 prompts to declare the field as `const`.

With the relocation to the Flash, however, the field can no longer be accessed directly. The access now happens, similar to the EEPROM, via a macro from pgmspace.h. The corresponding access to `uiAnaTemp` from Sect. 3.8.4 then looks like this:

```
if ((pgm_read_word(&(uiAnaTemp[j-1])) >= uiData) &&
              (pgm_read_word(&(uiAnaTemp[j])) < uiData))
(...)
```

The pointer to the content to be read is passed to `pgm_read_word` (analogous to `pgm_read_byte`, `pgm_read_dword`, `pgm_read_float`).

```
pgm_read_word(&(uiAnaTemp[j]));
```

Some pitfalls of this relocation are described in the cited references. In many cases, it is significantly justified and conserves memory space in the working memory.

References

1. Microchip: Reference Manual ATmega48/168. https://www.microchip.com/wwwproducts/en/ATmega88A. Accessed: 6. Jan. 2021.
2. Mikrocontroller.net. http://www.mikrocontroller.net. Accessed: 20. Dez. 2020.
3. NONGNU: AVR Libc. http://www.nongnu.org/avr-libc/user-manual/index.html. Accessed: Aug. 2020.
4. AVR Libc: http://savannah.nongnu.org/projects/avr-libc/ – last Accessed August 2020, also available at https://onlinedocs.microchip.com/ .
5. Meroth, A., & Tolg, B. (2008). *Infotainmentsysteme im Kraftfahrzeug*. Grundlagen, Komponenten, Systeme und Anwendungen. Vieweg.
6. Linear Technology: LTC1967 Precision Extended Bandwidth, RMS-to-DC Converter, data sheet (e.g. at https://www.mouser.de/datasheet/2/609/1967f-1271505.pdf. Accessed 20 Jan 2021).

Further Reading

7. Bernstein, H. (2020). *Mikrocontroller: Grundlagen der Hard- und Software der Mikrocontroller ATtiny2313, ATtiny26 und ATmega32–2*. Auflage Springer.
8. Schmitt, G. (2011). *Mikrocomputertechnik mit Controllern der Atmel AVR-RISC-Familie – 5.* Auflage De Gruyter.
9. Gaicher, H., & Gaicher, P. (2016). *AVR Mikrocontroller – Programmierung in C: Eigene Projekte selbst entwickeln und verstehen* (1. Aufl.). Tredition.
10. Salzburger, L., & Meister, I. (2013). *AVR-Mikrocontroller-Kochbuch* (1. Aufl.). Franzis.
11. Spanner, G. (2010). AVR-Mikrocontroller in C programmieren: Über 30 Selbstbauprojekte mit ATtiny13, ATmega8, ATmega32 (PC & Elektronik) (1. Aufl.). Franzis.
12. Williams, E. (2014). Make: AVR programming: Learning to write software for hardware – 1. Auflage, O'Reilly and Associates, Februar 2014.
13. Schäffer, F. (2014). AVR: hardware und Programmierung in C – Überarbeitete und erweiterte Neuauflage, Elektor, Dezember 2014.

Software Framework

Abstract

In this chapter, the basic architecture of a measurement system is presented, in particular the hardware abstraction.

After studying Chap. 3 you should be able to understand and operate important functions of the AVR microcontroller. This chapter now deals with organizing the code and modularizing it with the goal of readability and reusability. In addition, the chapter contains some considerations about the timing of the software, thus a simple multitasking scheme.

After the basic programming of the most important functions of a microcontroller has been described in the first chapters, it is now time to bring a little order into the program structure. This requires at least a rudimentary software architecture. Architecture is meant here as in IEEE 1471 as:

"Architecture is the fundamental organization of a system embodied in its components, their relationships to each other and to the environment and the principles guiding its design and evolution" [1].

An architectural design always has certain goals [2]. In the present case, the following goals are to be achieved first:

- Representation of self-contained modules
- Good reusability of the software modules
- Testability of the modules
- Good readability and comprehensibility

© The Author(s), under exclusive license to Springer Fachmedien Wiesbaden GmbH, part of Springer Nature 2023
A. Meroth, P. Sora, *Sensor networks in theory and practice*,
https://doi.org/10.1007/978-3-658-39576-6_4

- Combinability of the different solutions
- Hardware encapsulation or abstraction
- Independence from the operating system environment

Code efficiency, low power consumption, and optimal resource utilization are of course added, but will not be covered in too much depth in this book. Microchip/ATMEL has published its own application note [3] on this subject.

4.1   Views

A software architecture usually uses different views of the system. These are used to describe the relationship of the components with each other and with the outside world statically and dynamically. Some important views are:

- The static component view (Fig. 4.1), describes the relative relationship between the functions/components: Who accesses whom?

- The static distribution view (deployment): How are the functions/components distributed on physical devices in a distributed system?
- The context view: What does the interface look like to the outside world?
- Dynamic views: Here, the dynamic interaction of components (sequence chart) or the processes within a component (example: state machine) are described in temporal and logical dependence.

Some of these views are presented in more detail in the course of the chapter.

Figure 4.1 shows a simple software architecture for a measurement system. The sensors are connected either to an analog interface (temperature sensors, brightness sensors, proximity sensors, acceleration sensors with analog output), to a digital interface (presence and proximity sensors, limit or threshold switches) or to one of the serial interfaces described in the following Chaps. 7, 8, 9, 10 and 11. The hardware abstraction layer described in the following section serves to decouple the hardware from the application. The application finally takes over the processing of the data, whereby the preprocessing (for example filtering) can be outsourced to a separate sensor abstraction layer. The result of the processing is then either available on a bus system as a message (Chaps. 7, 8, 9, 10 and 11), is shown on a display (Chap. 25) or for an intervention with an actuator (lamps, motors, valves, thermal elements), as indicated in Chap. 3, the latter usually via a PWM control or by a digital output.

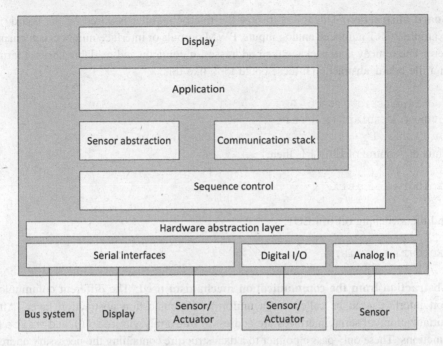

Fig. 4.1 Simple software architecture (static component view) for a measurement system

4.2   Hardware Abstraction

Hardware abstraction represents an important measure for improving the readability and portability of software and facilitates reuse in different projects. In this book, hardware abstraction refers to the following areas (Fig. 4.1):

- **Abstraction from the microcontroller family:** Here common concepts, for example timers, analog and digital inputs and outputs, EEPROM and others (see Chap. 3) are represented by own functions and modules (one also speaks of "wrapping", because the functionality is packed into own functions). Access is then only via these wrapper functions. For example, the following functions can be provided for a digital input/ output in a separate DIO module, which is exchanged from processor type to processor type:

```
- DIO_SetDirection(address)
- DIO_Write(address,value)
- DIO_Read(address)
```

- **Board abstraction:** This refers to how the microcontroller is wired on the board, for example PORT numbers, analog inputs, PWM outputs or interface numbers can change here. The remedy is to package their addresses in common and readable abstract terms, in a file board_abstraction.h these could look like this:

  – #define LED1OUT PORTB
  – #define LED1BIT (1 << PB4)

Thus the control of LED 1 is then:

```
LED1OUT |= LED1BIT;
```

and the switching off of LED1:

```
LED1OUT &= ~LED1BIT;
```

- **Abstraction from the communication mechanism used:** The different communication interfaces can be called via a uniform communication abstraction layer. After initialization, all serial interfaces are addressed uniformly via `read()` and `write()` functions. These only pass a pointer to a data structure containing the necessary address information and data to a ring buffer (Sect. 5.1.1 ff.) from which the data is then passed to the relevant interface in the background.
- **Abstraction from the sensors:** in the sensor **abstraction** layer, data structures and functions are available that behave like a generic sensor of a certain type. These are addressed by the actual functional program layer. The latter then does not need to be touched when a new sensor is introduced. Only the abstraction layer needs to be adapted.

Of course, the introduction of such abstraction layers always means a loss of performance and memory. At the same time, the power requirements of the processor may increase. Individual compromises must be found here.

4.3   Modularisation and Access to Modules

Modularization also requires tradeoffs between code efficiency (flash), memory efficiency, and readability or reusability. Beginners and teams working together on software are advised to keep readability and reusability in mind first. These involve combining functions that belong together into a module. The module consists of a .c file containing the sources and an .h file in which shared symbolic placeholders (#define) and the module interfaces are declared in the form of access functions. At this point, the use of variables that are global (`external`) beyond the module is initially discouraged (see also Sect. 2.8.5).

Although these provide an efficient means of accessing the module, they also have the disadvantage that their use is difficult to control. A simple alternative, which is not too expensive in terms of program memory, are access functions ("setter" and "getter" functions) to the global variables in the module.

Later, in Sect. 5.2, we will use a state machine. This uses a module-global variable ucState declared in the statemachine.c module. Other modules should be able to query and set this variable. For this purpose, we agree on the following setter and getter functions, whose task is easy to recognize:

```
uint8_t ucState ;
void set_ucState(uint8_t state)
{
    ucState = state ;
}
uint8_t get_ucState(void)
{
    return ucState;
}
```

This means that the variable no longer has to be published over module globally by extern. In addition, this procedure offers the advantage that a value range check can take place immediately on access or access can be prevented by interrupts, as the following example shows, in which a variable cPercentage is to be written, provided it remains in the value range:

```
#define ERROR_OUT_OF_BOUNDS 0
#define WRITE_SUCCESSFUL 1
char cPercentage;
char set_cPercentage(char value)
{
    char result;
    cli(); //disable all interrupts
    if ((value >= 0) && (value < 100))
    {
        cPercentage = value;
        result = WRITE_SUCCESSFUL;
    }
    else result = ERROR_OUT_OF_BOUNDS;
    sei(); //enable all interrupts again
    return result;
}
```

Furthermore, modules can be better protected against accidental overwriting of global variables in this way, because only the variables that can also be used publicly are known

via the .h file. Another example for accessing module-global variables is shown in the
following Sect. 4.4.

4.4   Time Control

Timing is usually essential for coordinated software execution. As described in Sect. 3.7.2,
we use a timer for global software flow control. This timer generates an interrupt that is
triggered in the shortest time that occurs in the system. All important system times are then
derived from this in the interrupt service routine. With a suitable abstraction, a small
operating system can be built from this, in which various tasks are regularly processed in
a time-controlled manner. In the following example, the times 10 ms, 50 ms and 100 ms are
required in the system. In addition, a "stopwatch function" is to inform how much time has
elapsed since the last call of the function in 100 ms. In a module Timer.c the following
global variables are declared:

```
uint8_t ucTimer1_Flag_10ms = 0; //becomes 1 when 10 ms have elapsed
uint8_t ucTimer1_Flag_50ms = 0; //becomes 1 when 50 ms have elapsed
uint8_t ucTimer1_Cnt_50ms = 0; //counts the 1ms until 50 ms has elapsed
uint8_t ucTimer1_Flag_100ms = 0; //becomes 1 when 100 ms have elapsed
uint8_t ucTimer1_Cnt_100ms = 0; //counts the 1ms until 100 ms has elapsed
unsigned long ulSystemClock = 0; //counts globally the 100 ms
```

Timer 1 is used in this example, but another timer can be used just as well.

Timer 1 is therefore initialized as described in Sect. 3.7 so that it triggers an interrupt
every 10 ms, which is possible with high accuracy for a 16-bit timer at 8 MHz if the crystal
frequency is divided by 64 and counted from 0 ... 1249:

```
void Timer1_Init(void)
{
    TCCR1B |= (1 << WGM12); //CTC Mode
    TCCR1B |= (1 << CS11) | (1 << CS10); //prescaler 64
    OCR1A = 1249; //Comparison value occurs every 10 ms
    TIMSK1 |= (1 << OCIE1A); //enable interrupt for compare registers
    sei(); //enable all interrupts
}
```

With 18.432 MHz crystals, the same result would be achieved by dividing by 1024 and
counting from 0 ... 179. This is then also possible with timer 0 or timer 2.

In the ISR of the timer the above mentioned time variables are now incremented. As
soon as they have reached their target value (which is 5 for the 50 ms variable, 10 for the
100 ms variable and 100 for the 1 s variable), the corresponding flags are set to 1 and
counting immediately starts again:

```
ISR(TIMER1_COMPA_vect) //is triggered by the timer every 10 ms
{
    ucTimer1_Flag_10ms = 1;
    ucTimer1_Cnt_1s++;
    if (ucTimer1_Cnt_1s == 100)
    {
        ucTimer1_Cnt_1s = 0;
        ucTimer1_Flag_1s = 1;
    }
    ucTimer1_Cnt_50ms++;
    if (ucTimer1_Cnt_50ms == 5)
    {
        ucTimer1_Cnt_50ms = 0;
        ucTimer1_Flag_50ms = 1;
    }
    ucTimer1_Cnt_100ms++;
    if (ucTimer1_Cnt_100ms == 10)
    {
        ucTimer1_Cnt_100ms = 0;
        ucTimer1_Flag_100ms = 1;
        ulSystemClock++;
    }
}
```

The evaluation of these "clocks" is done in the module Timer.c, shown here as an example for the 100 ms clock:

```
char Timer1_get_100msState(void)
{
    if (ucTimer1_Flag_100ms == 1)
    {
        ucTimer1_Flag_100ms = 0;
        return TIMER_TRIGGERED;
    }
    else return TIMER_RUNNING;
}
```

If the flag is set, it is cleared when the getter function is called and the value TIMER_TRIGGERED, which is different from 0, is returned. Otherwise 0 is returned. Of course, the following definitions must still be made in Timer.h for this:

```
#define TIMER_RUNNING 0
#define TIMER_TRIGGERED 1
```

The main program is now structured as follows:

```
int main(void)
{
    Timer1_Init();
    while(1)
    {
        if (Timer1_get_10msState()) Task10ms();
        if (Timer1_get_50msState()) Task50ms();
        if (Timer1_get_100msState()) Task100ms();
        IdleTask();
    }
}
```

As you can easily see, the `Task10ms()` function is now called every 10 ms, provided that the While loop is run in significantly shorter cycles. We therefore call this function the 10-ms task. The same applies to the other timed tasks. Only the `IdleTask();` is called in every loop pass and should not contain long code, at most you can toggle a pin there to check the load of the processor with the oscilloscope. In the final code it is mostly omitted.

With this simple tool, you can already build a simple, real-time "multitasking" operating system. The while loop can be called a *scheduler*. However, it is extremely important that all tasks together run faster than the shortest time in the system. In the above example, the processor time for all called tasks together must not be longer than 10 ms, otherwise the real-time is no longer maintained. This scheme is therefore also called *cooperative* multi-tasking, since all tasks must behave cooperatively and are not monitored. In contrast, in *preemptive multitasking* the tasks are interrupted when it is the turn of another task. This happens in a timer interrupt and guarantees a really exact adherence to the time constraints (real-time). However, considerable effort must be expended here to ensure the integrity of shared variables. For this reason, this book does not provide a more detailed explanation.

In the example above, a system clock was added. This counts the 100 ms as an `unsigned long` variable and would therefore only start again at 0 every 429,496,729.6 s (which is still over 119,000 h). But already counting the 10 ms or even the milliseconds, this value shrinks again below the threshold, where realistically an overflow can happen at runtime. To use this counter as a stopwatch, an additional variable `ulLastClock` is created globally in timer.c. Whenever the counter function is called, it is set to the current timer value and the difference between the current and the old value is output. If a (simple) range overflow occurred in the meantime, then the difference between the two numbers is counted and subtracted from the maximum of the value range. This can be determined by the macro `ULONG_MAX` from the includefile limits.h, which is loaded automatically. In this way, the stopwatch can of course also be implemented with shorter variables.

```
unsigned long TimerMeasure(void) //returns how much time has passed
                                 //between the last and the current call
{
    unsigned long diff;
    if (ulSystemClock > ulLastClock)
    {
        diff = ulSystemClock - ulLastClock;
    }
    else
    {
        diff = ULONG_MAX - (ulLastClock - ulSystemClock) ;
    }
    ulLastClock = ulSystemClock;
    return diff;
}
```

It must be clear that no lapping is intended here. If necessary, a "lap counter" must also be installed.

References

1. ISO/IEC/IEEE 42010 – Recommended practice for architectural description of software intensive systems. (2020). http://www.iso-architecture.org/ieee-1471. Accessed 3 Jan 2020.
2. Starke, G., & Hruschka, P. (2011). *Software-Architektur kompakt* (2. Aufl.). Spektrum Akademischer Verlag.
3. Microchip/ATMEL. (2016). AVR035: Efficient C coding for AVR. https://www.microchip.com/wwwAppNotes/AppNotes.aspx?appnote=en590906. Accessed 2 Apr 2021.

Memory Concepts and Algorithms

Abstract

This chapter introduces basic algorithms such as queuing, FIFO, and state machine that are necessary in many sensor projects.

In this chapter some important memory concepts and algorithms are presented. All examples used here are processor-independent and can also be used in other systems. Only a few of the algorithms described below are suitable for beginners, most likely the simple FIFOs mentioned in Sect. 5.1.1 and the first of the two state machines.

5.1   Important Storage Concepts

Although the possibilities of the processors of the AVR-family are limited, it is nevertheless worth to look at a few memory concepts, these are anyway independent of the used processor type. Especially the queue is important when it comes to data communication, especially when data packets occur irregularly (not isochronously). In this chapter lean implementations are presented, which are easily portable and usable.

5.1.1   Queues and Ring Buffers (FIFO)

Queues are used whenever two temporally independent, sometimes stochastic processes are coupled together. In the field of embedded systems of the size described in this book, queues can be used when, for example, a regular measurement is transmitted over an

© The Author(s), under exclusive license to Springer Fachmedien Wiesbaden GmbH, part of Springer Nature 2023
A. Meroth, P. Sora, *Sensor networks in theory and practice*,
https://doi.org/10.1007/978-3-658-39576-6_5

insecurely available radio interface or the data is to be transmitted via a bundled packet. Another use case is when asynchronous input variables from a network are to be processed in an isochronous process[1] (buffer). Due to the access sequence First in – First Out, queues are also called FIFO. Examples of applications can be found in Sect. 9.3.5.

A queue has two functions: `QueuePut()` and `QueueGet()`. Thess function push a new element into the queue (put) and take it out of (get).

A queue can be represented by an array or by a linked list. In an array with a defined length, two indices mark the state of the FIFO: `QueueIn` and `QueueOut`. Both are set to 0 after initialization, as can be seen in Fig. 5.1 on the left. If an element is now written into the FIFO, the pointer `QueueIn` is incremented until it finally arrives at the last element of the array. If an element is removed from the FIFO, the pointer `QueueOut` is incremented and thus the space in the array is released. You now have to make sure that `QueueOut` does not point beyond `QueueIn`, otherwise an invalid element, i.e. one that is not currently written in, would be output. The main problem, however, is that the array's field limits would quickly be exhausted and regular restoring of the contents would be too costly. FIFOs with arrays are therefore usually designed as *ring buffers*.

With a ring buffer, the two indices are each set back to zero when they reach the array size. This achieves a closed ring. The out-index always moves behind the in-index, when it has caught up with it, the FIFO is empty and no element can be removed. Conversely, if the in-index "overtakes" the out-index from behind, the FIFO is full and a decision must be made whether to overwrite old elements or stop filling.

We first define:

```
#define QUEUE_SIZE 10
#define QUEUE_FULL -1
#define QUEUE_EMPTY -2
#define QUEUE_NOERROR 0
int QueueIn, QueueOut;
```

We also need the contents of the FIFO, this can be a structure for example, or simply an elementary data type, usually a byte.

```
typedef uint8_t FIFO_t;
```

In the case of a bus message, it might also look like this:

```
typedef struct msg_s {
    long msg_addr;
    uint8_t data[8];
} FIFO_t;
```

---

[1] From Greek ισος = "equal": sampling processes with always equal sampling intervals.

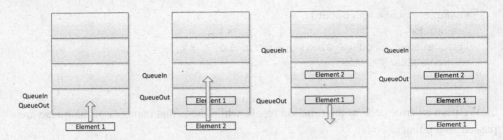

Fig. 5.1 FIFO principle

The actual queue is then:

```
FIFO_t Queue[QUEUE_SIZE];
```

In the simplest case, incrementing the in-index looks like this:

```
QueueIn++;
if (QueueIn+1 >= QUEUE_SIZE) QueueIn= 0;
```

With this a ring buffer is realized. The query is still missing whether the buffer is full. This is the case, if `QueueIn` is exactly before `QueueOut` or if `QueueIn` is just at the end of the array and `QueueOut` is at the beginning of the array. With this an important function of the FIFO, the write-in, is realized:

```
int QueuePut(FIFO_t new_element)
{
    if ((QueueIn +1 == QueueOut) ||
        (QueueOut == 0 && QueueIn == QUEUE_SIZE-1))
            return QUEUE_FULL;
    Queue[QueueIn] = new_element;
    QueueIn++;
    if (QueueIn >= QUEUE_SIZE) QueueIn = 0;
    return QUEUE_NOERROR; //no errors
}
```

Reading from the FIFO takes place in the same way:

```
int QueueGet(FIFO_t *result)
{
    if (QueueIn == QueueOut)
    {
        return QUEUE_EMPTY;
    }
```

```
*result = Queue[QueueOut];
QueueOut++;
if (QueueOut >= QUEUE_SIZE) QueueOut = 0;
return QUEUE_NOERROR;
}
```

We pass an error code as the function result, while the actual contents of the queue are passed as a pointer.

You can save the time-consuming query for the array boundary if you determine the index via a modulo operation [1].

```
QueueIn = (QueueIn + 1) % QUEUE_SIZE;
```

However, this has some disadvantages, for example, the modulo operation is "cheap" only if QUEUE_SIZE is a power of two, namely then it can be replaced by a bitmask [2]:

```
#define QUEUE_SIZE 16 //must be 2^n (8, 16, 32, 64 ...)
#define QUEUE_MASK (QUEUE_SIZE-1) //do not forget the parentheses
```

And the increment operation then looks like this:

```
QueueIn = ((QueueIn + 1) & QUEUE_MASK);
```

Besides this relatively simple example, there are many more examples on the internet, as you can see in [1, 2].

▶   If it is desired to read or write to the queue from an interrupt, it is important that all interrupts are disabled before accessing the queue!

5.1.2   Queue with Dynamic Data Structures

Alternatively, you can also implement a queue with so-called linked lists. The advantage of these dynamic data structures is that only as much memory is used as is actually needed, the disadvantage is that you have to be extremely careful not to fill the memory to the limit.

In a list, each element exists independently in memory. To move from one element to another in a list, each element has a pointer to its successor (see Fig. 5.2). Thus, without further management, the concatenation of elements is created, which can be traversed by remembering the first element and then jumping from element to element. Exactly one element in the list has no successor, more precisely, there the pointer to the successor points to NULL. This element is called the last element in the list.

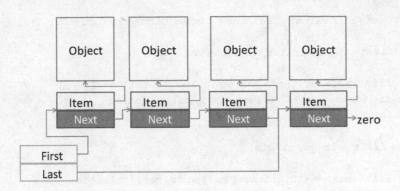

Fig. 5.2   Structure of a simple linked list

When a list is built, a new element is created and made known to its predecessor by reference. In a simple linked list, however, it does not know the predecessor. The successor relationship is therefore a simple, directed relationship. So in addition to the elements, we need least one other auxiliary, namely a reference to the first element of the list first. Conveniently, however, we will also store a reference to the last object last, since otherwise, when creating each new element, we would first have to transpose the entire list up to the last element.

If the pointer last pointed to the first element, we would again have a ring buffer or cycle. In Fig. 5.2 it is indicated that the list elements again point to the actual content. This may or may not be the case. In the example already used at the end of Chap. 3, a list element is constructed as a structure as follows:

```
struct sListItem {
    char i;
    struct sListItem *next;} ;
typedef struct sListItem tListItem;
tListItem *first;
tListItem *last;
```

The variable i stands for the content, which can of course be more complex.

To build a FIFO from a linked list, let's do the following thought experiment: Each new element pushed into the FIFO is simply appended to the last element. A function to append an element to a linked list must first check whether the list is still empty or whether it already contains elements. In the first case, the pointer first is assigned the address of the first element, which naturally becomes the last element. A new element is created using the malloc() function described in Sect. 3.11. If the creation fails, for example because there is no more heap memory available, the function must of course abort and output an error code. If there are already elements, the new element is appended to the previous last element and thus becomes the last element itself.

```c
char FIFOput (char value) //returns 0 if function was successful
{
    if (first==NULL) //list was empty until now
    {
        first = (tListItem*) malloc (sizeof(tListItem));
        if (first == NULL) return -1 ; //malloc failed
        last = first;
    }
    else //list already exists
    {
        last->next = (tListItem*) malloc (sizeof(tListItem));
        if (last->next == NULL) return -1 ; //malloc failed
        else last = last->next; //last must be adjusted
    }
    last->i = value; //assignment of the value
    last->next = NULL; //make sure that the last element points to NULL
    return 0;
}
```

The thought experiment continues: Each time an element has been taken from the FIFO, it can be deleted, i.e. released for further writing. This is done by deleting the first element and letting the working pointer first point to the second element. Here, of course, one must be careful not to accidentally lose a pointer, so an auxiliary pointer buf is created first:

```c
char FIFOget (char *value)
{
    tListItem *buf;
    buf = first; //save the pointer
    if (buf == NULL) //list was empty
    {
        return -1 ;
    }
    else first = first->next; //move the pointer
    *value = buf->i; //output the value
    free(buf); //...and free the heap
    return 0;
}
```

Of course, it is necessary to check whether the list contains an element at all. For this reason, and also because list elements can be (almost) arbitrarily complex, the contents of the element are output indirectly via a pointer (Sect. 2.9.5). The actual return value of the function is 0 if it was successful and there is a valid value in the passed range, or − 1 if the list was empty.

For example, you can now fill the FIFO from an ISR or a task:

```
FIFOput(val);
```

and then output it in another task via a serial interface (Sect. 5.2.1).

```
char res;
FIFOget(&res);
uartTransmitChar(res);
```

5.1.3 Multiple Queues in One Program

If it should actually happen that you want to use several queues in one program, then the functions must be rewritten. Each queue then gets a so-called "handle". In the case of the ring list, this is of course the pointer to the relevant array as well as the two working indices QueueIn and QueueOut:

```
typedef struct sFIFOH {
    int QueueIn;
    int QueueOut;
FIFO_t *Fifo;} FIFOH_t;
```

The initialization then looks as follows:

```
FIFOH_t MyFIFO;
FIFO_t Data[FIFOLENGTH];
QueueInit(&MyFIFO, Data);
```

Accordingly, the two working indices from the handle must then be used in the two access routines. The necessary sample code can be found in the supplementary material to the book.

The same way will be used with FIFOs of the linked list type, these are correspondingly simpler in structure and the handle contains only the pointers to the first and the last element:

```
typedef struct sFIFOH {
    tListItem *first;
    ListItem *last;
} FIFOH_t;
```

When accessing, the respective pointer of the handle is referenced instead of first and last:

```
char FIFOput (FIFOH_t *Handle; char value)
{
    if (Handle->first == NULL)
    {
...
```

A handle is created for each new queue.

```
FIFOH_t MyFIFO;
MyFIFO.first = NULL;
MyFIFO.last = NULL;
...
FIFOput (& MyFIFO, value);
...
```

5.1.4 Multiple Queues with Different Types

If you want to manage queues independently of the type of the contents, another concept must be taken into account. In this case, it is not possible to write directly to or read from the queue, because the access functions must take the type of the element into account (read as return parameter, write as input parameter). Here one helps oneself by an extension of the handle with pointers to functions written especially for it:

```
typedef void FIFO_t;
typedef struct sFIFOH {
    int QueueIn;
    int QueueOut;
    int QueueSize;
    void (*setfnct) (FIFO_t *element, int index);
    FIFO_t* (*getfnct) (int index);
    void *Fifo;
} FIFOH_t;
```

The type FIFO_t is a pointer to void, which must then be explicitly converted in the access functions.

When the queue is initialized, the access functions are also passed:

```
void QueueInit(FIFOH_t *Handle, int size, FIFO_t *Fifo,
                void (* setfnct)(FIFO_t *element,
                int index),FIFO_t* (*getfnct)(int index))
{
    Handle->QueueIn = 0;
    Handle->QueueOut = 0;
    Handle->QueueSize = size;
    Handle->Fifo = Fifo;
    Handle->getfnct = getfnct;
    Handle->setfnct = setfnct;
}
```

The getter and setter functions (access functions to the arrays) then simply replace the index brackets in the queue, so the queue functions put and get look like this:

```
int QueuePut(FIFOH_t *Handle, FIFO_t *new_element)
{
    if ((Handle->QueueIn + 1 == Handle->QueueOut) ||
            (Handle->QueueOut == 0 &&
             Handle->QueueIn == Handle->QueueSize-1))
        return QUEUE_FULL ;
    Handle->setfnct(new_element, Handle->QueueIn);
    Handle->QueueIn++;
    if (Handle->QueueIn >= Handle->QueueSize) Handle->QueueIn= 0;
    return QUEUE_NOERROR; //No errors
}


FIFO_t* QueueGet(FIFOH_t *Handle, int *errcode)
{
    FIFO_t *result;
    if(Handle->QueueIn == Handle->QueueOut)
    {
        *errcode = QUEUE_EMPTY;
        return NULL;
    }
    result = Handle->getfnct(Handle->QueueOut);
    Handle->QueueOut++;
    if (Handle->QueueOut >= Handle->QueueSize) Handle->QueueOut= 0;
    *errcode = QUEUE_NOERROR;
    return result;
}
```

The call could then look like this:

```
typedef struct a {
    int a;
    float b;
} a_t;
FIFO_t* get_a(int index)
{
    return &Queue_a[index];
}
void set_a(FIFO_t *element, int index)
{
    Queue_a[index]=*(a_t*)element;
}
int main()
{
    ...
    a_t Queue_a[5];
    FIFOH_t ha;
    QueueInit(&ha, 5, Queue_a,set_a, get_a);
    ...
}
```

5.2   State Machines

State machines are frequently used in everyday programming, especially in the field of embedded systems. Often one wants a scheme that is easy to change and requires few resources. Two of these schemes are described here, a very simple one, which has to be rewritten manually for each case, and a more complex one, which can be used especially for larger state machines and which is very readable.

5.2.1   General Consideration

Finite state machines are powerful concepts for the behavior of machines by modeling the memory-based response of a system to internal or external events. First, a finite set of states is defined in which – and only in which – the machine can be. Furthermore, there exists a list of events to which the machine reacts by changing from one state to another. This change may be associated with an externally visible action, i.e., an output; also, remaining in one state may be associated with an output. Important here is the "memory", thus a new state is not only dependent on the input but also on the previous state. So we can say that a state machine maps the set of states z_i^n at time n to a set of states z_i^{n+1} at time $n+1$ as a function of events or input variables e_j.

$$z_i^{n+1} = f\left(z_i^n, e_j\right) \tag{5.1}$$

and in addition, the set of current states and input variables is mapped to a set of reactions or output variables o_k:

$$o_k = f\left(z_i^n, e_j\right) \tag{5.2}$$

Figure 5.3 illustrates this relationship.

The changing between two states is called a transition. Outer transitions are transitions in which a state is left and a new state is entered. Inner transitions are those in which an event causes an activity but not a change of state.

A transition can occur asynchronously, in that a spontaneous event also triggers a spontaneous change. Mostly, however, state machines are executed synchronously, the event is stored temporarily and evaluated in the next step.

Events can be:

- Receiving a character or a character string via a serial interface
- A measured value changes or exceeds a defined upper or lower threshold
- The state of a digital input changes (edge), for example when a key is pressed.

Transitions can be restricted by conditions (guard), in which case the transition is only executed when the event occurs if the condition is also fulfilled.

Conditions may be, for example:

- A certain amount of time has elapsed since the last state change, or in general:
- A calculated variable exceeds a defined threshold

Particularly common use cases of state machines are:

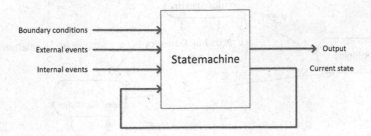

Fig. 5.3 General design of a state machine

- The control of sequential processes based on external events
- Human-machine interfaces: Responses to user input and machine events
- The realization of communication protocols

The external behavior of the state machine is associated with the states and the state transitions and is described by activities (or actions). We distinguish:

- Activities that are executed when the state machine enters the state (entry behaviour).
- Activities that are executed when the state machine leaves the state (exit behaviour).
- Activities that are executed while the state machine is in the state (doActivity).
- Activities that are carried out when a certain event occurs. (eng. Event behaviour)

5.2.2   Description of State Machines

For the description of state machines, either the state chart defined in UML (Unified Modeling Language) or SysML (Systems Modeling Language) [3–5] is used or the state transition table, also called state sequence table. Both techniques are presented here using the following example.

Example

HMI application: Often the parameters of a machine have to be set via a few input keys. A simple example is setting the time for an alarm clock, which knows the keys SET and + or -. By pressing SET the alarm clock expects the input (increment/decrement) first of hours, then of minutes. ◀

The state diagram looks as shown in Fig. 5.4; for the sake of readability, the actions have been omitted here. The most important language elements of the UML state diagram can

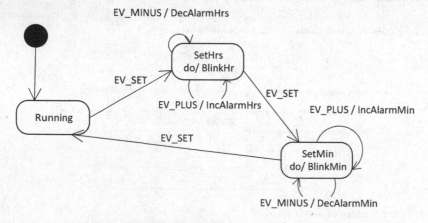

Fig. 5.4 Alarm clock state machine for time setting

Table 5.1 Status sequence table for the alarm clock

State	Event	Subsequent state	Action
Running	Set	SetHrs	
SetHrs	Plus	SetHrs	IncAlarmHrs
SetHrs	Minus	SetHrs	DecAlarmHrs
SetHrs	Set	SetMin	
SetMin	Plus	SetMin	IncAlarmMin
SetMin	Minus	SetMin	DecAlarmMin
SetMin	Set	Running	

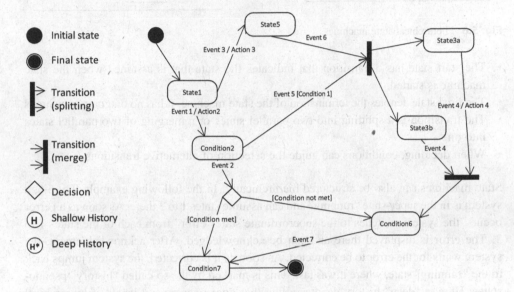

Fig. 5.5 UML language elements of the state diagram

already be seen in this diagram. The black filled circle is a so-called pseudo state, here the start state. After the reset, the automaton is in the start state and then immediately changes to the "running" state without any further event. The start state therefore has only one transition, which specifies the state in which the machine starts. The transitions are assigned the events that trigger the transition. In the example above, EV_SET is an event that triggers an outer transition, while EV_PLUS and EV_MINUS trigger inner transitions.

In a state sequence table, the same description looks like in Table 5.1:

For reasons of clarity, the actions have not been divided into do/entry/exit and event actions. A typical do action here would be to make the digit for the hours flash while the state machine is in the SetHrs state and correspondingly to make the digit for the minutes flash in the SetMin state (Fig. 5.4).

Figure 5.5 summarizes important language elements for the state diagram.

In particular, the following pseudo-states can be identified here:

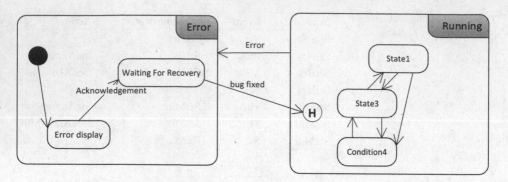

Fig. 5.6 Hierarchical state machines

- The start state has a transition that indicates the state that is assumed when the state machine is started.
- The final state denotes the termination of the state machine. It has no outgoing transitions
- The transition is a splitting into two parallel states or a merging of two parallel states into one.
- When deciding, conditions can guide the selection of alternative transitions

State machines can also be structured hierarchically. In the following example Fig. 5.6, a system is in the super-state "running" and can assume states 1 to 3 there. As soon as an error occurs, the system switches to the superordinate state "error" from each of the states 1 to 3. The error is displayed there and can be acknowledged. After acknowledgement, the system waits for the error to be corrected. As soon as it is corrected, the system jumps back to the "running" state, where it was last. This is marked by the so-called "history" pseudo-states. Here, a "deep" history denotes a flag that goes over several levels of hierarchical state machines, while the "shallow" history always considers only the next level.

5.2.3   Implementation of State Machines on Microcontrollers

In the alarm clock example above, three states and three events are defined, which can be mapped in C in this way, with a "non-event" EV_NOEVENT indicating that there is currently no event:

```
#define STATE_RUNNING 0
#define STATE_SETHRS 1
#define STATE_SETMIN 2
#define EV_SET 1
#define EV_PLUS 2
#define EV_MINUS 3
#define EV_NOEVENT 0
uint8_t state = STATE_RUNNING; //starting state
uint8_t event = EV_NOEVENT;
```

In the main loop of the program, the event is evaluated first, then the transition is executed or the transition activity is executed, then the do activity is executed:

```
while(1)
{
    event = ReadEvent();
    state = statemachine(event);
    PerformDoAction(state);
}
```

The actual state machine can be constructed in various ways. In the following code, it is based on nested `switch` statements, which can of course also be replaced by multiple branches with `if`. The advantage of `switch` is the fact that the compiler translates these statements as tables and thus can process them much more efficiently.

```
uint8_t statemachine(uint8_t event)
{
    switch (state)
    {
        case STATE_RUNNING: switch (event)
        {
            case EV_SET: state = STATE_SETHRS; break;
            default: break;
        } break;
        case STATE_SETHRS: switch (event)
        {
            case EV_SET: state = STATE_SETMIN; break;
            case EV_PLUS: IncAlarmHrs(); break;
            case EV_MINUS: DecAlarmHrs(); break;
            default: break;
        } break ;
        case STATE_SETMIN: switch (event)
        {
            case EV_SET: state = STATE_RUNNING; break;
            case EV_PLUS: IncAlarmMin(); break;
            case EV_MINUS: DecAlarmMin(); break;
            default: break;
        } break;
        default: break;
    }
    event = EV_NOEVENT;
    return state;
}
```

The transition actions, in this case the incrementing and decrementing of hours and minutes, for example, look like this:

```
void IncAlarmHrs(){
    alarmHrs++;
    if (alarmHrs == 24) alarmHrs = 0;
}
```

Instead of the sometimes confusing branches, a state table can also be programmed directly, in this case it is one with pointers to the transition actions. An entry looks like this:

```
struct sStateTableEntry{
    uint8_t state; //current state
    uint8_t event; //event
    uint8_t newstate; //new state
    void (*transition)(); //transition action
};
```

For the alarm clock, the example implementation of the table is:

```
#define STLEN 7
struct sStateTableEntry stm[] =
                     {{STATE_RUNNING,EV_SET,STATE_SETHRS,NULL},
                      {STATE_SETHRS,EV_SET,STATE_SETMIN,NULL},
                      {STATE_SETHRS,EV_PLUS,STATE_SETHRS,IncAlarmHrs},
                      {STATE_SETHRS,EV_MINUS,STATE_SETHRS,DecAlarmHrs},
                      {STATE_SETMIN,EV_PLUS,STATE_SETMIN,IncAlarmMin},
                      {STATE_SETMIN,EV_MINUS,STATE_SETMIN,DecAlarmMin},
                      {STATE_SETMIN,EV_SET,STATE_RUNNING,NULL}};
```

To run the automaton now, the function statemachine() from above is replaced by statemachine2():

```
uint8_t statemachine2(uint8_t ev)
{
    int i;
    if (ev == EV_NOEVENT) return state;
    for (i = 0; i < STLEN; i++)
    {
        if (state == stm[i].state && ev == stm[i].event)
        {
            state = stm[i].newstate;
            if (stm[i].transition != NULL) stm[i].transition();
            event = EV_NOEVENT;
            return state;
        }
    }
}
```

First, it checks whether a valid event has arrived, otherwise the state is retained. Then the algorithm searches in the `for` loop whether a state and an event correspond to the current combination and changes to the matching subsequent state. If a transition action was defined, it is still executed.

References

1. https://stackoverflow.com/questions/215557/how-do-i-implement-a-circular-list-ring-buffer-in-c. Accessed: 27. Dez. 2020.
2. https://www.mikrocontroller.net/articles/FIFO. Accessed: 27. Dez. 2020.
3. Booch, G., Rumbaugh, J., & Jacobson, I. (2006). *Das UML-Benutzerhandbuch, Aktuell zur Version 2.0.* Addison-Wesley. ISBN 978–3827325709.
4. Balzert, H. *UML2 in 5 Tagen*, 2008, W3L. ISBN 978.3937137612.
5. Alt, O. (2012). *Modellbasierte Systementwicklung mit SysML.* Carl Hanser Verlag GmbH & KG.

Theoretical Considerations for IoT Networks

Abstract

The chapter contains information about the ISO/OSI layer model and further about coding, physical layer, lines, transceivers, security, topologies, synchronization, MAC, media access, as well as protocols of the higher layers.

After the first preliminary considerations on programming and the software framework in the first part of the book, the second part deals with the networking of sensors. For this purpose, the ISO/OSI layer model is first briefly introduced in the sixth chapter. Since the book is limited to small embedded (8-bit) systems, the focus is on the implementation of the physical layer and the link layer. Nevertheless, a brief overview of protocols in the Internet of Things will be given to the reader. At the end of the chapter, the limitations to which sensor nodes in the IoT are subject are considered.

6.1 The ISO/OSI Layer Model

Communication takes place when data[1] is transmitted from at least one source (sender) to at least one sink (receiver) via a message channel and this data has the same meaning (semantics) for sender and receiver.[2] For this purpose, information must be provided in advance about the type of communication, specifically

---

[1] Only through interpretation by the recipient does it become information.

[2] This section was taken from the book Meroth, Tolg: "Bussysteme im Kfz" [1] and heavily revised.

© The Author(s), under exclusive license to Springer Fachmedien Wiesbaden GmbH, part of Springer Nature 2023
A. Meroth, P. Sora, *Sensor networks in theory and practice*,
https://doi.org/10.1007/978-3-658-39576-6_6

- the coding of the data by signals,
- the physical parameters of the signals and the message channel, and
- the protocol used

be known. These agreements about the communication process and the data exchanged are called protocols. They can refer to physical processes (e.g. establishing a physical connection), processes for communication control (e.g. connection setup, routing of data packets, diagnostics) and to the transmission of the user data itself. To enable communication participants from different manufacturers to use the same protocols, these are usually standardized. Most communication standards are based on the so-called OSI[3] layer model. It subdivides the procedures for data transmission into seven layers, each with specific tasks. This approach is based on the idea that software applications do not communicate directly with their counterpart application – how could they? -but via a service access point with a service that passes the message on to lower instances of the operating system and then via the hardware, if necessary via exchange switches, to the recipient computer, which in turn makes the message available to the application concerned (vertical communication vs. horizontal communication), as the following Fig. 6.1 shows.

This has considerable consequences for communication control. The layers, which can exist as independent instances, for example as part of an operating system or a runtime environment, require protocols for (virtual) horizontal communication with each other, but the use of the next lower layer for vertical communication also follows a protocol. The required protocol data (for horizontal communication) or interface data (for vertical communication) are appended before (header) or after (trailer) the actual payload. This increases the size of the data packet from layer to layer, whereby protocol data and user data of the next higher layer are interpreted as user data of the next lower layer (see Fig. 6.2). This is referred to as a protocol stack. When considering the data rate on a bus system, a careful distinction must be made between the gross data rate and the net data rate, which can differ significantly depending on the protocol efficiency.

The OSI model, whose roots go back as far as the 1970s, acts as a framework that provides a general approach to the design of new protocols.[4] In addition, there are the RFCs, *Request for Comments,* a series of technical and organizational documents on the Internet, which have been published since April 1969 [4]. Originally, these were literally proposals put up for discussion; today, discussion already takes place during the creation of the drafts, so that a published RFC can usually be regarded as a peer-reviewed technical specification.

The following Table 6.1 provides an overview of the layers of the OSI model; the following sections then present aspects from layers 1–4.

The primary difference between the wired networks described in this chapter is in the two lowest layers. Above these are different protocols, which are not described in this

---

[3] Open Systems Interconnection Reference Model.
[4] Specified in the ISO 7498-1994 standard [2].

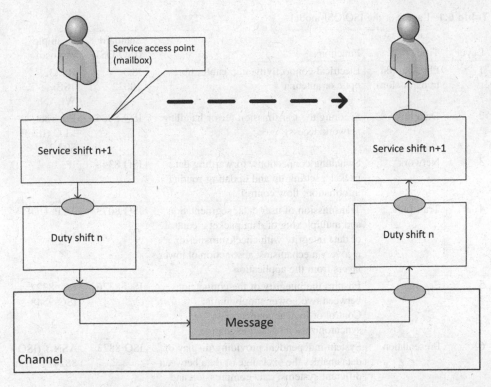

Fig. 6.1 Vertical and horizontal communication in the layer model

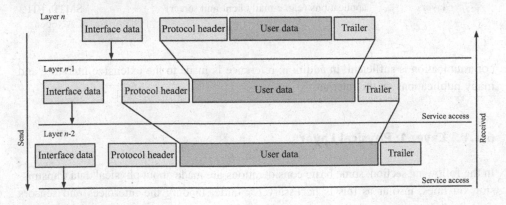

Fig. 6.2 Header and trailer as part of the protocol stack

book. However, there is a strong trend towards standardization of the upper layers in the direction of TCP/IP.

For the networks and software solutions described in this book, an understanding of layers 1 and 2 of the ISO/OSI layer model and a short excursion into TCP/IP

Table 6.1 Layers of the ISO/OSI model

Layer	Title	Functions	Standard (ISO-OSI)	Example protocol
1	Physical (bit transmission)	Electrical connectivity, e.g. cable, fibre optic or antenna	ISO 10022	RS232, 10Base-T, WLAN
2	Data link	Securing the transmission (error handling, network access)	ISO 8886	Ethernet, LLC (IEEE 802.2)
3	Network	Switching connections, forwarding data packets; setting up and updating routing information; flow control	ISO 8348	IP
4	Transport	Transmission of user data; segmentation and multiplexing of data packets; control of data integrity with checksums; error recovery mechanisms; abstraction of lower layers from the application	ISO 8073	TCP, UDP
5	Session	Ensures the integrity of the connection between two power supply units; Control of logical connections; synchronization of data exchange	ISO 8326	ISO 8327 ISO 9548
6	Presentation	System-independent providing/display of data enables the exchange of data between different systems; data compression and encryption	ISO 8823	ASN.1 (ISO 8824)
7	Application layer	Exchange of user data of special applications (e.g. e-mail client and server)	ISO 8649	HTTP, SMTP, FTP, CANopen

communication is sufficient; in addition, reference is made to the extensive literature and many publications on the Internet.

6.1.1 Layer 1: Physical Layer

In the following section, some basic considerations are made about physical data transmission on lines, insofar as this is necessary for understanding the interfaces and sensors described. A comprehensive description of bus system technology has been omitted; here it may be necessary to study the further literature described at the end of the chapter.

6.1.1.1 Wires

Electric lines transmit energy via the electric and magnetic fields surrounding them. The transmission is based on the principle that a changing electric field generates a magnetic field in its vicinity, which in turn generates an electric field, and so on. The consequence is that an electromagnetic wavefront propagates, analogous to an acoustic wavefront in a medium.

Since the propagation speed c is finite, the signal takes a certain time to travel from the transmitter to the receiver. c depends on the surrounding medium; in a vacuum, the propagation speed is equal to the speed of light c_0, i.e. $c = c_0 \approx 3 \cdot 10^8$ m/s; in real cables, it is mainly determined by the insulating material used and is approximately $c = 0.6c_0$. Thus, a transmission line represents a storage element for energy and for information. For a 5000 km long line with $c = 2 \cdot 10^8$ m/s, for example, the transit time is 25 ms, which means that at a data transmission rate of one Mbit/s, 25,000 bits are already stored in the line. For a comparable 10 m long line, the runtime is 50 ns. The absolute length of a line is less important than the relationship between length and data rate, in other words, even a short line is long for transmission if only the data rate is high enough.

However, due to conduction losses and polarization effects in the insulating material, energy is also dissipated, so that a real line represents, in a first approximation, a low-pass filter consisting of a series resistor and a parallel capacitance.[5] This causes the signal to be ground, so that signal regeneration is necessary for long line lengths. This is generally not yet relevant for the applications described here, but must be taken into account for greater distances to the sensors.

In addition to storage and damping, a line also has the property of radiating energy and absorbing energy from the surroundings of the line. This leads to *crosstalk* if several cables are routed in parallel next to each other, or to interference pulses if cables are routed directly next to loads through which high dynamic currents flow (electrical drives, switching actuators). On the other hand, cables with high data rates can act like antennas that directly influence other loads. Preventing this is the task of electromagnetic compatibility, EMC. For reasons of space, reference is made here to the literature [3]. In addition to the twisting and shielding of cables and the shielding of housings, EMC requirements go directly into the specification of the bus systems. Among other things, care must be taken to ensure that there are sufficient ground planes on the circuit board, and that signal and (switched) power lines are not routed in parallel or are shielded from each other via ground planes. This also applies to the juxtaposition of highly clocked signal lines.

However, one specific point should be addressed here: The interaction between electric and magnetic fields around the line establishes a constant, material-dependent ratio between voltage and current at each point on the line. This ratio is called wave impedance Z_L. It is independent of any circuitry at the ends of the line, since a wavefront travelling along the line does not "see" these ends.

It is now interesting to consider how a wave of amplitude U_{2h} or I_{2h} behaves when it reaches the end of the line and encounters a load resistor Z_V there. If the line is loaded (terminated) at the end with the impedance $Z_V = Z_L$, then the voltage/current ratio at the end

---

[5]This is true for lines where the propagation time is short compared to the rise time of the signals, otherwise the line forms a borderline infinitesimal chain of low passes.

corresponds exactly to that on the line. As a result, all the energy can be absorbed. This is called matching. If, on the other hand, the line is short-circuited, a voltage of 0 is forced. This leads to a compensation process between the non-zero voltage on the line and the voltage 0 at the end of the line, which continues as a returning wave of amplitude U_{2r} or I_{2r} towards the beginning of the line. On the other hand, for lines open at the end, a current 0 is forced and this also results in a returning wave. Between these extremes, there will be partial absorption of the wave by the load and partial reflection in the line.

The ratio of outgoing to returning wave is determined by the reflection factor r with

$$U_{2r} = r \cdot U_{2h} \tag{6.1}$$

$$I_{2r} = -r \cdot I_{2h} \tag{6.2}$$

which is calculated as follows:

$$r = \frac{Z_V - Z_L}{Z_V + Z_L} \tag{6.3}$$

Figure 6.3 shows the measurement of a pulse of 320 ns duration at terminated line start on a 100 m long coaxial cable with a wave impedance of 50 Ω. The upper curve in the figures shows the pulse at the end of the line, which obviously arrives 500 ns delayed. The flattening of the pulse due to the low-pass property of the line can be seen well. After another 500 ns the reflected pulse reaches the beginning of the line again (lower curve). In the case of the open line (left figure) it is obviously reflected positively, in the case of the short-circuited line negatively. The further flattening shows the renewed passage through the low-pass system. If the beginning of the line is also not properly terminated, the pulse wanders back and forth between the ends of the line, becoming increasingly blurred and finally disappearing.

However, it is easy to see that reflected bits on a line look like "real" bits, at least in an early phase, and therefore transmission errors can occur. Therefore, proper line termination is extremely important, even for short lines. As a rule, the terminating resistors are purely ohmic, but sometimes they also consist of an R-C element.

6.1.1.2 Transceiver

The term transceiver is a made-up word, composed of transmitter and receiver. Its task is signal generation or signal amplification on the transmitter side and measurement and thus detection of the signal on the receiver side. The following considerations play a role here:

- Transmission type: The transmission type determines how two bus participants can communicate. If one of the bus participants is only a sender and the other only a receiver, this is referred to as simplex or directional operation. If the communication can run in both directions, it is referred to as duplex operation, whereby a distinction is

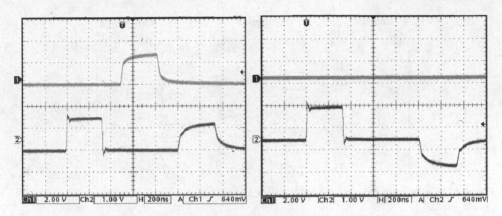

Fig. 6.3 Pulse of 320 ns duration on a 100 m long coaxial line (50 Ω); left: open end (upper signal in each case), right: short circuit at the end. Channel 1 (top) is measured at the end of the line, channel 2 (bottom) at the beginning of the line

made between full duplex and half duplex. Figure 6.4 illustrates the relationship using a simplified sequence diagram (see also Fig. 7.10). In full duplex operation, simultaneous transmission and reception is possible, usually through two signal lines (Tx for transmission, Rx for reception) or through two other transmission channels. In half-duplex operation, only either transmitting or receiving is possible.

- Connection type Point-to-Point or Multipoint: Many bus systems (Ethernet, CAN, LIN, I^2C/TWI) allow simultaneous access to the transmission medium, i.e. communication takes place not only between one transmitter and one receiver (point-to-point), but between several transmitters and several receivers (nodes). Electrically, this means that the transceivers must ensure that simultaneous transmission attempts by two bus nodes do not result in electrical damage (short circuit). Here, a circuit has become established which is called wired-AND or wired-OR, depending on the point of view: The bus is continuously held at the high level via a pull-up resistor. Each participant is connected to the bus with the open collector of its output transistor (open-collector). If any node sends a low signal (0 V), the overall bus level is low. Therefore the low level is called *dominant*, the high level is called *recessive*. Hence the designation wired-AND (Fig. 6.5), since a 1 can only be present if all nodes transmit a 1. Since the logic 0 is present at the collector when a 1 is present at the base of the transistor, the circuit is sometimes called wired-OR (negative logic).

- In this circuit, each node can read the bus level via its receiver side and, in the event of a discrepancy between its own transmitted high level and a measured low level, determine that another bus node is also attempting access. One consequence of this collision possibility is the requirement for access control, which will be described later.

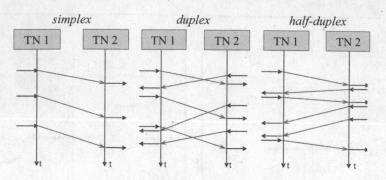

Fig. 6.4 Transmission types simplex, full duplex and half duplex

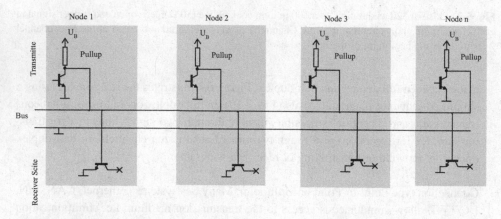

Fig. 6.5 Wired-AND circuit (open collector)

- Line type single-wire line or two-wire line: Depending on the reliability requirements, the signal lines are designed as single-wire or two-wire lines. The single-wire signal line (unbalanced interface) (Fig. 6.6) is the more cost-effective variant. The signal level is applied against the general ground. This makes the single-wire signal line susceptible to interference from electromagnetic induction[6] or by raising the ground level. It is therefore only used in bus systems with low data rates (e.g. LIN). With the two-wire signal line, an inverted signal is transmitted on each of two wires. A differential amplifier determines the signal level from the difference between the voltage levels independently of the ground potential, which is why we speak of differential transmission (example: CAN, RS485). By twisting the wires, this already robust cabling is even better protected against interferences, because these cancel each other out in the small loops that form the twists.

---

[6]Forward and return conductors form a loop with a very large loop area.

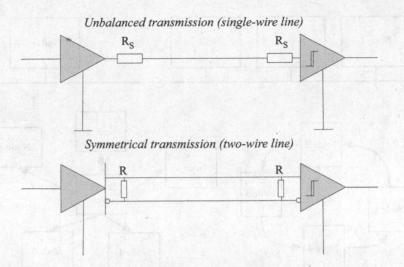

Fig. 6.6   Unbalanced and balanced transmission

6.1.1.3 Transmission Security

The transceiver is ultimately responsible for the reliable and low error transmission of the signal. On the transmitter side, this is achieved by pulse shaping suitable for the transmission medium. A steep-edged square-wave pulse is not always an advantageous solution, because the low-pass characteristic of the line smears it and causes it to trail. If a pulse train is transmitted, the trailing edges of the square-wave pulses overlap and can lead to *intersymbol interference* [5], i.e. mutual interference until the pulses become unrecognizable. An effective way to prevent this is to shape the pulses in such a way that their spectrum does not contain any significant frequency components beyond the cutoff frequencies given by the bandwidth of the channel (Nyquist criterion). For this purpose, so-called cosine roll-off pulses are usually used, which are generated from the original square-wave pulses by a filter of the same name.

On the receiver side, the difficulty is to decide whether a high level or a low level is present at a given measurement time. A digital signal overlaid by noise or affected by jitter (Sect. 6.1.2) complicates the choice of the measurement time on the one hand, but also the decision itself. The probability that a logical '1' was detected when a logical '0' was transmitted and vice versa can be calculated from the normal distribution and is called residual error probability. For this, please refer to the literature.

6.1.1.4 Network Topologies

A network with several nodes can be classified according to the structure principle. These are referred to as topologies. In local networks (LAN = Local Area Network), the following topologies (Fig. 6.7) have become established.

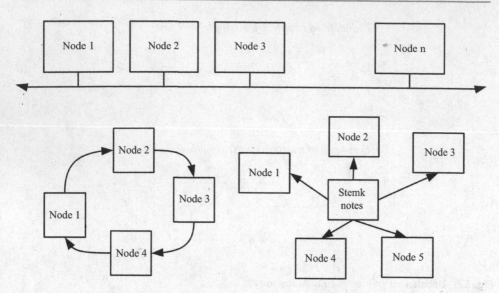

Fig. 6.7 Bus, ring and star topology

- *Bus*: The bus, (also line or line-bus) is the most common topology, represented for example by CAN and I^2C/TWI, based on point-to-multipoint access. All nodes transmit and receive over the same medium, so access control becomes necessary. If one node is dropped, the bus is generally still available to the other nodes; in the event of a line break, the bus splits into two partial buses.
- *Ring*: The ring consists of a closed sequence of point-to-point connections. As a rule, the nodes pass their data synchronously around the ring; a message to the predecessor of a node must pass through almost the entire ring before it is passed on to the desired recipient. Since the signal is regenerated in each node, ring topologies are robust; however, a line break will result in total system failure. Ring topology is used in MOST network which is described in [1].
- *Star*: In the star topology, as used in modern office LANs, all nodes communicate via a point-to-point connection with a star point. A passive star only distributes the amplified signal (repeater), if necessary, so this topology is logically comparable with the bus. With an active star, the star point assumes a master role or at least selects the data streams according to addressees.
- *Mesh*: In the mesh or network topology, the individual nodes are directly connected with several, in extreme cases with all nodes in the network. An example can be found in Chap. 13 for ZigBee.
- *Hierarchical topologies:* Hierarchical topologies are those in which individual network nodes represent the starting point for subnetworks (e.g., as gateways). One can represent these structures as a tree topology. Their advantage is that subordinate networks can operate locally autonomously and only system control takes place in the superordinate

network. This principle from automation technology is found, for example, in vehicle bus systems in the area of local subnets (LIN) and in the diagnostic interfaces.

6.1.1.5 Synchronization and Line Coding

The line coding has the task of defining the physical signal representation on the transmission medium (see Fig. 6.8). Level sequences, e.g. voltages, currents or light intensities, are assigned to the binary bit sequences. Various requirements are relevant for this:

- Synchronization and clock recovery: The reception of signals is a periodic measurement process that starts at a specified time and runs with a specified period duration (step size). The transmitter and receiver must be synchronized, i.e. they must change to the next transmitted character almost simultaneously. This is achieved either by transmitting a clock signal via an additional medium (additional line) (as with the I^2S- or I^2C-bus) or

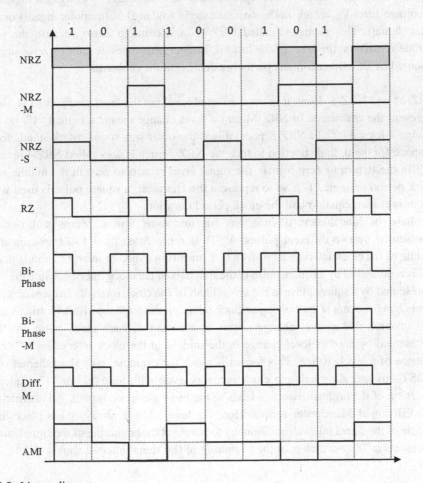

Fig. 6.8 Line coding

by cleverly selecting the line coding so that the clock is present in the signal and can be recovered from it. Only in the case of asynchronous transmission methods, such as those used in UART-based networks, for example the RS232 or RS485 interface or in the LIN bus, do the transmitter and receiver operate with free-running oscillators. Here, additional measures must be taken to ensure that the clock deviations do not exceed bit limits and thus lead to incorrect measurement (see Chap. 7).

- DC voltage-free: If the data transmission system has integrating behavior (low-pass), the DC voltage component of the signal must be 0 on average, otherwise the system will "charge up".
- Distinction between "signal 0" and "no signal": The question must be clarified how it can be determined whether a sequence of zeros is being transmitted or simply nothing at all.

In the simplest case, a logical "1" is represented by a voltage level V_1 (high), a logical "0" by a voltage level V_2, which in the special case is low or 0 V (unipolar signal) or $-V_1$ (bipolar signal). This format is called NRZ (Non Return to Zero) because the signal amplitude remains at the level of the last bit. Longer sequences of ones or zeros are thus represented as DC. Other formats partially circumvent the problem:

- NRZ-M and NRZ-S: Here, it is not the signal level but the level change that is used to represent the characters. In NRZ-M(ark) a level change means a logical "1", no level change a logical "0". In NRZ-S(pace) this is the other way round. Mark stands for the 1, space for the 0. In distinction to this, the NRZ format is also called NRZ-L(evel).
- RZ: In the Return to Zero format, the signal level returns to zero in the middle of the clock period (at logic "1"). So to represent the character, a square pulse is used whose width is chosen equal to half the clock period duration.
- Bi-phase or Manchester: designations for the same format. Zeros and ones are represented by two different pulses. A "1" is represented by a 1–0 transition in the middle of the clock interval, a "0" by a 0–1 transition. Another interpretation is that the "1" is represented by a square pulse in the first half of the clock interval, while the "0" is represented by a square pulse in the second half of the clock interval. This ensures that a level change occurs at least once per clock pulse. In the bi-phase M(ark) format variant, it is ensured that a level change occurs at the start of each clock pulse. A "1" is represented by another level change in the middle of the clock interval, a "0" by the absence of a level change. This format is used, for example, with the Ethernet or the MOST bus and allows simple clock recovery even with long "0" or "1" sequences. Inversion of the original sequence leads to the bi-phase space format. An alternative is the differential Manchester format: Here the level change always takes place in the middle of the signal interval, additionally a change at the beginning of the signal interval represents a "0". No change at the beginning of the signal interval shows "1".

Since exactly one bit (i.e. two states) can be transmitted with each transmission step, the formats mentioned above are also called binary formats. Among many other formats, the AMI format (Alternate Mark Inversion) should be mentioned, a pseudo-ternary format. Ternary means that the signal can assume three levels. But since the levels +1 and −1 each stand for a 1, the format is not really ternary after all. Each "1" is represented by an inverted polarity of the 1-level, the "0" by a 0-level. This ensures a level change even for long one sequences. This also has the consequence that the signal is DC-free (application: ISDN basic access).

Longer zero sequences are also considered in variants of the AMI format. These include HDB-n codes and the BnZS code. The basic principle is that longer zero sequences are replaced by a ternary string that specifically violates the principle of the AMI format.

Of course, it is possible to use even more signal levels for transmission, e.g. four levels (example: −2 V, −1 V, 1 V, 2 V), i.e. a quaternary format. Since $\log_2 4 = 2$ bits (i.e. four states) can thus be transmitted, the "felt" data rate doubles and we now distinguish between the step rate v_s. (unit: baud, i.e. signal change or measuring steps per second) and the data rate in bit/s. In real procedures, four, eight or sixteen amplitude steps are usually selected, corresponding to two, three or four bits transmitted simultaneously.

Nyquist had found out that the maximum channel capacity of the transmission medium c_c in bit/s

$$c_c = 2 \cdot B \tag{6.4}$$

where B is the bandwidth of the channel in Hz. This can be illustrated by the fact that one bit can be transmitted per half-wave of a sinusoidal oscillation. If N different amplitude levels are added, the channel capacity grows according to the above considerations to

$$c_c = 2 \cdot B \ \log_2 N \tag{6.5}$$

Unfortunately, with an increasing number of level steps, the influences of an ever constant noise level also become higher, so that according to Claude Shannon, the maximum channel capacity of the transmission medium c_c is limited to

$$c_c = 2 \cdot B \ \log_2(1 + \text{SNR}) \tag{6.6}$$

where SNR is the signal-to-noise ratio, i.e. the amplitude ratio between average signal power and average noise power.

At this point it should be noted that line coding is only part of the signal representation on the medium. Equally interesting is the question of how the voltage level is represented on the medium itself (modulation). The direct mapping of the signal level to the voltage level on the line is called baseband modulation. This reaches its limit at the latest when a line is no longer available, but must be transmitted via an air interface. In this case, a high-frequency carrier signal is usually modulated with the wanted signal. In the simplest case, the amplitude of the carrier signal varies with the signal level of the wanted signal (amplitude modulation or amplitude shift keying ASK) by multiplying the carrier signal with the wanted signal. Alternatively, the different signal levels can be represented as different frequencies (frequency modulation or frequency shift keying, FSK) or the phase position or several of these characteristics of the carrier signal are influenced as a function of the useful signal.

6.1.2 Layer 2: Data Link Layer

The physical layer and the data link layer together form the basis for data transmission in even the simplest point-to-point communication. When talking about network technologies, these two layers are usually mentioned in the same breath. The best-known example is the IEEE 802 family. This includes WLAN (802.11), Ethernet (802.3), mechanisms for establishing and terminating connections (LLC 802.2), but also Bluetooth® (802.15.1) or the data link layers underlying other wireless networks (e.g. 802.15.4 for ZigBee).

In the automotive sector, the basic CAN standard (ISO 11898 or Bosch specification) is also limited to these two layers [6, 7]. Network and/or transport protocols then sit on top of these protocols (e.g. TCP/IP or CAN-TP). Since the three main tasks of the link layer: frame generation, error handling and media access can be handled largely autonomously, the link layer is often subdivided into the three sub-layers:

- Framing Sublayer for Framing
- Media Access Control (MAC) sublayer for access control to the physical layer
- Logical Link Control (LLC) for communication control and, if necessary, error handling

6.1.2.1 Framing

The task of framing is on the one hand to inform the receiver where a message unit begins and where it ends, and on the other hand to append protocol information to a user data packet. Frames are usually appended to the beginning of a data packet (header), in which case the packet length is fixed or specified in the header. In some protocols, frame components (trailers) attached at the end are also used, for example, to transport checksums that were formed during data transmission. A basic distinction is made between

- Character-based transmission: Here only one byte is transmitted by the transmitter at a time, the start of the transmission is initiated in NRZ coding by a start bit, which fetches the bus level from the idle state (high) to the active state (low) and thus marks the start of the transmission. To ensure that the bus is back in the high state at the end of the transmission, one or two stop bits (high) are appended. Since the bit clock oscillator of the receiver (see above) can be resynchronized after each byte (character

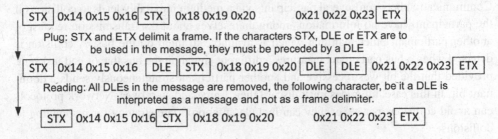

Fig. 6.9 Bit stuffing

| STX | 0x14 0x15 0x16 | STX | 0x18 0x19 0x20 | 0x21 0x22 0x23 | ETX |

Plug: STX and ETX delimit a frame. If the characters STX, DLE or ETX are to be used in the message, they must be preceded by a DLE

| STX | 0x14 0x15 0x16 | DLE | STX | 0x18 0x19 0x20 | DLE | DLE | 0x21 0x22 0x23 | ETX |

Reading: All DLEs in the message are removed, the following character, be it a DLE is interpreted as a message and not as a frame delimiter.

| STX | 0x14 0x15 0x16 | STX | 0x18 0x19 0x20 | 0x21 0x22 0x23 | ETX |

Fig. 6.10 Bytestuffing

synchronization), no clock has to be transmitted and the oscillators can be set up with a relative coarse tolerance of approx. 5%, e.g. as RC oscillator. (see Chap. 7)

- Bitstream-based transmission: In this transmission method, a data packet consisting of frame and user data is transmitted without further character synchronization. Bit synchronization is usually achieved by means of a suitable formatting procedure and clock recovery. If NRZ coding is nevertheless used (e.g. with CAN), it must be ensured that regular level changes are present, even if long "0" or "1" sequences are transmitted.

With CAN, this is done with the aid of the bit stuffing procedure (Fig. 6.9). As soon as the transmitter is to send more than five consecutive bits of the same value, it inserts a bit of the complementary value into the bit stream [6]. Similarly, when the receiver has received five consecutive bits of the same value, it can discard the subsequent bit if it has the complementary value or report a receive error if it has the same value. In bitstream-based transmission, data packets of 8 (CAN, see Chap. 10) to 1500 bytes (Ethernet) are transmitted without any overhead other than the protocol overhead.

An interesting question for this transmission method is how the beginning of a new frame is recognized. In many protocols this is solved in such a way that a special character is sent as the first character, which must not occur anywhere else in the message.

This runs counter to the requirement of having the entire representable code space available for user data (code transparency). The dilemma is solved, for example, by *bytestuffing* or *character stuffing* (Fig. 6.10). If the frame-initial character occurs in the code, it must be introduced by an escape character, which is introduced into the user data for this purpose. The escape character itself must of course also be introduced by an escape character if it occurs as a valid user data character. Alternatively, characters can be sent as frame delimiters in a bitstream-based system, which, for example, violate the bit stuffing rule and therefore cannot occur in the user data.

Other protocols use bus timing to detect a new frame: A transmission pause, for example, signals the start of a new frame in the RTU protocol of the Modbus (Chap. 11). If this pause is detected during the transmission of the frame, the frame transmitted up to that point must be discarded.

6.1.2.2 Media Access: MAC

Communication between several participants via a medium inevitably leads to collisions if the participants have no information among themselves about the transmission request of another participant other than that which is available to them on the bus. A collision is detected when a bus participant measures a different signal on the bus than it had just sent, provided that the bit was recessive and another participant simultaneously sends a dominant bit. In this case, all nodes must discard their frame and start over. Network protocols can avoid collisions per se, or they can include an appropriate strategy for responding to collisions.

With collision-avoiding protocols, there can be no collisions due to the transmission control:

- Master-slave protocols: For example, with the LIN bus, there is only one bus device (master) that can start data transmission. It regularly polls all other bus participants (slaves) for transmission requests according to a predefined schedule. The slaves only transmit after release by the master, even if they have messages to send to other slaves.
- Token passing principle: This rather rarely used procedure is based on a special message (token) that grants send authorization to the bus station that is in possession of the token. Once it has completed its transmission, it passes the token on to another bus subscriber (for example by counting up), so that it is passed on in turn.
- TDMA, Time Division Multiple Access: The TDMA method is based on exact timing. In a periodic time window, which is started by a synchronization message, time slots are reserved exclusively for a bus subscriber (static). If the subscriber has no transmission request, the time slot remains empty for this communication cycle. This method is used in the GSM mobile radio standard, in the MOST protocol and in FlexRay and is always suitable when isochronous measured values[7] have to be transmitted with low jitter.

Other protocols accept the fact that collisions are possible in principle and react with appropriate strategies to establish a new, ordered transmission sequence after a detected collision:

---

[7]Isochronous means that the signals are transmitted periodically with exactly the same period duration.

- ALOHA method: In this method developed in Hawaii, each node transmits as soon as there is a transmit request. A collision leads to a disturbed frame, which is detected by the receiver and acknowledged accordingly. If a collision has been detected, all bus nodes make a new transmission attempt after a certain time, which is determined by a random generator, in order to prevent the new transmission attempt from leading to collisions again.
- CSMA/CD procedure: The ALOHA method is not very efficient, which is why it is supplemented in Ethernet LANs, for example, by bus monitoring (CSMA stands for Carrier Sense Multiple Access), which prevents a transmission attempt from being made if another subscriber is already transmitting. Nevertheless, collisions can occur due to signal propagation times on the line, which the transmitter detects itself (CD stands for Collision Detection) and then also reacts after a random time with a new transmission attempt. If the number of messages on the medium increases, the participants extend the frame for the random time exponentially (exponential backoff), so that theoretically every message is transmitted after an arbitrarily long time.
- Binary countdown or CSMA/CA procedure: Reliability requirements, for example in motor vehicles, do not allow to have to wait arbitrarily long for certain messages. Therefore, in an extended procedure, which is used for example in the CAN bus, it is ensured that a priority is coded in the message headers. In the event of a collision, the node with the higher priority has the right to continue transmitting, while the node with the lower priority makes a new transmission attempt. The procedure thus avoids collisions and is therefore called CSMA/CA for collision avoidance (see Chap. 10).

6.1.2.3 Communication Control

Another task of the link layer is communication control. In principle, a distinction is made between

- connection-oriented protocols, i.e. protocols in which a medium is reserved exclusively for the partners for the duration of the communication[8] (circuit switching), the "old" dial-up procedure for the telephone works according to the principle of and
- connectionless protocols or datagram-oriented protocols, i.e. protocols in which each communication process is context-free and isolated, and thus for each communication process the protocol communicates between which partners the communication is to take place.

In any case, communication between potentially several different partners requires information about to whom a message is addressed (i.e., an address) and possibly information about the sender, at least if a response is expected. In networks, one distinguishes:

---

[8]Thus, the connection is contextual and the protocol must take care of providing the context, i.e., establishing and terminating the connection.

- Instance or device-oriented addressing: Here, the receiver itself is addressed, which can be an ECU or an application in the software of an ECU. The receiver exclusively evaluates the message, while the other bus nodes discard the rest of the data packet (example: Ethernet).
- Message-oriented addressing: Here, not a receiver but the message itself is addressed. All receivers who are interested in the message can evaluate it. This procedure is used in particular in protocols in which information must be sent to several participants, e.g. in CAN (multicasting). The addresses and the priority of the messages are defined during system development and recorded in a table in which all possible receivers are also noted. If the address or the format of a message is changed, the developers of all affected ECUs must be notified. The integration effort for message-oriented networks is correspondingly high. Some systems (e.g. LIN) have configurable interfaces with which the addresses of the messages can be changed subsequently.

6.1.2.4 Time Behaviour

An important criterion in the selection of a bus system is the timing. It is largely determined by the question of whether a message is sent with a determinate timing behavior, i.e. whether it is clear at any time when the message was sent and how up-to-date it was at that time. Deterministic networks determine this; in nondeterministic networks, for example, transmission may have been delayed because of a priority conflict. Furthermore, two quantities are decisive for the timing behavior:

- The *latency time*: This describes the time between an event (i.e. the sending process) and the occurrence of the reaction (the receiving process). It is composed of the time for preparing the transmission, the waiting time until the medium is released, the actual transmission time, which results from the packet size times the transmission rate (in bit/s), the propagation speed, which is only relevant for very long lines or satellite links, and the time required by the receiver to decode the message. If a response is expected, the roundtrip time is usually more interesting, which is about twice as long as the latency, plus the time the receiver needs to respond to the message.
- The *jitter* is a measure of the deviation of the actual period duration from the prescribed period duration for isochronous messages. It plays a role when measured values must be transmitted with a very reliable repetition rate. For multimedia data, which also require a low jitter, a buffer is used on the receiver side, which temporarily stores the received data packets and passes them on to the processing instance with a low jitter.

6.1.2.5 Error Handling

The last task of the data link layer to be described here is error handling. If data is sent via a message channel or even just read from a storage medium,[9] there is a risk of transmission

---

[9] A storage medium is also a high latency message channel in the broadest sense.

errors. Depending on the error, a distinction is made between single bit errors, bundle errors (whole words are transmitted incorrectly) or synchronization errors (the entire message is read incorrectly). Depending on where the error occurs (localization), a distinction is made between *user data errors*, which affect the actual message, and *protocol errors*, which affect the protocol information and can thus lead to misinterpretation of the message. The link layer is the first layer to be confronted with transmission errors. Its task is divided into two parts:

- Error detection and
- Error handling (error correction)

Error detection is based on the principle of redundancy. The basic idea is to deliberately reduce the information content of a single character or a group of characters within a message by adding additional characters to the message that do not contain any new information. This is then referred to as *redundancy*.

Behind this is the communication theory[10] or the coding theory. A simple and well-known example of a redundancy increase is the introduction of a parity bit: All ones in a codeword are counted. If the number is odd, a one is appended to the codeword, otherwise a zero (or vice versa). This ensures that each codeword contains an even (in the reverse case: odd) number of ones. Deviations from this rule are recognized as errors. This increases the number of digits by one bit, which is equivalent to doubling the number of possible codewords. However, since the information remains the same, the information per bit has decreased and half of all possible codewords are invalid. This procedure is used in simple communication systems, but it is not very efficient. A second error in the codeword completely cancels the error detection.

Other procedures form multi-digit checksums from larger message blocks or even the entire message with the aim of also being able to detect a second or more errors. The checksum must be reproduced in the receiver and compared with the transmitted sum. The most frequently used form of checksum formation is the CRC code[11] (polynomial code, cyclic code) because it can be easily reproduced by hardware and is very efficient. CRC codes are based on the assumption that a bit sequence (from now on called frame) of fixed length m can be understood as a polynomial of powers of two of degree r, for example the bit sequence 110001 as a polynomial of degree 5

---

[10]This large-scale mathematical construction goes back to the work of Claude Shannon, which he published in 1948.

[11]Cyclic redundancy check.

$$1 \cdot x^5 + 1 \cdot x^4 + 0 \cdot x^3 + 0 \cdot x^2 + 0 \cdot x^1 + 1 \cdot x^0$$
$$= x^5 + x^4 + x^0$$

In the so-called modulo-2 arithmetic, strongly simplified rules apply, for example $a + b = a - b$ (thus $1 + 1 = 0 = 1 - 1$, $1 + 0 = 1 = 1 - 0$, $0 + 1 = 1 = 0 - 1$), without remainder, thus an addition corresponds to a subtraction and a logical exclusive-or. A modulo division is correspondingly simpler: if one polynomial "fits" into another, the result is 1, otherwise 0. It can be mathematically proved that here is an algebraic body in which all algebraic operations can be mapped. A multiplication of a polynomial by a power of two x^n corresponds to a shifting of the bit sequence with appending zeros by n. A multiplication can only be done with 1 or 0, thus preserving the bit sequence or not. One advantage of the CRC is that it is very easy to implement the calculation in hardware, since the essential operation – the division – is performed with a shift operation that takes place automatically in the transmit shift register during transmission.

The procedure for calculating the checksum is:

- Choose a generator polynomial $G(x)$ of degree r with $r + 1$ bits
- The most significant bit and the least significant bit (MSB and LSB) must be 1.
- Append r 0 bits to the frame. The result is a polynomial $x^r \cdot M(x)$ with $m + r$ bit length
- Divide the bit string $x^r \cdot M(x)$ by $G(x)$ using modulo 2 division
- Subtract the remainder (always r bits or less) from $x^r \cdot M(x)$. The result is the polynomial $T_S(x)$, whose corresponding bit sequence is transmitted.

Mathematically written, the following transmission sequence is to be formed:

$$T_S(x) = x^r \cdot M(x) - \left[\frac{x^r \cdot M(x)}{G(x)} \right] \tag{6.7}$$

A receiver receives the bit sequence $T_E(x)$ which is composed of $T_S(x)$ and an error $E(x)$ which is additively superimposed according to the applicable physical laws.

$$T_E(x) = T_S(x) + E(x) = x^r \cdot M(x) - \left[\frac{x^r \cdot M(x)}{G(x)} \right] + E(x) \tag{6.8}$$

If now the receiver divides the received bit sequence again according to the same rules by the generator polynomial, the result, if the error is 0, logically must also be 0, because by subtracting the remainder no remainder can remain when dividing again (this is also valid for the usual arithmetic). But as soon as there is an additive error, the result becomes

$$\frac{T_E(x)}{G(x)} = \frac{x^r \cdot M(x) - \left[\frac{x^r \cdot M(x)}{G(x)}\right]}{G(x)} + \frac{E(x)}{G(x)} = \frac{E(x)}{G(x)} \tag{6.9}$$

That is, all errors $E(x)$ that are not multiples of the generator polynomial $G(x)$ are detected.
Example:

- The generator polynomial is chosen with $G(x) = x^4 + x^1 + x^0$ (bit sequence 10011, $r = 4$). There are various standardized generator polynomials that have proven their worth. One is defined for a protocol and is part of the protocol, i.e. it does not change and must of course be identical for receiver and sender.
- Let the frame in this example be $M(x) = x^3 + x^2 + x^1 + x^0$ (bit sequence 1111)
- The shift $x^r * M(x) = x^7 + x^6 + x^5 + x^4$ yields the bit sequence 11110000
- Now divide $x * M(x) \% G(x)$ modulo:

```
11110000 / 10011 = 1110 remainder 10
-10011
 011010
 -10011
 010010
 -10011
 000010
```

- Then the transmission polynomial is sent: $T_S(x) = x^r \cdot M(x) - [x^r \cdot M(x) \% G(x)] = 000011110010$

The receiver divides the received polynomial again by the generator polynomial, here:

```
11110010 / 10011 = 1110 remainder 0
-10011
 011010
 -10011
 010011
 -10011
 000000
```

Standardized generator polynomials are for example:

• IEEE802:	$x^{32} + x^{26} + x^{23} + x^{22} + x + x^{1612} + x^{11} + x^{10} + x^8 + x^7 + x^5 + x + x^{42} + x^1 + 1$
• CRC-12:	$x^{12} + x^{11} + x^3 + x^2 + x + 1$
• CRC-16:	$x^{16} + x^{15} + x^2 + 1$
• CRC-CCITT:	$x^{16} + x^{12} + x^5 + 1$
• CAN-CRC:	$x^{15} + x^{14} + x + x^{108} + x^7 + x^4 + x^3 + 1$

They are of course applied to larger frames, usually a few hundred bits. Section 12.1.4.2 shows a practical example of programming and hardware implementation of a CRC checksum.

For transmission networks that do not work with fixed codeword lengths, convolutional coders exist that generate a new, longer bit stream with the same information content but a lower average information content (per bit) from a theoretically infinitely long bit stream.

Since the necessary level of redundancy for a given target error rate depends on the error rate of the message channel, no procedure can be generalized for all message channels. Therefore, the measures for transmission error detection and avoidance are referred to as *channel coding*.

Two fundamentally possible measures for *error correction* must be distinguished:

- (Forward) error correction, also FEC (Forward Error Correction): The receiver detects errors in the code due to the redundancy and can reconstruct the original message because the faulty message (e.g. in the case of exactly one error) can be clearly traced back to the original message. For this purpose, higher redundancy is necessary even under good transmission conditions, but no back channel is needed and the correction is faster because it takes place in the receiver. However, there is a risk that if more than the allowed number of errors occur, the correction will lead to a wrong but valid message. FEC is used, for example, on CDs to compensate for scratches.
- Error control by error-detecting codes, also ARQ (Automatic Repeat Request): The receiver detects errors and reports them back to the transmitter so that the transmission can be repeated. However, this is only possible if a return channel is available; this is not the case with a CD or a satellite receiver, and in other applications the latency is too high (for example, on a Mars probe) for a return communication.

Different strategies can be distinguished for ARQ. A simple stop-and-wait protocol is based on the fact that the sender expects an acknowledgement of correct reception or, if necessary, an error message from the receiver in response to each message (handshake). If no acknowledgement message arrives after a certain time or if an error message arrives, the transmission of the message is repeated (timeout). This leads to a very inefficient transmission, since several new messages could be sent in the time it takes the receiver to process the message. Moreover, the acknowledgement or error messages may also be lost, so that the sender does not know which message to repeat if necessary. Two mechanisms in particular help here [9]:

1. The messages are provided with a running sequence number. To limit the necessary data field, one tolerates that the number repeats occasionally, e.g. after 256 messages. Acknowledgement or error messages always refer to this number.
2. A defined number of messages is temporarily stored in the sender and in the receiver and initially sent en bloc. The receiver now requests specific faulty messages (Selective Repeat) or it reports back the first faulty message and expects the new receipt of all data

sent since this faulty message (Go-back-N). Timeout is also used here. The receiver now puts all messages together in sequence based on the sequence number. As soon as sender and receiver agree that a message has been received safely, it is deleted from the cache of both communication participants. This procedure is known as sliding window.

Depending on the requirements for transmission rate and residual error probability (two requirements that contradict each other), error control strategies differ for each network.

6.1.3   Layer 3: Network Layer

The network layer has the task of routing data packets across network boundaries. A data packet of the data link layer always addresses a node in the network of the sending node. If you want to send data across networks (for example, from your own Wifi to the Internet), a data packet from the network layer is packed into the link layer payload and sent to a gateway required in the network. A gateway is always a network node that connects to multiple networks, regardless of their link layer protocol. Gateways that work with layer 3 protocols are called routers, but there are others (for example, CAN gateways work only with the CAN protocol, which is layer 2). Strictly speaking, however, the router's task is not to calculate a route (as a totality of all paths), but only to forward the data packet to the next router. This is done with an intended route, but ultimately it is up to each node the packet passes by to decide how the further route should look like, so that theoretically also endless loops are possible. A layer 3 protocol must therefore also contain mechanisms for looping.

The IP protocol is the best known layer 3 protocol and has become generally accepted. In IPV4, the network-independent address consists of four bytes, which are represented by four decimal numbers 0 ... 255 separated by a dot. A portion of the address space is reserved as local and is not routed externally. This address space is 10.x.x.x, 172.16.0.0 to 172.131.255.255 and 192.168.0.0 to 192.168.255.255. All other addresses may only occur once worldwide and are assigned centrally by the "Internet Assigned Numbers Authority" (IANA), which rules over the entire IP address space. It allocates IP address blocks to so-called "Regional Internet Registries" (RIR), which are responsible for regional IP address allocation. These in turn allocate IP address ranges to, for example, Internet Service Providers (ISPs). The IP addressing scheme is divided into the actual network address and a device address. The boundary between network address and device address is not fixed, but is determined by two mechanisms: Address classes and subnet masks.

In the early days of the Internet, the sheer mass of connected computers was not taken into account. Initially, the address classes listed in Table 6.2 were defined.

Most networks (companies, Internet providers, universities) have more needs than Class C but less than Class B, so the subnet mask was introduced as another mechanism.

Subnet masks identify the range of the IP address that describes the network and the subnet. This range is determined by ones (1) in the binary form of the subnet mask. This

Table 6.2 Address classes in IPV4

Name	Label	Address space	Scope
Class A	First bit of the first byte is 0	1.0.0.0 to 127.255.255.255	126 networks with 16 million nodes each
Class B	First two bits of the first byte are 10	128.0.0.0 to 191.255.255.255	16,384 networks with 65,534 nodes each
Class C	First three bits of the first byte are 110	192.0.0.0 to 223.255.255.255	2 million networks with 254 nodes each
Class D	First four bits of the first byte are 1110	Multicast address 224.0.0.0 to 239.255.255.255	Multicast address does not address individual nodes: 2^{28} networks
Class E	First five bits of the first byte are 11110	240.0.0.0 to 247.255.255.255	Reserved

makes it possible for several networks/services to share a Class B address. To make better use of the addresses, they are often assigned dynamically in the local network. One of the methods used for this is the DHCP protocol, where a node that logs on to the network is assigned an IP address. Private users are usually not assigned public addresses but use private address spaces. The router connected to the provider is assigned a dynamic IP, which can change regularly. The forwarding of data to an end device is done by layer 4 protocol, the NAT (Network Address Translation), see the next section.

In order to get a systematic approach to the flood of Internet addresses, the Domain Name System was introduced as early as 1983 in RFC882. This system divides the Internet hierarchically. The so-called top-level domains are assigned by the IANA and have developed from a purely American system (e.g. .com for companies, .gov for the government, .org for associations, .edu for universities) to a broad spectrum of country-specific domains. Each top-level domain is administered by a Network Information Center (NIC) or a Domain Name Registry, which in turn administers subordinate domains. In Germany, this is DENIC, which was initially located in Karlsruhe and now resides in Frankfurt (in Switzerland: SWITCH and in Austria nic-at).

The domain names are associated with IP addresses on so-called domain name servers (DNS) in a name directory. Since these can change, the domain name servers regularly exchange information about changes. If a name is not in a name directory, the DNS can query a higher-level DNS recursively. Thus, if a node on the Internet is addressed as a domain name, the sending node must first make a name request, after which it will get back an IP address. Since IP addresses can change constantly (for example, for load balancing in server farms or in the private sector), the domain name is the more stable addressing variant. For private individuals, a so-called DynDNS is available, here you get a domain assigned, the domestic router then reports every IP address change the new address to the DynDNS server, which in turn informs the other DNS in its environment.

Since the address space of 2^{32} addresses became scarce years ago, there is now a switch to the IPV6 protocol, which provides 128-bit addresses and thus $2^{128} = 3.4 \cdot 10^{38}$

addresses. This is so much that static address allocation is no longer necessary. Addresses are written in hexadecimal separated by a colon, for example: 2002:6dc1:ac2a:0: 1cce:7c9d:62b8:d6e0. In IPv6, the address is also divided into a network part and a node part.

So how do Layer 3 messages communicate over a Layer 2 network?

If a node wants to reach another node, inside or outside its own network, via an IP address, it sends an ARP (Address Resolution Protocol) message into the network and thus asks who owns this IP address. Either another node or the router that knows this address or has access to a routing table containing this address can reply. With the reply comes the local (MAC) address, to which the original sending node then sends its IP message wrapped in a layer 2 frame. The router unpacks the message and in turn sends it into the network, which is the best possible route to the addressed destination in terms of defined quality factors. To ensure that these routes are always up-to-date and optimal, all routers regularly exchange routing protocols (e.g. BGP – Border Gateway Protocol or RIP – Routing Information Protocol).

A route is optimal, for example, if it contains as few intermediate routers (hops) as possible, or has as little packet loss as possible, or the shortest possible transit times. The connection costs can also play a role. Which route is ultimately taken is decided by each router using a routing algorithm based on the aforementioned information from its neighboring routers.

The IP protocol therefore uses the store-and-forward method and is connectionless, since each IP packet is sent separately with full address information. Each hop stores the packet for a short time in order to then forward it. If confidential data is to be sent, it must therefore be encrypted end-to-end. An encrypted IPsec extension of the IP protocol therefore exists at layer 3.

6.1.4 Layer 4: Transport Layer

The transport layer has the task of separating the network from the application. The applications therefore deliver data to the transport layer and the latter prepares it so that it is sent correctly and delivered to the correct application on the other side. There are several mechanisms for this, which are implemented in the most well-known Internet transport layers, including:

- Segmentation: Messages that are too large can be split up and reassembled at the receiver (a mechanism that is also implemented in the CAN transport layer).
- Flow control: The transport layer controls the communication between asynchronous partners (especially if the receiver is slower than the sender) in such a way that data transmission is as continuous as possible without losses.

- Reliability (securing the communication): If the communication is based on unreliable lower layers (such as IP), the procedures described in Sect. 6.1.2.5 are used here to ensure that the data is transmitted without errors, without losses and in the correct order.
- End-to-end addressing: Layer 3 protocols usually connect computers. However, since communication takes place between applications (example: mail client to mail server, browser to web server, sensor to broker), another mechanism (on the Internet: port) is introduced that is equivalent to an address extension. An application is assigned to a port and all messages directed to that port go to that application. At the same time, a sender specifies the port for a reply. Known ports on the Internet, defined by common conventions, are 20 (FTP), 22 (Secure Shell), 25 (Mail), 80 (http) and so on.
- Connection control: In the Internet there are two very well known transport layers TCP and UDP, see below. The TCP protocol is connection-oriented, the UDP protocol is connectionless (see Sect. 6.1.2.3).

TCP (Transport Control Protocol) combines the above mechanisms. It is able to segment, builds a connection context, has handshake mechanisms to secure the data traffic, a checksum, flow control to name the most important. TCP is used by the http protocol, for example, and is always appropriate when larger data packets need to be sent securely.

UDP (User Datagram Protocol) is very simple and actually only an extension of IP by a port address and a checksum. No handshake mechanisms are provided and no error handling. On the other hand, the protocol is fast and is suitable for smaller data packets of up to 65 kByte, for which a possible packet loss is more acceptable than a costly resending.

On more complex operating systems (such as UNIX/Linux or Windows), the transport layer is provided in the form of sockets, a programming interface that can be used by the application. The process is as follows:

TCP Sockets (Stream Socket) On the client side:

- Create socket (with `socket()`)
- Connect the created socket to the server address from which data should be requested, first by querying the IP address from DNS or creating an IP address structure from the known IPv4 or IPv6 bytes: `inet_pton(AF_INET, "1.2.3.4", & srv.sin_addr);` then assign a port and connect: `srv.sin_port = htons(1234);`

  `connect(sockfd, (struct sockaddr*)& srv, sizeof(struct sockaddr_in));`

- Sending and receiving data `read(), write()`
- If necessary, shut down socket at the end (`shutdown(), close()`)
- Disconnect, close socket

Server-side:

- Create server socket (see above)
- Binding the socket to a port through which requests are accepted (bind())
- wait for requests (listen())
- Accept request and thus create a new socket pair for this client (accept())
- Processing the client request on the new client socket
- Close the client socket again.

UDP (datagram) sockets.
  Client-side:

- Create socket
- Send (send())

Server-side:

- Create socket
- Bind socket
- Waiting for packages

For servers there is still the speciality that waiting for a request blocks the software, as does waiting for a response at client and server. This can be circumvented in a higher operating system by moving the wait to a thread.

6.1.5 Application Protocols for IoT Networks

On the application layer side (layer 5–7 of the ISO-OSI layer model) the semantic aspects of the transmitted data are bundled.

On the Internet, a whole range of protocols are standardized to connect applications, the best known being:

- DoIP (Diagnostics over IP) – Transport protocol for vehicle diagnostics
- FTP (File Transfer Protocol) – File transfer
- HTTP (Hypertext Transfer Protocol, WWW)
- HTTPS (Hypertext Transfer Protocol Secure)
- IMAP (Internet Message Access Protocol) – Access to e-mails
- NTP (Network Time Protocol)
- POP3 (Post Office Protocol, Version 3) – E-mail retrieval
- PTP (Precision Time Protocol) – Time synchronization of clocks in a network

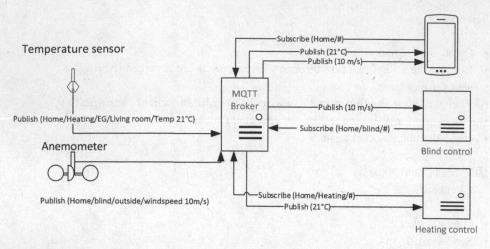

Temperature sensor

Subscribe (Home/#)
Publish (21°C)
Publish (10 m/s)

MQTT
Broker

Publish (Home/Heating/EG/Living room/Temp 21°C)

Publish (10 m/s)
Subscribe (Home/blind/#)

Anemometer

Blind control

Publish (Home/blind/outside/windspeed 10m/s)

Subscribe (Home/Heating/#)
Publish (21°C)

Heating control

Fig. 6.11 MQTT principle

- RDP (Remote Desktop Protocol) – Display and control desktops on remote computers (Microsoft)
- RTP (Real-Time Transport Protocol)
- SNMP (Simple Network Management Protocol) – management of devices in the network
- SMTP (Simple Mail Transfer Protocol) – Sending e-mails
- SSH (Secure Shell) – encrypted remote terminal[12]
- Telnet – Unencrypted login on remote computers (remote terminal)

The application system also contains various formats for sensor data, which should be briefly mentioned here:

- MQTT: Message Queuing Telemetry Transport (MQTT) is an open client-server protocol for machine-to-machine communication (M2M) that enables the transmission of measurement and telemetry data despite high delays or limited networks [8]. A client sends the server (the MQTT broker) a so-called topic, which is hierarchically named (for example, Home/Heating/EG/Living room/TemperatureIST). Other clients, e.g. controllers or end devices, can subscribe to these topics and receive them whenever they are changed. Wildcards are possible, e.g. Home/Heating/ + / + /TemperaturIST subscribes to all actual temperature values from all rooms, while Home/Heating/EG/# subscribes to all heating-related data from all rooms on the ground floor. A simplified example can be seen in Fig. 6.11.

---

[12] A pretty good list can be found in Wikipedia under the keyword "Internet protocol family".

- JSON: The JavaScript Object Notation is not a protocol but a compact data format in an easily readable text form and serves the purpose of data exchange between applications. JSON – defined in RFC 8259 – is independent of the programming language. Parsers and generators exist in all common languages. In JSON, data structures are exchanged that are structured hierarchically. Each entry consists of a name (key) and a value, whereby the latter can again consist of a structure. Arrays are also possible.

```
{ "city":
    { "name": "Heilbronn",
    "country": "DE",
    "coord":
        { "lon": 9.206321,
        "lat": 49.147134}
    },
"temp": [281.15, 285.31, 288.22, 290.51, 292.98]
}
```

JSON is currently replacing more and more XML, which is similar in structure but can often be notated ambivalently.

The description of the ISO/OSI stack ends at this point, since knowledge of the first two layers is sufficient for a further understanding of the serial interfaces shown. A more detailed overview of computer communication can be found in [9, 10], among others.

6.2   Requirements for IoT Networks

RFC 7228 (see [4] and [11]) describes the special requirements for IoT networks with constraint nodes. Constraint nodes are network participants where "some of the characteristics that are otherwise fairly self-evident for Internet nodes are not achievable at the time of writing, often due to cost constraints and/or physical limitations on characteristics such as size, weight, and available power and energy." (RFC7228).

The concepts presented in this book fulfill this definition insofar as they are implemented on 8-bit microcontrollers. Due to the portability of the presented codes, however, it should be noted that already an implementation with a microcomputer with LINUX operating system (e.g. Raspberry Pi) hardly needs to be called a constraint node, especially if a power supply from the network is available.

The essential properties of such networks are:

- Integration of the network into existing infrastructures that have a connection to the Internet at least in the gateway or in the backend.
- For the development and operation, well-known technologies and tools are used.

- New standards and protocols are being used for the "last mile", and the protocols are often open (not always!).
- Compared to the "normal" Internet, a sensor network consists of a large number of simple devices.
- These are often subject to physical and economic constraints.

Devices in the IoT are often resource constrained. This affects the

- Size and interfaces, especially the user interface, which is often missing, but also programming interfaces for updating the software, which are often missing or only accessible with physical proximity or limited bandwidth (in contrast to full remote or OTA (over-the-air) access for more complex nodes).
- Power supply, especially with battery operation or operation with energy harvesting, i.e. operation in which the energy supply is taken from the environment, for example with solar cells, wind generators, but also temperature differences, mechanical movement (vibration, pressure) or magnetic fields. For example, radio switches have existed for some time that use piezo crystals to generate enough energy to transmit their radio signal. RFID also uses this technology.
- Computational power and memory: Size, large number (cost) and low power also limit the computational power and memory, one reason why only concepts tested on 8-bit microcontrollers are presented in this book. The same applies to the code complexity, which has to be adapted due to the small memories.
- Network connection: Accordingly, bandwidth, data volume and latency times are limited and special protocols are required, a selection of which will be discussed in more detail in the following chapters.

Specifically, the networks are called constraint networks. Constraints may include [11]:

- low achievable bit rate/throughput (including limited sampling rates),
- high risk of packet loss and high variability of packet loss due to low transmission power and regulatory limitation of available transmission bands,
- strongly asymmetrical link properties, for example no return channel,
- high penalties for using larger packets (especially in the public ISM bands), which can lead to fragmentation[13] of the data packets (thus also risk of packet loss),
- Limits on accessibility over time (a significant number of devices could be turned off for energy reasons, but these are usually "woken up" periodically and can communicate for short periods of time); and
- the absence of (or severe limitations on) advanced services such as IP

---

[13] Fragmentation or segmentation is the process of breaking down a data packet into small subpackets that are transmitted independently of each other. By using suitable protocols, they are reassembled into the original packet at the receiver.

Table 6.3 Energy classes in the IoT [10]

Name	Type of energy restriction	Example of energy sources
E0	Limited energy per event, e.g. when a switch is actuated.	Event-based energy supply (harvesting), e.g. from the actuation energy of the switch or a magnetic field
E1	Time limited energy	Battery or accumulator with exchange or recharge cycles
E2	Energy available over the limited lifetime	Non-replaceable battery (e.g. in smoke detectors)
E9	No restrictions	Supplied with power supply unit

Multicast.

In summary, the limitations:

- exist on the cost side,
- can be caused by the nodes and their conditions of use,
- physical origin (environment, such as EMC, temperature, temperature change, weather exposure, vibration, water),
- regulatory reasons, as in the case of transmission in the ISM band, which is used by the radio standards presented in the following.

Moreover, in a growing IoT, it is also reasonable to assume that technologies and protocols of different maturity and development levels can coexist, and therefore downstream processing must be strictly compatible.

There are two characteristic variables for the energy supply, namely the average power consumption during operation, which determines, for example, the size of a solar cell or other energy converter, and the minimum energy that must be available to keep the device alive. The latter determines the size of any storage (battery, powercap).

RFC 7228 proposes four energy classes, which can be found in Table 6.3.

Depending on the energy class, different strategies can be used, see also Sect. 3.9 described in RFC 7228:

- Always on: No specific behaviors are prescribed here if it is not necessary due to the energy constraints. The device is simply always on, but it may make sense to design the hardware to optimize performance, to clock the processor slowly, to limit the number of transmissions or to implement other energy-saving measures. However, the device is always accessible in the network.
- Normally off: Here the device is normally switched off and is only activated when it has to transfer data. In doing so, it must reconnect to the network each time. The optimization goal is therefore to perform network dial-up as quickly and cheaply as possible (in terms of computing effort and the resulting energy consumption). If the switch-off time is long and only very little data is transferred, however, additional consumption due to dial-up can be accepted.

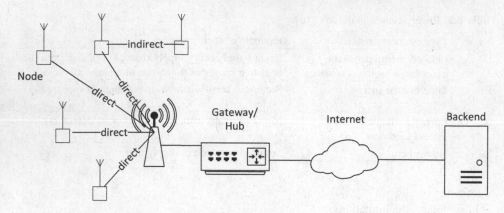

Fig. 6.12 Greatly simplified structure of an IoT network

- Low-power: Devices use this strategy if they still have to be accessible on the mains with the lowest possible energy consumption, if necessary with a slight delay (latency). In vehicles, these are usually control devices that are connected to the steady plus (terminal 30) and are woken up by bus activities. The optimization goal here is low energy consumption for the minimum activity on the network.

IoT networks use wireless or wired protocols optimized for these purposes, which will be discussed in Chaps. 10, 11, 12 and 13. Previously, in Chaps. 7, 8, and 9, the serial interfaces UART, SPI and I^{2}C inherent to most microprocessors are described, which represent the bridgehead to the outside world. The topology of these networks is often set up in such a way that the terminal devices (network nodes) implement simple connections according to the strategies described above and with the lowest possible data rates; if necessary (as in BlueTooth® Mesh), network nodes are also used as repeaters. The data packets are received by a front-end gateway, packed into IP packets and forwarded via the Internet to a back-end computer, from where data concentration and further processing is possible.

A highly simplified representation of such an architecture is shown in Fig. 6.12.

References

1. Meroth, A., & Tolg, B. (2008). *Infotainmentsysteme im Kraftfahrzeug. Grundlagen, Komponenten, Systeme und Anwendungen.* Vieweg.
2. ISO: ISO/IEC 7498-1: Information technology – Open systems interconnection – Basic reference model 1994, zugl. ITU-T Recommendation X.200 öffentlich. https://standards.iso.org/ittf/PubliclyAvailableStandards/s020269_ISO_IEC_7498-1_1994(E).zip. Accessed 31 Dez 2020.
3. Adolf, J. (2007). *Schwab.* Wolfgang Kürner: Elektromagnetische Verträglichkeit, Springer Verlag.
4. RFC Editor: RFC Index. https://www.rfc-editor.org/.Accessed 31 Dez 2020
5. Roppel, C. (2006). *Grundlagen der digitalen Kommunikationstechnik.* Carl Hanser.

6. NXP: CAN Bosch Controller Area Network (CAN) Version 2.0 PROTOCOL STANDARD http://www.nxp.com/assets/documents/data/en/reference-manuals/BCANPSV2.pdf.Accessed 1 Dez 2016
7. ISO 11898: Road vehicles – Controller area network (CAN) (6 Teile)
8. MQTT: Offizielle Webseite von MQTT. https://mqtt.org/.Accessed 1 Jan 2021
9. Andrew, S. T., & David, J. (2012). Wetherall: Computernetzwerke (5., aktualisierte Aufl.). Pearson
10. Riggert, W., Märtin, C., & Lutz, M. (2015). *Rechnernetze – Grundlagen, ethernet, internet (5., aktualisierte Aufl.)*. Hanser.
11. Bormann, N. et al. (2014). RFC 7228 terminology for constrained-node networks. https://www.rfc-editor.org/rfc/rfc7228.html.Accessed 31 Dez 2020.

Asynchronous Serial Interfaces

7

Abstract

This chapter describes the asynchronous interface of the AVR family and the principle of UART in general.

After the underlying theory of communication has been discussed in the sixth chapter, the following three chapters are dedicated to the interfaces that are installed in typical microcontrollers and sensors, or in network components. The classic is the UART interface, which is certainly a bit outdated but still used in many applications. In further chapters the SPI and the TWI/I^{2}C are described.

7.1 Universal Asynchronous Receiver/Transmitter (UART)

The UART interface (Universal Asynchronous Receiver/Transmitter) is a regular feature of microcontrollers and many protocols are based on this interface. For example, the somewhat outdated RS232 (TIA/EIA-232) is based on the UART, but also the protocols TIA/EIA-422, TIA/EIA-485 and partly also MODBUS and Profibus, which are still widely used in industry. Other single-wire bus systems, such as the LIN bus established in automobiles, are also based in principle on the UART interface and can be easily set up with a UART. Please refer to Chaps. 11 and 12. A UART driver is used for the hardware connection of various higher protocol layers and is interesting for sensor networks in that simple connections to PCs can be established with a "virtual COM interface". In addition, various wireless standards allow the transparent forwarding of serial protocols via a UART

© The Author(s), under exclusive license to Springer Fachmedien Wiesbaden GmbH, part of Springer Nature 2023
A. Meroth, P. Sora, *Sensor networks in theory and practice*,
https://doi.org/10.1007/978-3-658-39576-6_7

connection, for example the Bluetooth® protocol with the Serial Profile, which will be explained in more detail later, or various other wireless standards.

The basic idea of the UART is to establish a partner-to-partner (*peer-to-peer*) connection by clocking data packets of 5 to 9 bits from a shift register using the non-return-to-zero (NRZ) method, i.e. a logical 1 is represented by a particular voltage level, a logical 0 by a second one, usually 0 V. The term *asynchronous* comes from the fact that no clock information is transmitted and the receiver must sample the signal with its own clock generator. The step size (baud rate) and the length of the data packet must be agreed between receiver and transmitter before transmission. However, some UARTs also have an optional line for clock transmission, for example in the entire AVR family; they are then called USART (universal synchronous/asynchronous receiver/transmitter).

A clear disadvantage of the UART interface is the fact that the clock generators, for example due to temperature drift or component tolerances, no longer run synchronously for longer sequences, therefore the interface is only designed for relatively slow data rates and short frames. As can be seen in Fig. 7.1, the transmission begins with a so-called start bit complementary to the idle level of the bus. The remote station detects from this that a data word is following and starts scanning. After the transmission of the data word, which can be between five and nine bits long – usually eight bits are specified – first an optional parity bit and then one or two bits in the idle level (stop bits) are transmitted, this is a sign for the receiver that the transmission is finished before a new start bit is sent. In the figure the optional bits are marked with a square bracket.

The parity bit is used to determine correct transmission. Its introduction doubles the code space of the data to be transmitted,[1] so that only every second theoretically possible data frame carries a valid message. Thus, an odd number of erroneous bits can be detected per data frame. The determination rule[2] for parity is

$$P_{even} = d_{n-1} \oplus \cdots \oplus d_3 \oplus d_2 \oplus d_1 \oplus d_0 \oplus 0 \tag{7.1}$$

$$P_{odd} = d_{n-1} \oplus \cdots \oplus d_3 \oplus d_2 \oplus d_1 \oplus d_0 \oplus 1 \tag{7.2}$$

In other words: With even parity, the number of ones in the user byte is padded with a 1 to an even number, with odd parity to an odd number. Thus, the following parameters must be clarified in advance during data transmission: Data rate, number of transmitted bits, parity (odd, even, no) and number of stop bits (1 or 2).

Figure 7.1 additionally shows the minimum wiring for communication with a UART. Most UART interfaces are *full-duplex* capable, i.e. they have an output TxD for sending and an input RxD for simultaneously receiving a data word. If two partners are connected

---

[1] This is referred to as a "Hamming distance" of 2.
[2] The character $\oplus$ stands for an exclusive OR function.

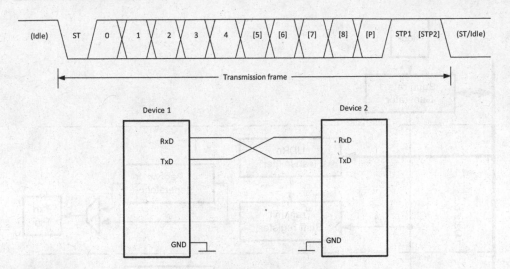

Fig. 7.1 Basic structure of a UART data frame and the circuitry

together, the two lines must of course be crossed, i.e. RxD must be connected to TxD and vice versa.

7.1.1   Hardware Connection in the AVR Family

In the AVR family there is at least one USART on each processor. At this point only the asynchronous mode shall be described. The basic internal circuitry is shown in Fig. 7.2. The USARTs are numbered consecutively in AVR, for simplicity we always use USART 0 here.

The data rate of the ATMega88 is determined in the baud rate generator by

$$BAUD = \frac{f_{osc}}{16 \cdot (UBBR0 + 1)}$$

where UBBR0 is the value of the baud rate register of the same name. With a given baud rate this must therefore be set to

$$UBBR0 = \frac{f_{osc}}{16 \cdot BAUD} - 1$$

can be set. At 9600 baud and a crystal frequency of 18.432 MHz, the value of UBBR0 = 119.

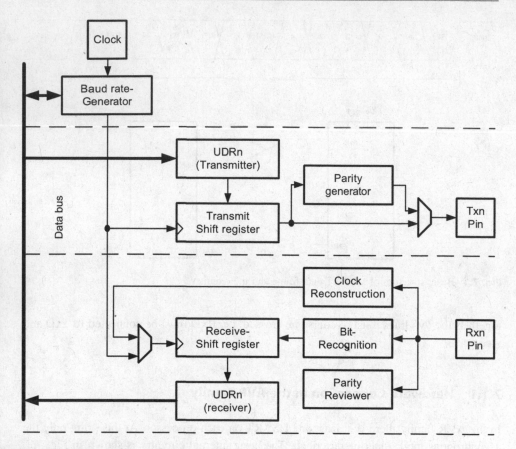

Fig. 7.2 Principle of the UART in asynchronous mode with the AVR family

7.1.2   UART Registers on the ATmega 88

In addition to the baud rate register, other registers are necessary for the configuration and operation of the USART interface. The transmit and receive registers of the interface can be addressed at the same address. Transmitting is done by writing to the UDR0 register, after a successful receive operation the received bits are to be read in this register. The interface is configured in the USART Control and Status registers (UCSR0A, UCSR0B, UCSR0C) (see Fig. 7.3).

In register UCSR0A, RXC0 is set by the processor when a new byte has been received. TXC0 is set when a transmit operation has been completed. UDRE0 is set, if UDR0 is empty. FE0 is set if an error was detected during reception. DOR0 is set if UDR0 has not been read and a new data word is pending at the input. UPE0 indicates a parity error. U2X0 doubles the transmission speed (description of this mode see ATMega88 manual). MPCM0 switches on the multiprocessor mode. In this mode several receivers are connected in parallel. First an address is transmitted and only the processor assigned to this address responds.

UCSR0A	Bit 7	Bit 6	Bit 5	Bit 4	Bit 3	Bit 2	Bit 1	Bit 0
	RXC0	TXC0	UDRE0	FE0	DOR0	UPE0	U2X0	MPCM0
Direction	R	R/W	R	R	R	R	R/W	R/W
Initial value	0	0	1	0	0	0	0	0

UCSR0B	Bit 7	Bit 6	Bit 5	Bit 4	Bit 3	Bit 2	Bit 1	Bit 0
	RXCIE0	TXCIE0	UDRIE0	RXEN0	TXEN0	UCSZ02	RXB80	TXB80
Direction	R/W	R/W	R/W	R/W	R/W	R/W	R	R/W
Initial value	0	0	0	0	0	0	0	0

UCSR0C	Bit 7	Bit 6	Bit 5	Bit 4	Bit 3	Bit 2	Bit 1	Bit 0
	UMSEL01	UMSEL00	UPM01	UPM00	USBS0	UCSZ01	UCSZ00	UCPOL0
Direction	R/W	R/W	R/W	R/W	R/W	R/W	R/W	R/W
Initial value	0	0	0	0	0	1	1	0

Fig. 7.3 The registers for configuring the UART in the ATmega88. (After microchip [1])

The UCSR0B register controls the interrupt behavior of the USART interface and enables transmit or receive. RXCIE0 switches on the Receive Complete Interrupt. TXCIE0 enables the Transmit Complete Interrupt. UDRIE0 turns on the USART Data Register Empty Interrupt. RXEN0 and TXEN0 switch receiving and transmiting on (Receive Enable and Transmit Enable, respectively), UCSZ02 together with UCSZ01 and UCSZ00 in the UCSR0C register specifies the number of data bits (see below). RXB80 and TXB80 contain the ninth bit in 9-bit mode.

UMSEL01 and UMSEL00 in register UCSR0C indicate the interface mode. With 0 in each case, the asynchronous mode is switched on. UMP01 and UPM00 select the parity according to the following scheme (Table 7.1):

USBS0 set means "two stop bits", not set means "one stop bit". Finally, the bits UCSZ02 ... UCSZ00 (USCR0B) indicate the number of data bits. 0 1 1 means "8 bits", the usual measure, see code example. Finally, UCPOL0 specifies the polarity of the clock in synchronous mode. The synchronous mode is described in Sect. 8.6.

Table 7.1 Setting the parity of the UART

UPMn1	UPMn0	Parity mode
0	0	Disabled
0	1	Reserved
1	0	Enabled, even parity
1	1	Enabled, odd parity

7.1.3   Initializing the UART Interface on the ATmega88

With a usual value of 8 data bits, no parity check and 2 stop bits and at 9600 baud (9600,8, N,2), a file uart.c will contain the following configuration (source: ATMega88 manual):

```
void UART_Init(uint16_t baud)
{
    /*set baud rate*/
    UBRR0H = (uint8_t)(baud >> 8);
    UBRR0L = (uint8_t) baud;
    //switch on receiver and transmitter as well as
    //Received Complete Interrupt
    UCSR0B = (1 << RXEN0) | (1 << TXEN0) | (1 << RXCIE0);
    /*frame format: 8data, 2 stop bit, no parity*/
    UCSR0C = (1 << USBS0) | (3 << UCSZ00);
}
```

In this case baud $= 119$ (in the case of a crystal frequency of 18.432 MHz and a target baud rate of 9600 baud) is passed during the call. This can be set, for example, with.

```
#define F_OSC 18432000uL
#define BAUDRATE 9600uL
#define BAUDPARAM (F_OSC/(BAUDRATE*16))-1
```

so that the value is calculated in the precompiler and transferred as a constant. If the baud rate does not have to change at runtime, this measure saves a lot of processor capacity. The suffix *uL* after the constants is important so that the precompiler knows the correct value range. A look at the translated code (assembler) shows the effectiveness of this measure, here only one constant is entered into the registers r24 and r25, which together result in the transfer parameter of type unsigned short:

```
    uartInit(BAUDPARAM);
70: 87 e7 ldi r24, 0x77 ; 119
72: 90 e0 ldi r25, 0x00 ; 0
74: f0 df rcall . -32 ; 0x56 < uartInit>
76: ff cf rjmp . -2 ; 0x76 <main+0x6>
```

However, one should be careful with the macro-based calculation of the baud rate parameters. The calculation in the integer always takes place with rounding off, at a crystal frequency of 18.432 MHz the common baudrates are integer, at other frequencies (e.g. 16 MHz) the rounding errors can produce considerable deviation in the real baudrate and a better result can be achieved if you calculate the value by hand, round up instead of down and use it directly as a constant. In the manual of the ATmega 88 there are detailed tables with the optimal values and their deviations from the ideal frequency.

7.1.4   Receiving Data

Basically, bit RXC0 is set in register UCSR0A when a new data word has been received. However, since data is received asynchronously, a receive function would have to wait blockingly for a new byte, which does not make sense. Alternatively, the receive bit can be polled in a task every 10 ms, for example, or the corresponding interrupt can be used. Furthermore, the driver should run in its own module. In general, it is possible in C to define global variables across module boundaries, but this can also lead to errors due to unintended use. If the overall scope of the program allows it, encapsulation should be used. If not, of course, the UDR data register of the UART can be accessed directly from anywhere in the program. The driver file uart.c contains the following code for encapsulation:

```
#define DATA_RECEIVED 1
#define DATA_WAITING 0
uint8_t ucRecBuf;
uint8_t ucRecFlag=0;
ISR (USART_RX_vect)
{
    ucRecBuf = UDR0;
    ucRecFlag = DATA_RECEIVED;
}
//returns DATA_RECEIVED if new data is present and resets ucRecFlag
//Serves to encapsulate the flags in the module
uint8_t UART_CheckReceived()
{
    if (ucRecFlag)
    {
        ucRecFlag=DATA_WAITING; //reset flag return DATA_RECEIVED;
    }
    else return DATA_WAITING;
}
/*serves to encapsulate the data in the module*/
uint8_t UART_GetReceived()
{
    return ucRecBuf;
}
```

In the interrupt it is ensured that a received byte is always fetched and written into a buffer (here a global variable ucRecBuf). The read buffer is thus ready to receive again and cannot be accidentally overwritten by another received data byte. However, even this solution does not exempt from emptying the own data buffer, in a task UART_CheckReceived() must be called regularly, for example with.

```
if (UART_CheckReceived()) {data=UART_GetReceived();}
```

If this is not possible with certainty, the data can also be written into a FIFO (Chap. 6), the interrupt service routine then changes accordingly:

```
ISR (USART_RX_vect)
{
    FIFOput(FIFOHandle,UDR0);
    ucRecFlag = DATA_RECEIVED;
}
```

Of course, a corresponding FIFO must first be set up. The application software can then regularly read out the FIFO (see Chap. 6).

Another important significance of the UART_CheckReceived() function is that it encapsulates the global variables. It allows the entire functionality of the UART interface to be offloaded to a single, precompiled module without the application software that uses it having to access global variables of that module.

7.1.5 Sending Data

Sending with the USART is done by simply writing to the USART Data Register UDR0. The transmission process takes some time, 9600 baud for example corresponds to $1.041 \cdot 10^{-4}$ s per bit and with an 8,N,2 frame (1 start bit, 2 stop bits, 8 data bits $= 11$ bits) the transmission time is 1.146 ms per transmission. Thus, it is recommended to move the transmission to a task that is called less frequently than the transfer time, or by checking the UDRE0 bit in the UCSR0A register, which indicates the end of the transfer operation. Alternatively, one can fill the bytes to be sent into a write buffer and flush it depending on the Transmit Complete interrupt. More on this later. The simplest one-byte transmit operation with blocking wait would be first:

```
void UART_TransmitChar(char data)
{
    /*wait until the send buffer is empty*/
    while ( ! ( UCSR0A & (1 << UDRE0)));
    /*and send*/
    UDR0 = data;
}
```

This is also suggested in the manual of the ATMega88. However, in the case of the above configuration, the processor remains blocked for 1.146 ms.

7.1.6   Implementation of UartWriteBuffer()

A practical way to send an entire data buffer is to use the ISR (USART_TX_vect). If the data are present in a byte array, they can easily be sent from the ISR by triggering the first send operation and incrementing a position index in the send buffer in the ISR until the content of the buffer at the current index is NULL (null terminated string) or a globally passed buffer length has been reached. If this abort criterion is missing, the controller sends continuously with the solution proposed here!

```
uint8_t ucSendIndex;
uint8_t ucBufLength;
uint8_t ucUART_Tx_Complete;
char *data;
#define TXCOMPLETE 1
#define TXNOTCOMPLETE 0
void uartWriteBuffer(char* databuf, uint8_t length)
{
    ucSendIndex=0;
    UDR0 = databuf[ucSendIndex++];
    ucBufLength = length;
    data=databuf;
    ucUART_Tx_Complete= TXNOTCOMPLETE;
}
ISR (USART_TX_vect)
{
    if (data[ucSendIndex] && ucSendIndex < ucBufLength)
        UDR0 = data[ucSendIndex++];
    else ucUART_Tx_Complete= TXCOMPLETE;
}
```

A simple way to check whether the data buffer has been completely transferred is to regularly test ucUART_Tx_Complete for TXCOMPLETE. Attention! In order to start the ISR USART_TX_vect, the corresponding byte TXCIE0 in register UCSR0B must of course be set! The array data[] must be defined globally to be visible in the ISR. The memory space must of course be reserved in the function that calls uartWriteBuffer().

7.1.7 UART Multiprocessor Mode

Multiple microcontrollers can be networked via UART as in Fig. 7.4 to form a daisy-chain connection. This can be converted to a ring topology by connecting the last microcontroller to the first, but communication is only unidirectional.

Bidirectional communication is enabled by the UART bus in Fig. 7.5, which operates according to the master-slave principle with a master defined by the hardware. Depending on their role, the bus nodes are connected to the two data lines with a specified communication direction. Each communication session, consisting of a master frame and possibly the frame response of the addressed slave, is initiated by the master. The UART-TX pin of the master is connected to the RX pins of all slaves. On its RX pin, the master can receive the messages sent by the slaves. The UART protocol as it is presented here does not provide for arbitration, so the slaves can only send their message when requested by the master. Data exchange between the slaves of this bus is excluded. Communication with the individual slaves can take place by means of explicit or content-based addressing. In the first case, each slave is assigned a system-wide (i.e. for all connected nodes) unique address, in the second case the messages are provided with an identifier. Using the identifier, the master can address a single slave or several at the same time. If several slaves are addressed simultaneously, the identifier must not trigger a response.

The structure of a frame could look like in Fig. 7.6. In the following example, the identifier is one byte in size and takes values in the range 0 ... 0x3F. The following byte

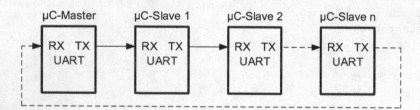

Fig. 7.4 UART daisy chain connection

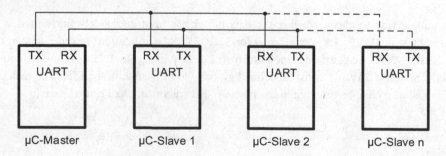

Fig. 7.5 UART – Networking of several microcontrollers

Identifier	Length of the data field	Data field	Checksum

Fig. 7.6 UART multiprocessor bus frame

specifies the number of data bytes and thus the communication can be made variable. The 1-byte sum of all bytes including the data bytes is added at the end of the frame as a checksum. The frames can be stored in a data structure uart_mp_frame as in the following program code:

```
typedef struct
{
    uint8_t ucID; //identifier
    uint8_t ucLength; //data field length
    uint8_t ucData[8]; //data vector for the maximum message length
}uart_mp_frame;
```

Networking the microcontrollers in this way via the standard UART has the disadvantage that all messages sent by the master must also be received by all slaves, which can lead to their overload.

7.1.7.1 Initialization of the Bus Devices in UART Multiprocessor Mode

To avoid overloading the slaves, the microcontrollers of the ATmega family can be integrated into a UART multiprocessor network. From the entire master frame, the identifier alone triggers a UART-RX interrupt at all slaves in multiprocessor mode. The other bytes of such a frame trigger an interrupt only at the addressed slaves. The initializations of the master and the slaves are different, but can be combined in a single function as illustrated in the following program code for an ATmega88. A global array ucBaudrate is filled with the values for UBBR0, the indexes could then look like this (for 18.432 MHz):

```
#define BAUDRATE_2400 0
#define BAUDRATE_4800 1
#define BAUDRATE_9600 2
#define BAUDRATE_14400 3
#define BAUDRATE_19200 4
uint16_t ucBaudrate[] = {479,239,119,79,59};
```

When calling this function, ucbaudrate – i.e. the index of the baud rate in the mentioned field – must be passed as parameter, as well as the role of the microcontroller in the network: Master or Slave. The same index is also used to set the timeout time. A timeout is required to signal a timeout during the transmission of the individual bytes and as a result to abort the reception of a message.

```
void UART_MP_Init(uint16_t ucbaudrate, uint8_t ucposition)
{
    UBRR0 = ucBaudrate[ucbaudrate]; //the baud rate is set
    UCSR0B = 1 << UCSZ02; //nine data bits
    UCSR0C = (1<<UCSZ00) | (1 << UCSZ01);
    UCSR0B |= 1 << RXEN0; //the receiver is switched on and
    UCSR0B |= 1 << RXCIE0; //receive interrupt enabled
    //the transmitter of the master is activated
    if (ucposition == MASTER) UCSR0B |= 1 << TXEN0;
    //the MPCM0 bit is activated for the slave to be able to receive IDs
    else if (ucposition == SLAVE) UCSR0A |= (1 << MPCM0);
    ucPosition = ucposition;
    UART_MP_Init_IdFilter(); //the receive filter is initialized
    //the timer 0 for the timeout is initialized
    TimeOut_Init(ucbaudrate);
}
```

The communication is initialized for nine data bits per UART frame and the reception of a data byte must trigger an RX interrupt for all bus participants. For the master the transmitter and for the slaves the multiprocessor mode is additionally activated. The function initializes a list of identifiers to which the slave responds and a timer to detect timeout errors during transmission. If the MPCM0 bit is set and 9-bit transmission is enabled, an RX interrupt is triggered at the receivers only if the received ninth bit is 1. To transmit an identifier, the master first sets the TX8B0 bit in the UCSR0B register and then, with the 8-bit identifier stored in the UDR0 data register, the transmission of the UART frame begins. The slaves compare the received identifier with their own list and if there is a match, they reset the MPCM0 bit. After the identifier is transmitted, the master resets the TX8B0 bit and sends the other bytes, which trigger a UART-RX interrupt only on the addressed slaves. When requested, the addressed slave enables its UART-TX and sends its frame, which can only be received by the master due to the wiring as described above. The slave then disables the UART-TX and re-enables the multiprocessor mode.

7.1.7.2 Receiving Data in UART Multiprocessor Mode

The entire reception of a message at higher transmission rates must take place in the UART-RX interrupt. Figure 7.7 shows an example of the flowchart of an interrupt service routine designed as a state machine. This routine receives a master message, stores it in a buffer and generates a code that shows whether the reception was error-free. A distinction is made between four states:

- **NO_ID** – the slaves that are waiting for an identifier are in this state. When an identifier from the own list is received, the state is switched to OWN_ID.
- **OWN_ID** – the byte received in this state indicates the data field length. If the length is zero, the state is switched to CRC, otherwise to DATA.

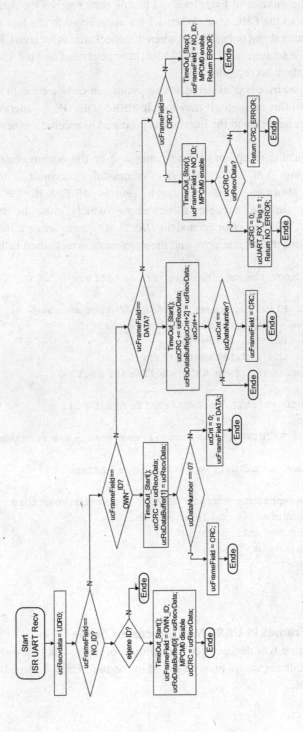

Fig. 7.7 State transition diagram for the UART-RX interrupt in UART multiprocessor mode

- **DATA** – when the number of bytes received in this state reaches the data field length, the system jumps to the CRC state. When a byte is received in the first three states, a timeout timer is started and only stopped when the checksum is received. If a byte is not received within the set time, a TIMEOUT_ERROR is generated and the interface is reset and prepared for the next reception.
- **CRC** – the received byte is considered as checksum and compared with the calculated sum of all bytes of this message. If they match, MESSAGE_OK is generated, otherwise CRC_ERROR is generated and the interface is prepared to receive the next message.

In the main function, the reception of a new message or the occurrence of an error is checked in polling. In case of a successful reception, the function UART_MP_Check_Message(&sRecFrame) returns MESSAGE_RECEIVED and stores the message in the variable sRecFrame which must be of data type uart_mp_frame. When an error occurs, the UART_MP_get_Error function returns the identifier of the last message in error and the number of errors when called.

```
uint8_t UART_MP_Check_Message(uart_mp_frame *sframe)
{
    if(!ucUART_Rx_Flag) return NO_MESSAGE; //error message
    ucUART_Rx_Flag = 0; //the receive flag is reset
    sframe->ucID = ucRxDataBuffer[0];
    sframe->ucLength = ucRxDataBuffer[1];
    for(uint8_t ucI = 0; ucI < ucRxDataBuffer[1]; ucI++)
    {
        sframe->ucData[ucI] = ucRxDataBuffer[ucI + 2];
    }
    return MESSAGE_RECEIVED; /*an error-free message was received*/
}
uint8_t UART_MP_Get_Error(uart_mp_errorframe *sframe)
{
    if(!ucFaultReception) return NO_ERROR; //no receive errors
    sframe->ucLastError = ucFaultReception;
    sframe->ucErrCnt = ucErrorCnt;
    ucFaultReception = NO_ERROR;
    return ERROR;
}
```

7.1.7.3 Sending Frames in UART Multiprocessor Mode

The uart_mp_frame was declared for a maximum of eight data bytes as an example. Before sending, the current values must be stored in the send frame (sSendFrame) as in the following example:

```
sSendFrame.ucID = 0x28; //store the current ID
sSendFrame.ucLength = 2; //two data bytes are sent
sSendFrame.ucData[0] = 'S'; //the first byte
sSendFrame.ucData[1] = '2';
UART_MP_Send_Message(&sSendFrame, SLAVE); //a slave sends the frame
```

The structure of the send function of a frame in UART multiprocessor mode is illustrated in the following program code. The address of the data structure that stores the elements of the frame and the role of the bus nodes: master or slave are passed as parameters. In the function the checksum is calculated and this is sent as the last byte of the frame.

```
void UART_MP_Send_Message(uart_mp_frame *sframe, uint8_t ucposition)
{
    //the variable ucCRC will store the checksum of the frame
    uint8_t ucCRC = 0;
    if(ucposition == SLAVE)
    {
        //the UART transmitter of the slave is switched on
        UART_MP_Set_On(SEND_ON);
        MPCM_SET_OFF; //the MPCM bit is reset
        TimeOut_Start(DELAY_YES);
        //it is waited until the line reaches a stable state. This is the
        //only way to ensure that the master recognizes the start bit
        while(TimeOut_Get_State() == TIMEOUT_RUNNING);
    }
    else if(ucposition == MASTER)
    {
        //the TXB80 bit is set to send the ID
        UART_MP_Set_On(TXB8_ON);
    }
    UART_MP_Send_Char(sframe->ucID); /*Master or slave send the ID*/
    ucCRC += sframe->ucID;
    //the bit is reset so that the master sends data
    //for the slave the value of this bit is irrelevant
    UART_MP_Set_Off(TXB8_OFF);
    UART_MP_Send_Char(sframe->ucLength); //the data field length is sent
    ucCRC += sframe->ucLength;
    for(uint8_t i = 0; i < sframe->ucLength; i++)
    {
        UART_MP_Send_Char(sframe->ucData[i]); /*the data are sent*/
        ucCRC += sframe->ucData[i];
    }
    UART_MP_Send_Char(ucCRC); //the checksum is sent
    if(ucposition == SLAVE)
```

```
{
        //the UART transmit part of the slave is switched off to avoid
        //bus collisions
        UART_MP_Set_Off(SEND_OFF);
        /*the MPCM bit is set so that the slave can receive the next ID*/
        MPCM_SET_ON;
    }
}
```

7.2   Connection of the Serial Interface to USB

For the communication of the processor with a PC the serial interface can be used by using the USB. For this the conversion of the UART to USB is necessary. The most common variant for this is the use of a device from the company Future Technology Devices International Limited (FTDI). From the diverse family of USB controllers, the FT232R controller was chosen here, which is addressed on the PC by a virtual COM driver and supports up to 12 MBit/s data rate. This allows the serial interface to be addressed directly as such in the PC, for example via a terminal program or via any other software that can access the COM interfaces. The drivers can be downloaded free of charge from FTDI [2].

Figure 7.8 shows the simplest possible wiring of the FT232R. In addition, LEDs can be connected to indicate the communication states and further digital inputs/outputs can be switched. Furthermore, the chip has the possibility to perform a hardware handshake.

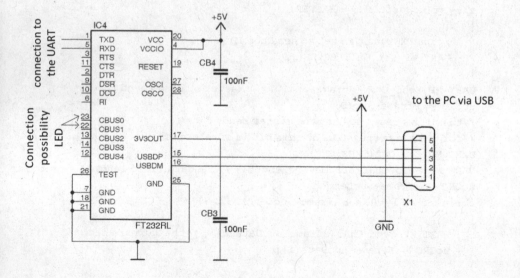

Fig. 7.8 Simplest wiring of the FT232R

7.3   A Simple Serial Protocol

This section describes a simple protocol that combines the following features, largely described in Chap. 6:

- Framing, i.e. a start and end byte
- One command addressing
- A packet counter for the detection of failed data packets
- A checksum

To keep the code manageable, only two bytes of data are transmitted, for example from the 16-bit AD converter. Protocols of this type are found in many sensors, which are described below. The protocol relies on the byte-by-byte transmission of control and content bytes. A frame always starts with an ASCII SOH (0x01) followed by an address. In the minimum case, the receiver therefore only has to wait for an SOH and read the address. If the address is valid, it is followed by a command. In this case, we have chosen only one command to avoid bloating the state machine: 0xC0. This value is chosen arbitrarily and means in our example "temperature follows". There could also be commands like: "Sleep", "Calibrate" or whatever is necessary.

Then, after exactly this command, two data bytes follow, for which it must of course be made clear whether the higher-value byte is transmitted first (Motorola format, big-endian, MSByte first) or the lower-value byte first (Intel format, little-endian, LSByte first). This is a topic in all protocols! We decide for big-endian.

Following the data bytes a message counter byte is sent, which of course starts from 0 after 256 messages. Here it should be left open what is done in case of a data loss, this depends on the application. Data bytes and message counter together form the checksum, in this case it is a simple arithmetic sum (instead of a CRC quite common), which is truncated after 8 bits. The frame is terminated by an ASCII EOT (0x04).

A message (data 0x12 0x34, message counter 0x56) could therefore read as follows
0x01 0xFF 0xC0 0x12 0x34 0x56 0x9C 0x04

The sender must provide a byte field of length eight, which is sent with the send function from Sect. 7.1.6. Only the data (bytes 4 and 5), the counter (byte 6) and the checksum (byte 7) are ultimately filled.

The receiver maps itself as a state machine that reacts to two events: A byte was received or an error was detected in the last step. The state machine is called via polling in the main loop (Fig. 7.9).

The receiver code can be implemented using the code described in Sect. 5.2.

For the message bytes we define in the receiver.

```
#define SOH 0x01
#define MY_ADDRESS 0xFF
#define READ_INTEGER 0xC0
#define EOT 0x04
```

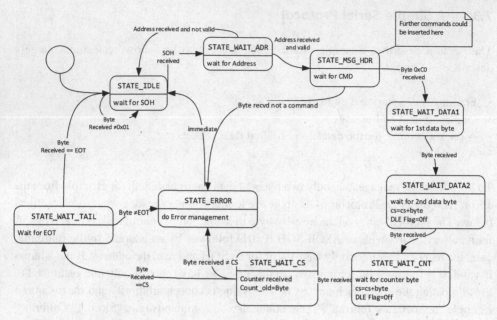

Fig. 7.9 State machine for a fictitious serial protocol

For the states.

```
#define STATE_IDLE 1
#define STATE_WAIT_ADR 2
#define STATE_MSG_HDR 3
#define STATE_WAIT_DATA1 4
#define STATE_WAIT_DATA2 5
#define STATE_WAIT_CNT 6
#define STATE_WAIT_CS 7
#define STATE_WAIT_TAIL 8
#define STATE_ERROR 9
```

And for the error codes.

```
#define ERROR_WRONG_CHECKSUM 3
#define ERROR_WRONG_COMMAND 2
#define ERROR_FRAME_ERROR 1
#define ERROR_NO_ERROR 0
```

We need some global variables.

```
static uint8_t ucState; //current state
static uint8_t ucChecksum; //checksum
static uint8_t ucError, ucSuccess; //error or success
short uiWordToReceive; //data to be received
```

In the main loop (after all variables and the serial port have been initialised of course, as explained in Sect. 7.1.3 and following), the state machine is executed. This is triggered either by an incoming byte or automatically if it has returned with an error from the last call:

```
while (1)
{
    if (uartCheckReceived() || error)
    {
        if (!error) ev = uartGetReceived();
        error = StateMachine(ev);
    }
    if (ucSuccess)
    {

        //message was received completely and without errors
        //add code here
    }
}
```

The state machine looks like this:

```
uint8_t StateMachine(uint8_t event)
{
    switch(ucState)
    {
        case STATE_IDLE:
            if (event != SOH) ucState = STATE_IDLE;
                // state 1 - idle remains
            else ucState = STATE_WAIT_ADR;
                //SOH received: wait for address
            ucSuccess = 0; //this is the beginning of a message
        break;
        case STATE_WAIT_ADR:
            if (event != MY_ADDRESS) ucState = STATE_IDLE;
            //address was not for us
            else ucState= STATE_MSG_HDR; //address ok, wait for a command
        break;
        case STATE_MSG_HDR:
```

```
        if (event != READ_INTEGER) //command unknown
        {
            ucError = ERROR_WRONG_COMMAND;
            ucState = STATE_ERROR;
        }
        else ucState = STATE_WAIT_DATA1; //command known,
                                         //wait for data
    break;
    case STATE_WAIT_DATA1:
        ucState = STATE_WAIT_DATA2; //data is read
        ucChecksum += event; //checksum is updated
        //write the first byte to the result
        uiWordToReceive = ((short)event) << 8;
    break; //big endian format
    case STATE_WAIT_DATA2:
        ucState = STATE_WAIT_CNT; //data are read
        ucChecksum += event; //checksum is updated
        //write second byte to result and wait for counter
        uiWordToReceive |= ((short)event);
    break;
    case STATE_WAIT_CNT:
        //if necessary, the handling of the counter is written here
        ucState = STATE_WAIT_CS; //counter read
        ucChecksum += event; //checksum is updated
        //wait for checksum
    break;
    case STATE_WAIT_CS:
        if (event != ucChecksum) //checksum false
        {
            ucError = ERROR_WRONG_CHECKSUM;
            ucState = STATE_ERROR
        }
        else ucState = STATE_WAIT_TAIL; //checksum correct
        ucChecksum = 0;
    break;
    case STATE_WAIT_TAIL:
        if (event != EOT) //EOT character false
        {
            ucError = ERROR_FRAME_ERROR;
            ucState = STATE_ERROR;
        }
        else
        {
            ucState = STATE_IDLE;
            ucSuccess = 1;
```

```
                    ucError = ERROR_NO_ERROR;
             }
             //EOT correct, message complete
        break;
        case STATE_ERROR : //do error handling
             ErrorHandling(); //do something with the error
             ucState = STATE_IDLE;
             return 1; //return 1, so that the SM again is triggered
                        // without a character being was received
        default: break;
    }
    return 0; //last character processed without errors
}
```

7.3.1   Establishing Code Transparency by Bytestuffing

The code becomes a bit more complicated if the command can be followed by any number of bytes. Then one should introduce a length field and if necessary also a byte stuffing, since the data could naturally also contain the value 0x01, which is reserved for the header. In this case one would have to insert a DLE (ASCII 0x10) as soon as the character 0x01 is to be sent in the data field and of course also when the DLE character itself is sent (see Chap. 6).

An example with bytestuffing is (data 0x01 0x23, message counter 0x21):

$$0x01\ 0xFF\ 0xC0\ 0x10\ 0x01\ 0x23\ 0x21\ 0x45\ 0x04$$

The byte stuffing is of course not used for the calculation of the checksum. In the sender the corresponding bytes are written without byte stuffing into a field of length 8. The byte stuffing takes place in the send interrupt service routine already introduced above, which is adapted accordingly for this purpose with the global variable stuffed:

```
ISR(USART_TX_vect)
{
    if ((datBuf[ucSendIndex]) && (ucDataLen > ucSendIndex))
    {
        if (datBuf[ucSendIndex] == 0x01 && !stuffed)
        {
            UDR0 = 0x10;
            Stuffed = 1;
        }
        else
```

```
        {
            UDR0 = datBuf[ucSendIndex++];
            Stuffed = 0;
        }
    }
}
```

The header byte is spared from this measure, since it does not have to be filled with DLE! With the above presented function `uartWriteBuffer()` you can send the previously completely created message and the stuffing is done automatically. In the receiver, of course, the DLEs must be removed again in the state machine. If there are no DLEs before a 0x01, the state machine could then interpret this as an incomplete frame being received and therefore jump back to the start of the protocol.

The state machine of the receiver is extended to the effect that in all states except `STATE_WAIT_ADR`, where the header byte was received, it is checked after receipt of a byte whether the incoming byte was a DLE. In this case the byte is discarded, the state is not left and a `StuffDetected` flag is set. With the next incoming byte and set flag, the byte is accepted again and the flag is reset.

For reasons of space, an implementation is left to the readers here.

Another serial protocol can be found in Chap. 11 in the Modbus.

References

1. Microchip: Reference Manual ATmega48/168. https://www.microchip.com/wwwproducts/en/ATmega88A. Accessed: 6. Jan. 2021.
2. FTDI: FT232R – USB UART IC, online. http://www.ftdichip.com/Products/ICs/FT232R.htm. Accessed: 5. Jan. 2021.

Serial Peripheral Interface (SPI)

8

Abstract

This chapter describes the SPI interface of the AVR family.

The synchronous serial interface SPI (Serial Peripheral Interface), sometimes also called Clocked Serial Interface (CSI), which is available in many types of processors, offers one of the most popular ways to access sensors on the PCB or outside the housing. Accordingly, many sensors are coming to market with an SPI connection. By transmitting the clock on a separate clock line (CLK, sometimes called SCK), it allows synchronization between the transmitter and receiver at much higher data rates than asynchronous interfaces. However, it should be noted that an increased EMC risk occurs with increasing length of the supply line. EMC measures may therefore have to be taken on the PCB or in the cable, for example by means of ground planes or appropriate shielding measures.

8.1   Structure and Mode of Operation

The SPI interface is always a fast and very convenient communication interface when it comes to the operation of a master CPU with several subordinate nodes (slaves). The clock always originates from the master, so it does not have to be configured at the slave. The respective slave is controlled via the slave select ($\overline{SS}$) connection. If this is low, the slave is active, otherwise not. When clocking, the master shifts the data from the master shift register into the slave shift register with either the rising or falling edge (selectable via register) and the data from the slave into the master with the same clock, around the ring.

A. Meroth, P. Sora, *Sensor networks in theory and practice*,
https://doi.org/10.1007/978-3-658-39576-6_8

205

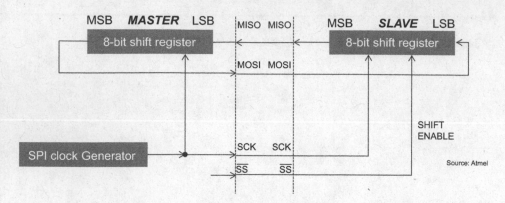

Fig. 8.1 Schematic drawing of the SPI interface. (According to Microchip/ATMEL)

This means that data is exchanged between master and slave (Fig. 8.1). The connection is made via the lines MISO (Master In Slave Out) as input of the master and MOSI (Master Out Slave In) as output of the master. Due to the functionality of the lines, MISO is always switched to input for the master and to output for all slaves. The slave select lines prevent several outputs from switching to the bus in parallel.

Processors of the AVR family have a slave select input, so that they can also be operated as slaves. If they are operated as masters, the corresponding port pins must be made available as outputs, depending on the number of slaves. Clock and data lines are connected in parallel.

Figure 8.2 shows how several sensors can be connected to a bus via an SPI interface. If a μC of the AVR family is operated as a slave, for example because it forms a digital bus participant with an analog sensor, its slave select pin can be operated on input.

It should be noted that there is no separate receive routine for the SPI interface! Each send operation of the master simultaneously reloads the buffer data of the slave into the buffer of the master. So, from the slave's point of view, the data must be made available before the master starts a new request. For this reason, sensor protocols between the microcontroller and the sensor often consist of several data words, often a command is clocked to the slave first, followed by a 0x00 or 0xFF, which then ensures that the sensor's data, which has been made available in the meantime, is transferred to the microcontroller. The following sections contain some examples of this procedure.

Fig. 8.2 Wiring several sensors via the SPI interface

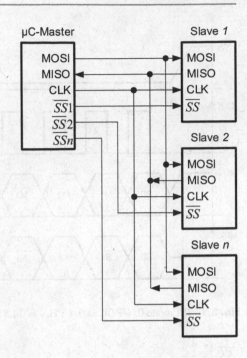

8.2   Configuration of the SPI Interface

To understand the configuration of the SPI interface, one must first be clear about the representation of a byte and the type of sampling. With sensors this is usually fixed, sometimes it can also be configured, in any case master and slave must agree on the same behavior. First of all, the order in which the individual bits of a byte are transmitted is determined. LSB first means that bit 0 (least significant bit) is transmitted first, MSB first means that the highest bit 7 (most significant bit) is transmitted first. Furthermore, there are four different modes, which are differentiated in

- The polarity of the clock (CPOL)
- The clock phase (CPHA) (cf. Figs. 8.3, 8.4, 8.5 and 8.6).

CPOL = 0 means that the clock has the level "0" (low) when idle, with CPOL = 1 the level is "1" (high) when idle. The clock phase indicates whether the signal is sampled on the first or the second edge after $\overline{SS}$ goes low. The following figures show the four possible modes that result from this.

When CHPA = 0, the master and slave start their data with the falling edge of $\overline{SS}$, after which, depending on CPOL, the rising or falling edge of the clock signal CLK is used to sample the signal present (master samples MISO, slave samples MOSI). The subsequent

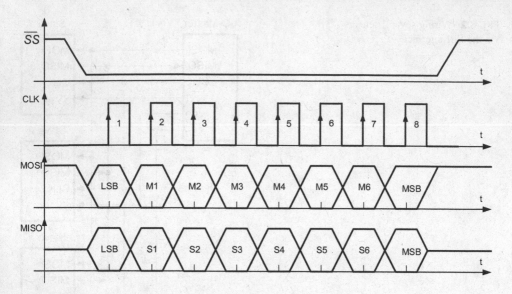

Fig. 8.3 SPI mode 0: CPOL = 0, CPHA = 0 LSB first

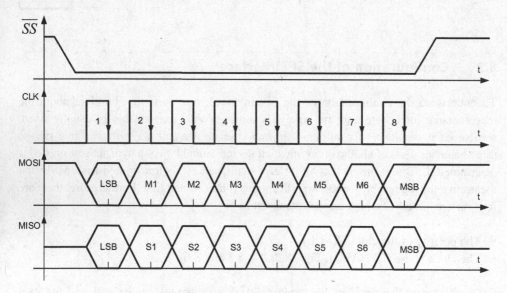

Fig. 8.4 SPI mode 1: CPOL = 0, CPHA = 1 LSB first

reversed edge causes the master and slave each to place the next bit on the data lines until the entire data word has been transmitted.

When CHPA = 1, sampling is first performed on the second clock edge and the first (rising in the case of CPOL = 0, falling in the case of CPOL = 1) edge after $\overline{SS}$ is low causes

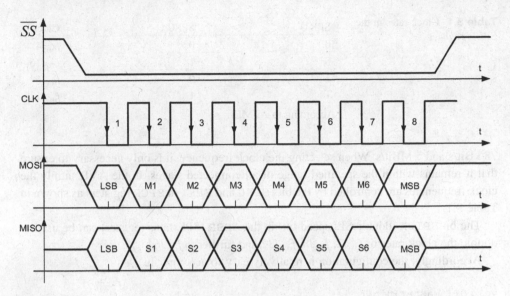

Fig. 8.5 SPI mode 2: CPOL = 1, CPHA = 0 LSB first

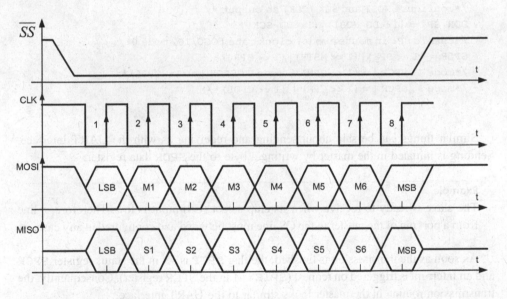

Fig. 8.6 SPI mode 3: CPOL = 1, CPHA = 1 LSB first

the slave to make its first bit available on the bus, the complementary edge again triggers sampling.

Since the clocking of the SPI interface depends exclusively on the master, it is not necessary to set exact clock rates. AVR microcontrollers provide a prescaler that allows clocking from $f_{osc}/2$ to $f_{osc}/128$. At a processor clock rate of 10 MHz, this means between

Table 8.1 Clock rates in the
SPCR register

SPR1	SPR0	Clock
0	0	$f_{osc}/4$
0	1	$f_{osc}/16$
1	0	$f_{osc}/64$
1	1	$f_{osc}/128$

According to ATMEL

78 kBit/s and 5 MBit/s. When selecting the clock frequency, it is only necessary to ensure that it remains within the specified range of all connected slaves. In the AVR family the clock frequencies are controlled by the bits SPR0 and SPR1 in SPCR registers as shown in Table 8.1.

The bit SPI2X (double SPI speed bit) in the SPSR (SPI status register) can be used to double the clock rates to $f_{osc}/2$, $f_{osc}/8$, $f_{osc}/32$ and further to $f_{osc}/64$.

Accordingly, the configuration is relatively easy to accomplish:

```
void SPI_MasterInit()
{
    /*configure MOSI and SCK (CLK) as output */
    DDR_SPI = (1<<DD_MOSI)|(1<<DD_SCK);
    /*enable SPI in master mode, clock rate fOSC/16, mode 0*/
    SPCR = (1 << SPE) | (1 << MSTR) | (1 << SPR0);
    /*mode 1: SPCR |= (1 << CPHA), mode 2: SPCR |= (1<<CPOL)*/
    /*mode 3: SPCR |= (1 << CPH) | (1 << CPOL)*/
}
```

Similar things can be said about sending and receiving as with the UART interface. Sending is initiated in the master by writing a byte to the SPDR data register.

Example

The slave is ready to receive when its chip select (CS) input is low. Therefore, a line from a port pin of the master to the CS line must be wired and configured in any case. ◄

As soon as the transmission is finished, the flag SPIF is set in the status register SPSR and an interrupt is triggered on request (SPIE = 1 in the SPCR register). Consequently, the transmission routine in the master looks similar to the UART interface.

```
void SPI_MasterTransmitChar(char cCharToSend)
{
    /*start transmission*/
    SPDR = cCharToSend;
    //Attention! Transmission continues even
    //if function has been terminated
}
```

Again, we have the problem of waiting until the transmit buffer is empty when sending successively. Analogous to the UART interface one can simply wait blocking (Attention! Such loops are often removed by the optimizer of the compiler), like here:

```
/*start transmission*/
SPDR = cCharToSend;
/*wait until transmission is finished*/
while(!(SPSR & (1 << SPIF)));
```

Alternatively, you can send from a time-controlled task or interrupt-controlled until the send buffer is empty. According to Fig. 8.1, the content of the slave transmit register is located in register SPDR after the transmit process. In this respect, there is no separate receive function for the SPI interface.

8.3   SPI Interface in Slave Mode

In slave mode it is strongly recommended to use the interrupt service routine of the SPI interface. This is illustrated by the following scenario (Fig. 8.7).

The right microcontroller is configured in slave mode:

```
void SPI_SlaveInit(void)
{
    /*set MISO as output, the others (CLK,/SS) as input*/
    DDR_SPI = (1 << DD_MISO);
    /*switch on SPI and enable interrupt*/
    SPCR = (1 << SPE) | (1 << SPIE);
    sei();
}
```

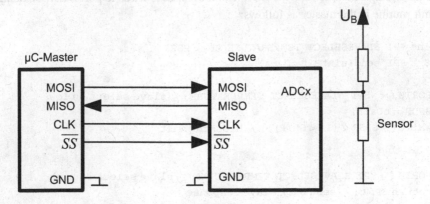

Fig. 8.7 Using a microcontroller with analog sensor in slave mode

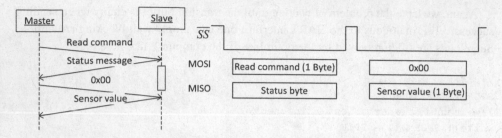

Fig. 8.8 Sequence of an SPI sequence with command byte

In the circuit example in Fig. 8.7, only one byte of the AD converter is to be used actively. The sequence is as follows: The master sends a command byte, for example to read out the current sensor value. In practice, an entire command set is used, in which, among other things, the sensor can be calibrated, an error register can be read out and cleared, the version and serial number of the sensor can be read out or measurement modes can be switched. Since SPI communication always involves an exchange, the sensor can return a status byte during this transmission, which contains, for example, whether valid data are present or whether an error memory entry exists. After the command word, the master sends eight zeros (0x00), after which the value of the sensor to be queried is ready in the SPDR register. To do this, the sensor must have set it in its own transmit register between the command word and the 0 sequence.

The principle sequence can be seen in Fig. 8.8. However, many SPI sensors are so complex that 16-bit sequences are transmitted, this is then analogous.

In the master, two bytes must therefore be sent in succession, with an appropriate pause in between, so that the slave can provide its data. This can be done in a time-controlled task (see Chap. 6) or by blocking waiting, which, however, is usually not permitted in real-time operation and also does not make sense. For simplicity, the master in the following code example shall at least wait in a blocking manner until the interface has sent the data byte. Also, this example assumes that the CS line of the sensor is wired to port B2. We extend the transmit routine in the master as follows:

```
#define SPI_SLAVESELECT_TEMPERATURE (1 << PB2)
uint8_t SPI_send(uint8_t cData)
{
    PORTB &= ~SPI_SLAVESELECT_TEMPERATURE; //slave select on low
    SPDR = cData;
    while(!(SPSR & (1 << SPIF))) //blocking wait
    {
    }
    PORTB |= SPI_SLAVESELECT_TEMPERATURE; //slave select on high
    return SPDR ; //send back slave response
}
```

In a 1 ms task the master code then looks like this:

```
//States of the SPI master
#define SPI_SENDCOMMAND 0x10
#define SPI_WAITANSWER 0x20
#define SPI_READSENSOR 0x15
#define SPI_VOID 0x00
(...)
if (SPI_SENDCOMMAND == masterstate)
{
    slavestate = SPI_send(SPI_READSENSOR);
    //here you can evaluate the status of the slave
    masterstate = SPI_WAITANSWER ; //wait now for response
}
else if (SPI_WAITANSWER == masterstate)
{
    sensorvalue = SPI_send(SPI_VOID);
    masterstate = SPI_SENDCOMMAND; //ready for the next transmission
}
(...)
```

If this code is called every millisecond, the master alternately sends the send command and the 0x00, which causes the slave to send the measured value. This corresponds to a state machine with the states SPI_SENDCOMMAND and SPI_WAITANSWER. By calling every millisecond one can save the blocking wait, because the transmission is finished for sure.

In the slave (i.e. the second microcontroller in Fig. 8.7), operation is controlled by the SPI interrupt. The slave also initially knows two states: Waiting for a command and sending the sensor value. For the following code it is assumed that a status byte unsigned char slavestate exists, into which the current status is always written.

```
ISR(SPI_STC_vect)
{
    SPI_recByte = SPDR;
    if (SPI_READSENSOR == SPI_recByte)
    {
        /*the sensor value is transferred at the next clock pulse*/
        SPDR = sensorvalue;
    }
    else if (SPI_VOID == SPI_recByte)
    {
        //sensor value was transferred and thus the starting
        //position is returned to jumped back
```

```
        SPDR = slavestate;
    }
    else
    {
        /*here you can trigger an error handling*/
    }
}
```

The further configuration of the SPI interface can be taken from the data sheet of the respective processor.

8.4   SPI Interface in a Sensor Network

For the effective programming of a microcontroller, which controls a SPI sensor network as master, a SPI software module can be created. The functions of this module can be accessed from the main function as well as from the corresponding software modules of the sensors. The module provides functions for initializing the SPI interface, for controlling the slave select line of a sensor and for byte exchange between master and slave. The module can also be used for other ATmega family microcontrollers by changing the definitions in the SPI.h header file that characterize the interface. The following example represents the definitions of an ATmegax8:

```
//MOSI signal
#define SPI_MOSI_DDR_REG DDRB
#define SPI_MOSI_PORT_REG PORTB
#define SPI_MOSI_BIT PB3
//MISO signal
#define SPI_MISO_DDR_REG DDRB
#define SPI_MISO_PORT_REG PORTB
#define SPI_MISO_BIT PB4
//Clock signal
#define SPI_CLK_DDR_REG DDRB
#define SPI_CLK_PORT_REG PORTB
#define SPI_CLK_BIT PB5
```

The following Fig. 8.9 shows the flow chart of a program for a microcontroller which, as a master, controls two slaves (sensors) with different modes via SPI.

After the general initialization, before the microcontroller is configured as SPI master, the dedicated slave select port pin must be configured as output. In order to keep the SPI module independent of the respective board configuration, the microcontroller connections for controlling the slave select inputs for each sensor are defined in the main file. For this

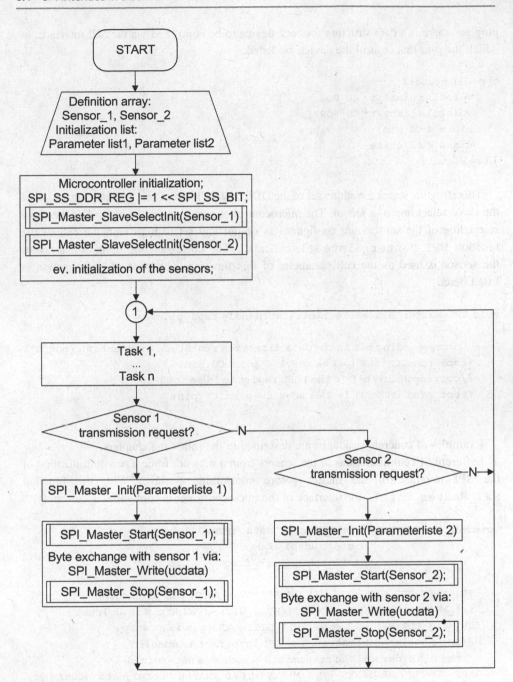

Fig. 8.9 Control of two sensors with different SPI modes

purpose, there is a data structure for each device to be controlled via the SPI interface, in which the pins that control the device are listed:

```
typedef struct{
    volatile uint8_t* CS_DDR;
    volatile uint8_t* CS_PORT;
    uint8_t CS_pin;
    uint8_t CS_state;
} tspiHandle;
```

This structure stores the addresses of the DDR and PORT registers and the connection of the slave select line of a sensor. The microcontroller port pins for the control of the slave select line of the sensors are configured as output and set to high with the call of the function SPI_Master_SlaveSelectInit. The corresponding definition array of the sensor is used as the call parameter of this function. The code of this function is listed here:

```
void SPI_Master_SlaveSelectInit(tspiHandle tspi_pins)
{
    //corresponding bit in the data direction register is set to high (output)
    *tspi_pins.CS_DDR |= 1 << sdevice_pins.CS_pin;
    //corresponding bit in the PORT register is set to high
    *tspi_pins.CS_PORT |= 1 << sdevice_pins.CS_pin;
}
```

Examples of concrete structures are described in the following chapters.

Different SPI configurations of the sensors from a network force a new initialization of the SPI interface of the master before controlling a slave. With the function SPI_Master_Init() the interface of the microcontroller is configured accordingly.

```
void SPI_Master_Init(uint8_t ucspi_data_order,
                     uint8_t ucspi_mode,
                     uint8_t ucspi_sck_freq)
{
    SPI_MOSI_DDR_REG |= 1 << SPI_MOSI_BIT; //MOSI pin is an output
    SPI_MISO_DDR_REG &= ~(1 << SPI_MISO_BIT); //MISO pin is an input
    SPI_CLK_DDR_REG |= 1 << SPI_CLK_BIT; //CLK pin is an output
    //µC is declared as master, the SPI interface is enabled
    //the bit order and the transmission mode are determined
    SPI_CONTROL_REGISTER = SPI_MASTER | SPI_ENABLE | ucspi_data_order |
                           (ucspi_mode << 2) | (ucspi_sck_freq % 4);
    //the desired SPI clock frequency is selected
    SPI_STATUS_REGISTER = ucspi_sck_freq / 4;
}
```

For better readability of the software, the following definitions can be used as call parameters of the initialization function. The call parameters are summarized in the flow chart as a parameter list.

```
//SPI Register
#define SPI_CONTROL_REGISTER SPCR
#define SPI_DATA_REGISTER SPDR
#define SPI_STATUS_REGISTER SPSR
//
#define SPI_ENABLE 0x40
#define SPI_MASTER 0x10
//DORD bit in the SPCR register (data order: LSB or MSB first)
#define SPI_LSB_FIRST 0x20
#define SPI_MSB_FIRST 0x00
//SPI mode is determined via 2 bits CPOL and CPHA in the SPCR register
#define SPI_MODE_0 0x00
#define SPI_MODE_1 0x01
#define SPI_MODE_2 0x02
#define SPI_MODE_3 0x03
// Clock frequency of the SPI interface
//FOSC = the clock frequency of the microcontroller)
#define SPI_FOSC_DIV_2 0x04
#define SPI_FOSC_DIV_4 0x00
#define SPI_FOSC_DIV_8 0x05
#define SPI_FOSC_DIV_16 0x01
#define SPI_FOSC_DIV_32 0x06
#define SPI_FOSC_DIV_64 0x02
#define SPI_FOSC_DIV_128 0x03
#define SPI_RUNNING (!(SPSR & (1 << SPIF)))
```

The master must be reconfigured before each transmission request. It starts the communication with the slave by setting the chip select input low with the call of the function SPI_Master_Start and stops it with the call of the function SPI_Master_Stop.

```
void SPI_Master_Start(tspiHandle tspi_pins)
{
    *tspi_pins.CS_PORT &= ~(1 << tspi_pins.CS_pin);
}
void SPI_Master_Stop(tspiHandle tspi_pins)
{
    *tspi_pins.CS_PORT |= (1 << tspi_pins.CS_pin);
}
```

With `SPI_Master_Write` the master sends a byte to the slave. This byte is passed as a parameter when the function is called. At the same time the master receives a byte from the slave, which is returned by the function. Thus the same function can be used for bidirectional communication.

```
uint8_t SPI_Master_Write(uint8_t ucdata)
{
    #define SPI_DATA_REGISTER SPDR //definition in the SPI.h file
    SPI_DATA_REGISTER = ucdata;
    while(SPI_RUNNING);
    return SPI_DATA_REGISTER;
}
```

8.5   3-Wire SPI Communication

SPI was basically developed as a 4-wire bus that allows synchronous data exchange between the master and a slave in full duplex mode. For cost reasons, some slaves are equipped with a 3-wire SPI interface. The data transfer takes place via a single pin. This saves an external connection, but communication can then only take place in half-duplex mode. The control of such a slave can be realized by a microcontroller via general I/O pins (see Fig. 8.10).

When controlling a 4-wire SPI bus with a dedicated interface, the software on the master side only has to ensure the control of the CS line, the storage of the byte to be sent in the `SPDR` register and the reading of the received byte. The clock generation as well as the bit transmission are realized by the hardware. With a minimal hardware modification, the above advantages can also be used to control slaves that have a 3-wire SPI interface. Figure 8.11 shows an SPI bus where Slave1 is connected to the bus using only three wires instead of four. The master's MISO pin is directly connected to the slave's SDI/SDO pin, while a resistor is connected between the MOSI and the SDI/SDO pin. Resistor R1 prevents a short circuit from occurring between the master's MOSI and slave's SDI/SDO pins while the master is transmitting a dummy byte to generate the clock for reception. At the same time it remains possible that the signal present at the SDI/SDO pin can be read via MISO.

Fig. 8.10   3-wire-SPI over I/O's

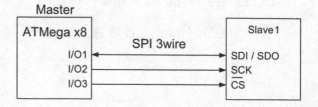

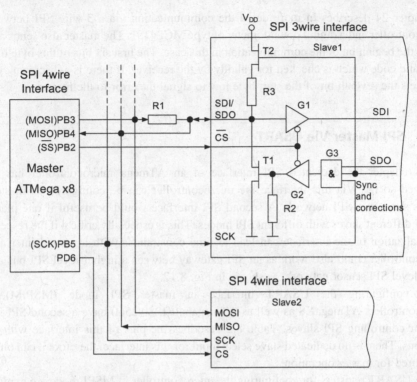

Fig. 8.11 3-wire SPI communication

Internally, such slaves have both signal lines SDI and SDO connected to the SDI/SDO pin as in Fig. 8.11. As usual, the internal transmit and receive buffers of the 3-wire interface are only activated when the CS signal goes low, otherwise the SDI/SDO pin is high impedance. The 3-wire SPI communication can only take place in half-duplex mode, therefore the addressed Slave1 sets the signal SDO to "0" at the beginning of a communication sequence. This switches it to receive mode and it waits for a valid command code that determines the communication direction. In case of a write command, the SDI/SDO port of the slave remains on input and receives the following data bytes. With a read command, the signal is switched synchronously with the clock signal according to the bits to be sent.

Sensors such as ADXL312 (Chap. 16) and L3GD20 (Chap. 17) are equipped with both an I^2C and an SPI interface for serial communication. The selection of the active interface is realized by an external circuit. An internal register flag can additionally be used to switch between the preset 4-wire and a 3-wire SPI communication. The master starts an SPI communication with such sensors with a start byte. The first bit of this byte specifies the transmission direction, while the last six bits transmit a register address. If the read/write bit is 0, the master overwrites the content in the slave's destination register, otherwise it requests the transfer of the addressed register from the sensor.

Chapter 24 describes in more detail the communication via a 3-wire SPI between a microcontroller and a digital potentiometer of type MCP4151. The master also sends a start byte at the beginning of the communication in this case. The first six bits of this byte form a command code which is checked for validity by the receiver. If there is a discrepancy, the slave sets the seventh bit of the start byte low to signal the error to the master.

8.6   SPI Master Via USART

In this chapter the dedicated SPI interface of an ATmega microcontroller has been described so far. Via this interface the microcontroller can be configured as master as well as slave in a SPI network. A second SPI interface would be useful if one plans to control different slaves with different SPI modes. This is especially critical if the respective re-initialization of the interface can lead to time constraints. With a second interface, a microcontroller could also work as an SPI gateway between a higher-level SPI bus and a lower-level SPI sensor network, as shown in Fig. 8.12.

By configuring the USART interface in master SPI mode (MSPIM), the microcontrollers ATmegaX8 as well as ATmega640/1280/2560 have a second SPI interface for controlling SPI slaves. Table 8.2 describes the pins of this interface with their functions. There is no dedicated slave select input for this interface, therefore it can only be configured for master operation.

The USART registers for configuring the microcontroller in MSPI mode are explained in Table 8.3. The functions of the active bits of the UCSR0A and UCSR0B registers for this operating mode are identical to those described in Chap. 7. By setting bits UMSEL00 and UMSEL01 of register UCSR0C, the interface is configured as master SPI. By setting bit

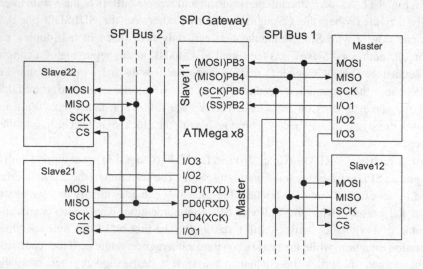

Fig. 8.12 ATmegaX8 as SPI gateway

Pin name	SPI function
TXD	Master out slave in
RXD	Master in slave out
XCK	Clock

Table 8.2 Port pins of the USART configured in MSPI mode

UDORD0, the LSB bit is transmitted first, and the combination of bits UCPOL0 and UCPHA0 determines the SPI mode.

A general software module is created here for this interface, which is structured as described in Sect. 8.4. For the control of the chip select inputs of the slaves, a data structure is defined in the header file of the software module in analogy to the above procedure:

```
typedef struct{
    volatile uint8_t* CS_DDR;
    volatile uint8_t* CS_PORT;
    uint8_t CS_pin;
    uint8_t CS_state;
} tmspimHandle;
```

According to [1], the interface must first be initialized before it can be used, such as in the following function:

```
void MSPIM_Master_Init(uint8_t ucspi_data_order,
                       uint8_t ucspi_mode,
                       uint8_t ucspi_sck_freq)
{
    MSPIM_BAUDRATE_REGISTER = 0;
    //CLK pin is declared on output; master mode is enabled
    MSPIM_CLK_DDR_REG |= 1 << MSPIM_CLK_BIT;
    //the microcontroller is configured as master
    // the bit order and the SPI transfer mode are determined
    MSPIM_CONTROL_REGISTER_C = MSPIM_MASTER | ucspi_data_order |
                                  ucspi_mode;
    //the USART as SPI interface is enabled
    MSPIM_CONTROL_REGISTER_B = MSPIM_ENABLE;
    //the desired SPI clock frequency is set
    MSPIM_BAUDRATE_REGISTER = ucspi_sck_freq;
}
```

By enabling the transmitter and the receiver by setting the bits TXEN0 and RXEN0 in the UCSR0B register, the corresponding pins TxD and RxD are automatically configured to output and input, respectively. Explicit configuration of the two pins is not necessary. The call parameters of the function MSPIM_Master_Init() for a microcontroller ATmegaX8 are defined in the header file of the module as follows:

Table 8.3 The registers for configuring the USART in MSPI mode

UCSR0A	Bit 7	Bit 6	Bit 5	Bit 4	Bit 3	Bit 2	Bit 1	Bit 0
	RXC0	TXC0	UDRE0	–	–	–	–	–
Direction	R	R/W	R	R	R	R	R	R
Initial value	0	0	0	0	0	1	1	0
UCSR0B	Bit 7	Bit 6	Bit 5	Bit 4	Bit 3	Bit 2	Bit 1	Bit 0
	RXCIE0	TXCIE0	UDRIE0	RXEN0	TXEN0	–	–	–
Direction	R/W	R/W	R/W	R/W	R/W	R	R	R
Initial value	0	0	0	0	0	1	1	0
UCSR0C	Bit 7	Bit 6	Bit 5	Bit 4	Bit 3	Bit 2	Bit 1	Bit 0
	UMSEL01	UMSEL00	–	–	–	UDORD0	UCPHA0	UCPOL0
Direction	R/W	R/W	R	R	R	R/W	R/W	R/W
Initial value	0	0	0	0	0	1	1	0

According to Microchip

```
//MSPIM Register
#define MSPIM_DATA_REGISTER UDR0
#define MSPIM_CONTROL_REGISTER_A UCSR0A
#define MSPIM_CONTROL_REGISTER_B UCSR0B
#define MSPIM_CONTROL_REGISTER_C UCSR0C
#define MSPIM_BAUDRATE_REGISTER UBRR0
//
#define MSPIM_ENABLE 0x18 //bits RXEN0 and TXEN0 in register UCSR0B
#define MSPIM_MASTER 0xC0 //bits UMSEL00 and UMSEL01 in register UCSR0C
//UDORD0 bit into the UCSR0C register (data order LSB or MSB first)
#define MSPIM_LSB_FIRST 0x04
#define MSPIM_MSB_FIRST 0x00
//SPI mode is determined by 2 bits UCPOL0 and UCPHA0 in the UCSR0C register
#define MSPIM_MODE_0 0x00 //UCPHA0 = 0, UCPOL0 = 0
#define MSPIM_MODE_1 0x02 //UCPHA0 = 1, UCPOL0 = 0
#define MSPIM_MODE_2 0x01 //UCPHA0 = 0, UCPOL0 = 1
#define MSPIM_MODE_3 0x03 //UCPHA0 = 1, UCPOL0 = 1
//clock frequency of the SPI interface
//FOSC = clock frequency of the microcontroller)
#define MSPIM_FOSC_DIV_2 0x00
#define MSPIM_FOSC_DIV_4 0x01
#define MSPIM_FOSC_DIV_8 0x03
#define MSPIM_FOSC_DIV_16 0x07
#define MSPIM_FOSC_DIV_32 0x0F
#define MSPIM_FOSC_DIV_64 0x1F
#define MSPIM_FOSC_DIV_128 0x3F
//sending a byte in progress
#define MSPIM_SENDING (!(UCSR0A & (1 << UDRE0)))
//receiving a byte in progress
#define MSPIM_RECEIVING (!(UCSR0A & (1 << RXC0)))
//Clock-Pin
#define MSPIM_CLK_DDR_REG DDRD
#define MSPIM_CLK_PORT_REG PORTD
#define MSPIM_CLK_BIT PD4
```

After initialization the USART works like the SPI interface. When a byte is stored in the data register UDR0, the individual bits are shifted in the set sequence onto the MOSI line synchronously with the clock generated at pin XCK. At the same time, the bits sent by the activated slave are loaded into the input buffer. The data exchange between the master and a slave takes place by calling the following function.

```
uint8_t MSPIM_Master_Write(uint8_t ucdata)
{
    while(MSPIM_SENDING);
    MSPIM_DATA_REGISTER = ucdata;
```

```
    while(MSPIM_RECEIVING);
    return MSPIM_DATA_REGISTER;
}
```

Unlike asynchronous transmission, the bit rate for synchronous transmission is calculated using Eq. 8.1.

$$Baud = \frac{f_{osc}}{(UBRRn + 1) * 2} \tag{8.1}$$

With UBRRn $= 0$ the maximum transmission rate of $f_{osc}/2$ is reached, where f_{osc} is the clock frequency of the microcontroller. The set transmission rate should be lower than the maximum transmission rate of all slaves connected to the bus.

Reference

1. Microchip: Reference Manual ATmega48/168. https://www.microchip.com/wwwproducts/en/ATmega88A. Accessed: 6. Jan. 2021.

The I²C/TWI Interface

Abstract

This chapter describes the I²C/TWI interface.

After the SPI interface was described in the eighth chapter, the third fundamental serial interface still missing is the interface called in Microchip/ATMEL jargon "Two wire" (TWI). In many applications it is known as I²C-Bus and is actually more than a serial interface. In contrast to the aforementioned, the I²C interface is, firstly, easily scalable and, secondly, because of its minimalist cabling (only one data and one clock line, no select lines or similar), it can also be used over somewhat longer distances, for example in a control cabinet. However, the hardware does not allow transmission over long distances; a "real" bus system must be used for this.

The I²C-Bus[1] (or I2C) was developed by Philips (now NXP)([2–4]) and allows networking of microcontrollers, sensors and other devices such as: RAM and EEPROM memory, A/D, D/A converters, port extensions, LCD and LED display controls, real-time clocks, radio and TV receivers, I²C-bus extensions and many others. Based on this interface, other protocols have been developed such as: CBUS, System Management Bus (SMBus), Power Management Bus (PMBus), Intelligent Platform Management Interface (IPMI), Advanced Telecom Computing Architecture (ATCA), and Display Data Channel (DDC). These protocols extend the original interface, improve it or adapt it to specific application areas. They allow, for example, the dynamic assignment of slave addresses, the forcing of I2C communication by the slaves via an additional interrupt line, or the detection

---

[1] I²C – Inter-Integrated Circuit.

© The Author(s), under exclusive license to Springer Fachmedien Wiesbaden GmbH, part of Springer Nature 2023
A. Meroth, P. Sora, *Sensor networks in theory and practice*,
https://doi.org/10.1007/978-3-658-39576-6_9

Table 9.1 I²C modes

Mode	Transmission rate	Transmission direction
Standard	<100 kBit/s	Bidirectional
Fast-mode	<400 kBit/s	Bidirectional
Fast-mode plus	<1 Mbit/s	Bidirectional
High-speed-mode	<3.4 Mbit/s	Bidirectional
Ultra-fast- mode	<5 Mbit/s	Unidirectional

and release of a "hang-up" of a slave blocking the bus. Conditionally, it is even possible for a microcontroller as master to control slaves with different protocols within a single bus. Depending on the transmission rate and direction, the following modes are defined (Table 9.1).

The I²C bus operates according to the master-slave principle. The master is the bus node that initiates and terminates the I²C communication, provides the clock for synchronizing the data transmission and detects and solves communication problems according to the programmed software. Each slave has an address that uniquely identifies it on a bus. The address can be seven or ten bits in size and allows accordingly up to 128 respectively 1024 different slaves to be addressed on the same bus. Devices with different address lengths can coexist on the same bus.

Before a master transmits or receives data, it must address a slave with a previously agreed address. However, the address groups 0000 xxx and 1111 xxx are reserved and should not be assigned to the slaves. The protocol defines that a master can address all slaves in a bus simultaneously using the reserved address 0000 000 in order to send them a message at the same time. This is called a broadcast message. Attention, not all I²C components have this function implemented!

I²C devices such as memory or some sensors offer, in addition to the permanently programmed device type address, a device chip address that can be set via external connections. If n is the number of address connections and m is the number of possible logic states, then up to n^m devices of the same type with different addresses can be identified in a bus. In general, m is equal to 2. A terminal can be connected to ground or to the supply voltage. Taking the M24C64 memory as an example, which has three address inputs and whose device type address is 1010 xxx, there are eight possible address combinations listed in Table 9.2. An address input connected to V_{cc} sets the corresponding address bit to 1. Rarely, as in the case of the TMP175 temperature sensor, a third state can be assigned to an unconnected pin. The address corresponding to each input combination can be found in the data sheet of the device.

Table 9.2 Addressing of a M24C64 device

Address inputs			Chip address							
E2	E1	E0	A6	A5	A4	A3	A2	A1	A0	R/$\overline{W}$
GND	GND	GND	1	0	1	0	0	0	0	1/0
GND	GND	V$_{CC}$	1	0	1	0	0	0	1	1/0
GND	V$_{CC}$	GND	1	0	1	0	0	1	0	1/0
GND	V$_{CC}$	V$_{CC}$	1	0	1	0	0	1	1	1/0
V$_{CC}$	GND	GND	1	0	1	0	1	0	0	1/0
V$_{CC}$	GND	V$_{CC}$	1	0	1	0	1	0	1	1/0
V$_{CC}$	V$_{CC}$	GND	1	0	1	0	1	1	0	1/0
V$_{CC}$	V$_{CC}$	V$_{CC}$	1	0	1	0	1	1	1	1/0

9.1 I²C-Bus Configuration

In the minimum configuration shown in Fig. 9.1, the master is connected to the slave via the bidirectional bus lines: SDA (data line) and SCL (clock line). These bus lines are connected to the supply voltage via pull-up resistors (or bus resistors), resulting in AND wiring of the devices. Additional devices (masters or slaves) can be connected to the bus by connecting the dedicated connectors SDA and SCL to the corresponding bus lines. In this example, both the master and the slave are supplied with the same voltage, which is not always the case.

I²C is designed as a true multi-master bus in which any connected microcontroller can temporarily function as a master or slave. In such a bus, any device configured as a master can take control of the communication when the bus is free. A bus is considered free when both bus lines are high for a specified time. Any collisions that may occur are detected by arbitration and resolved without interrupting communications or losing data. Arbitration, implemented hardware-based in a master, intervenes when two masters attempt to take control of the bus quasi-simultaneously. A master compares each bit it sends with the logical state of the SDA line. If there is a discrepancy (a "1" was sent but the SDA line is low), it loses arbitration and relinquishes control of the bus through the software. If arbitration is lost in the addressing phase, the master must immediately switch to slave mode in order to be able to react if it is to be addressed itself.

Devices with different technologies can be networked via I²C (bipolar, NMOS, CMOS), which are supplied with different voltages or operate with different maximum clock frequencies. To make this possible, the minimum output voltage and the maximum input voltage are standardized:

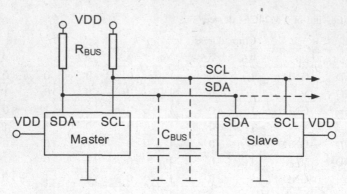

Fig. 9.1 I²C – minimum configuration

$$V_{IL} = 0.3 \times V_{DD}$$
$$V_{OH} = 0.7 \times V_{DD}, \qquad\qquad (9.1)$$

where V_{DD} is the supply voltage of the device. The maximum number of bus nodes and the length of the bus lines are not standardized, but are limited by the specified maximum bus capacitance of 400 pF. The bus capacitance is the total electrical capacitance between a bus line and ground. It is composed of the parallel connection of the input capacitances of all devices connected to the line, the line capacitance to ground and the capacitance of the terminals. Since all these capacitances are connected in parallel, they simply add up. Increasing the number of bus devices or the line length may cause the recommended maximum bus capacitance to be exceeded. In this case, the maximum clock frequency can no longer be guaranteed. The clock frequency of the bus is usually based on the slowest device. A disadvantage of the I²C protocol is that a slave has no way of reporting the availability of new data to the master. The master must always poll the data.

To design an I²C network and create the software of a master, the slave addresses and the clock frequency of the bus must be known. A master does not have to make any agreements with the slaves, which simplifies the extension of a bus.

In terms of hardware, only the bus resistor must be determined. The outputs of the bus nodes can be connected together because they have an open-collector or open-drain output. During communication, the bus capacitance discharges quickly across a conducting transistor, but charges slowly across the bus resistor when all output transistors are blocking. When charging, the voltage across the bus capacitance follows the equation:

$$V_{Cbus} = V_{DD}\left(1 - e^{-\frac{t}{\tau_{Bus}}}\right) \qquad\qquad (9.2)$$

with the time constant:

$$\tau_{Bus} = R_{Bus} \cdot C_{Bus} \tag{9.3}$$

The rise time of the voltage V_{Cbus} between the limits given in Eq. 9.1 must be less than the specified rise time t_r for each I^2C mode. With these considerations, the maximum value [1] is obtained for the bus resistance:

$$R_{Bus(max)} = \frac{t_r}{0{,}8473 \cdot C_{Bus}} \tag{9.4}$$

To limit the maximum current through the output stages to the specified value, the minimum bus resistance is calculated according to [1]:

$$R_{Bus(min)} = \frac{V_{DD} - V_{OL(max)}}{I_{OL}} \tag{9.5}$$

In the data sheets of the components, the output voltage V_{OL} and the output current I_{OL} are specified for the conductive state of the output transistor. A small bus resistor enables communication over longer lines, but causes higher energy losses. When choosing the bus resistor, the supply voltage, clock frequency, bus capacitance and energy consumption must be taken into account. The choice can also be based on the recommendations of the component manufacturers.

9.2 Bus Extension

In order to design complex I^2C buses dynamically, to enable longer bus lines, to exceed the maximum bus capacity, or to connect devices that are supplied with different voltages, have the same address, or operate with different clock frequencies, bus extensions are required. A bus extension is a device which is also controlled via I^2C. The exact conditions under which one can use these bus extensions are described in [3, 4]. All bus extensions generate a propagation delay of the signals, which must be taken into account.

9.2.1 I^2C-Repeater

An I^2C repeater has bidirectional drivers for the bus lines and splits the bus into two bus sections as in Fig. 9.2. The EN input can be used to isolate the right bus section from the left. Capacitively, the two sections are separated by the repeater so that the capacitance of each section can reach 400 pF. The sections can be supplied with different voltages. The repeater is supplied with the low voltage (usually 3.3 V), but its inputs are 5 V tolerant. If the master from the example can work with up to 400 kBit/s, but some devices only in standard mode, then the devices can be separated by bus clock: in the left bus section the

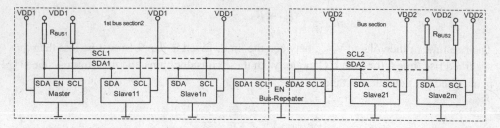

Fig. 9.2 I²C-bus extension with a repeater

master with the fast devices, in the right the slow ones. Before the master initiates the communication with 400 kBit/s, it must isolate the right bus section via the repeater. The EN input of the multiplexer must not be switched during communication.

9.2.2 I²C-Hub

An I²C hub is an extension of an I²C repeater. It can be used to network several bus sections. These can be supplied with different voltages, communicate via different clock frequencies and each bus section can reach 400 pF. The master can isolate bus sections via several EN inputs.

9.2.3 I²C Multiplexer

In order to control devices with the same address without conflicts occurring, an I²C multiplexer is used. The master selects a device or a bus section via the multiplexer and thus forms a new configuration of the bus. In this case, the bus capacity limit is calculated for the bus created via the selected path. An I²C multiplexer makes it possible to supply each bus section with a different voltage without additional level converters.

9.2.4 I²C-Switch

An I²C switch allows more complex configurations of a bus than a multiplexer because the master can be networked with several bus sections simultaneously. Each bus section can be supplied with a different voltage, the bus capacity limit applies to the active bus.

9.3   TWI in the AVR Family [5]

TWI^2 is an interface implemented by default in the ATmega family microcontrollers and compatible with I^2C. A microcontroller of this family can operate as a master or as a slave in Fast-mode (up to 400 kHz) with a 7 bit addressing and is multi-master capable. In this section, we present a sample implementation for AVR microcontrollers acting as participants of a single-master TWI bus. This implementation can of course be modified and it also exposes some basic mechanisms of the TWI communication. The communication in our example is byte-oriented and the interface is interrupt-based. Eight bits (bytes) are always transmitted, with the most significant bit (MSB) first. The end of a TWI communication process can be detected by the microcontroller by blocking waiting, or with the TWI interrupt. The transmission of one byte (8 data bits + acknowledge bit) in Fast-mode (400 kHz) takes about 22.5 µs. The clock line is held low by the master after the transmission of a byte until the interrupt has been serviced to prevent bus takeover by another master. The master generates eight bit clocks for bit synchronization and a ninth clock for receive acknowledgement. If the SDA line is pulled low by the data receiver during the ninth clock, it is referred to as Acknowledge (ACK), otherwise it is referred to as Not Acknowledge (NACK). A slave address sent by the master that is not present on the bus is acknowledged with NACK. A slave also sends a NACK if it cannot accept any more data due to internal problems. Finally, the master software must respond to the last byte sent by a slave with a NACK.

9.3.1 TWI Register at the ATmega 88

The TWI interface has a set of registers that are necessary for the configuration and operation of the interface. The registers belonging to it are: Data register, Baud rate register, Status register, Control register and for slave operation an address register TWAR (TWI Slave address register) and a register for masking this address TWAMR (TWI Slave address mask register).

The transmit and receive registers TWDR (TWI data register) of the TWI interface can be addressed under the same name. In transmit mode the byte to be transmitted is written into the register, in receive mode the register contains the received byte.

The clock frequency f_{SCL} of the communication is determined by

$$f_{SCL} = \frac{f_{OSC}}{16 + 2 \cdot TWBR \cdot TWI_{Prescaler}} \qquad (9.6)$$

---

[2]TWI – two wire interface.

Table 9.3 Setting the TWI prescaler

TWPS1	TWPS0	TWI prescaler
0	0	1
0	1	4
1	0	16
1	1	64

where f_{OSC} is the clock frequency of the microcontroller. The value of TWBR (TWI bit rate register) is set according to Eq. 9.7 and the value of TWI prescaler is selected according to Table 9.3. TWPS1 and TWPS0 belong to the status register.

Solving Eq. 9.6 for TWBR, we obtain:

$$TWBR = \left(\frac{f_{OSC}}{f_{SCL}} - 16\right) / (2 \cdot TWI \ Prescaler) \tag{9.7}$$

With the TWI prescaler setting equal to 1, values of the TWBR register for clock frequencies f_{SCL} from approximately 40 kHz at $f_{OSC} = 20$ MHz can be determined approximately with the following macro:

```
#define F_CPU 20000000UL //clock of the board µC in Hz
//desired TWI clock frequency in Hz;
#define TWI_SCL_FREQ 400000UL
//value of the TWBR register
#define TWI_MASTER_CLOCK (((F_CPU / TWI_SCL_FREQ) - 16 ) / 2 +1)
```

The TWI communication is controlled via the TWCR register (TWI control register):

Bit 7	TWINT – is set by the hardware when the following conditions exist: after execution from the master of a START or RESTART sequence, after sending or receiving a byte, after recognizing its own address as a slave, after losing arbitration, after an unauthorized START or STOP operation. The bit must be cleared by software before each TWI operation by writing a 1. If the TWIE bit and the I bit in the SREG status register of the microcontroller are set, setting TWINT generates an interrupt. Otherwise, the state of this bit can be determined by polling.
Bit 6	TWEA – The setting of this bit by the software generates an ACK after each byte received and additionally in slave mode after the recognition of the own address.
Bit 5	TWSTA – By setting this bit from the software in master mode, a START sequence is generated and communication is initiated.
Bit 4	TWSTO – If the master sets this bit, a STOP sequence is generated and communication is terminated.
Bit 3	TWWC – Is a collision flag set by hardware when attempting to write to the data register while the TWINT bit is "0". If this bit is "0", it may mean that a byte transfer is not yet complete.

(continued)

Bit 2	TWEN – Setting this bit enables the TWI interface and takes control of the dedicated ports SDA and SCL without additional settings in the I/O registers.
Bit 1	Is reserved.
Bit 0	TWIE – Setting this bit enables the TWI interrupt. In the corresponding interrupt service routine the bit TWINT must be reset.

The status register TWSR contains a code in the bits 7:3 after each TWI operation, which indicates whether the operation was successful or not. This code must be evaluated by the software and taken into account for the next step. Bits 1:0, TWPS 1:0 are used for prescaler selection (see Table 9.3). In order to evaluate the status code, the status register must be masked so that the prescaler bits mentioned above are hidden:

```
#define TWI_STATUS_REGISTER (TWSR &0xF8)
```

9.3.2   Initializing the TWI Interface

In a single master TWI bus, a microcontroller is configured either as a master or as a slave. The setting parameters should ideally be compiled in a data structure in the header file of the TWI module, as illustrated in the following example:

```
typedef struct{
     uint8_t ucDevice; //TWI_MASTER or TWI_SLAVE
     uint8_t ucTwiClock; //transfer rate for the master
     uint8_t ucSlaveAddress; //slave address
     uint8_t ucGenAddress; //enable/disable general call
     uint8_t ucAddressMask; //ev. address mask for the slave
} TWI_InitParam;
```

The individual parameters are defined in the main file (.c file). The role of the bus device is determined by the first parameter: Master or Slave. If the first parameter is TWI_MASTER, the second parameter must be the coded value of the clock frequency. The other parameters are not relevant for the master.

For a slave the parameter clock frequency is not relevant, instead the 7 bit address and possibly the address mask must be entered as a 7 bit number. In addition, CGA_ENABLE (general call enable) should be passed as the fourth parameter to determine whether the slave can be addressed via the general address 0x00.

The initialization function enables the interface. If the processor is configured as a slave, the TWI interrupt is additionally enabled as follows:

```
void TWI_Init (TWI_InitParam sinit_param)
{
    if(sinit_param.ucDevice == TWI_MASTER)
    {
        TWCR = (1 << TWEA) | (1 << TWEN);
        TWBR = sinit_param.ucTwiClock;
        //the clock frequency is calculated in the main file
    }
    else if(sinit_param.ucDevice == TWI_SLAVE)
    {
        //the address of the slave is determined and the
        //general call address 0x00 is enabled
        TWAR = (sinit_param.ucSlaveAddress << 1) |
               sinit_param.ucGenAddress;
        //save the address mask
        TWAMR = sinit_param.ucAddressMask << 1;
        TWCR |= (1 << TWEN) | (1 << TWEA) | (1 << TWIE);
    }
}
//The function is called as follows:
TWI_Init(sTWI_InitParam);
```

9.3.3   TWI Communication

In terms of hardware, TWI communication takes place on AVR controllers exactly as described for I²C above. In our example, all necessary functions are combined in a TWI software module and are structured in such a way that the microcontroller can be operated in master or slave mode by initialization. For the single master mode (a TWI network with only one master), we present in the following sections the basic functions used to initiate, terminate, and transfer a byte in both directions (master-slave and slave-master) of a TWI communication. Because the master has control over the timing of the communication in this mode, there is a blocking wait for the end of a TWI sequence. These simple functions provide a good basis for programming higher-level functions for controlling various I²C slaves. Numerous examples in the following chapters are based on exactly these functions.

The microcontroller in slave mode does not know the time of a communication. So that the microcontroller is not unnecessarily loaded with polling in this mode, functions with blocking waiting have been omitted. All slave communication is controlled by the TWI-ISR.

The following data structure is proposed as the TWI message frame:

```
typedef struct{
    uint8_t ucAddress; //for the slave: private or general address
    uint8_t ucTWIData[4];
    uint8_t ucTWIDataLength;
}twi_frame;
```

The structure component ucAddress is only relevant for the slave as recipient. It codes the addressing type of the message as private (only intended for this slave) or general (broadcast message). The maximum number of data bytes of a frame ucTWIDataLength can be adapted as required for the specific application.

9.3.4   The Microcontroller ATmega as TWI Master

Calling the function TWI_Init() with the following parameter initializes a microcontroller as TWI master:

```
TWI_InitParam sTWI_InitParam = {/*ucDevice*/ TWI_MASTER,
                                /*ucTwiClock*/ TWI_MASTER_CLOCK,
                                /*ucSlaveAddress*/ 0x80,
                                /*ucGenAddress*/ CGA_DISABLE,
                                /*ucAddressMask*/ 0x00};
```

When the bus is idle, the SDA and SCL lines are high. The master initiates the communication with a START sequence by setting the SDA line low before the SCL line. By calling the function TWI_Master_Start().

```
void TWI_Master_Start(void)
{
    TWCR = (1 << TWINT) | (1 << TWSTA) | (1 << TWEN);
    //start: set SDA line to Low before SCL line
    while(!(TWCR & (1 << TWINT))); //wait until SDA is set
}
```

the microcontroller, now in master mode, starts a START sequence as soon as the bus is free. After the START, the bus is considered busy. After a successful start, the status register of the master contains the code 0x08 and the software initiates the addressing of a slave. The master transmits a byte whose bits 7:1 form the address of the desired slave; bit 0 (read/write) is used to determine the direction of the transmission. On the ninth clock pulse, the addressed slave confirms the recognition of its own address with an ACK. During the entire transmission, a bit on the SDA line may only change while the SCL line is low. It must maintain a stable state throughout the SCL pulse, as outlined in Fig. 9.3. This requirement is implemented in the protocol's hardware link layer and does not need to be implemented by the user software.

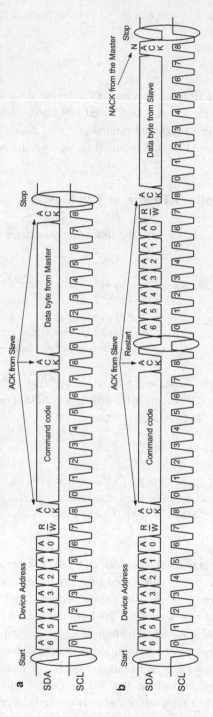

Fig. 9.3 TWI communication

9.3.4.1 The TWI Master as Transmitter

The master is in transmitter mode when it sets the read/write bit to 0 in the addressing phase. The time course of an exemplary data transmission from master to slave is shown in Fig. 9.3a. After successful transmission of the slave address, the status register contains the code 0x18, and one byte can be sent at a time using the TWI_Master_Transmit function. To avoid collisions (this would set the bit TWWC), the byte ucdata is first written in the data register and then the bit TWINT is cleared.

```
void TWI_Master_Transmit(uint8_t ucdata)
{
    TWDR = ucdata;
    TWCR = (1 << TWINT) | (1 << TWEN);
    while(!(TWCR & (1 << TWINT)));
    //the hardware sets the bit TWINT when the byte ucdata has been
    //completely transmitted
}
```

The addressed slave acknowledges each received byte with ACK, resulting in code 0x28 in the status register. After the last byte, the master terminates communication with the slave by a STOP sequence. In hardware, this sequence means setting the SCL line high before the SDA line, which can be done in software by calling the function TWI_Master_Stop(). After the STOP, the bus is considered free after a delay time specified in the standard and can be claimed by another master.

A simple function for the master to send a frame to the microcontroller slave under the above conditions is as follows. The frame to be sent is filled in the master file and addressing can be private, multicast or broadcast. The completion of each TWI sequence is checked. If errors occur, the function is interrupted and an error code is returned to the call location.

```
uint8_t TWI_Send_Frame(uint8_t ucdevice_address,
                       twi_frame *sframe)
{
    uint8_t ucDeviceAddress;
    ucDeviceAddress = (ucdevice_address << 1);
    ucDeviceAddress |= TWI_WRITE; //write mode
    TWI_Master_Start(); //start
    if((TWI_STATUS_REGISTER) != TWI_START) return TWI_ERROR;
    //send device address
    TWI_Master_Transmit(ucDeviceAddress);
    if((TWI_STATUS_REGISTER) != TWI_MT_SLA_ACK)
    {
        TWI_Master_Stop();
        return TWI_ERROR;
    }
```

```
for(uint8_t i = 0; i < sframe->ucTWIDataLength; i++)
{
    TWI_Master_Transmit(sframe->ucTWIData[i]);
    if(TWI_STATUS_REGISTER == TWI_MT_DATA_NACK)
    {
        TWI_Master_Stop();
        return TWI_NACK;
    }
}
TWI_Master_Stop(); //stop
return TWI_OK;
}
```

9.3.4.2 The TWI Master as Receiver

Because most I²C devices have a complex structure and therefore output a wide variety of data or can be broadly configured, a master can usually only access the desired data indirectly via registers or functions. Before it switches to receiver mode, it must transmit the key to the data as a sender as described in the last section. For a memory block, this can be the address of the desired byte or a function code. In order not to lose control of the bus, the master generates a so-called RESTART sequence by a renewed START and thereby continues to reserve the bus, so that a possibly further master cannot access it. A new addressing of the slave follows, this time with the R/W bit set. If the slave confirms the reception with ACK, the status register in the master contains the code 0x40. From now on the transmission direction changes, the slave transmits and the master receives the data. The master continues to generate the clock frequency and must respond to the ninth clock pulse of a byte with ACK (code 0x50 in the status register) or with NACK (code 0x58 in the status register). The last requested byte is acknowledged with NACK, which signals the end of communication to the slave. Finally, a STOP sequence is generated. This procedure is shown graphically in Fig. 9.3b. The master initiates communication, addresses the slave in write mode, and transmits a command code. After RESTART and re-addressing in read mode, the slave pushes out a byte on the SDA line. A RESTART sequence is generated with both bus lines high before the bus can be considered free.

In the following two functions for the reception of a byte are presented, depending on whether the master confirms the byte reception with ACK or NACK:

```
//the master responds with ACK to the received byte
uint8_t TWI_Master_Read_Ack(void)
{
    TWCR = (1 << TWINT) | (1 << TWEN) | (1 << TWEA);
    while(!(TWCR & (1 << TWINT)));
    return TWDR;
}
```

```
//the master responds with NACK to the received byte
uint8_t TWI_Master_Read_NAck(void)
{
    TWCR = (1 << TWINT) | (1 << TWEN);
    while(!(TWCR & (1 << TWINT)));
    return TWDR;
}
```

Both functions return the received byte to the call location. Numerous examples of write and read functions are presented in the following chapters on sensor examples.

The communication of the master as receiver with a microcontroller slave is simpler and is illustrated below using the `TWI_Read_Frame()` function. The master addresses the slave only once privately with the R/W bit set and continues to establish the clock for data reception. The data bytes are stored in a data structure of type `twi_frame`. The master acknowledges every byte except the last one with ACK. Different functions can also be reached via different addresses in this case.

```
uint8_t TWI_Read_Frame(uint8_t ucdevice_address,
                       twi_frame *sframe)
{
    uint8_t ucDeviceAddress, i;
    ucDeviceAddress = (ucdevice_address << 1);
    ucDeviceAddress |= TWI_READ; //write mode
    TWI_Master_Start(); //start
    if(TWI_STATUS_REGISTER != TWI_START) return TWI_ERROR;
    TWI_Master_Transmit(ucDeviceAddress); //send device address
    if(TWI_STATUS_REGISTER != TWI_MR_SLA_ACK)
    {
        TWI_Master_Stop();
        return TWI_ERROR;
    }
    for(i = 0; i < (sframe->ucTWIDataLength - 1); i++)
    {
        sframe->ucTWIData[i] = TWI_Master_Read_Ack();
        if(TWI_STATUS_REGISTER == TWI_MR_DATA_NACK)
        {
            //if the slave answers with NACK, the master
            //terminates the communication
            TWI_Master_Stop();
            return TWI_NACK;
        }
    }
    //the master responds to the last byte with NACK,
    //and terminates the communication
```

```
sframe->ucTWIData[i] = TWI_Master_Read_NAck();
TWI_Master_Stop(); //stop
return TWI_OK;
}
```

9.3.5   The Microcontroller ATmega as TWI Slave

To configure the microcontroller as a slave, the function TWI_Init() is called with the
following parameter sTWI_Init_Param as an example:

```
TWI_InitParam sTWI_InitParam = {/*ucDevice*/ TWI_SLAVE,
                                /*ucTwiClock*/ TWI_MASTER_CLOCK,
                                /*ucSlaveAddress*/ 0x10,
                                /*ucGenAddress*/ CGA_ENABLE,
                                /*ucAddressMask*/ 0x0B};
```

The structure is defined in the main file. The component ucTwiClock is not relevant
for the slave and can assume any value.

In the initialization function the TWI interface and the TWI interrupt are activated. Thus
the hardware takes control of the pins SCL and SDA and the internal data flow is controlled
by interrupt.

Each microcontroller that is to operate in slave mode is assigned a 7 bit address by
software, which may occur once in the TWI bus. This address is shifted one bit to the left
and stored in the address register TWAR. By setting the least significant bit of the TWAR
register, the slave can receive broadcast messages. These messages are sent by the master to
the general address 0x00. The 7 bit addresses of the form 1111 xxx are reserved for later
applications.

The slave compares the address received via TWI with its own and, if it matches, it
acknowledges the addressing sequence with ACK. During this comparison, the bits marked
with a 1 in the mask register TWAMR are neglected. With each set bit in the mask register,
the address range of a slave increases by a factor of two as shown in Table 9.4.

Table 9.4 Determination of
the slave addresses

TWAR	TWAMR	Confirmed address
0001 0000	0000 0000	0001 000x
0001 0000	0000 0010	0001 000x 0001 001x
0001 0001	0000 0010	0000 0000 0001 000x 0001 001x

When the slave is addressed as a receiver, the bit denoted by "x" in Table 9.4 is set to 0; when the slave is addressed as a transmitter, it is set to 1. Broadcast and multicast addressing are only useful for slaves in receiver mode.

Address masking opens up new possibilities for communication with microcontroller slaves. This means that each write or read function of a slave can be called directly via one of the addresses. This leads to a shortening of messages and simplification of master functions.

The microcontroller initialized as a slave is in unaddressed mode. The R/W bit in the addressing sequence switches it to receiver or transmitter mode.

9.3.5.1 The TWI Slave as Receiver

A TWI slave switches to receiver mode when it is addressed with the R/W bit equal to 0. In this mode a point-to-point or broadcast communication is possible. By clever use of address masking, multicast communication is also possible. In this mode, the master can address a slave group simultaneously as shown in Table 9.5.

Setting three bits in the TWAMR register increases the number of private addresses of the two slaves to eight each. The last four addresses listed are identical for the two slaves, which allows the master to address only these two slaves together (multicast mode) in receiver mode.

The following TWI events cause an interrupt for the slave in receiver mode:

- Detection of an address (private or general); the code 0x60 is written into the status register when receiving a private address, or 0x70 when receiving the general address. According to this address, a code is stored in twi_frame- > ucAddress in our example implementation. This code is used for the later assignment of the message.
- Receiving a data byte leads to storing the code 0x80 into the register TWSR; the TWI interface does not offer the possibility to buffer the received data bytes. Therefore, they are stored from the data register TWDR in a receive buffer of type twi_frame and a counter is incremented.

Table 9.5 Settings of two slaves for a multicast communication

	Slave 1	Slave 2
TWAR	0010 0000	0001 0000
TWAMR	0001 0110	0010 0110
Confirmed addresses	0010 0000	0001 0000
	0010 0010	0001 0010
	0010 0100	0001 0100
	0010 0110	0001 0110
	0011 0000	0011 0000
	0011 0010	0011 0010
	0011 0100	0011 0100
	0011 0110	0011 0110

- Detection of a stop or a new start sequence. After the occurrence of one of these events the code 0xA0 is stored in the status register. This signifies the end of a TWI frame and the number of bytes received (the contents of the counter) is stored in `twi_frame-> ucDataLength`.

In receiver mode the slave acknowledges all received bytes with ACK.

If the processing of the messages takes different amounts of time, it makes sense to store the frames in an array:

```
#define MAX_REC_BUFFER 5
twi_frame sTWI_RecBuffer[MAX_REC_BUFFER];
```

This array is treated as a FIFO ring buffer and managed in the interrupt service routine. The maximum number of components is defined globally according to the concrete application. A write control variable indicates the array index at which the next frame is stored. A read control variable points to the oldest unread frame. For further discussion and alternatives to FIFO ring buffers, see Sect. 5.1.1.

```
case 0xA0: /*a stop or a new start was detected; a new frame was received*/
    sTWI_RecBuffer[ucTWI_Buff2Save].ucTWIDataLength = ucIndex;
    ucTWI_Buff2Save++; /*control variable for the frame to be saved*/
    if(ucTWI_Buff2Save == MAX_REC_BUFFER) ucTWI_Buff2Save = 0;
    if(ucTWI_Buff2Save == ucTWI_Buff2Read)
    { /*the case occurs when the first write buffer was overwritten but not
        read*/
        ucTWI_Buff2Read++; /*control variable for the frame to be read*/
        if(ucTWI_Buff2Read == MAX_REC_BUFFER) ucTWI_Buff2Read = 0;
    }
    TWCR |= (1 << TWINT) | (1 << TWEA); //switch to unaddressed slave mode
break;
```

When storing a new message, if the store variable and the read control variable of the FIFO buffer are the same, an overflow from the FIFO takes place and the oldest, unread message is overwritten.

After initialization of the TWI interface, both control variables `ucTWI_Buff2Save` and `ucTWI_Buff2Read` are zero. Later, the inequality of the control variables `ucTWI_Buff2Save` and `ucTWI_Buff2Read` indicates that there are unread messages. This check can be implemented in polling mode using the `TWI_Check_Message()` function. This function returns `MESSAGE_RECEIVED` if there are new messages and in this case copies the oldest unread message into the data structure passed as parameter and updates the read control variable.

```
uint8_t TWI_Check_Message(twi_frame *sframe)
{
    if(ucTWI_Buff2Save == ucTWI_Buff2Read)
    {
        return NO_MESSAGE; //no new messages
    }
    else
    {
        sframe->ucAddress = sTWI_RecBuffer[ucTWI_Buff2Read].ucAddress;
        sframe->ucTWIDataLength =
                sTWI_RecBuffer[ucTWI_Buff2Read].ucTWIDataLength;
        for(uint8_t i = 0;
                i < sTWI_RecBuffer[ucTWI_Buff2Read].ucTWIDataLength; i++)
        {
            sframe->ucTWIData[i] =
                    sTWI_RecBuffer[ucTWI_Buff2Read].ucTWIData[i];
        }
        ucTWI_Buff2Read++; //the read control variable is updated
        if(ucTWI_Buff2Read == MAX_REC_BUFFER) ucTWI_Buff2Read = 0;
    }
    return MESSAGE_RECEIVED; //there was an unread message
}
```

9.3.5.2 The TWI Slave as Transmitter

The microcontroller is set to transmitter mode as a slave when the R/W bit in the addressing sequence is set to 1. In this mode, the broadcast and multicast communications are meaningless. Address masking allows slave actions to be triggered directly in this mode as well.

A transmit buffer of type twi_frame is declared for the slave in transmitter mode. The structure component ucAddress is not relevant in this mode. The higher-level user software takes care of updating this buffer from which the data bytes are sent, controlled by the ISR. This can be implemented as in the following example:

```
void TWI_Fill_TransmitBuffer(twi_frame *sframe)
{
    sTWI_TransmitBuffer.ucTWIDataLength = sframe->ucTWIDataLength;
    for(uint8_t i = 0; i < sTWI_TransmitBuffer.ucTWIDataLength; i++)
    {
        sTWI_TransmitBuffer.ucTWIData[i] = sframe->ucTWIData[i];
    }
}
```

The time of data transmission is determined by the master, which initiates the transmission. Updating the send buffer during communication leads to corruption of the message. To prevent this, a check is made beforehand to see whether the send buffer is empty.

The events that lead to an interrupt for the slave in transmitter mode are:

- When the own address is detected, the code 0xA8 is stored in the status register. In the ISR the first data byte is sent from the transmit buffer.
- Sending a data byte that is acknowledged by the master changes the code in the status register to 0xB8. The next data byte is sent and the control variable is incremented.
- Sending a data byte that was acknowledged by the master with a NACK results in the code 0xC0 in the status register. The slave evaluates the code as end of communication, switches to unaddressed slave mode and stores the value TRANSMIT_BUFFER_EMPTY in the variable ucTWI_TransmitBuffer.
- When the last data byte is sent, the master responds with ACK and the code 0xC8 is stored in the status register. The slave switches to non-addressed slave mode and reports the transmit buffer as empty.

References

1. Meroth, A., & Tolg B. (2008). Infotainmentsysteme im Kraftfahrzeug. Grundlagen, Komponenten, *Systeme und Anwendungen*. Friedr. Vieweg & Sohn Verlag Wiesbaden.
2. NXP Semiconductors. UM10204 – I²C-bus specification and user manual- Rev.6–4 April 2014. www.nxp.com.
3. NXP Semiconductors. Application Note AN255-02 – I²C/SMBus Repeaters, Hubs and Expanders, 2015. www.nxp.com.
4. NXP Semiconductors. Application Note AN262_2 – PCA954x Family of I²C/SMBus Multiplexers and Switches, 2015. www.nxp.com.
5. Microchip: Reference Manual ATmega48/168. https://www.microchip.com/wwwproducts/en/ ATmega88A. Accessed: 6. Jan. 2021.

CAN Bus

10

Abstract

In this chapter the CAN bus is explained and a library implementation is presented.

After the fundamental serial interfaces for networking sensors and other peripheral components have been presented in the last chapters, the following chapters are dedicated to some selected wired networks between control units. The CAN bus, which originated in the automotive industry but has since become indispensable in automation technology, is the first of these.

CAN (Controller Area Network) is the classic sensor network in the automotive industry and has established itself as a standard not only in vehicles but also in automation technology in particular due to its simplicity, its reasonably high data rates (up to 1 Mbit/s at lengths of up to 40 m) and its robustness. Originally specified by Bosch [1], it is now internationally standardized as ISO 11898 (parts 1–6) [2]. Part 1 reflects the original specification, Part 2 and Part 3 define the physical layer for high-speed and low-speed CAN, respectively, and Part 4 is the extension to time-triggered synchronous communication. Parts 5 and 6 refer to the low-power mode and selective wakeup. An overview of other standards in the CAN environment is given in [3, 4].

A. Meroth, P. Sora, *Sensor networks in theory and practice*,
https://doi.org/10.1007/978-3-658-39576-6_10

10.1   CAN Basics Compact

CAN is usually designed as a balanced two-wire line in Wired-AND technology, but is also specified as a single-wire line. The data transmission is bit stream oriented and uses bit stuffing after five equal bits (see Sect. 6.1) to prevent long 0 and 1 sequences.

CAN uses the CSMA/CA technique for collision avoidance. As long as one station is transmitting, the other stations remain quiet. Initially, CAN transmits an 11-bit message identifier (CAN2.0A) after a start bit (dominant $= 0$), i.e. it is message-oriented. In the CAN2.0B specification, the identifier can be extended to 29 bits. The message frame is then called "extended" instead of "standard". Each identifier is bound to a message of a participant in a mandatory and unique way. Since all participants on the bus "listen in", no one makes a transmission attempt as long as another is transmitting. If two participants attempt to transmit at the same time, the participant that has sent a recessive ($=1$) bit (logical 1) but has a dominant bit on the bus (logical 0) will be the first to notice the problem. It must then immediately interrupt the transmission. Consequently, the message with the lower address prevails undisturbed. Since the messages are relatively short, the chance of getting transmission capacity after a short time is high.

In Fig. 10.1 a corresponding course can be seen, here four stations try to send messages with the CAN IDs 0x64C, 0x658, 0x704, 0x6DC messages. As soon as a station detects that the signal it sends on the bus has been overridden by a dominant bit, it aborts the transmission. The transmission phase of the identifiers is also called the competition phase, because this is where it is decided which message will prevail on the bus.

CAN uses a protocol frame (Fig. 10.2) with up to 64 bits of user data and a header of 19 bits in length for the CAN2.0A specification or 37 bits for CAN2.0B and a trailer of 25 bits in length. In addition, there are at least 3 bits of waiting time after which a node is allowed to make a transmission attempt itself after a successful transmission (Interframe Space, IFS), and a total of up to 15 stuff bits. There is space for a maximum of eight data bytes between the header and the trailer. In the worst case, the protocol efficiency, i.e. the ratio between the user data and the total packet length, is just under 50%. At 500 kBit/s bit rate, the transmission time (latency) is 225 µs for the short header and 260 µs for the long header, which corresponds to a user data rate of 34 kByte/s and 30 kByte/s, respectively. In addition to the message identifier, the header also consists of:

- **SOF**: Start of Frame: Dominant bit that signals the start of a message if the bus has previously been at recessive level for at least 11 bits (see Acknowledge Delimiter, EOF and IFS below).
- **RTR**: Remote Transmission Request: A CAN station that expects a certain message sends a frame with the message identifier of this message, with recessive (1) RTR bit but without data field. The node that normally generates this message recognizes the remote frame and then completes the message with the corresponding data. In this way, a simple client-server model can be established.

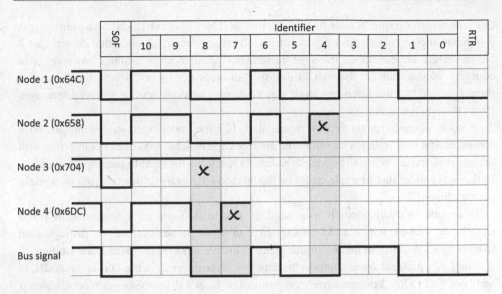

Fig. 10.1 CSMA/CA competition period

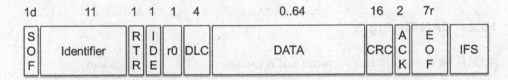

Fig. 10.2 CAN frame

- **IDE**: Identifier Extension: If it is recessive, it indicates that the identifier is extended to 29 bits.
- **r0**: Reserved (will be set dominant).
- **DLC**: Data Length Code: Shows the length of the following data field in bytes.
- **DATA**: User data: Can comprise zero to eight bytes.
- **CRC**: CRC checksum (see Sect. 6.1.2.5).
- **ACK**: Acknowledge: Consists of the acknowledge slot sent recessively by the sender and a specified recessive delimiter bit (acknowledge delimiter). A bus station that has received the message consistent with the checksum sets the acknowledge slot to dominant level.
- **EOF**: End of Frame: The transmission is terminated with seven recessive bits.

ISO 11898 also specifies an error handling procedure that provides for the possibility of a node disconnecting itself from the bus if it is the cause of frequent errors. For this purpose, a node that detects an error sends a message with six dominant bits (error frame), which thus violates the bit stuffing rule and is detected by every other node. These also respond to this with an error frame. A node that detects that an error has occurred during a transmission it initiated increments an error counter (transmit error counter). A node that detects that it has received an erroneous message increments another error counter (receive error counter). Nodes that are the first to discover an error get more "error points" because there is a risk that the discovery itself was an error. Correctly sent or received messages lower the counter again.

A node whose error counter is greater than 127 may no longer send dominant error messages, but only indicate the error with six recessive bits, but thus does not interfere with the network traffic. When the counter reaches 255, it loses the right to participate in traffic. In this way, nodes that generate errors on the bus too frequently or detect errors incorrectly are switched off.

If a node is overloaded, it may send an overload frame in a transmission pause (interframe space), which looks exactly like an error frame, but can be distinguished from it by its position in the interframe space. This prevents other nodes from sending.

The CAN protocol corresponds to the data link layer (layer 2) in the OSI layer model. In addition, ISO 15765-2 defines a transport protocol on layer 4 in the OSI layer model, which is used in diagnostics, for example. A detailed description can be found, for example, in [3, 4].

10.2   CAN Timing

A CAN bit consists of four segments and is measured in TQ (Time Quantum):

- The sync segment is used to give all nodes the necessary time to detect the start of the bit. It is one TQ long
- The propagation segment compensates for delays caused by propagation effects from the line. It is 1 … 8 TQ long
- The phase segments PS1 and PS2 represent the actual bit. PS1 is between 1 and 8 TQ long, PS2 is at least 2 TQ and up to 8 TQ long. The sample point, i.e. the time at which the value of the bit is measured, is at the transition between PS1 and PS2. Thus PS2 is also responsible for processing (Information Processing Time).

Figure 10.3 shows the structure of a bit graphically. The total bit time is therefore

$$t_{Bit} = t_{SyncSeg} + t_{PropSeg} + t_{PS1} + t_{PS2} \tag{10.1}$$

This then results in the nominal bit rate.

$$NBR = \frac{1}{t_{Bit}} \tag{10.2}$$

All CAN controllers in a network must have the same nominal bit rate so that the bits can be read safely. Fluctuations of the respective oscillators are compensated via a PLL.

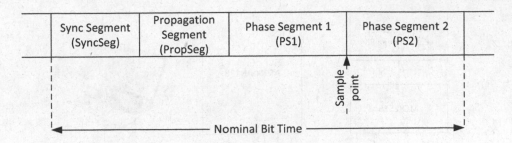

Fig. 10.3 Bit timing in CAN

10.3   Use of CAN with Processors of the AVR Family

With the AT90CAN128, the AVR family has a processor that directly has a CAN controller on board. Only a component for the adaptation of the electrical transmission is necessary, here for example a PCA82C251 or a TJA1050 offers itself in each case from NXP. To get other processors to communicate via CAN, an MCP2515 from Microchip is often used. This also needs a physical transceiver, both components are available as adapter boards for about 2€. In this case the communication with the processor is done via SPI. In the following section, only the rough basics of using the MCP2515 are briefly described. There are ready CAN drivers available on the net, among others the free software of the Roboterclub Aachen, which supports both variants and furthermore the CAN controller SJA1000 via corresponding precompiler switches [5]. Here we present our own library, which can be freely downloaded as source code from the book's website.

Figure 10.4 shows a CAN network in which three nodes access the CAN bus with different wiring. Here you can see the different configurations depending on the devices used.

10.3.1 CAN Controller MCP2515

The CAN controller MCP2515 from Microchip [6] is one of the most successful CAN controllers on the market. It can be connected via SPI and provides signals for CAN Receive (CAN Rx) and Transmit (CAN Tx) at the output, which must be electrically connected to the bus system via a transceiver (here: TJA1050). An overview of the architecture can be seen in Fig. 10.5.

The controller has three transmit buffers, two receive buffers, two receive masks and a total of six receive filters. It performs all error and overload handling itself. A complete description is beyond the scope of this book, so please refer to the very comprehensive and instructive data sheet [6]. The transmit buffers contain the messages, the CAN identifier and other control data. They can be addressed as registers via SPI. The block uses the

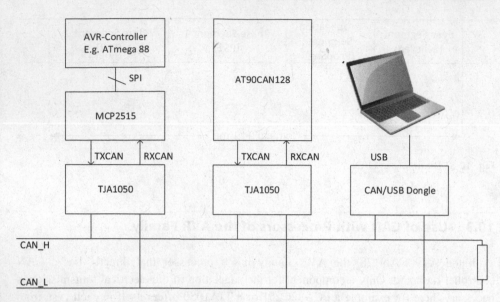

Fig. 10.4 CAN network with three nodes

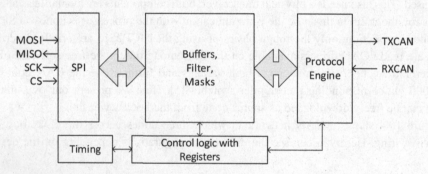

Fig. 10.5 Block diagram_of_the_MCP_2515. (After [6])

CSMA/CA method to transmit the contents of these buffers, which can be prioritized among each other. In this respect, the actual processor is completely unloaded as soon as the messages are delivered in the transmit buffer. The protocol engine assembles the CAN frame and handles the relatively complex timing and checksum formation.

On the receive side, a Message Assembly Buffer (MAB) takes incoming messages from the Protocol Engine and distributes each message that passes through the receive filters and masks into one of the two receive buffers RXB0 and RXB1. RXB0 is assigned two filters and one mask, RXB1 is assigned the second mask and four filters. If a message is received whose identifier matches the filter criteria for both buffers, it is stored only in RXB0. As long as an unread message is in the buffer, it is blocked for further messages. However, if a message arrives that is intended for RXB0 although an unread message is already stored there, the controller can store the old message in RXB1 and receive the new message (rollover operation), this must be configured specifically. As soon as a buffer contains a new message, the corresponding pin RXB0 or RXB1 can optionally be pulled high to low, so that the MCP2515 can trigger an interrupt (pin change interrupt) at the processor.

The mask controls whether the corresponding bit in the filter is effective, i.e. if a mask bit is set to 0, the bit of the message identifier is accepted in any case, otherwise it is only accepted if it matches the corresponding filter bit. A message passes through the filters if all bits of the identifier enabled by the mask are accepted. Thus, a mixture of masks can be used to pass individual messages and entire message groups. Filters and masks can be described using registers, as can the entire CAN configuration that controls timing, error behavior, and power management. Figure 10.6 gives an overview of the structure of the receiving system.

10.3.1.1 Initialization

The initialization of the MCP2515 requires a precise knowledge of the CAN timing. Finally, depending on the crystal oscillator used, one sets a prescaler (BRP) that determines the time quantum and then sets the TQs for the individual segments.

$$TQ = \frac{2 \cdot BRP}{f_{osc}} \tag{10.3}$$

Our library does this with a simple call

```
//CAN Baud rate in MCP2515_HHN.h
#define BAUDRATE_20_KBPS 0
#define BAUDRATE_50_KBPS 1
#define BAUDRATE_100_KBPS 2
#define BAUDRATE_125_KBPS 3
#define BAUDRATE_250_KBPS 4
#define BAUDRATE_500_KBPS 5
#define BAUDRATE_1_MBPS 6
//call
MCP2515_Init(MCP2515_1, BAUDRATE_250_KBPS);
```

Before this, the management structure for the SPI interface is set up once as described in Sect. 8.4. This measure makes it possible to control other sensors or one or more other CAN buses with different $\overline{SS}$ outputs in addition to the CAN bus.

```
MCP2515_pins MCP2515_1 ={{
    /*CS_DDR*/ &DDRB,
    /*CS_PORT*/ &PORTB,
    /*CS_pin*/ PB2,
    /*CS_state*/ ON}};
```

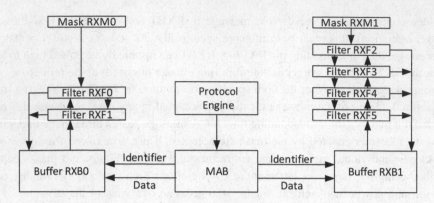

Fig. 10.6 Structures of the receive buffers in the MCP2515. (After [6])

This structure `MCP2515_pins` contains the structure of type `tspiHandle` known from Sect. 8.4. Therefore, the two curly brackets must also be opened and closed.

10.3.1.2 Sending Messages

The central data structure for the use of CAN is the CAN message. In the CAN library it is defined for standard or extended CAN messages as follows:

```
typedef struct
{
    uint8_t EIDE_Bit; //STANDARD_ID or EXTENDED_ID
    uint8_t RTR_Bit; //DATA_FRAME or REMOTE_FRAME
    uint8_t RB0_Bit; //Reserved bit: 0x00 or 0x01
    uint8_t RB1_Bit; //Reserved bit: 0x00 or 0x02
    uint32_t ulID; //11 or 29 bit ID
    uint8_t ucLength; //number of data bytes
    uint8_t ucData[8]; //data vector for the maximum message length
}can_frame;
```

The ID field with 32 bits is suitable for extended identifier messages (29 bits) as well as for standard identifier messages. Sometimes a time stamp (16 bits are usually sufficient) is also stored, but this is not supported by the hardware of the MCP2515 and must then be sent as part of the message.

The sending of a message by the function `MCP2515_Send_Message()` runs as follows:

- First the function checks if one of the three transmit registers is free by sending the code for READ STATUS via the SPI interface, this is 0xA0. Sending any value again (usually 0x00 or 0xFF) pushes the current status into the SPI receive register.

- The status of the three transmit registers is represented in bits 2, 4 and 6 (TXREQ0 ... 2) of the buffer status register, here n means the number of the transmit register. If the respective bit = 1, there is still a message to be sent in the corresponding register. If the bit = 0, the message can be written into this register.
- Writing is done by sending the command Load TX Buffer (0x40) together with the address of the TX buffer followed by transmission of the identifier, the length byte, possibly together with an RTR (see [6]) and the corresponding data.
- The transmission process is then started with the Request to Send (RTS) command. This consists of a 0x80 ORed with the address of the TX buffer (0x01, 0x02 or 0x04).

The call in our library is simple. The example shows the sending of a 16 bit message with a temperature value in little-endian format. First the message is compiled and then sent with a call.

```
can_frame sSendFrame;
sSendFrame.EIDE_Bit = STANDARD_ID;
sSendFrame.RTR_Bit = DATA_FRAME;
sSendFrame.ulID = 0x19; //please enter accordingly!!!
sSendFrame.ucLength=2;
/*******************************/
sSendFrame.ucData[0]=(uint8_t) MyTemperature;
sSendFrame.ucData[1]=(uint8_t) (MyTemperature >> 8);
//bracket is important, because typecast has higher prio than shift
/*******************************/
MCP2515_Send_Message(MCP2515_1, &sSendFrame);
```

A simple sensor circuit that sends a sensor value every 100 ms only has to transmit a send command with the corresponding data in a 100 ms task.

10.3.1.3 Receiving Messages
Receiving messages can be done in several ways:

- If the interrupt lines RX0BF and RX1BF are wired, a new message in one of the two receive buffers triggers an interrupt that can fetch the contents of the registers. In our library, we do not use interrupt operation, with which we have not had good experiences. Indeed, at high bus loads, stack overflows can occur in the library used in [5] if more messages arrive than interrupts can be processed. This leads to a total crash of the system, which the authors experienced.
- If the interrupt is not used, the Read Status (0xB0) command can be used to query the receive status of the two buffers (see [6]) and the receive status and the filter numbers used to enable reception are obtained (polling mode). In both cases, the message must then be retrieved anyway by sending the Read Rx-Buffer command and then reading out the buffer contents.

In general it is recommended to use the filter function for high bus loads, in any case the query is quite simple (in this case two frames with the IDs 0x10 and 0x15 are read):

```
can_frame sRecFrame;
if(MCP2515_Check_Message(MCP2515_1,&sRecFrame))
{ //here you can now evaluate the frame:
    switch(sRecFrame.ulID)
    {
        case 0×10: Call_function_1(&sRecFrame); break;
        case 0×15: Call_function_2(&sRecFrame); break;
        default: break;
    }
}
```

The simplest way to implement a filter operation is to statically preallocate the filters, which is provided in our library.

The `MCP2515_Set_Filter_Mask` function is used to store the desired receive filters and masks in coded form in the controller registers. These registers can only be accessed in the controller's configuration mode [6], which requires that no messages are sent during the setting. The settings are stored in the main file in a data structure that is passed as a parameter when the function is called.

```
typedef struct
{
    uint8_t Rec_Buff0_Rollover; //ROLLOVER_ON / ROLLOVER_OFF
    uint8_t RecBuff_ID[2]; //STANDARD_ID or EXTENDED_ID
    uint32_t RecBuff_Mask[2]; //filter masks of the receive buffers
    uint32_t RecBuff_Filter[6]; /*the first two filters are for the receive
                                  buffer 0*/
}can_filter;
```

Storing the constant `ROLLOVER_ON` in `Rec_Buff0_Rollover` enables rollover operation for receive buffer 0. The second element of the data structure is used to determine the frame type (standard or extended) for each receive buffer. The filter masks that are stored in `RecBuff_Mask` must correspond to the selected frame type. The six receive filters (the first two for receive buffer 0 and the other 4 for receive buffer 1) are stored in the array `RecBuff_Filter`. This allows up to six individual ID's or – depending on the mask – parts of them to be filtered out.

In the following code two filters are used which allow messages with IDs 0x123 and 0x789 to pass, by setting the masks to 0x7FF all used bits of the default identifier are covered.

```
can_filter sFilter = {
    /*Rec_Buff0_Rollover*/ ROLLOVER_ON,
    /*RecBuff_ID*/ {STANDARD_ID, STANDARD_ID},
    /*RecBuff_Mask*/ {0x7FF, 0x7FF},
    /*RecBuff_Filter*/ {0x123, 0x7FF, 0x789, 0x7FF, 0x7FF, 0x7FF}};
```

The receive filters that are not used are set to 0x7FF for the standard identifiers, or 0x3FFFF for the extended identifiers. These identifiers may no longer be used in the bus.

If you also want to receive messages with the identifier 0x121, this can be entered in the filter list. Since 0x123 and 0x121 only differ by bit 1, the filter list does not have to be changed; instead, the receive mask of receive buffer 0 is set to 0x7FD. Setting n bits of a receive mask to 0 increases the number of filtered identifiers to $m \cdot 2^n$, where $m = 2$ for receive buffer 0 and $m = 4$ for receive buffer 1.

10.3.2 AT90CANxx

The AT90CAN128 has its own CAN controller, which performs the following tasks according to the CAN specification:

- it implements the Logical Link Control (LLC) sublayer of the link layer,
- the Medium Access Control (MAC) sublayer
- the physical layer implements the Physical Signalling (PLS) sublayer

Physical Medium Attach (PMA) and Medium Dependent Interface (MDI) are not supported, which must be implemented by a transceiver as mentioned above.

The CAN controller is able to process all types of frames (data, remote, error and overload) and achieves a bit rate of 1 Mbit/s. As far as it corresponds to the concept described above, it works completely different:

The controller contains a "mailbox" that can contain up to 15 messages. It is divided into an 8x15-byte (120-byte) buffer, which contains the user data, and a register set, in which all the data required for sending/receiving are stored for 15 so-called message objects (MOb). Figure 10.7 shows the architecture described in [7].

The CAN controller scans the mailbox for messages to be sent or for free message objects in which received messages are stored.

To be able to send or receive data, you have to initialize the MOb. There are five modes:

- In Disabled mode the MOb is not in use (free)
- In Tx-Data/Remote-Frame mode the message must be stored in a free MOb with all fields (i.e. DLC, RTR, IDE, reserved bits, identifier and the data), Sect. 10.1. and a request to send. Then the two CONMOB bits are set with the transmit command 01b. The controller then starts sending the message and acknowledges with a TXOK flag or interrupt.

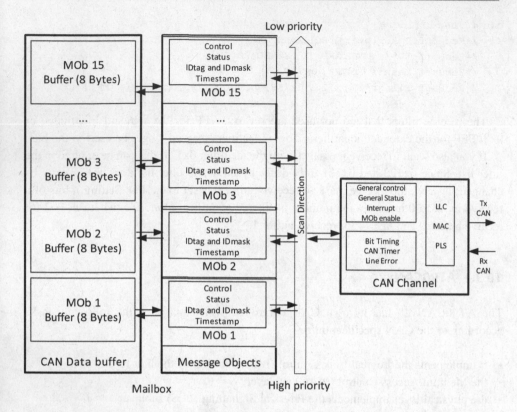

Fig. 10.7 Architecture of the internal CAN controller on the AT90CANx. (According to [7])

- In Rx Data/Remote Frame mode, the complete message (except for the data that is to be received) and an acceptance filter mask must also be written to the MOb. As with the MCP2515, a 1 at the corresponding bit of the filter mask means that this bit in the ID must match the corresponding ID bit in the message to be received. In addition, the CONMOB bits must be set to 10b. If a frame has been received on the bus, the controller scans all available MOBs and writes their data to the MOb with the highest priority that matches the message with ID and mask. Success is acknowledged with an RXOK flag and an interrupt. The handling of RTR requests must be done manually.
- In the Automatic-Reply mode it is possible to preassigned the data of RTR receive messages. As soon as an RTR message is received on the bus, the controller automatically completes the message with the data. The condition is that the Reply valid (RPLV) bit was set. After the successful reply, this bit is automatically cleared. It is therefore very easy to provide asynchronous data here that is requested by a remote client via an RTR message.
- If several messages are to be received on the bus, the frame buffer receive mode can be used. Here all messages are collected and an RXOK or interrupt is only triggered when all have been received.

To write your own CAN library for the AT90CANxx you can't avoid a careful study of the manual [7], this section is only meant to contribute to the first understanding. However, the libraries from [5] or our own library (on the website of the book at the publisher for download) are also helpful here. After good experiences with [5], we decided to take the step of developing our own, because our library is somewhat more oriented towards the specific requirements of the processors, is better adapted for multiple SPI participants, and also, in our opinion, uses the processor somewhat more efficiently, while the one from [5] unifies various controllers and thus does not necessarily exploit them completely. On the other hand, it is faster to use, especially by beginners, and well documented.

10.3.2.1 Implementation on the AT90CANx

The CAN controller of the microcontroller AT90CANxx is switched to standby mode after the POR phase[1] and the message objects (MOb) are initialized randomly. Before the CAN controller is enabled, the MOb must therefore be explicitly initialized!

10.3.2.1.1 Initialization

By calling the function AT90CAN_Init, all MOb are switched to disabled mode and the CAN channel is set to "normal operation" mode after an internally caused time delay. Subsequently, the baud rate is set.

```
uint8_t AT90CAN_Init(uint8_t ucbaud)
{
    AT90CAN_Init_AllBuffer();
    AT90CAN_Set_OpMode(ENABLE_MODE);
    //the normal operation mode is set
    //it waits until this mode is also set
    while(!(AT90CAN_Read_Status() & (1 << ENFG)));
    AT90CAN_Set_Baudrate(ucbaud); //the baud rate is set
    return CAN_OK;
}
```

Each individual MOb can be set as a transmit or receive buffer in one of the modes described above.

```
#define TX_BUFFER_1 0 //Mob 0 is called TX_BUFFER_1
#define RX_BUFFER_1 5 //Mob 5 is called RX_BUFFER_1
#define AUTOREPLAY_BUFFER_1 10 //Mob 10 is called AUTOREPLAY_BUFFER_1
```

The message data structure is defined as follows:

---

[1] Power-On-Reset (see Chap. 3).

```
typedef struct
{
    uint8_t EIDE_Bit; //STANDARD_ID or EXTENDED_ID
    uint8_t RTR_Bit; //DATA_FRAME or REMOTE_FRAME
    uint8_t RB0_Bit; //Reserved bit: 0x00 or 0x01
    uint8_t RB1_Bit; //Reserved bit: 0x00 or 0x02
    uint32_t ulID; //11 or 29 bit ID
    uint8_t ucLength; //Number of data bytes
    uint8_t ucData[8]; //Data vector for the maximum message length
}can_frame;
```

10.3.2.1.2 Sending Messages
The initialization of MOb 0 as the first transmit buffer could look like this:

```
sSendFrame.EIDE_Bit = STANDARD_ID;
sSendFrame.RB0_Bit = 0x00;
sSendFrame.RTR_Bit = DATA_FRAME;
sSendFrame.ucLength = 0x04;
sSendFrame.ulID = 0x77;
AT90CAN_Init_Buffer(TX_BUFFER_1, &sSendFrame);
```

where sSendFrame is of data type can_frame. These elements of the data structure are stored in the MOb page memory. Before the message is sent, the send data is updated and sent with the call:

```
AT90CAN_Send_Message(TX_BUFFER_1, & sSendFrame);
```

the message is ready to be sent.

10.3.2.1.3 Receiving Messages
For a better overview, a data structure sRecframe of the same data type can_frame is declared for data reception. Similar to the initialization of the transmit buffer, MOb 5 is also initialized for the reception of messages with the default identifier 0x101.

```
sRecFrame.EIDE_Bit = STANDARD_ID;
sRecFrame.RB0_Bit = 0x00;
sRecFrame.RTR_Bit = DATA_FRAME;
sRecFrame.ucLength = 0x04;
sRecFrame.ulID = 0x101;
AT90CAN_Init_Buffer(RX_BUFFER_1, &sRecFrame);
```

The stored parameters can be changed with a new initialization. A filter mask is used to enable the reception of messages with multiple identifiers. The mask as a data structure

must first be defined and then saved by calling the function
AT90CAN_Set_ReceiveMask. With the following settings, the buffer
RX_BUFFER_1 receives the CAN messages with the standard identifiers 101 and 103.

```
sMask.EIDE_Bit = STANDARD_ID;
sMask.IDE_MaskBit = 1; //a mask is used
sMask.RTR_MaskBit = 0; //no RTR message
sMask.ulMask = 0x7FD; //with this mask the buffer can receive 2 ID's
//setting the mask reception of the first message is enabled
AT90CAN_Set_ReceiveMask(RX_BUFFER_1, &sMask);
AT90CAN_Enable_Reception(RX_BUFFER_1);
```

The receipt of a new message in a mailbox is signaled by setting the RXOK bit in the
CANSTMOB register. A CAN interrupt can be triggered if it has been set accordingly. The
function AT90CAN_Receive_Message checks the bit RXOK of the desired mailbox
and if it is set, it returns CAN_RECEIVED and stores the new message in a data structure of
type can_frame.

```
if(AT90CAN_Receive_Message(RX_BUFFER_1, &sRecFrame) == CAN_RECEIVED)
{
    //evaluation of the message
}
```

10.3.3 Implementation with the CAN Library of the Robot Club Aachen

With this library [5], the CAN controller to be used must first be selected in config.h via
#defines. Here we show the procedure using the MCP2515. The initialization is done with
the call

```
typedef enum {
    BITRATE_10_KBPS = 0,
    BITRATE_20_KBPS = 1,
    BITRATE_50_KBPS = 2,
    BITRATE_100_KBPS = 3,
    BITRATE_125_KBPS = 4,
    BITRATE_250_KBPS = 5,
    BITRATE_500_KBPS = 6,
    BITRATE_1_MBPS = 7,
} can_bitrate_t;
can_init(BITRATE_250_KBPS);
```

The message data structure is defined as follows:

```
typedef struct
{
    uint16_t id; //ID of the message (11 bit)
    struct {
        int rtr : 1; //Remote-Transmit-Request-Frame?
    } flags;
    uint8_t length; //number of data bytes
    uint8_t data[8]; //data of the CAN message
} can_t;
```

For extended identifier messages (29 bits) the identifier is of type `unsigned long` and a further flag `int extended : 1` is added to the flag structure. Sometimes a timestamp (16 bits are usually sufficient) is also stored, but this is not supported by the MCP2515 hardware and must then be sent as part of the message.

```
typedef struct
{
    unsigned long id; //ID of the message (11 bit)
    struct {
        int rtr : 1; //Remote-Transmit-Request-Frame
        int extended : 1; //the ID is extended (29 bit)
    } flags;
    uint8_t length; //number of data bytes
    uint8_t data[8]; //data of the CAN message
    unsigned long timestamp //time stamp
} can_t;
```

Sending a message is done as follows:

```
can_t mymsg ;
mymsg.length = 4;
mymsg.id = 0x123;
mymsg.data[0] = 0;
mymsg.data[1] = 1;
mymsg.data[2] = 2;
mymsg.data[3] = 3;
can_send_message(&mymsg);
```

Here, a message with the identifier 0x123 and a length of four bytes with the content 0, 1, 2, 3 is sent. A simple sensor circuit that sends a sensor value every 100 ms only has to transmit a send command with the corresponding data in a 100 ms task.

Receiving messages in polling mode is the simplest form of message reception here:

```
char res;
if (can_check_message())
{
    res=can_get_message(&recmsg); //FALSE if no message
                                  //is present, otherwise
                                  //the filter code by which the
                                  //message has been accepted
    if (res)
    {
        //here you can process the message
    }
}
```

The simplest way to realize a filter operation is to preassign the filters statically. In the cited library there is the possibility to write all filter values and those of the two masks one after the other into an array and to use this for programming the filters. In the following code two filters are used, which allow messages with the IDs 0x123 and 0x789 to pass, by setting the masks to 0x7FF all used bits of the standard identifier are covered.

```
const uint8_t can_filter[] PROGMEM =
{
// Group 0
    MCP2515_FILTER(0x123), //filter 0
    MCP2515_FILTER(0x0), //filter 1
//Group 1
    MCP2515_FILTER(0x789), //filter 2
    MCP2515_FILTER(0x0), //filter 3
    MCP2515_FILTER(0x0), //filter 4
    MCP2515_FILTER(0x0), //filter 5
    MCP2515_FILTER(0x7ff), //mask 0 (for group 0)
    MCP2515_FILTER(0x7ff), //mask 1 (for group 1)
};
can_static_filter(can_filter); //sets the filter
```

The macros MCP2515_FILTER only help here to distribute the bytes of the identifiers correctly (the three MSB of the identifier must be in a separate byte). The function can_static_filter() requires that this array is in the flash (PROGMEM, see Sect. 3.12).

Setting individual filters at program runtime requires somewhat more program code, since more case distinctions are necessary here for programming. The function can_set_filter() is available for this purpose. It is called, as shown in the following example, with the number of the filter and a pointer to a structure can_filter_t. Static and dynamic filtering also work together.

```
can_filter_t cf;
(...)
cf.id = 0x654;
cf.mask = 0x7ff;
can_set_filter(3, &cf);
```

The receive function `can_get_message()` returns the number of the filter from 1 ... 6, while the quoted filter function numbers the filters from 0 ... 5. This may have to be taken into account.

10.4   CAN Transport Protocol

The CAN transport protocol according to ISO 15765-2 represents a transport layer (see Sect. 6.1.4) whose tasks are as follows

- Transmission of large data blocks (segmentation),
- Flow Control,
- Abstraction of the transport layers from the application layers.

These tasks are solved relatively simply in the CAN-ISO-TP, as it is also called. The idea is that the application sends large data blocks (i.e. larger than eight bytes) like a single message, the segmentation is handled by the ISO-TP layer.

For this purpose, the protocol provides four different message types within the data frame of eight bytes: Single Frame, First Frame, Consecutive Frame, Flow Control Frame.

Basically, the message identifier is used to code whether the message is a "normal" Layer 2 CAN message or an ISO TP message, i.e. the sender and receiver must agree on this beforehand. In an ISO-TP message, the first four bits are used to encode the frame type. Figure 10.8 shows the currently specified frames. If these four bits are 0, the message is interpreted as a single frame in which seven user data bytes are still available. In the lower four bits of the message the number (1–7) of user data bytes is encoded, whereas in all ISO TP messages the DLE byte is set to eight.

If a segmented message is now to be sent, the sender initiates the transmission with a "First Frame", in whose first byte the uppermost four bits carry the number 0001 (binary) and the lower four bits together with the second byte determine the total length of the message. Since the length field has 12 bits, messages up to a length of 4095 can be segmented with it. The slave responds with a flow control frame, which is provided with 0011 (binary) in the upper nibble (=four bits) of the first byte. The second nibble encodes the flow state, this can contain Continue to Send (0x00) or Wait (0x01). The second byte contains the Block Size (BS = 1 ... 255), i.e. the number of further CAN messages that can be received immediately (buffer size of the receiver). BS = 0 means any number of blocks. The Separation Time specifies the minimum time in milliseconds that the transmitter must wait between two messages.

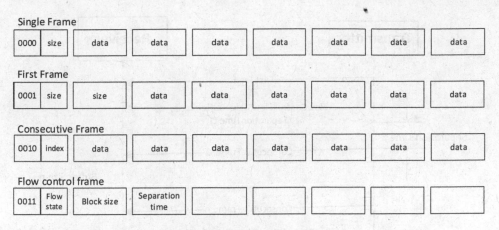

Fig. 10.8 Frames of the ISO transport protocol. (After [4, 8])

After the receiver response, the transmitter now sends the Consecutive Frames according to the agreements now made, here a 0010 is coded in the upper nibble and a sequence number in the lower one, which can be used to ensure the sequence correctness of the messages after reception (also in the case of error handling as described in Chap. 6. After the agreed number of messages (block size), the transmitter waits until the receiver acknowledges with a new flow control frame and then continues the transmission. Figure 10.9 summarizes the communication flow as a ULM sequence diagram).

According to the standard, data blocks that do not fill the CAN message must be padded to avoid long zero sequences.

Example:

The sender wants to send the 24 byte long message: 0x01 0x02 0x03 0x04 0x05 0x06 0x07 0x08 0x09 0x0A 0x0B 0x0C 0x0D 0x0E 0x0F 0x10 0x11 0x12 0x13 0x14 0x15 0x16 0x17 0x18.

The First Frame is: 0x10 0x18 0x01 0x02 0x03 0x04 0x05 0x06, where the 0x10 0x18 at the beginning means: First Frame of length 0x018 (so 24).

The receiver responds with 0x30 0x00 0x00 (i.e. Flow Control Frame Continue, any number of messages, no minimum waiting time).

Then the sender sends: 0x21 0x07 0x08 0x09 0x0A 0x0B 0x0C 0x0D (i.e. first consecutive frame) with the corresponding next data. The other messages are:

0x22 0x0E 0x0F 0x10 0x11 0x12 0x13 0x14 and 0x23 0x15 0x16 0x17 0x18 0xAA 0xAA 0xAA (i.e., the following frames counted on). The padding is applied in the last frame.

The ISO-TP is part of every diagnostic system in the vehicle. It is extended by alternative transport protocols (e.g. VW's own). It is an integral part of the widely used AUTOSAR middleware [9], lean implementations can be found on github [10].

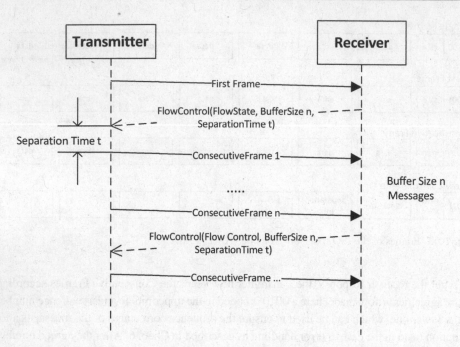

Fig. 10.9 Sequence of sending a segmented message in the ISO-TP. (According to [4, 8])

10.5   CANopen in Industrial Control Technology

For industrial control technology, there is a communication protocol called CANopen based on the CAN data link layer [11]. CANopen was developed as a standardized embedded network with highly flexible configuration options as an open standard, especially for medium-sized companies. It was originally designed for motion-oriented machine controls, such as handling systems. Today, it is used in various application areas, including medical devices, off-road vehicles, marine electronics, railway applications or in building automation or elevator technology.

The CiA (CAN in Automation) association, based in Nuremberg, is responsible for defining and updating the standards, each of which bears the name CiA followed by a number. CANopen provides communication objects, which are divided as follows:

- Service data objects (SDO) for the parameterization of object dictionary entries,
- Process data objects (PDO) for the transport of real-time data,
- Network Management Objects (NMT) for controlling the state machine of the CANopen device and for monitoring the nodes,
- other objects such as synchronization object (SYNC), time stamp and error messages (EMCY).

The Object Dictionary (OD) mentioned here is the heart of a CANopen stack. It contains all communication objects and all user objects. In the CANopen device model, the OV is the link between the application and the CANopen communication unit. This means that the application works with process parameters, sends them to the protocol stack, which generates a CAN message with the help of the OV. Conversely, the protocol stack will decode a received message using the object dictionary and pass it on to the application as a process parameter. In addition, the OV also contains the entire parameterization of the communication process.

The NMT master ensures that all devices in the network are logged on and ready for operation. It usually also sends a heartbeat (a recurring message) to synchronize the participants. In the meantime, a number of open CANopen stacks exist that can be downloaded from github (for example [12, 13]). A self-development is usually not worthwhile. Since the stack is quite data-intensive, it should be used on a larger system (ideally an embedded Linux system).

In [14, 15] an application for elevators (CiA417) is described.

References

1. NXP: CAN Bosch Controller Area Network (CAN) Version 2.0 PROTOCOL STANDARD. http://www.nxp.com/assets/documents/data/en/reference-manuals/BCANPSV2.pdf. Accessed 1 Apr 2021.
2. ISO 11898: Road vehicles – Controller area network (CAN) (6 Teile).
3. Etschberger, K. (Ed.). (1994). *CAN-Controller area network – Grundlagen, protokolle, bausteine, anwendungen*. Hanser.
4. Zimmermann, W., & Schmidgall, R. (2014). *Bussysteme in der Fahrzeugtechnik – Protokolle, Standards und Softwarearchitektur* (5. Aufl.). Springer.
5. Greif, F., & Aachen, R. (2008). Universelle CAN Bibliothek. http://www.kreatives-chaos.com/artikel/universelle-can-bibliothek. Accessed: Jan. 2021.
6. Microchip Technology Inc. MCP2515 Datasheet. https://www.microchip.com/wwwproducts/en/en010406. Accessed 10 Jan 2021.
7. Microchip Technology Inc. 8-bit Microcontroller with 32K/64K/128K Bytes of ISP Flash and CAN-Controller. (2015). https://www.microchip.com/wwwproducts/en/AT90CAN128. Accessed 10 Jan 2021.
8. ISO: *ISO 15765-2:2016*. 15765-2:2011 Road vehicles – Diagnostic communication over Controller Area Network (DoCAN) – Part 2: Transport protocol and network layer services 2016. Auflage. https://www.iso.org/standard/66574.html. Accessed 9 Jan 2021.
9. AUTOSAR: Classic Platform. https://www.autosar.org/standards/classic-platform/. Accessed 10 Jan 2021.
10. Peplin, C., & Boll, D. (Ford): ISO-TP (ISO 15765-2) Support Library in C. https://github.com/openxc/isotp-c/tree/master/src/isotp. Accessed 23 Jan 2021.
11. CiA: CANopen Protocol. https://www.can-cia.org/canopen/. Accessed 1 Feb 2021.
12. CAN Festival: CAN Festival free CANopen framework. https://canfestival.org/. Accessed 1 Feb 2021.
13. https://github.com/CANopenNode/CANopenNode. Accessed 1 Feb 2021.

14. Sußmann, N., & Meroth, A. (2017). "Model based development and verification of CANopen components" 22nd IEEE International Conference on Emerging Technologies And Factory Automation ETFA 2017, September 12–15, Limassol, Cyprus.
15. CANopen Lift. http://de.canopen-lift.org. Accessed 1 Feb 2021.

The Modbus

11

Abstract

This chapter describes some aspects of Modbus, which has become widely used as an industrial fieldbus.

The MODBUS standard specifies the application layer (the 7th layer of the OSI model) and partially the link layer of the communication protocol between networked devices [1]. Because of its versatility, it has been widely used in industry. The Modbus organization advises and shapes the standards around Modbus.

Communication can take place either via TCP/IP (IEC 61.158), or via a serial, asynchronous interface such as TIA[1]/EIA[2]-232 or TIA/EIA-485 [2]. Recently, a secure variant of the Modbus/TCP protocol based on Transport Layer Security (TLS) has also been published, which uses digital X.509v3 certificates to authenticate server and client, and role-based access control is also proposed.

Control devices with higher performance are usually connected via Modbus, so that the TCP/IP variant is usually preferable. This is omitted in this book due to the low performance of the processors of the AVR family described here. Nevertheless, we would like to present the variant with ASCII transmission here, since the protocol can be used to connect sensors with lower computer performance in a Modbus system and to feed the frame of the application layer transparently into the system via an intermediate gateway.

---

[1] TIA – Telecommunications Industry Association.

[2] EIA – Electronic Industry Alliance.

© The Author(s), under exclusive license to Springer Fachmedien Wiesbaden GmbH,
part of Springer Nature 2023
A. Meroth, P. Sora, *Sensor networks in theory and practice*,
https://doi.org/10.1007/978-3-658-39576-6_11

For the serial protocol, MODBUS defines the data link layer of the OSI model. A master-slave bus with up to 247 slaves is specified. Only the master can initiate a communication session with a slave, or address all slaves with a broadcast using the global address 0x00. The bus can also be operated as a multi-master under the condition that only one master is active at any time. Data transfer between a master and a single slave that is not too far away spatially (point-to-point connection) at a low transmission rate can take place via TIA/EIA-232, otherwise via TIA/EIA-485. In the scenario briefly described here, a microcontroller operates with one or more local sensors connected via I^2C, for example, preprocesses the data and transmits it in a MODBUS protocol frame via TIA/EIA-232 to the master, which in turn is connected to the control or process control level via the TCP/Modbus protocol.

11.1 TIA/EIA-485 as Physical Layer for MODBUS

The industrial standard TIA/EIA-485, also known as RS-485, was developed as a successor to RS[3]-232 (officially TIA/EIA-232) to achieve higher data rates (<50 Mbit/s), to bridge longer distances (<1200 m), to be less sensitive to interference than an interface based on voltage levels, and to allow networking of several devices on one physical channel. RS-485 is a serial interface that implements the first layer of the OSI model (see Sect. 6.1.1). Bit transmission is in reverse mode (full duplex) or alternating mode (half duplex) and is balanced and asynchronous. Balanced transmission uses two signal lines, which should be twisted and may be shielded. The voltages $U_A(t)$ and $U_B(t)$, with $U_B(t) = -U_A(t)$ are transmitted. The receiver evaluates the voltage $U_{AB} = U_A - U_B$ and reconstructs both the bit clock and the bit sequence. The amplitude of the differential voltage U_{AB} must be greater than 200 mV. An interference voltage $U_q(t)$ has an additive effect on the two signal voltages during transmission. Balanced transmission largely suppresses the common-mode interference (Fig. 11.1) that occurs on the signal lines and improves the bit error rate.

RS-485 theoretically allows the connection of several devices or sensors that do not have a common reference potential (ground). In practice, connecting the grounds via an electrical line is recommended. It is also recommended that all devices are connected directly to the signal bus (daisy-chain topology), or by means of short stubs as in Fig. 11.2. The data bus should be terminated with terminating resistors (RT). These resistors correspond to the wave impedance of the lines used. They prevent electrical reflections from occurring at the end of the line, which could lead to data corruption. Further protective measures against overvoltage, open circuit and short circuit can be found in [3].

---

[3] RS – Recommended Standard.

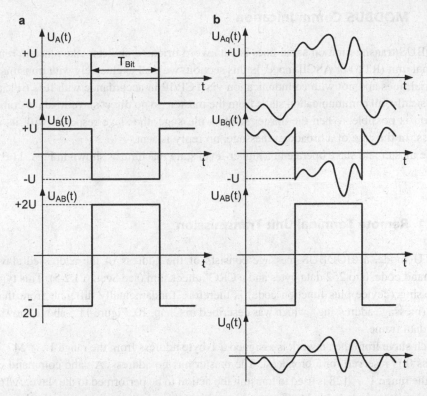

Fig. 11.1 (a) Balanced transmission without interference; (b) Balanced transmission with common mode interference

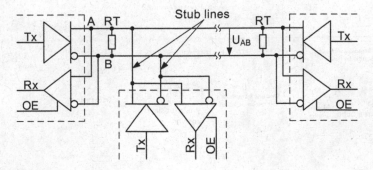

Fig. 11.2 MODBUS network topology

11.2   MODBUS Communication

MODBUS transmission on serial lines is data word oriented and can take place in remote terminal unit (RTU) or ASCII mode. In this section, we deal exclusively with transmission on serial lines and not with communication via TCP/IP in accordance with IEC 61158.

Basically, all communication starts from the master, so no direct slave-to-slave communication is possible. When the master sends a message, the slave responds with its own address. In the case of a broadcast message, no reply is sent.

The master and slave operate in a higher-level state machine as shown in Figs. 11.3 and 11.4 [2].

11.2.1   Remote Terminal Unit Transmission

In RTU mode, a MODBUS message consists of the address of the addressed slave, a command code, 0 to 252 data bytes and a CRC[4] checksum (see Sect. 6.1.2.5). This type of addressing (device plus function code) is therefore fundamentally different from that of CAN (message addressing), which was described in Chap. 10. Figure 11.5 above shows the RTU data frame.

Each slave from the network is assigned a 1-byte address from the range 1 ... 247. The address 0 is reserved for a broadcast. The master has no address. A valid command code from the range 1 ... 128 is used to transmit the action to be performed to the slave. A 16-bit CRC checksum is appended to the message, which, in addition to the parity check, is used to determine the data integrity.

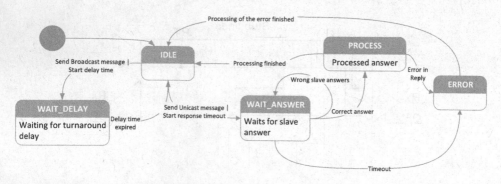

Fig. 11.3 State machine of the MODBUS master

---

[4]CRC – cyclic redundancy check.

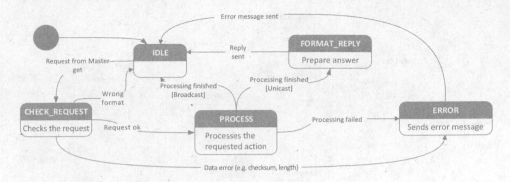

Fig. 11.4 State machine of the MODBUS slave

Slave-Address	Function code	Data	CRC low	CRC high	
1 Byte	1 Byte	1 Byte ... 252 Bytes	1 Byte	1 Byte	RTU Frame

Start :	Slave-Address	Function code	Data	LRC	End CR - LF	
1 char	2 char	2 char	0... 2 x 252 char	2 char	2 char	ASCII Frame

Fig. 11.5 Modbus frames in serial mode (top RTU, bottom ASCII)

A data word is always 11 bits in size in RTU mode and is similar in structure to that of the UART interface (Chap. 7). The frame starts with a start bit, followed by eight data bits, a parity bit and a stop bit. The transmission of the data byte always starts with the least significant bit first. The default setting is even parity. The MODBUS specification requires that odd parity should also be adjustable. If the parity bit is to be omitted completely, the data frame must be terminated with two stop bits. The parity bit is calculated in the same way as in Chap. 7 and can thus be executed via UART hardware.

Alternative structure of the data word in RTU mode.

Start	Bit 1	Bit 2	Bit 3	Bit 4	Bit 5	Bit 6	Bit 7	Bit 8	Parity	Stop
Start	Bit 1	Bit 2	Bit 3	Bit 4	Bit 5	Bit 6	Bit 7	Bit 8	Stop	Stop

The baud rate and the structure of the data word must be set the same for all bus participants of a network so that communication can take place. In RTU mode, the master can send a message if there has been no bus activity for at least 3.5 bytes beforehand. If this waiting period between two messages is not observed, the slaves regard the new message as an erroneous continuation of the previous one. A frame start byte is therefore not provided,

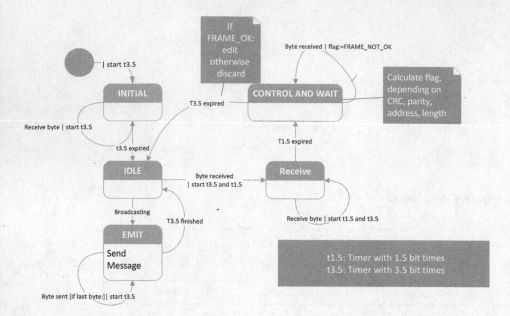

Fig. 11.6 State machine with timings for the sequence of master and slave communication

since the frame start results from this bus timing (3.5 byte lengths of waiting time before the frame start), thus eliminating the need to establish code transparency (see Sect. 6.1.2.1).

When transmitting a message, a waiting time is allowed between the individual bytes, which must not be greater than the transmission time of 1.5 bytes, otherwise an abort occurs and the frame is discarded (FRAME_NOT_OK in Fig. 11.6). After a broadcast, the master does not wait for a response, but delays sending the next message to allow the slaves to decode and execute the command. A directly addressed slave responds to the master with a message of similar structure. It sends its own address first to identify itself, followed by the received command code or an error code. The master checks whether the response of the slave it addressed comes back and the integrity of this message (parity plus checksum). In case of error, it can send the message again Fig. 11.6.

11.2.2 ASCII Transmission

In ASCII mode, the message to be transmitted is formed as in RTU mode. Instead of the CRC checksum, however, a 1-byte longitudinal parity checksum[5] is calculated and appended. Detailed examples of the calculation of the two checksums are in [2]. In this mode, each byte of the basic message is assigned two ASCII characters ('0', '1'...'9',

---

[5]LRC – longitudinal redundancy check.

'A'...'F') from the 7-bit ASCII table. These characters encode the hexadecimal values
(0 ... F) of the high-order and low-order nibble of a byte. Example: Byte 0x5B is encoded
as two characters: 0x35 and 0x42 (0x35 = "5", and 0x42 = "B" in ASCII). This doubles
the number of bytes to be transmitted and the net data rate becomes smaller than with RTU
mode. The code of the higher order nibble is transmitted first. A data word always consists
of ten bits and contains only seven data bits (ASCII).

Alternative structure of the data word in ASCII mode.

Start	Bit 1	Bit 2	Bit 3	Bit 4	Bit 5	Bit 6	Bit 7	Parity	Stop
Start	Bit 1	Bit 2	Bit 3	Bit 4	Bit 5	Bit 6	Bit 7	Stop	Stop

The following program excerpt converts the basic message, consisting of the slave
address, uiLength data bytes and the LRC checksum, stored in the variable
ucBasicFrame into an ASCII encoded message.

```
uint8_t ucTemp;
for(uint8_t ucI = 0; ucI < (uiLength + 2); ucI++)
{
    //the higher-order nibble of the current byte is encoded
    ucTemp = ucBasicFrame[ucI] & 0xF0;
    ucTemp = ucTemp >> 4;
    //if the value is less than 10, it is coded with the ASCII code of the digit
    if(ucTemp < 0x0A) ucCodedFrame[2 * ucI] = ucTemp + 0x30;
    //otherwise, with the letters from A to F
    else ucCodedFrame[2 * ucI] = ucTemp + 0x37;
    //the least significant nibble of the current byte is encoded
    ucTemp = ucBasicFrame[ucI] & 0x0F;
    if(ucTemp < 0x0A) ucCodedFrame[(2 * ucI) + 1] = ucTemp + 0x30;
    else ucCodedFrame[(2 * ucI) + 1] = ucTemp + 0x37;
}
```

A colon character ':' is sent as the start character of a new message. The coded basic
message is transmitted character by character and terminated with the character sequence
'CR'[6] and 'LF'.[7] The waiting time between the transmission of two characters shall be
adjustable (see also Fig. 11.5 below). Both the master and the slave implement a similar
state diagram in ASCII mode as in RTU mode. However, no timings are specified by using
the frame start (:) and frame end identifier (CR LF) (see Fig. 11.7). The requirement for
code transparency does not apply, since only the hexadecimal ASCII characters are used
for data transmission.

---

[6]CR – carriage return = 0x0D (carriage return).

[7]LF – line feed = 0x0A (line break).

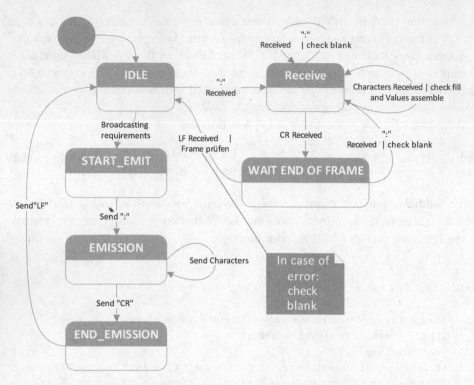

Fig. 11.7 State machine master and slave on byte level (ASCII mode)

Figure 11.8 also shows a flow chart of such a data transmission. After a message is completely received, the parity of each character is checked. By decoding the message, the basic message is restored, the address and the longitudinal parity checksum are compared.

Finally, the formation of the LRC or length parity checksum should be mentioned. This is calculated by modulo adding successive bytes (except the ":" and the CR LF) in the message, i.e. all carries are discarded. This corresponds to a continuous XOR of all transmitted bytes, where a byte consists of 7 bits of value and 1 bit of parity. In principle, the LRC is a simple parity check of all respective first, second, third, etc. bits of a message. Consequently, when this sum is added to the message, a zero must result when the LRC is recalculated. The combination of parity bit of a byte and longitudinal parity can therefore be used to locate and correct a single bit error. In MODBUS, the LRC is transmitted in two's complement. The specification proposes the following code to form the LRC, it returns the LRC. `alsoMsg` contains the message without ":" and CR LF.

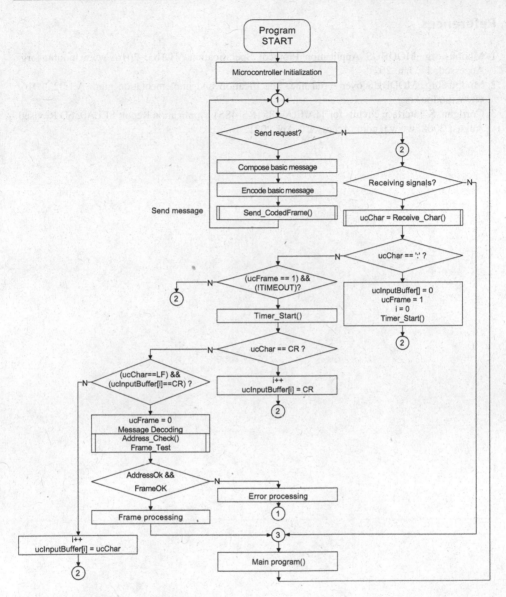

Fig. 11.8 MODBUS flow diagram of ASCII mode

```
static uint8_t LRC(alsoMsg, usDataLen)
uint8_t *alsoMsg ; /* message (message buffer)*/
unsigned short usDataLen ; /* length of message */
{
    uint8_t uchLRC = 0; /* LRC is initialized */
    while (usDataLen-) /* pass through message buffer */
        uchLRC += *alsoMsg++ ; /* add the characters */
    return ((uint8_t)(-((char)uchLRC))); /* two's complement*/
}
```

References

1. Modbus.org. MODBUS Application Protocol Specification V1.1b3, 2016. www.modbus.org. Accessed: 11. Jan. 2021.
2. Modbus.org. MODBUS over serial line. Specification and implementation guide V1.02, 2016. www.modbus.org
3. Corrigan, S. Interface circuits for TIA/EIA-485 (RS-485). Application Report SLLA036D Revised August 2008. www.ti.com. Accessed 1 Apr 2021.

Single-Wire Bus Systems

12

Abstract

This chapter describes other single-wire bus systems that can be used for sensor networks.

Single-wire bus systems are becoming increasingly popular especially for subsystems (sub-buses) in industrial technology, building services and vehicles due to their very simple and cost-effective wiring. They are partly based on the UART. We have selected the 1-Wire® bus and UNI/O® systems here for a more detailed description and mentioned the LIN bus at least in such a way that a basic understanding can be acquired; more detailed information can be found in the further literature.

The single-wire bus systems described here are asynchronous master-slave buses that require only one data line for communication, since no clock is transmitted and communication is bidirectional in half-duplex mode. A single bus master controls the data flow in a network of this type. The bus master initiates all communication, which usually takes place between the master and the addressed slave. Some protocols also allow communication between the master and all slaves, but direct data exchange between slaves is usually not supported. The components that implement such protocols are easy to integrate in terms of circuitry thanks to their simple design. Many slaves require only three connections: Ground, supply voltage and data line. On some buses, the bus master can even supply the slaves with power through the data line.

This simple hardware configuration makes it difficult or impossible to distinguish between identical slaves connected to the same bus – in this case, additional pins would

A. Meroth, P. Sora, *Sensor networks in theory and practice*,
https://doi.org/10.1007/978-3-658-39576-6_12

have to be provided for coding on the hardware side. For these slaves, the address must be burned into the components during the manufacturing process, or the use of slaves of the same type is not possible. Some sensors are delivered from the catalog with different addresses (as is the case with I^2C/TWI). Asynchronous transmission requires tighter time constraints from the master when building up the individual bits.

12.1   1-Wire®-BUS

1-Wire® is a proprietary bus protocol designed by Dallas Semiconductor, now Maxim Integrated [1], for the transmission of smaller amounts of data with longer transmission pauses for short distances, for example for operation in control cabinets. The protocol allows half-duplex communication between a master with up to 2^{64} slaves. To make the bus as simple as possible, the slaves do not require any external components and are identified by an internal, 64-bit long, unchangeable identification code (ID). The ID consists of an 8-bit family code, a 48-bit serial number (unique device ID) and an 8-bit CRC checksum, so that all connected nodes can be uniquely identified (see Sect. 12.2.4.2). The transmitted pulses are unipolar and NRZ coded, the communication takes place on a single data line. The bus line is connected to the supply voltage via a pull-up resistor. Sensors, A/D converters, EEPROMs, etc. have been developed for this bus.

12.1.1 Network Topology

A minimal configuration of a 1-Wire circuit is shown in Fig. 12.1a. The slave is supplied via the VDD connection, and data is exchanged via pin DQ. The master should preferably have an open-drain or a bidirectional connection for this bus. An internal wiring of the slaves allows a so-called "parasitic" power supply. In this case, an additional supply line is omitted as in Fig. 12.1b or even the supply pin as in Fig. 12.1c. During the recessive phase of the bus (high level), the internal buffer capacitor can be charged via the bus resistor. The size of the bus resistor must be adjusted to the power requirements of the slaves. In the case of parasitic supply, for processes that require more energy, such as storing in an EEPROM, or keeping the voltage constant over a longer period of time (drift behaviour of sensors!), the supply voltage is switched to the data line for a short time via a digital switch, as in Fig. 12.1b,c, in order to freshly charge the capacitor. During this time, no communication can take place via the bus.

The wiring of the individual slaves is very simple, as they do not require any additional components, not even a block capacitor on the supply pin. The 1-Wire nodes can be arranged as a star or bus with stub lines up to 3 m long [2]. Communication with reduced transmission rate with slaves supplied with 5 V also works over longer lines (see [2]). This enables low-cost networking of 1-Wire sensors in large rooms or even in an entire house. In order for the network communication to function without interference, attention must be

Fig. 12.1 (a) Minimum 1-Wire configuration, (b) "Parasitic" supply of a slave with supply pin, (c) "Parasitic" supply of a slave without supply pin

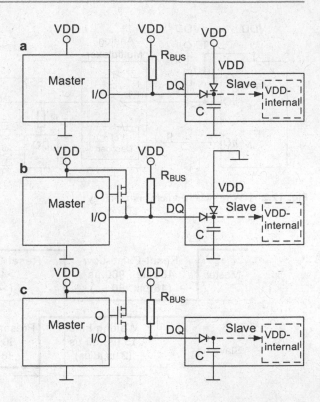

paid to the exact design of the network. Different lengths of the data lines to the slaves lead to different reaction times of the slaves, e.g. during bus initialisation (see Sect. 12.1.2) or when searching for nodes. With star topology, reflections at the end of the line can also falsify the signals.

Complex, mixed networks often lead to transmission errors. Figure 12.2 shows a switched network topology [2] that realizes a good compromise between network complexity and transmission reliability. The master switches one network branch at a time through the decoder, which in this case has a simple bus topology. The master needs one I/O pin for data communication and $q = \log_2 n$ outputs for selecting the n branches. The slaves can be supplied either via VDD, or "parasitically". Resistors R connected in series with the slaves are recommended to minimize the influence of reflections at the end of the line.

12.1.2 Initialization of the Bus

Before each communication session, the master must initialize the slaves that are connected to the bus. This process is shown in Fig. 12.3. The master generates a Reset Pulse Low for at least 480 µs and then releases the bus. For the next 480 µs, the master waits for the slaves

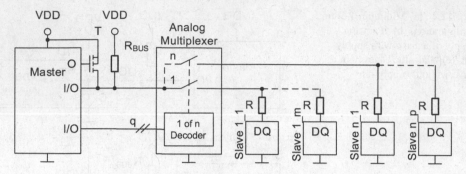

Fig. 12.2 1-Wire switched network topology

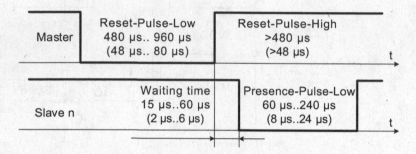

Fig. 12.3 1-Wire bus initialization

to respond. For an ATmega family microcontroller, the bus can be enabled by switching the controlling pin from output to input and still acting on the internal pull-up resistor, which is turned on by setting the PORT register when the corresponding DDR register is set to input. This allows the bus capacitance to be charged through the pull-up resistor. After the rising edge of the Reset Pulse High, the operational slaves generate a so-called Presence Pulse Low after a waiting period and then also release the line. Taking into account the speed with which the slaves respond and the distance to the master (propagation time), a time span of 15 µs ... 60 µs is provided for the waiting time and a time span of 60 µs ... 240 µs for the Presence Pulse Low. For the slaves that enable a faster transmission rate, the times from the brackets apply.

A possible initialization routine can use either a timer or wait functions tuned to the frequency of the microcontroller' quartz frequency for timing. This routine should report the initialization as unsuccessful for slaves with a slow data rate:

- no Presence Pulse Low was detected during the Reset Pulse High;
- the waiting time before the Presence Pulse Low is not in the range 15 µ ... 60 µs;
- the duration of the Presence Pulse Low is not in the range 60 µs ... 240 µs.

12.1.3 1-Wire Bit Transmission

The transmission of a byte always starts with the least significant bit first. The 1-Wire bus protocol provides a synchronous time-division multiplexing for the bit transmission. The duration of a bit is fixed at 60 µs for the normal transmission speed. In the first time window with a duration of 15 µs, the slaves can place their bit on the bus, the second time window with a duration of 45 µs is reserved for the bit transmitted by the master. With a recovery time – waiting time between two bits – of at least 1 µs, the transmission rate can reach a maximum of 16.3 kBit/s. The master synchronizes the transmission with a Start Pulse (low level) of at least 1 µs. This initiates the transmission of each bit. After the Start Pulse, the master normally releases the bus. The signal sequences of the individual bits depend on their logical values and the direction of transmission, as shown in Fig. 12.4. The times indicated in the brackets relate to transmission at increased speed, which in the ideal case can reach 142 kBit/s. This high transmission speed is not supported by all components. The deviating times, which are noted in the data sheets of the respective components (e.g [4].), must also be taken into account.

The master starts the transmission of a "1" with a Start Pulse, after which it leaves the bus line free again (recessive high). The bus capacitance is charged via the pull-up resistor within the first time window and the slaves sample the signal in the second (Fig. 12.4a). To send a logical "0", the master continues to hold the bus line low for the entire bit duration after the Start Pulse (Fig. 12.4b). The master requests the slaves to send data by so-called ROM commands (see Sect. 12.1.4.1). The master must initiate each requested bit with a Start Pulse. For the transmission of a logical "1", the slave does not access the data line after the Start Pulse, so that it remains immediately in the high level (Fig. 12.4c). If a slave wants to send a "0", it holds the signal low after the Start Pulse until the end of the first time window and then releases the bus line (Fig. 12.4d). As can be seen in Fig. 12.4, the bus line is predominantly high, which leads to the charging of the internal capacity of the slaves and favors the parasitic supply. This is referred to as the protocol not being dc-free, which is normally to be avoided but is useful here. Problematic in the case of a parasitic supply is the transmission of long zero sequences in both directions, since this discharges the internal capacitances and thus cuts off the slave from the supply.

The master must ensure that the signal level remains constant throughout the slave time window. It is recommended that the master samples the bus signal in the last third of the master time window to avoid errors due to a too high time constant (settling) of the bus.

Any I/O pin of an ATmega family microcontroller can be used to control a 1-Wire bus. A possible hardware abstraction for pin 2 of port B which should control a 1-Wire bus is:

```
#define _1_WIRE_DDR_REG DDRB
#define _1_WIRE_PORT_REG PORTB
#define _1_WIRE_PIN_REG PINB
#define _1_WIRE_BIT PB2
```

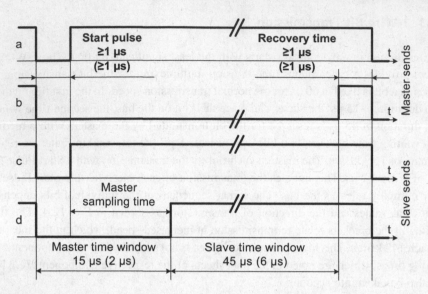

Fig. 12.4 1-Wire bit timing. (**a**) Master sends a "1", (**b**) Master sends a "0", (**c**) Slave responds with a "1", (**d**) Slave responds with a "0"

With the presented hardware abstraction the master can realize the control of the bus, respectively the sampling of the bus signal as follows:

```
//initialization of the PORT register
_1_WIRE_PORT_REG &= ~(1 << _1_WIRE_BIT);
    //the port is declared on output; with the corresponding bit
    //in the PORT register set to 0, the bus level is set to low
_1_WIRE_DDR_REG |= 1 << _1_WIRE_BIT;
    //the port is declared on input which puts it in the high impedance
    //state. This enables the bus and charges the bus capacitance
_1_WIRE_DDR_REG &= ~(1 << _1_WIRE_BIT);
    //with the following instruction the bus signal is sampled
ucBit = _1_WIRE_PIN_REG & (1 << _1_WIRE_BIT);
```

12.1.4 Communication Session

The communication of the master with the slaves generally takes place either with so-called ROM commands or with function commands. The function commands cause the slave to perform an action or to send a response message. The ROM commands have the task of addressing the slaves (a function command follows) or reading the ID of the slaves.

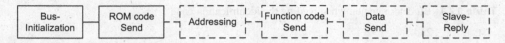

Fig. 12.5 1-Wire communication session

As shown in Fig. 12.5, the master sends a ROM command immediately after successful bus initialization (see Sect. 12.1.2).

In a bus with several nodes, the master must know their IDs in order to be able to address them individually. These IDs may not be known, especially during commissioning, and must then first be determined in a broadcast call.

An example with temperature sensors distributed in a control cabinet is intended to illustrate the problems involved in commissioning the bus. The sensors are installed at different locations in the control cabinet and have unique IDs, but it is not easy to see which ID belongs to which sensor. Since the temperature must be associated with a location, it makes sense to install the sensors one after the other and to teach them to the master.

12.1.4.1 ROM Commands
A distinction is made between general and specific ROM commands. The general commands, which are described briefly below, are implemented in all 1-Wire slaves, while the specific ones are only implemented in certain families of components.

12.1.4.1.1   Search ROM (Code 0xF0)
This ROM command requests the nodes to send their identification codes. After successful bus initialization and sending of the Search ROM code, the master initiates sending of each bit and all nodes send simultaneously, starting with the least significant bit. Simultaneous transmission of recessive ("1") and dominant ("0") bits, due to the wired-AND circuitry of the slaves, results in bus collisions. Unlike CAN, the 1-Wire protocol does not rely on collision avoidance, but on collision detection. The solution of these collisions takes place in a complex, repetitive process, which is described in detail in [1]. To detect the transmission collisions, the master requests for each participant to send each of the 64 bits twice. The first time the participants send the real bit and the second time they send the inverted bit. The result of this double result is listed in Table 12.1.

The process continues until all 64 bits have been read and is repeated (including bus initialization and sending of the Search ROM code) until all identification codes have been determined.

12.1.4.1.2   Read ROM (Code 0x33)
This ROM command requests the only slave connected to the bus to send its identification code. The bits are determined in a normal reading procedure.

Table 12.1 Bit determination when reading the identification code

1st scanning	2nd scanning	
0	0	Because some slaves have a 0 at this position and the others have a 1, both sample values are evaluated as 0 and thus a bus collision is detected. With the first collision for this bit position, the master stores a 0 for the current slave at the current bit position and sends a "0" to the slaves for confirmation. The participants whose bit has been confirmed will continue transmitting, the others will stop until the next bus initialization. The master remembers the bit location where the collision occurred. If all IDs with a 0 at this position have been detected, the master stores a 1 for the current slave and sends a "1" for confirmation in order to be able to detect the others.
0	1	All slaves have sent a "0" the first time and a "1" the second time; for the current slave a 0 is stored at the current bit position and the master sends a "0" for confirmation.
1	0	All slaves have sent a "1" the first time and a "0" the second time; for the current participant a 1 is stored at the current bit position and the master sends a "1" for confirmation.
1	1	Error; the entire process is aborted and repeated

12.1.4.1.3   Match ROM (Code 0x55)

This ROM command is followed by the sending of the identification code of a subscriber and a function command. Only the addressed slave responds to this function command.

12.1.4.1.4   Skip ROM (Code 0xCC)

In a network with a single slave, this ROM command can be used to skip the addressing phase. In a network with the same slaves of the same type – for example temperature sensors – the master uses this ROM command to inform the slaves that a broadcast function command follows. This command addressed to all slaves must not be a read command in order to avoid bus collisions.

To the ROM commands the manufacturer counts further commands, which are however not necessarily implemented in all components:

- **Alarm Search** causes a search in a 1-Wire bus for the temperature sensors of type DS18S20 that have triggered a temperature alarm;
- **Overdrive Skip ROM** and **Overdrive Match ROM** put all bus nodes into this mode, which are able to communicate with increased bit rate. After this command, all bus communication will take place with increased transmission rate and the other bus nodes will switch to inactive mode. This state is terminated with a Reset Pulse.

12.1.4.2 Addressing

If a slave is to execute a function, the master must address it via its ID. Addressing is only necessary after the Match ROM command. The identification code of all 1-Wire slaves is 64 bits (eight bytes) long and has the configuration shown in Table 12.2. The family code has the least significant position and identifies the components with the same function. For example, the DS18B20 series temperature sensors have the family code 0x28. The serial number uniquely identifies the 1-Wire slaves within a component family. A CRC checksum over these seven bytes is calculated and all together are burned into the component during the manufacturing process. The master must determine the identification codes of all bus nodes during power-up in order to be able to address them. The checksum is recalculated after receipt and compared with the received one in order to detect transmission errors.

The checksum is calculated as the remainder of the polynomial division $v(x)$: $g(x)$ where $v(x)$ satisfies Eq. 12.1.

$$v(x) = a_n x^{n-1} + a_{n-1} x^{n-2} + \ldots + a_2 x + a_1 \tag{12.1}$$

where $n = 64$, a_0 is the least significant bit of the family code and a_n is the most significant bit of the received checksum. See Sect. 6.1.2.5 and the book [5] for a more detailed description of the CRC checksum generation. The equation maps the polynomial representation of the identification code. The polynomial $g(x)$ is called the generator polynomial and for this procedure is of the form $x^8 + x^5 + x^4 + 1$. In the case of an error-free transmission, the remainder of the polynomial division is zero. The circuit shown in Fig. 12.6 implements the Euclidean division algorithm. The CRC byte is initialized with 0x00 before the calculation.

The following function calculates the checksum of a byte array according to the method of the Maxim company and returns this value to the calling location.

```
uint8_t _1_wire_CRC_Maxim(uint8_t *ucbytefield, uint8_t ucnumber)
{
    uint8_t ucCRC = 0, ucMask, ucPolynomial = 0x18, ucTest;
    for(uint8_t j = 0; j < ucnumber; j++) //byte number of the array
    {
        ucMask = 0x01; //bit pointer, starts with the LSB
        for(uint8_t i = 0; i < 8; i++)
        {
            //it is checked whether 0 or 1 is pushed to MSB
            if(ucbytefield[j] & ucMask) ucTest=(ucCRC & 0x01)^0x01;
            else ucTest = (ucCRC & 0x01) ^ 0x00;
            if(ucTest)
            {
                ucCRC = ucCRC ^ ucPolynom;
                ucCRC = (ucCRC >> 1) | 0x80;
            }
```

```
        else
        {
            ucCRC = ucCRC >> 1;
        }
        ucMask = ucMask << 1;
    }
}
    return ucCRC;
}
```

12.1.4.3 Function Commands

The master sends a function command after the Skip ROM command or after Match ROM followed by the address of the desired slave. The first group of commands triggers an internal action in the slaves, such as the start of a new measurement or the data transfer from a volatile to a non-volatile memory area. After transmitting such commands, the master terminates the communication session. If these commands affect all bus nodes of the same type, they are sent as a broadcast message after the Skip ROM command. With a second group of commands, the master can rewrite either registers or memory areas of the slaves. The master sends this data set directly after the function command. Broadcast messages are also allowed in this case if the slaves do not have to respond to the message. The third group of commands requires a response from a single bus device and in this case the slave must be addressed with its identification code. Broadcast messages are not allowed in this case because of the risk of collision.

12.1.5 Software Structure of the 1-Wire Bus Communication

The general case of an ATmega microcontroller which controls several 1-Wire buses as bus master is considered. Different slaves can be connected to each bus. For the bus communication a general software module is created, which provides the basic functions for the bit

Table 12.2 Composition of the identification code

CRC checksum	Serial number	Family code
1 byte (8 bits)	6 bytes (48 bits)	1 byte (8 bits)

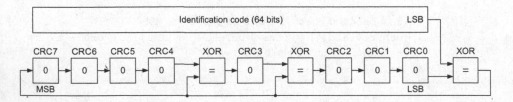

Fig. 12.6 CRC calculation according to [3]

generation and the general ROM functions. In the header file of the software module
_1_wire.h, a software structure is defined that abstracts the bus connection.

```
typedef struct{
    volatile uint8_t* DQ_DDR_REG;
    volatile uint8_t* DQ_PORT_REG;
    volatile uint8_t* DQ_PIN_REG;
    uint8_t DQ_pin;
} t1wireHandle;
```

To make the module hardware independent, the microcontroller connections for each
bus are defined in the main file. An example of a bus connection to pin 2 of port B is shown
in the following program section.

```
t1wireHandle _1wire_Bus_1 = {{/*DQ_DDR_REG*/ &DDRB,
                            /*DQ_PORT_REG*/ &PORTB,
                            /*DQ_PIN_REG*/ &PINB,
                            /*DQ_pin*/ PB2}};
```

The functions of the module are structured hierarchically and can be called from the
software modules of the individual slaves as well as from the main file. This structure
makes it easy to program the specific functions of the individual slaves. The master controls
the bus line to high or low as follows:

```
void _1wire_Set_High(t1wireHandle sdevice_pins)
{
    *sdevice_pins.DQ_DDR_REG &= ~(1 << sdevice_pins.DQ_pin);
}
void _1wire_Set_Low(t1wireHandle sdevice_pins)
{
    *sdevice_pins.DQ_DDR_REG |= 1 << sdevice_pins.DQ_pin;
}
```

The functions that serve the master to generate the individual bits:
_1wire_Write_Zero or _1wire_Write_One, are constructed for communication
with slow slaves due to the timing shown in Fig. 12.4. These primitive functions are used to
program more complex functions – such as transferring a byte – or to program entire
sequences.

```
void _1wire_Write_Zero(t1wireHandle sdevice_pins)
{
    //1-Wire data line is set low, transmission of one bit is initiated
    _1wire_Set_Low(sdevice_pins);
    //timer is started
```

```
    _1wire_Delay_Start(ucDelayTime[TIME_60_MICROSEC]);
    while(_1wire_Delay_Get_State() == _1WIRE_DELAY_RUNNING);
    _1wire_Set_High(sdevice_pins); //the µC releases the data line
}


void _1wire_Write_One(t1wireHandle sdevice_pins)
{
    //1-Wire data line is set low, transmission of one bit is initiated
    _1wire_Set_Low(sdevice_pins);
    _1wire_Set_High(sdevice_pins); //the µC releases the data line
    //timer is started
    _1wire_Delay_Start(ucDelayTime[TIME_60_MICROSEC]);
    while(_1wire_Delay_Get_State() == _1WIRE_DELAY_RUNNING);
}
```

12.1.6 Control of a 1-Wire Temperature Sensor of the Type DS18B20

The DS18S20 [6] device belongs to a family of temperature sensors with adjustable resolution. The large address range of the 1-Wire devices enables their use for temperature monitoring or temperature control in rooms with many measuring points. The measuring accuracy over the entire measuring range − 55 °C ... +125 °C is ±2 °C, over the range − 10 °C ... +85 °C even ±0.5 °C. The adjustable bit resolution of 9 to 12 bits enables a temperature resolution of 0.5 °C to 0.0625 °C. Increasing the resolution by one bit doubles the measuring time, which reaches 750 ms with 12 bits. The device has an alarm function that is activated when the temperature exceeds or falls below the set temperature limits.

12.1.6.1 Memory Organization
The device has three bytes of non-volatile memory in which the temperature limits and bit resolution can be permanently stored. In addition, the bus master can access a nine-byte RAM (scratchpad) with the structure shown in Table 12.3.

Bytes 2, 3 and 4 are initialized with the values stored in the EEPROM after the device has been started up. At the end of a temperature measurement, the measured value is stored in bytes 0 and 1, the checksum is recalculated and stored in byte 8.

12.1.6.2 Temperature Coding
The sensor stores the temperature values in two's complement (see Fig. 12.7) in bytes 0 and 1, which can only be read by the master. In Fig. 12.7, bits 11 to 15 are zero for positive temperature values and one for negative temperature values. The bits labeled x are undefined for the corresponding resolutions. Bytes 2 and 3 store temperature limits in two's complement as an integer. If the measured temperature value reaches or exceeds these limits, a flag is set internally which is reset as soon as the temperature returns to

Table 12.3 Structure of the RAM for DS18B20

Byte 0	Temperature value LSB
Byte 1	Temperature value MSB
Byte 2	Upper temperature limit
Byte 3	Lower temperature limit
Byte 4	Configuration register
Byte 5	Reserved (0xFF)
Byte 6	Reserved
Byte 7	Reserved (0x10)
Byte 8	CRC

Configuration-Register		Resolution [Bit]	Temperature register																	
			MSB (Byte 1)								LSB (Byte 0)									
Bit 6 R1	Bit 5 R0		Bit 15	Bit 14	Bit 13	Bit 12	Bit 11	Bit 10	Bit 9	Bit 8	Bit 7	Bit 6	Bit 5	Bit 4	Bit 3	Bit 2	Bit 1	Bit 0	
1	1	12	±	±	±	±	±	2^6	2^5	2^4	2^3	2^2	2^1	2^0	2^{-1}	2^{-2}	2^{-3}	2^{-4}	°C
1	0	11	±	±	±	±	±	2^6	2^5	2^4	2^3	2^2	2^1	2^0	2^{-1}	2^{-2}	2^{-3}	x	°C
0	1	10	±	±	±	±	±	2^6	2^5	2^4	2^3	2^2	2^1	2^0	2^{-1}	2^{-2}	x	x	°C
0	0	9	±	±	±	±	±	2^6	2^5	2^4	2^3	2^2	2^1	2^0	2^{-1}	x	x	x	°C

Fig. 12.7 DS18B20 coding of the temperature values

within the permissible range. The bit resolution is determined by bits 5 and 6 of the configuration register (byte 4) as shown in Fig. 12.7. The other bits of this register have fixed values that cannot be changed by the master.

12.1.6.3 Function Commands DS18B20

12.1.6.3.1   Start Temperature Measurement (0x44)
The temperature sensor that receives this command starts a new measurement.

12.1.6.3.2   Write RAM (0x4E)
With this command, followed by three data bytes, the bus master can change the temperature limits (bytes 2 and 3) and the bit resolution (byte 4) of a slave.

12.1.6.3.3   Read RAM (0xBE)
The entire RAM area (byte 0 to 8) of a temperature sensor can be read in one communication session, starting with byte 0. After reading, the checksum can be formed over the first

eight bytes – e.g. with the presented function _1_wire_CRC_Maxim - and then compared with byte 8 to detect any transmission errors. The master can terminate the transmission by sending a Reset Pulse if only part of the data is desired. The function DS18B20_Read_Scratchpad can be used to read the entire RAM area.

```
uint8_t DS18B20_Read_Scratchpad(t1wireHandle sdevice_pins,
                                uint8_t ucrom_command,
                                uint8_t *ucbytefield,
                                uint8_t *ucromfield)
{
    uint8_t ucMask;
    //initialization
    if(_1wire_Init_Comm(sdevice_pins) == INIT_NOK) return INIT_NOK;
    _1wire_Write_Byte(sdevice_pins,ucrom_command);
    //the ROM command is transferred
    if(ucrom_command == MATCH_ROM)
    {
        for(uint8_t i = 0; i < 8; i++)
        {
            _1wire_Write_Byte(sdevice_pins,ucromfield[i]);
        }
    }
    //the command code is transferred
    _1wire_Write_Byte(sdevice_pins,READ_SCRATCHPAD);
    for(uint8_t i = 0; i < 9; i++) //all nine bytes are read
    {
        ucMask = 0;
        ucbytefield[i] = 0;
        for(uint8_t j = 0; j < 8; j++) //the eight bits of a byte are read
        {
            ucbytefield[i] += _1wire_Read_Bit(sdevice_pins) << ucMask;
            ucMask ++ ;
        }
    }
    return INIT_OK;
}
```

If the identification code of a temperature sensor is stored in the array ucROM, its RAM area can be read into the vector ucScratchpad with the call: DS18B20_Read_Scratchpad(_1wire_Bus_1,MATCH_ROM, ucScrachpad, ucROMCode).

Save Data to EEPROM (0x48) – This command causes bytes 2, 3, and 4 to be permanently saved from the RAM area to the EEPROM.

Retrieve data from **EEPROM (0xB8)** – bytes 2, 3 and 4 from the RAM area are overwritten with the values permanently stored in the EEPROM.

Query Supply Type (0xB4) – After receiving this command, the DS18B20 slaves, which are supplied parasitically, switch the data line to low in response.

12.2   UNI/O® Bus

UNI/O® is a proprietary bus protocol [7] developed by Microchip Technology to realize bidirectional communication between a master and several slaves via a single data line (SCIO), similarly to 1-Wire protocol. The bus specification mentions as possible slaves: EEPROMs, temperature sensors and A/D converters. The master initiates the communication and establishes it with the individual slaves via a permanently stored or programmable, 8-bit or 12-bit address. Slaves with 8-bit and 12-bit addresses can coexist on a bus without causing collisions. According to the specification, communication can take place at a transmission rate between 10 kbps[1] and 100 kbps. The supply voltage and the manufacturing processes of the slaves are not specified, but the total bus capacitance is limited to 100 pF.

12.2.1 Network Topology

To ensure the advantages of 1-line transmission, the UNI/O slaves are connected via three connections as shown in Fig. 12.8. While slaves 1 to n use the V_{DD} line for their own supply, the slave in Fig. 12.11 is supplied "parasitically" via the data line. This requires additional external components: a fast rectifier diode with low voltage drop and a buffer capacitor.

12.2.2 Bit Coding

Similarly to what is described in Chap. 6, the individual bits are coded with a unipolar Manchester code, as can also be seen in Fig. 6.8. In the first half of the bit the logical "0" is at high level, in the second at low level (the first bit of the preamble). The logical "1", on the other hand, is at low level in the first half of the bit and at high level in the second half (the second bit of the preamble). Ultimately, therefore, a 1–0 transition takes place in the middle of the bit to encode a "0" and vice versa for the "1". The resulting data stream favors clock synchronization because of the level switching in the middle of each bit.

---

[1] kbps – kilobits per second.

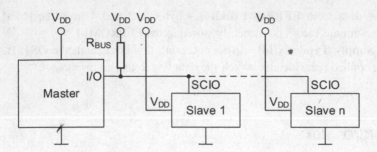

Fig. 12.8 UNI/O bus configuration

12.2.3 UNI/O Frame

A frame consists of one byte (eight bits), the acknowledgement of the master and the response of the slave. The byte is sent by the master or the addressed slave, starting with the most significant bit (MSB). The master then sends an acknowledgement bit and waits for the slave response. With a logical "1" (MAK = Master Acknowledge) the master signals the continuation of the communication, with a logical "0" (NoMAK = Not Master Acknowledge) its termination. With the recognition of its own address, the slave responds with a logical "1" (SAK = Slave Acknowledge) after each acknowledgement (MAK or NoMAK) from the master. The slaves connected to the bus respond with NoSAK (Not Slave Acknowledge) before successful addressing. This signal with the duration of one bit differs in its structure from SAK. An already addressed slave also responds with NoSAK in the event of a transmission error. The double acknowledgement from the master and the slave reduces the net data rate, but increases the security of the communication.

12.2.4 Communication Session

After the POR[2] phase or after an error, the slaves switch to the so-called idle mode in which they do not take any digital signals into account and wait to be switched back to standby mode. The slaves switch to standby mode if the data line is held high for longer than 600 μs. In this mode, power consumption is low and the slaves are ready to process digital signals. It is recommended to use a pull-up resistor for the bus line to ensure the standby mode even in a time when the master has no control over the bus.

12.2.4.1 Initialization of Communication

A UNI/O communication is initiated by the master with a Standby Pulse (Fig. 12.9), or also terminated prematurely. The next communication phase is initiated with a low pulse with a

---

[2]POR – Power On Reset; subcircuit of a complex function block which ensures that a defined state is reached after switching on or after a reset.

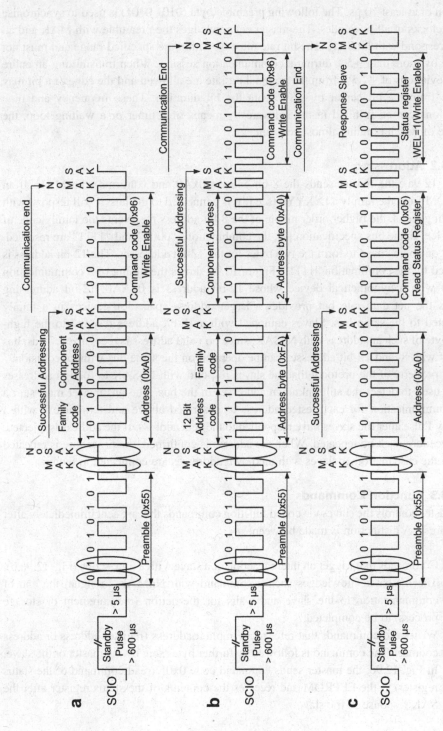

Fig. 12.9 UNI/O communication structure

duration of at least 10 µs. The following preamble byte (0101 0101) is used to synchronise the bit clocks of all bus nodes. The master acknowledges the preamble with MAK and all slaves respond with NoSAK. The bit rate must be within the specified range and must not change by more than ±5% during a communication session. When transmitting an entire byte, deviations of ±7.5% from the nominal bit rate are allowed and the edge of a bit may deviate from the bit center by ±8% from the bit duration. These frequency and time restrictions can be fulfilled in the software by means of a timer or a waiting loop, the passage of which remains almost constant.

12.2.4.2 Addressing

In Fig. 12.9a,c the master sends the 8-bit address 0xA0 and confirms with MAK. If an EEPROM from the family 11XXYY0 (see [8]) is connected to the bus, it will respond with SAK. In general, the higher order nibble of the 8-bit address represents the family code of the device. In the bus specification [7], the combinations "0000" and "1111" are reserved, the last one being used to form the two bytes for the 12-bit addresses. The 12-bit address is preceded by the combination "1111". Figure 12.9b shows the timing of a communication session with a hypothetical device whose 12-bit address is 0xAA0. 12-bit addressing reduces the net data rate but provides a larger address space for devices in a family connected to the same bus. Slaves equipped with an 8-bit address respond to the high-order byte of such an address with NoSAK, since no valid address is recognized. Thus, bus devices with 8- and 12-bit addresses can be operated on the same bus without collision.

The property of the protocol that the slaves respond with NoSAK to foreign addresses can be used to detect the still unknown addresses in the bus. The bus master must start a new communication for each tested address as described above and terminate it with a Standby Pulse after the slaves have responded to the sent address. If the address is rejected, the procedure must be repeated. When the address is confirmed, the procedure is repeated only in the case that other slaves with unknown addresses are connected to the bus.

12.2.4.3 Function Commands

The master controls the slaves via coded function commands that are sent immediately after the address. A distinction is made between:

(a) Commands that trigger an internal action of a slave with a time scope (Fig. 12.9a,b). The master acknowledges such a command with NoMAK to signal the end of communication to the slave and waits for the action (measurement or storage process) to be completed.

(b) Write/read commands that refer to an implicit address (register address or address counter). The command is followed by further bytes sent by the master or the slave. In Fig. 12.9c, the master sends command code 0x05 (read command of the status register) to the EEPROM and receives the contents of the status register after the SAK response of the slave.

(c) Write/read commands that refer to an explicit address. The bus master sends the complete address of a memory cell after the command code and, depending on the command, further bytes are sent by the master or the slave.

The UNI/O protocol does not provide for broadcasting.

12.2.5 Software Structure of UNI/O Bus Communication

To operate a bus as shown in Fig. 12.8, the master ideally has an I/O pin. The microcontrollers of the ATmega family have such pins that meet the requirements and can be used to operate several such buses. In general, the bus connection can be abstracted in a UNI_O software module as follows:

```
typedef struct{
    volatile uint8_t* SCIO_DDR_REG;
    volatile uint8_t* SCIO_PORT_REG;
    volatile uint8_t* SCIO_PIN_REG;
    uint8_t SCIO_pin;
} tuni_oHandle;
```

One notices the similarity with the software structure of a 1-Wire bus. Similarly the microcontroller pin is also defined in the main file in this case.

Functions are defined in the software module for the following actions:

- Sending of a Standby Pulse;
- Sending a byte
- Receiving a byte.

A further structuring of the functions on bit level (as with 1-Wire) was omitted in order to comply with the time restrictions of the UNI/O bus. In order to design the necessary times for the communication, a delay function is used, which was programmed in assembler. This function allows transmission rates up to approx. 16 kBit/s. To achieve higher transmission rates, the functions would have to be designed as state machines, which are controlled via a timer interrupt.

The function UNI_O_Write_Byte is used to send the preamble byte, the command codes and the data bytes. The function, whose flowchart is shown in Fig. 12.10, is called with the byte to be transmitted, ucByte2Send and the master acknowledgement (MAK or NoMAK) as parameters. At the beginning of the function, the microcontroller sets the bus connection to output and starts sending the byte with the most significant bit first. The generation of a bit is initiated by setting the output high or low depending on the bit significance. The output remains in this state for the duration of one half-bit (TA). The

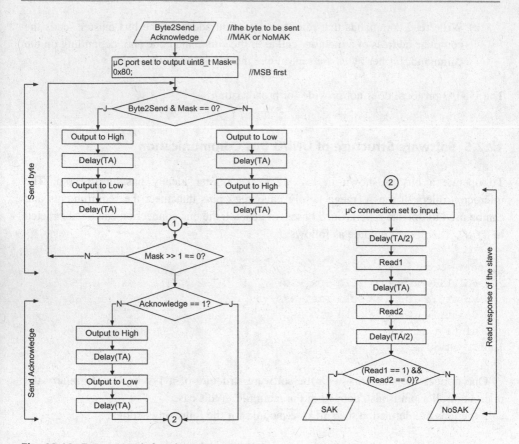

Fig. 12.10 Byte transmit function of the UNI/O bus

output is then toggled and it is waited for TA microseconds. After the eight bits are sent and acknowledged, the bus port is switched to input to read the response from the addressed slave. This response is determined in two steps by reading in the middle of the half bits, hence the delay of TA/2 microseconds at the beginning and end of the read. This response is returned by the function to determine the further transmission steps. The successful byte reception is answered by the slave, starting with the recognition of its own address with SAK.

An example function for reading a byte is the function UNI_O_Read_Byte. The individual bits are determined step by step like the slave response in the previous example and assembled to a byte. The function stores the read byte at the desired address (*ucbyte2read), sends the master confirmation and receives the slave response. The received byte is processed by a higher-level function only in the case of an SAK response, otherwise the byte is discarded and communication is terminated with a Standby Pulse.

```
uint8_t UNI_O_Read_Byte(tuni_oHandle sdevice_pins,
                        uint8_t *ucbyte2read,
                        uint8_t ucack)
{
    uint8_t ucByte2Read = 0, ucBit2Read = 0;
    uint8_t ucAck = 0;
    // the data line is set to input
    *sdevice_pins.SCIO_DDR_REG &= ~(1 << sdevice_pins.SCIO_pin);
    Delay(TA_HALF); //wait time to determine the middle of the bit half
    for(uint8_t i = 0; i < 7; i++)
    {
        // read a bit (1...7)
        if(*sdevice_pins.SCIO_PIN_REG & (1 << sdevice_pins.SCIO_pin))
            ucBit2Read = 2;
        Delay(TA);
        if(*sdevice_pins.SCIO_PIN_REG & (1 << sdevice_pins.SCIO_pin))
            ucBit2Read |= 1;
        Delay(TA);
        switch(ucBit2Read)
        {
            case 1:
                ucByte2Read |= ucBit2Read;
                ucByte2Read = ucByte2Read << 1;
            break;

            case 2:
                ucByte2Read = ucByte2Read << 1;
            break;

            default:
                return RECEIVE_NOK;
            break;
        }
        ucBit2Read = 0;
    }
    //reading the 8th bit
    if(*sdevice_pins.SCIO_PIN_REG & (1 << sdevice_pins.SCIO_pin))
        ucBit2Read = 2;
    Delay(TA);
    if(*sdevice_pins.SCIO_PIN_REG & (1 << sdevice_pins.SCIO_pin))
        ucBit2Read |= 1;
    Delay(TA_HALF);
    switch(ucBit2Read)
    {
        case 1:
```

```
            ucByte2Read |= ucBit2Read;
        break;

        case 2:
        break;

        default:
            return RECEIVE_NOK;
        break;
    }
    *ucbyte2read = ucByte2Read;


    *sdevice_pins.SCIO_DDR_REG |= 1 << sdevice_pins.SCIO_pin;
    if(ucack == MAK)
    {
        //the master sends MAK
        //Set_Low
        *sdevice_pins.SCIO_PORT_REG &= ~(1 << sdevice_pins.SCIO_pin);
        Delay(TA);
        //Set_High
        *sdevice_pins.SCIO_PORT_REG |= 1 << sdevice_pins.SCIO_pin;
        Delay(TA);
    }
    else
    {
        //the master sends NoMAK
        //Set_High
        *sdevice_pins.SCIO_PORT_REG |= 1 << sdevice_pins.SCIO_pin;
        Delay(TA);
        //Set_Low
        *sdevice_pins.SCIO_PORT_REG &= ~(1 << sdevice_pins.SCIO_pin);
        Delay(TA);
    }
    //the data line is set to input
    *sdevice_pins.SCIO_DDR_REG &= ~(1 << sdevice_pins.SCIO_pin);
    //the response of the slave is read SAK/NoSAK
    Delay(TA_HALF);
    if(*sdevice_pins.SCIO_PIN_REG & (1 << sdevice_pins.SCIO_pin))
        ucAck = 2;
    Delay(TA);
    if(*sdevice_pins.SCIO_PIN_REG & (1 << sdevice_pins.SCIO_pin))
        ucAck |= 1;
    Delay(TA_HALF);
    if(ucAck == SAK) return SAK;
    else return NOSAK;
}
```

12.2.6 Control of a 11XXYZ-EEPROM

The devices with the designation 11XXYZ are serial EEPROMs that are controlled via the UNI/O bus [8]. Because of their small memory capacities (128 bytes to 2 kbytes), they are only suitable for storing small amounts of data, but require little space and only one wire for their control. The internal architecture differs from the devices described in Chap. 21 only in the special features of the communication interface. The memories are organized in 16-byte memory pages. The letters XX code the permissible temperature range and the supply voltage:

- AA are the devices specified for the industrial temperature range ($-40\,°C \ldots +85\,°C$) and a supply voltage of 1.8 V to 5.5 V;
- the devices marked LC cover the automotive temperature range ($-40\,°C \ldots +125\,°C$), but must be supplied with a narrower tolerated voltage ($+2.5\,V \ldots +5.5\,V$).

The two-digit number YY encodes the memory size in kilobits.

12.2.6.1 Memory Write Protection of the 11XXYZ Devices

Due to the reduced number of pins, the memory can only be write-protected by software. The status register, which also controls the memory protection, is structured as follows:

- **Bits 7:4** – are reserved and return 0 when read;
- **Bits 3:2** – BP1, BP0 determine the read-only memory area as explained in Table 12.4;

- **Bit 1** – WEL (Write Enable Latch); when this bit is 0, write access to both the status register and the whole memory is disabled; the bit is reset to 0 after the POR phase and after all writes that complete successfully;
- **Bit 0** – WIP is a read-only bit that is internally set to 1 only during a save operation.

12.2.6.2 Addressing the 11XXYZ-EEPROMs

The eight-bit address is stored in the module during production and can no longer be changed. The digit "Z" in the device name codes the device address, which is "0000" for $Z = 0$, or "0001" for $Z = 1$. Together with the family code "1010", the addresses 0xA0, or 0xA1 result. A special feature of the 11AA02E48 and 11AA02E64 devices [9] is that a 48- or 64-bit unique identifier is stored in the last six or eight bytes of the memory area during

Table 12.4 Write-protected memory areas of the 11XXYZ devices	BP1	BP0	Write-protected memory area
	0	0	None
	0	1	The top quarter
	1	0	The top half
	1	1	Total memory area

the manufacturing process. To prevent this identifier from being accidentally overwritten, the corresponding memory area is write-protected by setting the BP0 bit in the status register. Unlike the 1-Wire devices, these EEPROMs cannot be addressed via this identifier, but only identified.

12.2.6.3 Function Commands of the 11 XXYYZ EEPROMs

The control functions of these EEPROMs are part of a software module with the name _11XXYYZ which is based on the described UNI/O functions.

12.2.6.3.1   Commands that Trigger an Internal Action

Write enable (0x96) – sets the bit WEL in the status register to 1 and thus changing the status register and the non-write protected memory is enabled. A corresponding function is listed further on:

```
uint8_t _11XXYZ_Write_Enable(tuni_oHandle sdevice_pins,
                             uint8_t ucstdby_pulse,
                             uint8_t ucaddress)
{
    if(UNI_O_Start_Header(sdevice_pins, ucstdby_pulse) == SAK)
        return START_HEADER_NOK;
    if(UNI_O_Write_Byte(sdevice_pins, ucaddress, MAK) != SAK)
        return SLAVE_ADDR_NOK;
    if(UNI_O_Write_Byte(sdevice_pins, WREN, NOMAK) != SAK)
        return COMMAND_NOK;
    return OK;
}
```

By calling the function UNI_O_Start_Header the master initiates the communication, then it addresses the desired block and transfers the command code. Only if one communication phase is successfully completed, the next one is started, otherwise the process is aborted and the function returns an error code that helps with error identification. This function must be called before each write operation.

Write inhibit (0x91) – resets bit WEL to 0 and inhibits writing to the status register and to all memory.

Clear memory (0x6D) – sets all bytes of the memory to 0x00.

Set memory (0x67) – sets all bytes of the memory to 0xFF.

12.2.6.3.2   Commands Indicating an Implicit Address

Read status register (0x05) – reads the contents of the status register and stores it in a variable. Reading the status register and testing the WIP bit can be used to determine when a write operation is completed.

Write status register (0x6E) – after successful write enable, this command changes the contents of the status register.

Read memory from current address (0x06) – this command can be used to read any number of memory cells from the current address. The address counter, which stores the current address, is initialized with a random value after the POR phase. For this reason, the function should not be called until the address has been uniquely determined. In general, the address counter is incremented after reading a memory cell within the entire memory area. After reading from the highest memory address, the address counter jumps to address 0.

12.2.6.3.3 Commands that Refer to an Explicit Address

Read memory (0x03) – the command can be used to read any number of memory cells. The starting address must be transmitted as a 2-byte number (with the most significant byte first) after sending the command code.

Write memory (0x6C) – with this command, as illustrated in the following example function, the contents of up to 16 memory cells (one memory page) can be changed. The start address is transmitted in the same way as for memory read. The bytes to be stored are first buffered in a volatile 16-byte memory buffer. From the buffer, the transferred bytes are permanently stored in the EEPROM. This process is started with the error-free completion of the communication after receiving the NoMAK acknowledgement from the master and takes approx. 5 ms. To be sure that the function works error-free even if the slave is parasitically supplied, the bus line is set high at the end of the function. After writing a byte, the address counter is only incremented within the memory page. After reaching the highest address of a page, the address counter jumps to the start address of this page.

```
uint8_t _11XXYZ_Write_Memory(tuni_oHandle sdevice_pins,
                       uint8_t ucstdby_pulse,
                       uint8_t ucdev_address,
                       uint16_t uimemory_address,
                       uint16_t uin_bytes,
                       uint8_t *uctargetaddress)
{
    uint16_t i;
    uint8_t ucMemoryLowAddress = uimemory_address;
    ucMemoryHighAddress = (uimemory_address >> 8);
    if(_11XXYZ_Write_Enable(sdevice_pins,ucstdby_pulse,ucdev_address)
        != OK) return NOK;
    if(UNI_O_Start_Header(sdevice_pins, ucstdby_pulse) == SAK)
        return START_HEADER_NOK;
    if(UNI_O_Write_Byte(sdevice_pins, ucdev_address, MAK) != SAK)
        return SLAVE_ADDRESS_NOK;
    if(UNI_O_Write_Byte(sdevice_pins, WRITE, MAK) != SAK)
        return COMMAND_NOK;
    if(UNI_O_Write_Byte(sdevice_pins, ucMemoryHighAddress, MAK)
        != SAK) return MEMORY_ADDR_NOK;
```

```
if(UNI_O_Write_Byte(sdevice_pins, ucMemoryLowAddress, MAK) != SAK)
    return MEMORY_ADDRESS_NOK;
for(i = 0; i < (uin_bytes - 1); i++)
{
    if(UNI_O_Write_Byte(sdevice_pins, uctargetaddress[i], MAK)
        != SAK) return RECEIVE_NOK;
}
if(UNI_O_Write_Byte(sdevice_pins,uctargetaddress[i], NOMAK) != SAK)
    return RECEIVE_NOK;
UNI_O_Set_SCIO_High(sdevice_pins);
return OK;
}
```

12.2.6.4 Parasitic Supply of an 11XXYZ-EEPROM

The parasitic supply of such an EEPROM is useful if the omission of the supply wire is more important than the more complex circuitry, as shown in Fig. 12.11. The choice of the diode and the sizing of the capacitor are described in [10]. In the following, the networking of a microcontroller ATmega88 with a UNI/O slave of type 11AA02, which is parasitically supplied, is considered. First, it is checked whether the requirements of a parasitic supply are fulfilled. For a supply voltage of the bus master $V_{MDD} = 5$ V the following information can be found in the data sheet [11]:

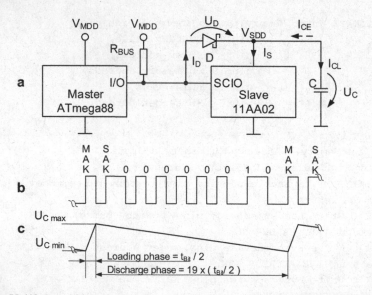

Fig. 12.11 Uni/O-Bus with parasitic wirirng of the slave. (**a**) Wiring; (**b**) Time curve during the transmission of a byte; (**c**) Voltage curve at the capacitor during the reading of a byte

- maximum high output current $I_{MOH\ max} = 40$ mA;
- minimum high output voltage $V_{MOH\ min} = 4.2$ V @ $I_{MOH} = 20$ mA;
- minimum high input voltage $V_{MIH\ min} = 0.6 \times V_{MDD} = 3$ V.

For the 11AA02 module, the following information can be taken from the data sheet [8]:

- minimum high output voltage $V_{SOH\ min} = V_{SDD} - 0.5$ V, where V_{SDD} is the supply voltage of the slave;
- Supply current during reading $I_{S\ Read} = 3$ mA @ $V_{SDD} = 5.5$ V;
- Supply current during storage $I_{S\ Write} = 5$ mA, $t_{Write} = 5$ ms.

A Schottky diode is the first choice at this point because of the negligible switching times and low forward voltage. The Schottky diode BAT48 has a forward voltage U_D of 0.35 V [12] with a forward current of 40 mA. The capacitor is to buffer the electrical energy and provide it to power the EEPROM. Figure 12.11b shows an example of the transfer of one byte. Because of the Manchester coding, the bit sequence of the byte does not matter because half of each bit is high.

There are two phases of operation of the capacitor. In the first phase, the microcontroller connection is switched to output and high, the diode conducts and the capacitor charges. In the second operating phase, the connection is switched to output and low, or to input. In this second phase, the diode blocks and the capacitor discharges through the slave's supply pin. Reading out a larger area of memory is the worst case, which will continue to be considered. When reading out a byte (ten bits), the master only sends the acknowledgement bit MAK, so that the capacitor can only be charged over 5% of the byte duration. During the charging phase, the output of the microcontroller can be considered as a constant current source that can supply a maximum of 40 mA and the capacitor with the current:

$$I_{CL} = I_{MOH\ max} - I_{S\ Read} = 40\ \text{mA} - 3\ \text{mA} = 37\ \text{mA} \tag{12.2}$$

charges up. At the end of this phase, the capacitor voltage reaches the maximum value:

$$U_{C\ max} = U_{SDD} = V_{MOH\ min} - U_D = 4.2\ \text{V} - 0.35\ \text{V} = 3.85\ \text{V} \tag{12.3}$$

and the supplied electric charge can be calculated as follows:

$$Q_L = I_{CL} \cdot t_{Bit}/2 \tag{12.4}$$

In the discharge phase, the reverse current of the diode and the input current of the microcontroller (both about 1 µA) can be neglected and the discharge current $I_{S\ Read} = 3$ mA can be assumed to be constant. The capacitor voltage reaches the value $U_{C\ min}$ at the end of this phase and the electrical charge drawn is:

$$Q_E = I_{CE} \cdot 19 \cdot (t_{\text{Bit}}/2) \tag{12.5}$$

A steady state is achieved by the capacitor voltage reaching the same value after each byte read, for this to happen the electrical charge supplied and the electrical charge removed must be the same. From the equations Eqs. 12.4 and 12.5 the maximum discharge current can be calculated:

$$I_{CE\ \text{max}} = \frac{I_{CL}}{19} = \frac{37}{19} \approx 1.95\ \text{mA} \tag{12.6}$$

The calculated current is smaller than the $I_{S\ \text{Read}}$ specified in the data sheet. However, the supply current decreases at lower supply voltages of the EEPROM (see [8]) and must be tested in practice. If the condition:

$$I_{S\ \text{Read\ real}} \leq \frac{I_{\text{MOH\ max}} - I_{S\ \text{Read\ real}}}{19} \tag{12.7}$$

is not valid, the prerequisite for parasitic supply is not fulfilled. The stationary state is not reached and communication errors occur after a few bytes.

In the next step, the capacitance of the capacitor is calculated at a bit rate of 10 kBit/s assuming that the condition in Eq. 12.7 is satisfied. In the limiting case, error-free communication is guaranteed if:

$$V_{\text{SOH\ min}} = V_{\text{SDD\ min}} - 0.5\ \text{V} \geq V_{\text{MIH\ min}}. \tag{12.8}$$

From $V_{\text{SDD\ min}} = U_{C\ \text{min}}$, follows that

$$U_{C\ \text{min}} \geq V_{\text{MIH\ min}} + 0.5\ \text{V} = 3\ \text{V} + 0.5\ \text{V} = 3.5\ \text{V} \tag{12.9}$$

In the discharge phase, the capacitor emits the maximum electrical charge Q_E

$$Q_E = I_{S\ \text{Read}} \cdot 19 \cdot t_{\text{Bit}}/2 = \frac{3 \cdot 10^{-3} \cdot 19\ \text{A} \cdot \text{s}}{2 \cdot 10000} = 2.85\ \mu\text{As}$$

and the voltage drop across the capacitor is:

$$\Delta U_C = U_{C\ \text{max}} - U_{C\ \text{min}} = 3.85\ \text{V} - 3.5\ \text{V} = 0.35\ \text{V}.$$

The charge delivered is proportional to the capacitance and voltage drop of the capacitor:

$$Q_E = C \cdot \Delta U_C \tag{12.10}$$

$$\text{Resulting from}: \quad C = \frac{Q_E}{\Delta U_C} = \frac{2.85 \ \mu \ \text{As}}{0.35 \ \text{V}} = 8.1 \ \mu F.$$

Experimentally, the readout of a data packet of 128 bytes was exmined using a ceramic capacitor with a nominal capacitance of 1 µF to buffer the parasitic supply of the EEPROM. Figure 12.12 shows time segments of this transfer in steady state. Figure 12.12a shows the data waveform on the bus and Fig. 12.12b shows the voltage waveform on the buffer capacitor. During the reading of a byte, the voltage at the capacitor varies between 4.2 V and 4.6 V. This voltage is sufficient for the safe supply of the EEPROM.

The experiment shows that even at smaller capacitances than those calculated by Eq. 12.10 and at higher frequencies, data transmission with parasitic supply works without errors, but it must be tested experimentally.

12.3 LIN Bus

The Local Interconnect Network (LIN) is referred to as a sub-bus because it is specified for minimum cost, minimum resource consumption and consequently to only low data rates. The specification document [13] the book [14, 15] describe the system in detail. Until 2019, the last specification 2.2 A was freely available on the internet, some vendors still keep it available for download.

In this book we describe the LIN bus for the sake of completeness, but not in its implementation.

LIN is based on the asynchronous UART interface, which is integrated in every common microcontroller. In contrast to a full-duplex UART interface, however, it is adapted to the voltage levels in the vehicle and designed as a half-duplex system in Wired-AND technology (open collector) with an unbalanced single-wire interface. Since LIN is also intended to be run by small computers (intelligent lamps, switches or actuators) without their own quartz oscillator if necessary, the specified limit of the data rate is 20 kBit/s. Depending on the version, the LIN transceiver contains the level matching, power management components (wake-up), monitoring circuits (watchdog) and the power supply for the microcontroller. In this way, very compact circuits can be realized.

Like CAN, LIN is message-addressed, i.e. messages are not sent between instances but the address refers to the meaning of the message, so that it can be read by different bus nodes if required. In communication technology, this is referred to as the producer-consumer model.

For bus control, LIN also uses a master-slave protocol: Each communication on the bus is initiated by a master node. The master sends, according to a more or less fixed schedule, a message header, which is headed by a 13 bits long synchronization field (transition from a recessive to a dominant bit), during which the connected slaves can prepare reception.

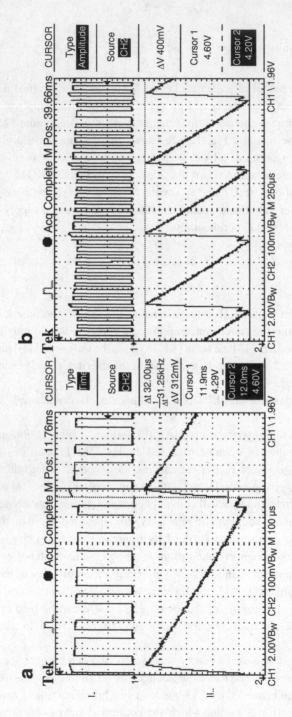

Fig. 12.12 Parasitic supply of a UNI/O-EEPROM

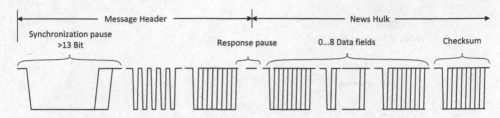

Fig. 12.13 LIN frame

From this point on, communication is taken over by UART byte by byte. First, a hexadecimal 0x55, i.e. a binary 0101 0101, is sent so that the slaves can synchronize to the clock rate. The byte synchronization is performed by the UART's own start and stop bits. Then a 6-bit identifier (ID0 ... ID5) with a 2-bit checksum is also sent via UART, so that a total of 64 messages can be addressed (see Fig. 12.13). This identifier is also called PID.

The 2-bit checksum consists of two parity bits:

$$P0 = ID0 \oplus ID1 \oplus ID2 \oplus ID4 \tag{12.11}$$

$$P1 = ! \ (ID1 \oplus ID3 \oplus ID4 \oplus ID5) \tag{12.12}$$

Now the following cases are distinguished:

- The message is to be sent from the master to one or more slaves: In this case, the master continues the UART transmission with the data field, which may consist of eight bytes. The slaves that are interested in the message read along.
- The message is to be sent by a slave. In this case, the master stops the transmission after sending the identifier including the checksum and the slave continues the transmission with the data field. For capacity reasons, so-called event-triggered frames are also permitted, i.e. several slaves could send a response to an identifier if the corresponding send reasons are present. In this case there is a collision, which is detected by the master and rectified by individual polling with separate identifiers, for which a further standard frame must be made available in each case.
- At the end of the transmission, an 8-bit checksum is sent. It is calculated as with the Modbus as LRC according to the following formula:

$$\text{Checksum} = \text{INV} \ (\text{Databyte } 1 \oplus \text{Databyte } 2 \oplus \ldots \oplus \text{Databyte } 8) \tag{12.13}$$

To form the checksum, the individual data bytes are added (XOR) using modulo-256 arithmetic and then inverted bit by bit.

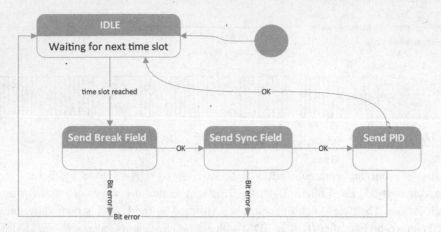

Fig. 12.14 State machine of the LIN master. (According to [13])

Ultimately, two state machines run on a LIN master, that of the master for sending the message header and that of a slave for sending or receiving a message. Each slave has a table with the identifiers that are intended for it. This table refers to the intended direction (Rx frame = the slave reads, Tx frame = the slave completes the message with data) and to the number of bytes to be sent or received in the message. The sequence is outlined in Figs. 12.14 and 12.15.

Since each LIN node can read the messages it sends due to the open-collector connection to the bus, a transmission error can be identified in this way.

The messages with the identifiers 0x60 and 0x61 are reserved for configuration and diagnostic messages. Here a superimposed transport protocol takes care of the segmentation of the messages, which are usually larger than 8 bytes. For example, message identifiers can be modified compared to the default settings, the node status can be read out, the node can be put to sleep, and so on. The procedure corresponds in simplified form to the UDS protocol (Unified Diagnostic Service) whose description would go beyond the scope of this book.

The LIN specification not only defines the protocols and the physical layer, but also the software interfaces of the protocol stack (API) and, above all, a development concept. This simplifies the integration of LIN components considerably. For example, a standardized description of its signals and messages must be supplied with each LIN component (Node Capability File NCF). The descriptions of all components in the system result in the so-called LIN Description File (LDF), from which the message identifiers can be determined. The LIN diagnostic protocol requires that the identifiers can be changed dynamically by configuration in the nodes. In this way, the high quantities required for the price segment can be realized while facilitating integration into the overall system. Nevertheless, LIN is of course not a plug-and-play network, because the functional diversity is deliberately not standardized at the transport level.

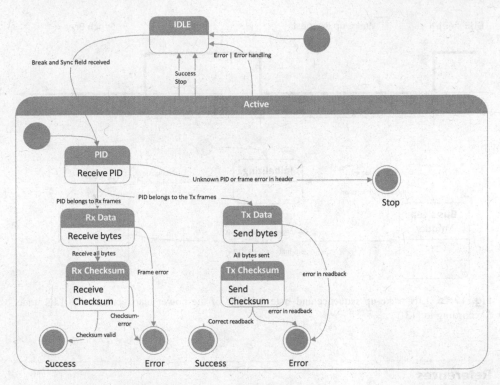

Fig. 12.15 State machine of the LIN slave (According to [13])

Finally, power management is also implemented in LIN. It is assumed that all sensors or actuators connected to LIN are permanently supplied with power (in the vehicle this is the so-called terminal 30). If the system is put into a power saving mode (sleep mode), the power supply unit (in the vehicle this is the DC/DC converter from the on-board power supply to the 5 V or 3.3 V supply voltage of the electronics) is put into a state of minimum power consumption. This means that operation of the processor is not possible or only possible to a limited extent (see Chap. 3), but the LIN device continues to respond to the bus. This sleep mode is reached when the master control device does not send any requests for 4 . . . 10 s (bus inactive) or explicitly sends a goto sleep request with the message 0x60. To wake up, the master must send a dominant pulse of at least 250 μs to 5 ms to the bus and then wait at least 100 ms for the slave to complete its boot process. The wake-up pulse causes the LIN controller to turn the power back on and perform a power-on reset of the processor. After that, normal communication according to the schedule table begins. Figure 12.16 illustrates this process.

LIN is integrated by default in AUTOSAR, an open and standardized software architecture for electronic control units in vehicles. In addition, there are some open source stacks that are more or less completely implemented.

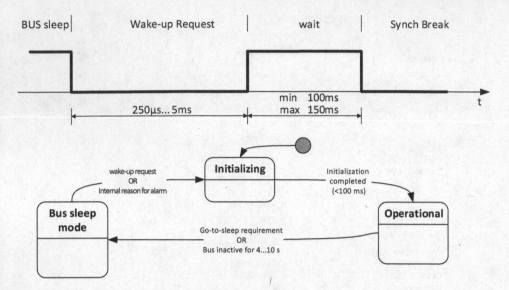

Fig. 12.16 LIN wake-up sequence and state machine of the power management of a LIN node. (According to [13])

References

1. Maxim Integrated Products. AN937: Book of iButton® standards. www.maximintegrated.com. Accessed: 20. März 2021.
2. Maxim Integrated Products. AN148: Guidelines for reliable long line 1-wire networks. www.maximintegrated.com. Accessed: 27. März 2021.
3. Maxim Integrated Products. AN27: Understanding and using cyclic redundancy checks with maxim 1-wire and iButton products. www.maximintegrated.com. Accessed: 21. März 2021.
4. Maxim Integrated Products. DS1972 data sheet. 1024-Bit EEPROM iButton. www.maximintegrated.com. Accessed: 19. März 2021.
5. Werner, M. (2008). *Information und Codierung.* Vieweg + Teubner.
6. Maxim Integrated Products. DS18B20 data sheet. Programable resolution 1-wire digital thermometer. www.maximintegrated.com. Accessed: 9. Febr. 2021.
7. Microchip Technology Inc. UNI/O® bus specification. www.microchip.com. Accessed: 31. Mai 2021.
8. Microchip Technology Inc. 1K-16K UNI/O® Serial EEPROM family data sheet. www.microchip.com. Accessed: 26. Okt. 2021.
9. Microchip Technology Inc. 2K UNI/O® Serial EEPROMs with EUI-48™ or EUI-64™ node identity. www.microchip.com. Accessed: 13. Mai 2021.
10. Microchip Technology Inc. AN1213 Powering a UNI/O® bus device through SCIO. www.microchip.com. Accessed: 13. Mai 2021.
11. Microchip. Reference Manual ATmega48/168, https://www.microchip.com/wwwproducts/en/ATmega88A. Accessed: 6. Jan. 2021.
12. STMicroelectronics. BAT48 – Small signal Schottky diode. www.st.com. Accessed: 7. Aug. 2021.

13. ISO: ISO 17987 Part 1–7:2016 Road vehicles – Local Interconnect Network (LIN).
14. Grzemba, A., & von der Wense, A. (2005). *LIN-Bus: Systeme, Protokolle, Tests von LIN-Systemen, Tools, Hardware, Applikationen* (1. Aufl.). Franzis.
15. Microchip Technology Inc. Local Interconnect Network (LIN) Bus. https://microchipdeveloper.com/lin:start. Accessed: 2. Febr. 2021.

Wireless Networks

13

Abstract

This chapter describes wireless networks, especially in the ISM bands around 433 MHz and 868 MHz, as well as in the ISM band 2.4 GHz. There are countless components for this, the use of a few selected ones is explained here.

The use of microcontrollers and sensors in areas where laying communication wires is difficult favors radio transmission. The same applies to networks with a large number of nodes or with sporadic data traffic, where wired networking is uneconomical. The technical development of radio transmission correlates with the price reduction of the required hardware, the increase of data protection mechanisms and transmission security allows the entry of radio networks in the modern car and in applications such as "Industry 4.0", "Smart Home", "Smart Metering" or "Internet of Things".[1] Wired transmission is more secure and reliable than radio transmission, but radio networks can be expanded more easily.

13.1 Basics of the Radio Interfaces

Radio communications in Germany are subject to the regulations of the Federal Network Agency for Electricity, Gas, Telecommunications, Post and Railway. The Federal Network Agency determines the maximum transmitting power for each frequency range for short-

---

[1] IoT = Internet of Things.

© The Author(s), under exclusive license to Springer Fachmedien Wiesbaden GmbH, part of Springer Nature 2023
A. Meroth, P. Sora, *Sensor networks in theory and practice*,
https://doi.org/10.1007/978-3-658-39576-6_13

313

range radios (SRD)[2] [7]. This limitation contributes to the interference immunity of radio communications. For industrial, commercial, medical, domestic or similar applications (ISM),[3] unlicensed frequency ranges are allocated and can be used freely with few restrictions [9]. The 2.4 GHz ISM band, to which the frequency range 2.4 GHz to 2.5 GHz is allocated in Germany, is defined worldwide and forms the selection space for the transmission channel of numerous radio protocols.

13.1.1 Multiplexing

Two radio transmitters that are within range can interfere with each other. To avoid this, it has to be ensured that the transmitters use different carrier frequencies or do not transmit simultaneously.

Frequency Division Multiplexing (FDMA)[4]
In frequency division multiplexing, a different radio channel is assigned to each transmitter. The frequency ranges of the radio channels used should not overlap.

Time Division Multiplex (TDMA)[5]
Time division multiplexing allows multiple use of the same radio channel. The transmission time is divided into time frames of fixed duration and each time frame is divided into time slots. In the synchronous procedure, the number of time slots is equal to the number of transmitters and each device may only transmit in the time slot assigned to it. The asynchronous method provides for a dynamic allocation of the time slots, depending on the transmission demand of the devices [10].

Code Division Multiplex (CDMA)[6]
Code division multiplexing belongs to the spread spectrum technique. Each bit of a message is multiplied before modulation by a special binary spreading code whose units are called chips. After demodulation, the message can be decoded only by the device that knows the spreading code. This process results in a broadband spectrum and increased information redundancy. This redundancy is used to reduce the transmit power.

---

[2] SRD – Short Range Device.
[3] ISM – Industrial, Scientific and Medical.
[4] FDMA – Frequency Division Multiplex Access.
[5] TDMA – Time Division Multiplex Access.
[6] CDMA – Code Division Multiplex Access.

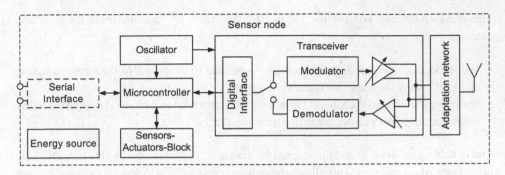

Fig. 13.1 Radio node – block diagram

13.1.2 Sensor Nodes

A simple wireless node that contains sensors and actuators, as shown in Fig. 13.1, is called a sensor node [1]. The microcontroller represents the digital core of the node and performs the following tasks:

- It initializes the sensors statically (once at program start) or dynamically, according to the requests received from outside;
- It reads in the measured values, assembles them into a data packet and performs channel coding;
- It sets the radio channel, the transmission power and the sensitivity of the receiver;
- It controls the actuators;
- It exchanges digital data with the transceiver;
- It implements the upper layers of the ISO/OSI layer model according to the selected protocol.

The transceiver is responsible for radio transmission and implements the physical bit transmission layer. It covers a defined frequency range and generates an adjustable carrier frequency. The transceiver consists of a modulator with adjustable output power, a demodulator with adjustable input sensitivity and an RF switch. In addition, some transceivers can perform link layer tasks (e.g., [2–4]) such as:

- Calculation of a CRC checksum for each data packet;
- Detection and possible correction of transmission errors;
- automatic sending of an acknowledgement of receipt;
- Retransmission of data packets that have been reported as faulty;
- Encryption and decryption of the data packets.

The transceivers can provide the following parameters by measurements on the radio channel:

- the RSSI[7] as analog or digital signal is proportional to the power of the received signal;
- the LQI[8] maps the reception quality;
- the energy of the radio channel; this allows the transmission power to be adjusted to the minimum that just ensures error-free transmission.

The matching network adapts the internal resistance of the transceiver to the characteristic impedance of the antenna. This allows the right antenna to be selected for each application and also optimizes energy consumption.

The electrical energy to power all the electronics is usually supplied by a battery or is generated locally by converting mechanical, kinetic, solar or thermal energy (energy harvesting).

The sensor node presented in Fig. 13.1 can be used for point-to-point communication with a simply constructed software, with the application layer as the focus. The development of proprietary software for networking multiple sensor nodes, with the application layer in the background, is costly. A flexible and inexpensive solution is shown in Fig. 13.2. The upper layers of the ISO/OSI model are still implemented on the microcontroller. The transceiver, as part of the radio module, usually integrates the bit transmission and the data link layers, while a protocol stack with the other layers is stored on the host controller. Some radio modules offer additional digital I/Os, possibly also analog inputs, in order to read in sensors directly or to control actuators directly.

13.2   Radio Transmission in the 433 MHz and 868 MHz ISM Bands

RFM12B is a radio transceiver manufactured by HOPERF Microelectronics [14], which uses the FSK[9] digital frequency modulation and covers the 433 and 868 MHz ISM frequency bands approved in Europe. Under the same name, ready-made modules are produced, which, in addition to the transceiver, includes a quartz crystal, interference suppression capacitors, possibly an analog filter and an antenna matching network. The transceiver is set via a four-wire SPI interface. The data to be transmitted is provided via SPI or as a PCM data stream via the FSK/DATA pin. The received data can be read via SPI or as a PCM data stream together with the reconstructed clock via the FSK/DATA, or DCLK pins. When the data is transmitted via SPI, the air transmission rate can reach 115.2 kBit/s, while it can reach 256 kBit/s for PCM transmission.

---

[7]RSSI – Received Signal Strength Indication.

[8]LQI – Link Quality Indication.

[9]FSK – Frequency Shift Keying.

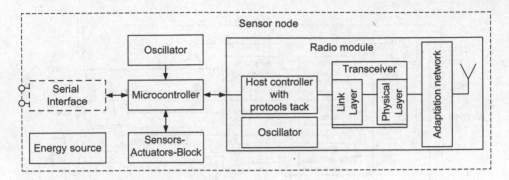

Fig. 13.2 Radio node with radio module

13.2.1 Structure of the RFM12B

A block diagram of the transceiver with the connections relevant to the radio module can be seen in Fig. 13.3. The module has two communication channels: a wired interface (SPI) for communication with the higher-level microcontroller and a radio channel for communication with other radio modules. The transceiver consists of a HF[10] part and a NF[11] part. The HF part contains a transmitter, which works with the FSK modulation method, and an I/Q[12] demodulator. Due to the low power consumption the transceiver can be integrated into battery powered devices. To optimize the power consumption and to avoid a voltage drop when switching between transmitter and receiver operation, which could lead to an interrupt, the corresponding circuit parts can also be switched on and off only partially. In the NF part, the I/O control logic takes over the pin control and the SPI communication.

A switchable oscillator clocks the entire RF section. An adjustable frequency divider can be used to generate the desired frequency at pin CLK to clock the host microcontroller.

When the device is switched on, a phase begins in which the POR circuit puts the chip into a safe state. According to [14], this phase should last a maximum of 100 ms. The POR circuit also becomes active after a hardware or software reset. No communication via SPI shall take place during this phase. The end of the phase is signaled by setting bit 14 in the status register.

Changes to the settings stored during the POR phase are made using commands transmitted via SPI. The settings are stored in the register block.

A two-byte transmit buffer can temporarily store the bytes to be transmitted before they are modulated and transmitted by radio.

---

[10] HF – High frequency.

[11] NF – Low Frequency.

[12] I/Q – In Phase & Quadrature (demodulation method).

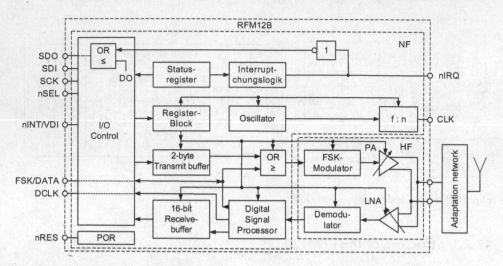

Fig. 13.3 RFM12B block diagram

A DSP (Digital Signal Processor) processes the demodulated data and either passes it into a 16-bit FIFO buffer or outputs it at pin FSK/DATA as a PCM data stream together with the reconstructed clock signal at pin DCLK.

13.2.2 Wiring of the RFM12 Radio Module

A possible control of the radio module with an ATmega88 microcontroller can be seen in Fig. 13.4. The user data is transmitted via SPI in this configuration. In addition to the SPI connections: SDI, SDO, SCK and nSEL, the microcontroller controls the reset pin nRES and evaluates the interrupt output nIRQ of the transceiver. The transceiver returns the contents of the status register on SPI request on pin SDO, otherwise it returns the inverted state of interrupt pin nIRQ, see Fig. 13.3. A pull-up resistor was provided on pin FSK/DATA to ensure full functionality on the RFM12S Rev. 3.0 modules tested. Without this resistor, an interrupt is triggered with the sending of the first byte and pin nIRQ is permanently switched low.

13.2.3 The SPI Communication

The RFM12B transceiver is preset as an SPI slave. The wired communication with the controlling microcontroller takes place via a four-wire SPI interface in mode 0. The most significant bit of a byte is transmitted first. The clock frequency of the communication must be less than a quarter of the crystal frequency. With a crystal oscillating at 10 MHz, the transmission rate can therefore reach a maximum of 2.5 Mbit/s.

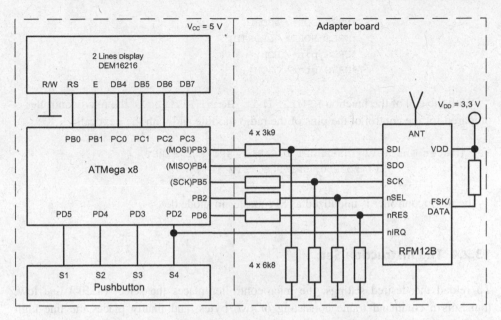

Fig. 13.4 Wiring of a radio module RFM12B

A software module for controlling the transceiver provides functions for managing the following pins: nSEL, nIRQ, nRES and nINT/VDI. The transceiver can be switched from the standby to the active state with a low pulse at pin nINT, which is configured as an input. As output, the pin VDI (Valid Data Indicator) is set high by the transceiver if the signal strength of the received signal and/or the reception quality exceed the set threshold values.

These pins are coded as part of a data structure, which are initialized in the main.h file according to the circuit shown in Fig. 13.4:

```
RFM12_pins RFM12_1 ={/*NSEL_DDR*/ &DDRB,
                     /*NSEL_PORT*/ &PORTB,
                     /*NSEL_pin*/ PB2,
                     /*NSEL_state*/ ON},
                     /*VDI_DDR*/ &DDRD,
                     /*VDI_PORT*/ &PORTD,
                     /*VDI_PIN*/ &PIND,
                     /*VDI_pin*/ PD6,
                     /*VDI_Pin_Dir*/ VDI_OUT,
                     /*VDI_state*/ OFF,
                     /*NIRQ_DDR*/ &DDRD,
                     /*NIRQ_PORT*/ &PORTD,
                     /*NIRQ_PIN*/ &PIND,
                     /*NIRQ_pin*/ PD2,
                     /*NIRQ_state*/ ON,
```

```
/*NRES_DDR*/ &DDRD,
/*NRES_PORT*/ &PORTD,
/*NRES_pin*/ PD6,
/*NRES_state*/ ON};
```

With the call of the function RFM12_Init_Hard(RFM12_1) the microcontroller is prepared for the control of the pins of the radio module and with the next call:

```
SPI_Master_Init(SPI_INTERRUPT_DISABLE, SPI_MSB_FIRST,
                SPI_MODE_0, SPI_FOSC_DIV_16);
```

the microcontroller is initialized as SPI master in mode 0.

13.2.4 The Instruction Set

To reload the desired settings, the microcontroller places the select (nSEL) line low, transmits a command frame consisting of two bytes, and finally places the line high again. A command frame consists of a command code that occupies the high-order bits of the frame, followed by setting parameters. The example function RFM12_Write_CMD allows the microcontroller to transfer individual commands passed as parameters to the transceiver. The function uses the SPI module introduced in Chap. 8.

```
void RFM12_Write_CMD(RFM12_pins sdevice_pins, uint16_t uicmd)
{
    uint8_t ucCmdLow, ucCmdHigh;
    ucCmdLow = uicmd;
    ucCmdHigh = uicmd >> 8;
    SPI_Master_Start(sdevice_pins.RFM12spi); //nSEL on Low
    SPI_Master_Write(ucCmdHigh); //operation code is transferred
    SPI_Master_Write(ucCmdLow); //the parameter bits are transferred
    SPI_Master_Stop(sdevice_pins.RFM12spi); //nSEL to High
}
```

The structure of the commands and a brief description are explained below.

Configuration Setting Command is the command used to set the transceiver.

- **Bits 15:8** – map the command code: 1000 0000 (0x80).
- **Bit 7** – EL = 1 validates the transmit buffer
- **Bit 6** – EF = 1 validates the FIFO receive buffer and switches off the PCM mode.
- **Bits 5:4** – these two bits select the frequency band: 01–433 MHz, 10–868 MHz, 11–915 MHz.

- **Bits 3:0** – these four bits are used to set the capacitance for the crystal oscillator. The 16 combinations encode the values between 8.5 pF and 16 pF in steps of 0.5 pF.

Power Management Command

- **Bits 15:8** – Command code: 1000 0010 (0x82).
- **Bit 7** – ER = 1 switches on the oscillator and the entire receiver.
- **Bit 6** – EBB = 1 switches only the demodulator on.
- **Bit 5** – ET = 1 switches on the oscillator and the entire transmitter and initiates transmission from the transmit buffer. When initializing the transceiver, this bit should not be set.
- **Bit 4** – ES = 1 switches only the FSK modulator on.
- **Bit 3** – EX = 1 switches only the oscillator on.
- **Bit 2** – EB = 1 validates the undervoltage detector.
- **Bit 1** – EW = 1 validates the internal timeout timer.
- **Bit 0** – DC = 0 enables the clock output.

Frequency Setting Command   The command is used to set the carrier frequency for the transmitter or receiver.

- **Bits 15:12** – Command code: 1010 (0xA).
- **Bits 11:0** – F11:0 encode the carrier frequency. The 12-bit number F must be between 96 and 3903, otherwise the change will not be implemented. Depending on the selected frequency band, frequencies between 430.24 MHz and 439.7575 MHz, between 860.48 MHz and 879.515 MHz and between 900.72 MHz or 929.2725 MHz can thus be set. The carrier frequency can be set with a resolution of 2.5 kHz in the 430 MHz band, or 5 kHz in the 860 MHz band. For the frequency bands permitted in Europe, the frequency code F of a valid carrier frequency f_0 in MHz can be calculated using Eq. 13.1, or the carrier frequency if the code is known.

$$
\begin{aligned}
\text{430 MHz Band:} \quad & F = 400 \cdot (f_0 - 430) \\
& f_0 = {}^F\!/_{400} + 430 \\
\text{860 MHz Band:} \quad & F = 200 \cdot (f_0 - 860) \\
& f_0 = {}^F\!/_{200} + 860
\end{aligned}
\tag{13.1}
$$

Data Rate Command   This command sets the transmission rate in the air.

- **Bits 15:8** – Command code: 1100 0110 (0xC6).
- **Bit 7** – CS is a parameter for the calculation of the bit rate.
- **Bits 6:0** – R6:0 encode the bit rate. For a given bit rate **BR** in kBit/s, the bit rate code **R** is calculated as an integer using Eq. 13.2

$$R = \frac{10000}{29 \cdot BR \cdot (1 + CS \cdot 7)} \tag{13.2}$$

To keep the deviation between the desired and the real bit rate small, set the parameter CS to 1 for bit rates between 4.8 kBit/s and 115.2 kBit/s.

Receiver Control Command
- **Bits 15:11** – Command code: 1001 0.
- **Bit 10** – P16 determines the direction of the pin nINT/VDI. If the bit is 0, the pin is an interrupt input.
- **Bits 9:8** – D1:0 determine the output of the VDI pin.
- **Bits 7:5** – I2:0 with valid binary values between 001 and 110 the bandwidth of the receiver is set to values between 400 kHz and 67 kHz.
- **Bits 4:3** – G1:0 set the gain factor of the input amplifier LNA. Values between 00 and 11 set gain factors of 0 dB, −6 dB, −14 dB and −20 dB.
- **Bits 2:0** – R2:0 set the RSSI threshold. The receiver demodulates and decodes the received signal if the signal strength is above the set RSSI threshold. With values between 000 and 101, the RSSI threshold can be set between −103 dB and −73 dB in steps of 6 dB.

Data Filter Command
- **Bits 15:8** – Command code: 1100 0010 (0xC2).
- **Bits 7:6** – AL, ML; the two bits affect the clock recovery. It is recommended to set both bits to zero to set a 2-byte preamble.
- **Bits 5:3** – are undocumented and should both be set to 1.
- **Bit 4** – S = 0 activates the internal digital filter for SPI mode.
- **Bits 2:0** – F2:0 determines the threshold for the data quality detector to be greater than four.

FIFO and Reset Mode Command
- **Bits 15:8** – Command code: 1100 1010 (0xCA).
- **Bits 7:4** – F3:0 with setting this 4-bit number to 1000_2 an interrupt is triggered after receiving a byte. The microcontroller must read a byte from the FIFO to prevent the receive buffer from overflowing.
- **Bit 3** – SP determines the length of the synchron pattern before the data bytes. If SP = 1, the synchron pattern consists of the byte 0x2D, otherwise of two bytes 0x2DD4.
- **Bit 2** – AL = 0 allows the receive buffer to be filled only after the synchron pattern has been recognized, otherwise all decoded bytes are stored in the buffer.
- **Bit 1** – FF = 1 validates the filling of the receive buffer after the synchron pattern has been recognized. When the FF bit is reset, saving to the input buffer is interrupted.

- **Bit 0** – if this bit is reset, a hardware reset is triggered when the supply voltage drops below 1.6 V and additionally enables a software reset. If the bit is set, the threshold of the supply voltage for triggering the hardware reset is lowered to 0.25 V.

Synchronous Pattern Command
- **Bits 15:8** – Command code: 1100 1110 (0xCE).
- **Bits 7:0** – with these eight bits the command can replace the least significant byte of the synchron pattern.

AFC[13] Command
- **Bits 15:8** – Command code: 1100 0100 (0xC4).
- **Bits 7:6** – the bits determine the AFC mode. With the combination 00 the automatic frequency correction (AFC) is switched off, which led to good results.
- **Bits 5:4** – if you do not want to limit the frequency offset register, these two bits are to be reset.
- **Bit 3** – by setting this bit the last measured frequency deviation is stored.
- **Bit 2** – FI = 1 enables better accuracy in determining the frequency deviation.
- **Bit 1** – OE = 1 validates the frequency offset register.
- **Bit 0** – EN = 1 enables the calculation of the frequency deviation.

TX Configuration Control Command
- **Bits 15:9** – Command code: 1001 100.
- **Bit 8** – with MP = 0 the higher transmission frequency codes the logical "0" and the lower the logical "1".
- **Bits 7:4** – (M3:0 + 1) × 15 kHz determines the frequency deviation as the amount of the difference between the carrier frequency and the frequency of the modulated signal.
- **Bit 3** – = 0.
- **Bits 2:0** – the relative output power of the transmitter is coded via the three bits. The output power at a matched antenna serves as a reference. Combinations of 000 to 111 are used to encode the relative output power from 0 dB to −17.5 dB in steps of −2.5 dB.

PLL Setting Command
- **Bits 15:8** – Command code: 1100 1100 (0xCC).
- **Bits 7:0** – this byte is set to the value 0x77 after the POR phase and is recommended by the manufacturer.

Wake-Up Timer Command
- **Bits 15:13** – Command code: 111.
- **Bits 12:8** – R4:0 a parameter that can take values between 0 and 29.

---

[13] AFC Automatic Frequency Control.

- **Bits 7:0** – M7:0.

By setting bit EW with the Power Management command, an interrupt is triggered after the time T_Z has elapsed:

$$T_{Z/ms} = 1.03 \cdot M \cdot 2^R + 0.5 \qquad (13.3)$$

To ensure that the interrupt is triggered repeatedly after the time T_Z, bit EW must be reset and set again with the Power Management command.

Low Duty-Cycle Command In order to save energy, the transceiver as receiver can be periodically switched from stand-by mode (the ER bit must be reset with the Power Management command) to active mode. The period duration of this cycle T_Z is determined using the Wake-up Timer command and the duration of the active mode T_{aktiv} is determined using the Low Duty-Cycle command. Even if the wake-up timer is active, no interrupt is triggered by the validation of this operation.

- **Bits 15:8** – Command code: 1100 1000 (0xC8).
- **Bits 7:1** – D6:0 represent a parameter for the calculation of the duty cycle according to Eq. 13.4.
- **Bit 0** – EN = 1 validates this cyclic operation.

$$\frac{T_{aktiv}}{T_Z} = \frac{2 \cdot D + 1}{M} \cdot 100 \ \ in \ \% \qquad (13.4)$$

where M is the parameter of the wake-up timer.

Low Battery Detector and Microcontroller Clock Divider Command
- **Bits 15:8** – Command code: 1100 0000 (0xC0).
- **Bits 7:5** – when the DC bit = 0, (see the Power Management command) these bits determine the frequency of the clock output between 1 MHz and 10 MHz.
- **Bit 4** – = 0.
- **Bits 3:0** – V3:0 these bits set the threshold U_S for the undervoltage detector according to Eq. 13.5.

$$U_S = 2.25 + 0.1 \cdot V \ in \ \ Volt \qquad (13.5)$$

13.2.5 The Status Register

The status register is a 16-bit read-only register that stores information about the state of the transceiver. When an interrupt is triggered, the interrupt output nIRQ is set low, the SPI data output SDO is set high, and a bit from the 15:10 range in the status register is set, identifying the source of the interrupt. Clearing most interrupts is done by reading the status register. This sets the output nIRQ high and resets the corresponding interrupt bit.

When the transceiver detects a logical zero as the first bit of an instruction, it shifts bitwise on the SDO line the 16 bits of the status register, which have the following meaning:

- **Bit 15** – when set, indicates that the transmitter is ready to receive via SPI another transmit byte; the interrupt is cleared by transmitting a byte or after the last byte of a frame by resetting bit ET with the Power Management command. When this bit is set at the receiver, it indicates that the received number of bits in the FIFO buffer has reached the set threshold (usually 8 bits). In this case the interrupt is cleared by reading one byte from the buffer.
- **Bit 14** – is set after successful completion of the POR phase.
- **Bit 13** – is set when the send or receive buffers overflow.
- **Bit 12** – becomes 1 when the wake-up timer overflows, provided the interrupt has been enabled by setting bit EW with the Power Management command.
- **Bit 11** – detecting a low pulse at input nINT triggers an interrupt and sets this bit.
- **Bit 10** – becomes 1 when the undervoltage detector is activated and the supply voltage falls below the set value.
- **Bit 9** – is set when the receive buffer is empty.
- **Bit 8** – is set when a sufficiently strong signal has been detected at the transmitter antenna, or the received signal strength on the receiver side is above the set RSSI threshold.
- **Bit 7** – Output of the signal quality detector.
- **Bit 6** – = 1 if the reconstructed clock signal is latched.
- **Bit 5** – is toggled with each AFC measurement cycle of the received carrier frequency.
- **Bit 4** – maps the sign of the difference between the carrier frequencies of the transmitter and receiver.
- **Bits 3:0** – correspond to the least significant bits of the measured difference between the carrier frequencies. This signed difference can be added to the carrier frequency code on the receiver side.

13.2.6 Initialization of the Transceiver

After the initialization process it should be ensured that the transceiver is ready to perform its task. By calling the function RFM12_Init the line nRES is set low for 10 ms and thus a hardware reset is initiated. A wait of at least 100 ms is made and then the status of the POR phase is queried by testing bit 14 from the status register. When this phase is successfully completed, the current configuration parameters, stored in an array, are transferred and stored in the register block. As a transfer parameter, the function requests the variable ucsens to determine whether the transceiver is operating as a transmitter or receiver, as well as to set the coded values for the radio rate and carrier frequency. The pinout of the transceiver is passed as a parameter using the sdevice_pins data structure. The function returns the value INIT_NOK if it was unsuccessful and must be called again in this case.

```c
uint8_t RFM12_Init(RFM12_pins sdevice_pins,
                   uint8_t ucsens, //ucsens = TX or RX
                   uint8_t ucbaudrate, //ucbaudrate=the coded radio rate
                   uint16_t uicenter_freq) //carrier frequency
{
    uint16_t uiInitStatus;
    uint16_t uiCmd_RX[14] = {0x80D8, 0x8289, 0x9420, 0xC22C, 0xCA81,
                             0xCED4, 0xC407, 0x9850, 0xE000, 0xCC77,
                             0xC800, 0xC000, (0xC600 | ucbaudrate),
                             (0xA000 | uicenter_freq)};
    uint16_t uiCmd_TX[14] = {0x80D8, 0x8209, 0x9420, 0xC22C, 0xCA81,
                             0xCED4, 0xC4C7, 0x9850, 0xE000, 0xCC77,
                             0xC800, 0xC000, (0xC600 | ucbaudrate),
                             (0xA000 | uicenter_freq)};
    //the reset input of the transceiver is set to low
    RFM12_Set_NRESLow(sdevice_pins);
    TimeOut_Start(DELAY_YES); //the waiting time is started
    while(!TimeOut_Get_State()); //10 ms are waited for
    //the reset pin of the transceiver is set to high
    RFM12_Set_NRESHigh(sdevice_pins);
    //the transceiver needs at least 100 ms after the reset to reach
    //a stable state
    for(uint8_t i = 0; i < 20; i++)
    {
        TimeOut_Start(DELAY_YES);
        while(!TimeOut_Get_State());
    }
    //the status register will be read
    uiInitStatus = RFM12_Read_StatusReg(sdevice_pins);
    if(uiInitStatus & 0x4000) //if true, then successful POR phase
    {
        for(uint8_t i = 0; i < 14; i++)
        {
            //the new settings are transferred
            if(ucsens == TX)
                RFM12_Write_CMD(sdevice_pins, uiCmd_TX[i]);
            else RFM12_Write_CMD(sdevice_pins, uiCmd_RX[i]);
        }
        return INIT_OK;
    }
    return INIT_NOK;
}
```

If more flexible setting options are desired, then the transfer parameters should be defined in a structure, as in the following example, and the initialization function should be adapted accordingly.

```
typedef struct{
    uint8_t ucFreqBand; //Freq. band: 315 MHz, 430 MHz, 866 MHz, 915 MHz
    uint8_t ucSense; //direction transmitter or receiver
    uint16_t uiFreq; //Carrier freq.: code for the carrier frequency;
                     // may contain values between 96 and 3902.
                     // This value is calculated as follows:
                     //433 MHz band: uicenter_freq = (F / MHz-430) : 0.0025
                     //868 MHz band: uicenter_freq = (F / MHz-860) : 0.005
    uint8_t ucBaudrate; //radio rate
    uint8_t ucRecv_BW; //Bandwidth
    uint8_t ucLNA_gain; //LNA gain factor
    uint8_t ucRSSI_Tresh; //RSSI threshold
    uint8_t ucModparam; //Modulation parameter
    uint8_t ucOutPow; //OutPower
} RFM12Param;
```

13.2.7 Send Data

The transceiver is set as a transmitter by setting the ET bit and resetting the ER bit with the Power Management command. In SPI transmit mode, the bytes stored in the transmit buffer are transmitted via radio. This buffer is released by setting the EL bit (see Configuration Setting Command). The microcontroller first transmits the command code 0xB8 and then the bytes to be sent, which are stored in the buffer. The sending of the first byte of a message begins when setting the ET bit with the Power Management command. The transceiver signals to the microcontroller that a byte has been completely sent by setting the output nIRQ low, or the output SDO high. The microcontroller transmits another byte, resetting the triggered interrupt. After the last byte is transmitted, with the reset of the bit ET the transmission is terminated and the interrupt is cleared. To send a sequence of data bytes, the microcontroller must transmit a send frame as in Table 13.1 via SPI to the transceiver.

The preamble consists of two or more bytes with alternating bits (0xAA) and is used for clock synchronization on the receiver side. If the receiver has enabled the synchron pattern detection, only the bytes starting from the payload segment are stored in the FIFO buffer. The trailer segment consists of dummy bytes and is needed to ensure that all payload bytes have been completely sent when the bit ET is reset after the SPI transmission of the last trailer byte.

Table 13.1 Transmit frame

Send command code	Header segment		Payload segment	Trailer segment
0xB8	Preamble 0xAAAA	Synchron Pattern e.g. 0x2DD4	n data bytes + possible CRC checksum	e.g. 0xAAAA

The function RFM12_Send_Array is used to send an entire payload segment. The address of the array that stores the payload bytes is passed to the function as call parameter together with the length of the array.

```
void RFM12_Send_Array(RFM12_pins sdevice_pins,
                      uint8_t *ucarray2send,
                      uint8_t ucarray_length)
{
    uint8_t ucSendHeader[4] = {0xAA, 0xAA, 0x2D, 0xD4}; //serves to
        //synchronize the receiver and transmit the
        //synchronization bit pattern for filling FIFO buffer
    uint8_t ucSendTrailer[2] = {0xAA, 0xAA}; //used to send the last
        //payload bytes from the send buffer
    //sending starts with the setting of the bit ET
    RFM12_Write_CMD(sdevice_pins, (POWER_MANAGMT_CMD | EX | DC |ET));
    //it waits for the interrupt
    while((RFM12_Get_NIRQState(sdevice_pins)));
    //nSEL is set to low
    SPI_Master_Start(sdevice_pins.RFM12spi);
    //the header is transferred
    SPI_Master_Write(0xB8); //transmission of the send command code
    for(uint8_t i = 0; i < 4; i++)
    {
        SPI_Master_Write(ucSendHeader[i]);
        while(RFM12_Get_NIRQState(sdevice_pins));
    }
    //the payload bytes are transferred
    for(uint8_t i = 0; i < ucarray_length; i++)
    {
        SPI_Master_Write(ucarray2send[i]);
        while(RFM12_Get_NIRQState(sdevice_pins));
    }
    //the trailer is transmitted
    for(uint8_t i = 0; i < 2; i++)
    {
        SPI_Master_Write(ucSendTrailer[i]);
        while(RFM12_Get_NIRQState(sdevice_pins));
```

```
}
SPI_Master_Stop(sdevice_pins.RFM12spi); //nSEL is set to High
//the transmitter is deactivated
RFM12_Write_CMD(sdevice_pins, ((POWER_MANAGMT_CMD | EX | DC) &
                               ET_NOT));
}
```

13.2.8 Reading the Received Data

To receive messages, a transceiver RFM12B must be set as receiver by setting the bit ER
with the Power Management command. In addition, the FIFO receive buffer must
be enabled and the same synchron pattern must be set as on the transmitter side. The
storage of the received bytes is validated only after the synchron pattern is detected and the
F3:0 parameter is set to eight with the FIFO and Reset Mode command. With the setting
of the FF bit, each received payload byte is stored and an interrupt is triggered during the
process.

The microcontroller monitors the nIRQ line and if it is set low, the microcontroller reads
the status register and checks if the reception of a byte has triggered the interrupt. The
microcontroller transmits the FIFO Read (0xB0) instruction code over SPI followed by a
dummy byte to read a byte from the FIFO buffer and prevent an overflow. This clears the
interrupt and returns the nIRQ line high. This process is repeated until all bytes of the
payload segment are read. At the end, the bit FF is reset to disable the storage to the buffer.

The described procedure is implemented in the function RFM12_Read_Array.

```
void RFM12_Read_Array(RFM12_pins sdevice_pins,
                      uint8_t *ucarray2rcvd,
                      uint8_t ucarray_length)
{
    uint16_t uiStatusReg; //content of the status register
    uint8_t ucInd = 0; //control variable
    //storage in the FIFO buffer is enabled
    RFM12_Write_CMD(sdevice_pins, FIFO_RESET_MODE_CMD |
                                  FIFO_INT_LEVEL | FF));
    while(ucInd < ucarray_length)
    {
        //waiting for an interrupt
        while(RFM12_Get_NIRQState(sdevice_pins));
        //reading the status register
        uiStatusReg = RFM12_Read_StatusReg(sdevice_pins);
        if(uiStatusReg & 0x8000) //true, if a payload byte was received
        {
```

```
/*the payload byte is stored and the interrupt is cleared*/
ucarray2rcvd[ucInd] = RFM12_Read_RecvByte(sdevice_pins);
ucInd++;
    }
}
//storage in the FIFO buffer is disabled
RFM12_Write_CMD(sdevice_pins,
          ((FIFO_RESET_MODE_CMD|FIFO_INT_LEVEL)& FF_NOT));
}
```

13.3   Radio Protocols in the 2.4 GHz ISM Band

13.3.1 Bluetooth®

Bluetooth is a radio protocol based on the IEEE 802.15.1 standard [8], which specifies the protocol's physical layer and media access control. The aim of its development was to replace the wired transmission between closely located devices with a radio transmission. The number of network nodes should be low, but the communication should reach the quality of wired transmission in terms of bit rate and security.

The protocol has evolved over the years and Bluetooth has established itself as a standard that has found its way into mass-market products such as computers of all kinds and cell phones.

As of version 4 of the protocol, the LE mode[14] was defined in addition to the BR/EDR mode.[15] In LE mode, the transmission rate and at the same time the energy consumption were drastically reduced in order to make the protocol attractive for battery-powered devices. The establishment of a communication in the older versions takes place in seconds, which is acceptable for long communications. An entire communication session (establishment, short transmission, and termination) under version 4 is said to be possible within 3 ms [11]. No connections between devices operating with different modes can be realized.

Bluetooth defines *profiles* as a catalog of rules and protocols to allow devices from different manufacturers to communicate with each other. The profiles are standardized and do not have to be implemented on every device. In addition, *services* are defined as data structures that are used to characterize a function or property of a device.

---

[14]LE Low Energy.
[15]BR/EDR Basic Rate /Enhanced Data Rate.

13.3.1.1 Physical Layer [12, 15]

The transmission takes place in the 2.4 GHz ISM band between 2402 MHz and 2480 MHz. In BR/EDR mode, 79 radio channels are defined with a bandwidth and spacing of 1 MHz. In BR mode, data is modulated with a Gaussian frequency shift keying[16] and a gross transmission rate of 1 Mbit/s is specified. To achieve higher bit rates of 2 Mbps or even 3 Mbps at the same frequency channel spacing, phase shift keying[17] is used. The devices are manufactured in three performance classes:

• Class 1	+20 dBm (100 mW), range < 100 m;
• class 2	+4 dBm (2.5 mW), range < 50 m;
• class 3	0 dBm (1 mW), range < 10 m

By measuring the signal quality, the output power can be adapted to the environment.

In LE mode, 40 radio channels with a bandwidth of 2 MHz are defined and Gaussian frequency shift keying is used for signal modulation. Three of these radio channels are reserved for advertising, a procedure used by a device to send unidirectional messages on a regular basis. The other 37 radio channels are data channels. Version 4 of the protocol specifies a transmission rate of 1 Mbit/s (LE 1 M) for the LE mode.

From version 5 onward, further transmission rates and a further power class 1.5 with 10 dBm (10 mW) are specified for the LE mode. In LE 2 M mode, data can be transmitted at 2 MBit/s, or in LE Coded mode at 500 kBit/s or 250 kBit/s. The transmission rate can now be switched depending on the requirements. Reducing the transmission rate enables a reduction in performance while maintaining the same range. In order to detect possible transmission errors, a 24-bit CRC checksum is added to the Bluetooth messages in all versions. By introducing error-correcting data coding, the range in LE-coded mode can be increased while maintaining the same transmission power. As of version 5.2, it is also possible to adjust the transmit power in LE mode.

To avoid interference with other protocols from the 2.4 GHz ISM band, Bluetooth uses an adaptive "frequency hopping" procedure from version 1.2. This frequency hopping procedure provides for a continuous change of frequency channels. The frequency hopping is synchronized according to a 625 µs clock. This clock is provided to the network by the master, i.e. the device that initiated the connection. The pseudo-random frequency hopping sequence is derived from the address of the master, which is communicated to the slaves when the connection is established. The method is called adaptive because the master has the possibility to exclude from the frequency hopping table those radio channels on which it measures high radio activity that could lead to interference of the communication in its own network. Data traffic between master and slaves is controlled using the time-division multiplexing method. In the first half of a clock period only the master can start sending a message, in the second half only the addressed slave. For longer transmissions, the radio

---

[16] Frequency Shift Keying (FSK).
[17] Phase Shift Keying (PSK).

channel is changed after five clock periods at the latest. There are 1600 frequency jumps per second.

13.3.1.2 Communication Topologies

Bluetooth was developed to enable point-to-point communication. This also applies to the star-shaped Bluetooth networks called Piconet, which consist of a master and up to seven slaves. The number of slaves is limited to seven because they are addressed via three bits after the connection is established and the address "000" is reserved for broadcast. The connection between devices within range can take place ad-hoc, i.e. without an appointment. The following rules apply when setting up piconetworks:

- there may only be one master in each piconet;
- one device may be part of two piconets;
- a device may be master only in one piconet.

13.3.1.2.1 BR/EDR Topologies

In a BR/EDR piconet, all nodes communicate via the same physical transmission channel. They use a common clock to synchronize the communication and a common frequency hopping sequence. Communication can only take place between the master and the addressed slave. Communication between the slaves as well as broadcasting are not provided.

The participation of a device in two or more piconets creates a scatternet as in Fig. 13.5 but not a meshnet because the specification of the protocol does not allow the forwarding of the data flow from one piconet to the other. If desired, data exchange between piconets can be regulated at a different level. Different piconetworks can coexist in close proximity, but must use different transmission channels. Up to version 5 of Bluetooth, up to 255 inactive (parked) slaves could be used in addition to the seven active slaves, but they were

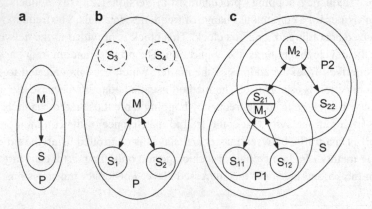

Fig. 13.5 Bluetooth BR/EDR network topologies. (a) Piconet with one slave, (b) Piconet with two active (S_1 and S_2) and 2 inactive (S_3 and S_4) slaves, (c) Scatternet (S) consisting of piconets P1 and P2

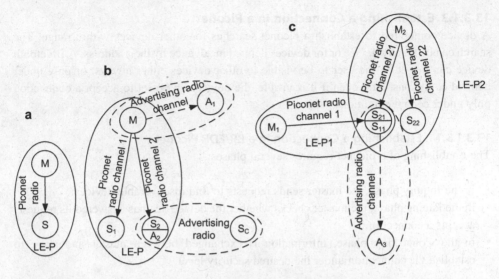

Fig. 13.6 Bluetooth LE network topologies. (**a**) Piconet with one slave, (**b**) Piconet with two slaves, one advertiser and one scanner, (**c**) Scanner network, consisting of the LE-P1 and LE-P2 piconets and one advertiser

synchronized with the master. The master could change the state (active or inactive) of the slaves synchronized with it. As of version 5, the "park" state has been removed from the specification.

13.3.1.2.2   LE Topologies [15]
Figure 13.6 shows the LE topologies corresponding to Bluetooth version 5.2. In an LE piconet, each slave communicates with the master over a different transmission channel. The master establishes a point-to-point connection with a slave, or reaches all slaves of the piconet with a broadcast. The transmission of data between piconetworks is allowed from version 5.1, so that Bluetooth measuring networks with a large range are possible.

Figure 13.6b shows an LE piconet consisting of one master and two slaves. Communication takes place via different radio channels. The master wants to extend the piconet. In the role of initiator, it listens (scans) on an advertising radio channel for advertisers (in this case A_1) in range that broadcast their willingness to connect. The slave S_2 transmits as advertiser and its messages are received by the scanner S_C. A connection between the two devices does not exist and is not intended.

Figure 13.6c shows an LE scatter network consisting of the piconets LE-P1 and LE-P2. The slave S_{21}/S_{11} is part of both piconets and aims to establish a piconet as master with the advertiser A_3, which has signalled its readiness to connect.

13.3.1.3 Establishing a Connection in a Piconet

A device configured to establish a piconet searches for other devices within range. The search can be for any device or for devices defined in advance by their address. A Bluetooth device can be set by the user to be visible to other devices either always, or only under defined conditions. And even if it is visible, the device can be set to accept a connection only under certain conditions.

13.3.1.3.1   Establishing a Connection in a BR/EDR Piconet

The establishing of a piconet requires several phases:

- in the Inquiry phase, the master sends requests to find discoverable devices;
- in the Paging phase, the master checks whether the device that has answered its requests accepts a connection;
- in the Connecting phase, information is exchanged between master and slave to establish the connection under the desired security level.

Pairing is the exchange of a connection key between devices that have previously authenticated each other.

13.3.1.3.2   Establishing a Connection in a LE Piconet

- **Advertising** is a procedure used by a BLE[18] device to send unidirectional data packets on an advertiser radio channel at defined time slots, with a predefined data structure. The procedure can also be used to establish a connection with a device as initiator. The initiator is a device that attempts to connect to a connection-enabled advertiser. An initiator can be a master, a slave, or a device that does not yet have a connection.
- **Scanning** is a procedure used by a BLE device to receive advertising data packets. A scanner can request additional data transmitted by the advertiser.
- **Discovering** is a procedure used by a BLE device to search for other devices in range or to signal its presence to other devices. The devices use the advertising or scanning procedure for this.
- **Connecting** is the procedure by which the connection between two BLE devices is established. During this procedure, one device is a scanner while the other is an advertiser. The advertiser must send a connection ready message.

13.3.1.4 Security of Communication

Bluetooth provides several mechanisms for securing communication such as: Authentication, Authorization and Encryption. The protocol defines the following security modes:

---

[18] BLE – Bluetooth Low Energy.

- Mode 1 (non secure mode). The device does not use any security mechanisms in this mode.
- Mode 2 (service level enforced security). The selected security mechanisms are used after the connection request has been accepted.
- Mode 3 (link level enforced security). In this mode, authentication is required during the connection setup, but encryption of the communication remains optional.

The security mechanisms are based on:

- Uniqueness of the Bluetooth address of the device;
- An adjustable 4 to 8 digit PIN number;
- Random numbers that are agreed between master and slave to secure the communication.

Version 4 of the protocol [12] extends the PIN to 16 alphanumeric characters and defines a fourth security mode in which devices use an authenticated connection key to make the connection more secure.

13.3.2 ZigBee

ZigBee was developed by the ZigBee Alliance (now Connectivity Standards Alliance) as a radio protocol for networking sensor nodes. The protocol allows complex network structures with up to 2^{16} nodes. Low radio power together with short messages allow to supply a node with a battery for years, simple nodes can even be supplied via energy harvesting.

13.3.2.1 ZigBee Devices

Depending on the complexity of the devices, ZigBee distinguishes between FFD[19] – and RFD[20]-devices (see Fig. 13.7). An FFD device fully implements the protocol functions, can build and manage a network and can communicate with all types of devices. An RFD device equipped with less memory implements only part of the protocol functions and can only communicate with other RFD devices.

A ZigBee device can have the following role in a network:

- A **coordinator** is an FFD device that sets up and manages an entire network.
- An **End-Device** is an RFD or FFD device that only partially implements the MAC layer requirements to reduce energy consumption.

---

[19] FFD – Full Function Device.
[20] RFD – Reduced Function Device.

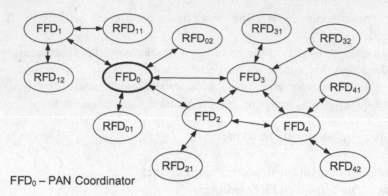

FFD₀ – PAN Coordinator

Fig. 13.7 ZigBee network

- A **router** is an FFD device that increases the range of a network.
- A **gateway** is an FFD device that connects a ZigBee network to another network (LAN, WLAN, etc.).
- **Trust center** is a ZigBee coordinator that centrally organizes network security.

The complexity of the implemented functions as well as the role in a network determine the energy consumption of a device. A distinction is made between:

- Green Power Devices: These are devices with extremely low energy consumption, which can be covered by energy harvesting or by a lifetime battery.
- End devices are predominantly in sleep mode and are powered by a replaceable battery.
- Router/Coordinator must be active all the time to forward the messages and are powered by the mains.

13.3.2.2 Protocol Structure

Figure 13.8 depicts the layer model of ZigBee version 3 [13]. The physical layer and media access control of the protocol are specified by the IEEE 802.15.4 standard, while the upper layers are specified by the ZigBee Alliance. The focus is on standardization of the protocol at all levels, as well as energy consumption.

13.3.2.3 Physical Layer

The ZigBee protocol defines 27 communication channels (see Table 13.2). The first channel is valid only in Europe, another 10 are valid only in the USA and channels 11 to 26 are defined in the 2.4 GHz ISM band. To ensure coexistence with other radio standards using the same radio space, ZigBee uses code division multiplexing. This method allows communication within the network at a transmission power of 1 mW (0 dBm). Despite the low transmission power, a large range can be achieved because the protocol provides for message forwarding. The devices have a globally unique address with which they must identify themselves when joining the network. They are then assigned a 16-bit network-internal address to reduce energy consumption.

Fig. 13.8 ZigBee 3 – Layer Model

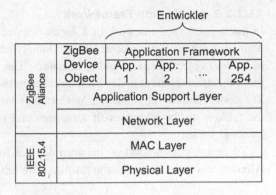

Table 13.2 ZigBee frequency bands

Frequency band	0	1	2	...	9	10	11	12	...	25	26
Frequency/MHz	868	906	908	...	922	924	2405	2410	...	2475	2480
Carrier spacing/MHz	–	2					5				
Bitrate/kBit/s	20	40					250				

13.3.2.4 Network Layer

The network layer of the protocol contains the necessary functions for setting up, administration, maintenance and security of a network. Each ZigBee network is built and managed by a single coordinator. The coordinator initializes the network, determines the communication channel and the name of the network. The network name is used to identify ZigBee networks that share the same radio space. Because the ZigBee protocol does not support "frequency hopping", the coordinator must search for a radio channel during the initialization phase of the network whose low radio activity ensures undisturbed communication. To do this, the coordinator measures the radio energy of a channel and searches for any ZigBee networks that are already active on this channel. The protocol provides for the networks to organize themselves. New devices can join or leave a network. Each new node is assigned an internal, unique 16 -bit network address. The coordinator assigns itself the network address (for example, 0x0000). A logical 16-bit address allows a large number of radio nodes in a network and, compared to the physical 64-bit identification number, leads to a shortening of messages and thus to a reduction in energy consumption. The routers continuously create and update a list of addresses of devices that are in range. The networks can be organized in a star, tree or mesh structure.

13.3.2.5 ZigBee Device Object

The ZigBee Device object is an important part of the protocol, which includes management functions related to setting up the network, updating the table of nodes in range, finding the appropriate endpoints (Application Endpoints) and managing the logical connections (bindings).

13.3.2.6 Application Framework

In the application framework of a ZigBee device up to 254 application endpoints can be defined. Each application object is individually addressable and describes the access to the individual objects: sensors and/or actuators. The logical connection between the application objects of different nodes from the same network is defined in a table (bindings table). The coordinator or the routers, as devices that are permanently supplied with power, store the table with the addresses of all nodes and their application objects, as well as the bindings table.

To ensure communication between devices from different manufacturers, the ZigBee Alliance has defined application profiles. The following profiles are already standardized:

- Building Automation (monitoring and control of commercial buildings);
- Home Automation (monitoring and control in the private sector);
- Health Care (transmission and monitoring of clinical parameters in the medical and fitness sector);
- Input Device (wireless connection of input devices such as mouse, keyboard, etc. to computer);
- Light Link (control of complex lighting systems);
- Remote Control;
- Retail Service (e.g. cashless payment);
- Smart Energy (transmission of energy measurement values and control of energy-relevant systems);
- Telecom Sevices (information transmission).

In the application layer of a ZigBee node, application objects can be defined from different application profiles.

13.4 Bluetooth Communication with the Serial Profile

The typical use case chosen in this example is the wireless connection of two microcontrollers. When choosing the protocol for the wireless connection between two microcontrollers, factors such as cost, distance, data rate, hardware and software complexity, interference and data security play an important role. A good solution for distances below 100 m is offered by Bluetooth, which has been specified for such applications. Various manufacturers offer inexpensive radio modules which, in addition to a 2.4 GHz transceiver, include a host controller that implements a Bluetooth version, a clock source and possibly an matched antenna. The modules provide various Bluetooth profiles and several serial interfaces such as USB, UART, SPI, PCM for communication with the host microcontroller (see Fig. 13.9).

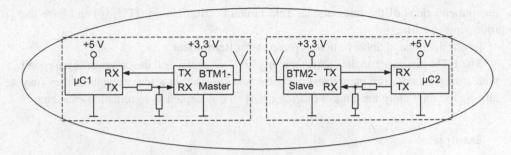

Fig. 13.9 Piconet with two microcontrollers

The SPP[21] profile, which emulates a serial RS-232 interface, is suitable for duplex communication. It should be ensured that the Bluetooth versions of the two modules BTM1 and BTM2 are compatible. For this example, modules from the Rayson company from the BTM series (e.g. BTM230) and, for simplicity, the UART interface were chosen for communication with the microcontroller. The communication parameters are preset to eight data bits, no parity, one stop bit and 19.200 kBit/s, but can be changed. The control of the radio modules and the entire communication between the microcontrollers run exclusively via UART and for this the functions presented in Chap. 7 can be used.

13.4.1 Operating Modes

The BTM modules can operate in data or command mode. After the successful pairing procedure, the modules are in data mode and all bytes received via UART are sent by radio.

In command mode, the module's settings can be changed and stored in non-volatile memory. When the module is in data mode, the command mode can be activated by the microcontroller sending a sequence consisting of three '+' characters followed by 'CR' via UART. The radio module will return the message 'OK' after approximately 1 s if the changeover is successful.

13.4.2 Command Set

The implemented command set is used to set the role of the module as master or slave, the settings of the UART interface, the connection type (automatic or manual), the Bluetooth address of the communication partner, etc. The complete command set can be found in the data sheet of the used module [5]. AT commands are used, which were first used to control modems. They consist of a sequence of ASCII characters, are based on the

---

[21] SPP – Serial Port Profile.

recommendations of the International Telecommunication Union (ITU) [6] and have the following structure:

[prefix][command abbreviation] <parameter> [terminator].

The BTM modules use the character string "AT" as prefix and the commands are coded with a single letter. The parameters are command-specific, in a few exceptions they may also be missing. Only the control character 'CR'[22] is accepted as terminating character.

Example	
"ATL1\r"	– Changes the UART transfer rate to 9.6 kBit/s.
"ATO\r"	– Switches the module from command mode to data mode.
"\r" stands for 'CR' and \n for 'LF' in the C programming language.	

The syntactically correct commands are usually executed in about 100 ms and acknowledged with "OK", otherwise the microcontroller receives the message "ERROR". The two messages are followed by "CR" and "LF". The establishment of a connection is acknowledged with "CONNECT", its interruption with "DISCONNECT". These messages are followed by the address of the communication partner.

13.4.3 Initialization of the Radio Module

With the function BTM_UART_Init the UART interface of the microcontroller is initialized for the communication with the radio module, respectively the communication parameters (baud rate, parity and number of stop bits) are changed according to the new settings.

```
void BTM_UART_Init (uint16_t uibaudrate,
                    uint8_t ucparity,
                    uint8_t ucstopbit)
{
    /*set baud rate */
    UBRR0 = uibaudrate;
    /*enable receiver, transmitter and receive complete interrupt*/
    UCSR0B = (1<<RXEN0) | (1<<TXEN0) | (1<<RXCIE0);
    /*set frame format*/
    UCSR0C = (1<<UCSZ00) | (1 << UCSZ01) | ucparity | ucstopbit;
}
```

---

[22] Coded in ASCII code with 0x0D.

For communication with a module with factory settings, the function is called as follows:

```
BTM_UART_Init(BAUD_19200, PAR_NO, STOP_BIT_1);
```

with:

```
#define BAUD_19200 59 //F_CPU = 18.432 MHz
#define PAR_ODD 0x30
#define PAR_EVEN 0x20
#define PAR_NO 0x00
#define STOP_BIT_1 0x00
#define STOP_BIT_2 0x08
```

In the following, the initialization of two BTM modules is explained as an example, which are to form a piconet together. The address of the master is "00126F250A6A" and of the slave "00126F250A6E". Saving the address of the communication partner prevents the module from being embedded in a scatternet. The radio connection is to be established manually. The UART_WriteBuffer function presented in Chap. 7 is used to transfer the AT commands.

13.4.3.1 Initialization of the Bluetooth Master

The one-time call of the function BTM_Master_Init initializes a BTM module as master and saves the settings in a non-volatile memory, provided the module is in command mode. The string "+++" activates the command mode, if this is already active, it is acknowledged with the message "ERROR" and the following commands are executed.

```
void BTM_Master_Init(void)
{
    delay(2000); //2 s wait before first command
    //the radio module is set to command mode
    UART_WriteBuffer((uint8_t*) "+++\r\0",5);
    delay(1500); //1.5 s wait to execute the previous command
    //the echo on the part of the radio module is switched off
    UART_WriteBuffer((uint8_t*) "ATE0\r\0",6);
    delay(100); //100 ms wait to execute the previous command
    //automatic establishment of the radio connection is deactivated
    UART_WriteBuffer((uint8_t*) "ATO1\r\0",6);
    delay(100); //100 ms wait to execute the previous command
    //the module is set as master
    UART_WriteBuffer((uint8_t*) "ATR0\r\0",6);
    delay(3000); //3 s wait to execute the previous command
    //the address of the slave with which the connection
    //is to be established is stored
```

```
    UART_WriteBuffer((uint8_t*) "ATD=00126F250A6E\r\0",18);
    delay(100); //100 ms wait to execute the previous command
}
```

The pairing procedure can now be initiated by calling the function BTM_Master_Connect. The master attempts to establish the connection with the slave whose address it has stored within 60 s. The slave is then connected to the master. If the slave is switched off or not in range, "Time out, Fail to connect" is reported to the microcontroller and the procedure must be restarted. The ·suggested waiting times are empirical values and may differ for other modules.

```
void BTM_Master_Connect(void)
{
    delay(100);
    //the pairing procedure is initiated
    UART_WriteBuffer((uint8_t*) "ATA\r\0",5);
}
```

If the command "ATO1" in the initialization routine is replaced with the "ATO0", the radio connection is automatically established as soon as the Master and the Slave are switched on and in range. The function BTM_Master_Connect does not have to be called anymore.

13.4.3.2 Initialization of the Bluetooth Slave

In the presented piconet the slave can be initialized as follows. With these settings the local echo is switched off and the slave allows a connection only with the master whose address it stores.

```
void BTM_Slave_Init(void)
{
    delay(2000); //2 s wait before first command
    //the radio module is set to command mode
    UART_WriteBuffer((uint8_t*) "+++\r\0",5);
    delay(1500); //1.5 s wait to execute the previous command
    //the echo on the part of the radio module is switched off
    UART_WriteBuffer((uint8_t*) "ATE0\r\0",6);
    delay(100); //100 ms wait to execute the previous command
    //the module is set as slave
    UART_WriteBuffer((uint8_t*) "ATR1\r\0",6);
    delay(3000); //3 s wait to execute the previous command
    //the address of the master that established the connection is saved
    UART_WriteBuffer((uint8_t*) "ATD=00126F250A6A\r\0",18);
    delay(100); //100 ms wait to execute the previous command
}
```

References

1. Yang, S.-H. (2014). *Wireless sensor networks. Principles, design and applications.* Springer.
2. Microchip: AT86RF231 Low Power 2.4 GHz Transceiver for ZigBee, IEEE 802.15.4, 6LoWPAN, RF4CE, SP100, Wireless HART and ISM Applications. https://www.microchip.com/wwwproducts/en/AT86RF231 resp. http://ww1.microchip.com/downloads/en/devicedoc/doc8111.pdf. Accessed: 4. Apr. 2021.
3. Nordic semiconductor. nRF24L01+ single Chip 2.4 GHz transceiver. Product Specification v.1.0. https://www.nordicsemi.com/Products/Low-power-short-range-wireless/nRF24-series. Accessed: 4. Apr. 2021.
4. Microchip Technology Inc. MRF24J40 data sheet. IEEE 802.15.4™ 2.4 GHz RF transceiver. www.microchip.com respectively. https://ww1.microchip.com/downloads/en/DeviceDoc/39776C.pdf. Accessed: 4. Apr. 2021.
5. Rayson Technology co. Ltd. BTM-230 data sheet. BC04-EXT Class1 module BTM-230. http://www.rayson.com/rayson/en/upload/prod_file/NP0019_1.pdf. Accessed: 4. Apr. 2021.
6. International Telecommunication Union. ITU-T V.250. Serial asynchronous automatic dialing and control. https://www.itu.int/rec/T-REC-V.250/de. Accessed: 4. Apr. 2021.
7. Vfg 30/2014, changed with Vfg 36/2014, changed with Vfg 69/2014. https://www.bundesnetzagentur.de/DE/Sachgebiete/Telekommunikation/Unternehmen_Institutionen/Frequenzen/Allgemeinzuteilungen/allgemeinzuteilungen-node.html. Accessed: 4. Apr. 2021.
8. IEEE Std 802.15.1™-2002. Wireless medium access control (MAC) and physical layer (PHY) specifications for wireless personal Area Networks (WPANs). http://ieeexplore.ieee.org. Accessed: 4. Apr. 2021.
9. Bundesnetzagentur. Vfg 76/2003. Allgemeinzuteilung von Frequenzen in den Frequenzteilbereichen gemäß Frequenzbereichszuweisungsplanverordnung (FreqBZPV), Teil B: Nutzungsbestimmungen (NB) D138 und D150 für die Nutzung durch die Allgemeinheit für ISM-Anwendungen. www.bundesnetzagentur.de. Accessed: 4. Apr. 2021.
10. Beuth, K., Breide, S., Lüders, C,. Kurz, G., Hanebuth R. (2009). Nachrichtentechnik. Vogel Industrie Medien GmbH & Co. KG, Würzburg.
11. Heydon, R. (2013). *Bluetooth low energy – The developer's handbook.* Prentice Hall.
12. Bluetooth. (2014). Specification of the bluetooth system core 4.2. www.bluetooth.com. . Accessed: 4. Apr. 2021.
13. ZigBee 3: Präsentation. (2014). https://zigbeealliance.org/solution/zigbee/. Accessed: 4. Apr. 2021.
14. HOPE MICROELECTRONICS CO., LTD. RFM12B – Universal ISM Band FSK Transceiver. www.hoperf.com. . Accessed: 10. Juli 2021.
15. Bluetooth SIG. Bluetooth core specification, revision v5.2. www.bluetooth.com. Accessed: 15. Febr. 2021.

Sensor Technology System Considerations

14

Abstract

In this chapter some system-related terms like sampling, quantization and digital filters are clarified. The chapter also presents some software tricks for abstracting the hardware pins and for using integer arithmetic.

A digital sensor usually consists of one or more analog transducers, a low-noise amplifier and possibly a voltage regulator. A transducer converts a physical quantity into an electrical one. This electrical, analog signal is amplified and made available for measurement, further processing or transmission. Additional processing such as linearization of the characteristic curve, compensation of the DC offset or the influence of other variables such as temperature, air pressure or humidity are easier to implement if the analog signal is converted into a digital signal beforehand. The digitization should take place as close as possible to the sensor in order to minimize the interferences. If the sensor and the digitization form one unit, it is called a digital sensor. It is important to understand the processing path of the digitization at least in rudimentary form in order to be able to assess the quality and significance of the processed signal.

Modern technologies, in particular microsystem technology (MEMS) and the high integration of different semiconductor structures, make it possible to integrate an ever increasing number of sensors in more and more devices or to integrate them into a single housing. Digital sensors have emerged to take advantage of digital processing and transmission, to reduce power consumption and circuit complexity, and to reduce the load on the host microcontroller. An example block diagram of such a sensor is shown in Fig. 14.1.

© The Author(s), under exclusive license to Springer Fachmedien Wiesbaden GmbH, part of Springer Nature 2023
A. Meroth, P. Sora, *Sensor networks in theory and practice*,
https://doi.org/10.1007/978-3-658-39576-6_14

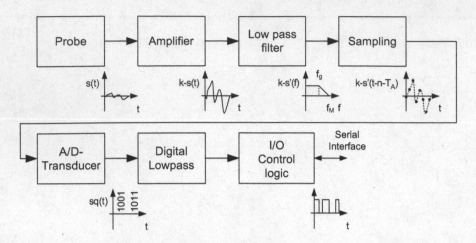

Fig. 14.1 Digital sensor – block diagram

Any interference on the analog side of the sensor is reduced to a minimum because of the extremely short lines. The measured values are made available in digital form with selectable resolution on a data bus which serves to network several circuit components.

14.1 Sampling

An analog signal whose transmission is susceptible to interference and whose processing is difficult also contains irrelevant and redundant components. Such a continuous-time and continuous-value signal $s(t)$ is completely described by a discrete-time and continuous-value sequence $s(t - n \cdot T_a)$ if the spectrum of the signal $s(f)$ is limited to a frequency f_M such that:

$$2 \cdot f_M \leq f_a = \frac{1}{T_a}, \qquad (14.1)$$

where n is a natural number and f_a is the sampling frequency. The requirement described by Eq. 14.1 is known as the sampling theorem or Nyquist criterion, and the frequency $f_a/2$ is referred to in the literature as the Nyquist frequency. In practice, this value is hardly tenable, i.e. one will usually sample considerably above this frequency, usually about seven to ten times higher than the cut-off frequency of the signal to be measured. To limit this cutoff frequency to the relevant spectral range of the analog signal, the signal is limited with a low-pass filter before sampling, which removes the irrelevant parts of the analog signal. This process is irreversible, and the removed parts cannot be recovered later. Band limiting is usually implemented with active filters. As the complexity (order) of the filter increases, so does the slope with which the amplitude response of the filter falls off from the

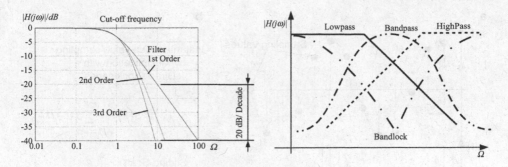

Fig. 14.2 Butterworth filter – Bode diagrams. (From [1])

cutoff frequency f_g. A higher order filter allows the cutoff frequency to be chosen close to the Nyquist frequency as illustrated in Fig. 14.2. The figure plots the magnitude of the transfer function $|H(j\omega)|$ in dB for first, second and third order Butterworth filters with unitary gain as a function of the normalized angular frequency $\Omega = \omega/\omega_g$ with $\omega = 2\pi f$. These filters exhibit an amplitude drop of $1/\sqrt{2}$ at the cutoff frequency and an attenuation of $n \cdot 20$ *dB/Decade* above the cutoff frequency, where n is the order of the filter. By calculating back to linear values, we obtain an attenuation factor of 10^n/*Decade*.

Sampling an analog signal produces a PAM[1] signal from which the original baseband signal can be reconstructed without error if the requirement of the sampling theorem is met. The reconstruction is implemented with a low-pass filter. Figure 14.3a shows the sampling of a harmonic signal using different methods. While in theory the samples are calculated by multiplying the original signal by a Dirac comb, in practice the sampling is realized using FET[2] transistors as electronic switches as shown in principle in Fig. 14.3b. The sliding and instantaneous sampling Fig. 14.3 II, III are used to transfer the samples of several channels to one data line in time-division multiplexing. The sample and hold method Fig. 14.3 IV is used in all analog/digital converters to ensure a constant voltage during conversion. The sampling theorem must also be considered when selecting the conversion rate of a microcontroller's analog-to-digital converter.

The spectrum $s(f)$ of an analog bandlimited signal $s(t)$ is shown in Fig. 14.4. The mathematical expression of the Fourier spectrum $s_a(f)$ of the sampled signal is [3]:

$$s_a(f) = \frac{1}{T_a} \cdot \sum_{n=-\infty}^{+\infty} s(f - nf_a) \tag{14.2}$$

Equation 14.2 shows that:

---

[1] PAM – Pulse Amplitude Modulation.
[2] FET – Field Effect Transistor.

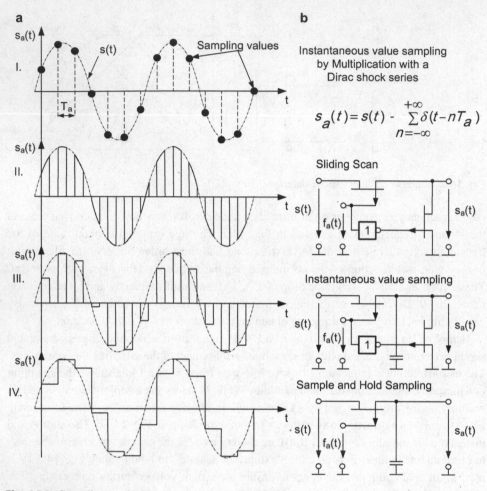

Fig. 14.3 Sampling method

- the spectrum of the baseband signal repeats around the sampling frequency and its multiples;
- the sampling frequency and its multiples are not present in the spectrum of the sampled signal.

Figure 14.4b shows an image of the real spectrum when the requirement of the sampling theorem is fulfilled. It is easy to see that with a steep low pass filter (reconstruction low pass filter) the original spectrum can be reconstructed.

If the input filter is missing, has a too high cut-off frequency or is not steep enough, an overlap of the partial spectra occurs as in Fig. 14.4c, which is called aliasing. The spectral components with frequency f_x with $f_M > f_x > f_a/2$ that satisfy the condition $|n \cdot f_a - f_x| < f_a/2$ are mirrored into the reconstructed signal and cannot be removed. These mirrored

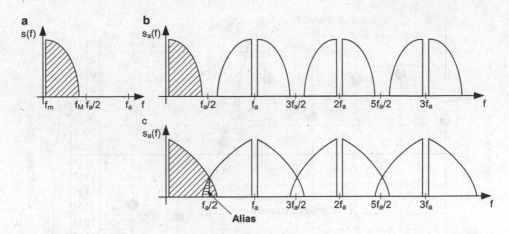

Fig. 14.4 Spectrum of the bandlimited signal $s(f)$ and the sampled signal $s_a(f)$ for $2f_M < f_a$ and $2f_M > f_a$

frequencies are called alias components. For example, if the input filter of a transmission system operating at a sampling frequency of 8 kHz limits the signal spectrum to 5 kHz, then a 4.5 kHz spectral component will appear in the reconstructed signal with a frequency of 3.5 kHz. The input filter is called an antialiasing filter. The occurence of the alias components is also the basis for the Nyquist criterion mentioned above. Since the spectrum is mirrored as shown in Fig. 14.4, it stands to reason that it must be limited to the range $\pm f_a/2$.

14.2 Quantization

Quantization as a subsequent step of sampling converts the discrete-time and continuous-value PAM signal into a discrete-time and discrete-value signal. The resulting quantized values are encoded over a number of n bits. A distinction is made between linear and nonlinear quantization. In linear quantization, the full scale is divided into 2^n intervals of equal size, as shown in Fig. 14.5 as an example for $n = 4$. Each quantization interval is assigned a binary code as an integer from the range $[0; 2^n - 1]$. For sensors that need to convert an AC voltage, such as acceleration, angular velocity, or magnetic field sensors, among others, the quantization intervals of the A/D converters are coded in two's complement to facilitate later data processing. For sensors that measure air pressure, temperature, or humidity, the encoding may be unsigned. Quantization is a comparison method in which each analog sampled value is assigned the corresponding code of the interval in which the sampled value falls. This assignment also eliminates the redundant components from the signal. The process is reversible and the sampled signal can be recovered except for an error. The error signal is called quantization noise and is caused by the rounding that is done during quantization. If u_q is the amplitude of a quantization interval and one refers to

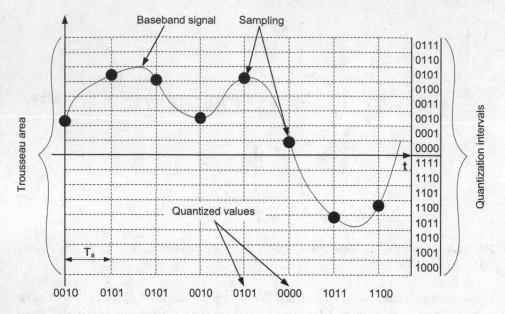

Fig. 14.5 Linear quantization

the middle of the interval [5], then the quantization noise takes values in the range $[-u_q/2;$ $+u_q/2]$. A measure for this noise is the signal-to-noise ratio [4]:

$$\frac{S}{N_q}\Big|_{dB} = 10 \ \log \frac{S}{N_q} \approx n \cdot 6 \ \ \text{in} \ \ \text{dB} \tag{14.3}$$

where S is the signal power, N_q is the quantization noise power and n is the bit resolution. When the modulation range is exceeded, we talk about overdriving, an effect that greatly affects the signal-to-noise ratio. An underdrive of 50% reduces the signal-to-noise ratio by about 6 dB, which from this point of view corresponds to quantization with $(n-1)$ bits.

Non-linear quantization is used in audio technology to ensure a constant signal-to-noise ratio over large dynamic ranges.

14.3 Digital Filtering

A digital filter is used as shown in the block diagram of a digital sensor Fig. 14.1. Before sampling, the spectrum of the signal to be measured is limited with a low-order antialiasing filter to a frequency f_M which is much larger than the measurement frequency. To satisfy the Nyquist criterion under these conditions, the signal must be oversampled. After quantization, the signal is digitally filtered.

The digital filter replaces a complex analog filter that is difficult to integrate, parameterize and reproduce. The new cut-off frequency of the digital filter depends on the desired measuring rate. The achieved slope of the filter is very high. The filter characteristics can be changed during operation and are not affected by temperature and operating voltage fluctuations. The behavior of a digital filter can be accurately simulated already in the design phase.

14.3.1 Finite Impulse Response (FIR) Filter

FIR filters are finite impulse response filters with the transfer function [3]:

$$H(z) = \sum_{k=0}^{N} h[k] \cdot z^{-1})$$ (14.4)

N is the filter order, $h[k]$ are the filter coefficients which can be calculated by sampling the impulse response of the filter [3] and z^{-1} signifies a time delay.

$$h[k] = \frac{1}{2} \cdot si\left(\frac{\pi}{2} \cdot \left(k - \frac{N}{2}\right)\right) \text{ with } i = 0 \text{ N.}$$ (14.5)

The si function, or sample function, is defined as $si(x) = sin(x)/x$. The structure of an FIR filter is shown in Fig. 14.6. For the calculation of the filtered value y_n, only the N previous values are considered in addition to the current sample value:

$$y_n = \sum_{k=0}^{N} x[n-k] \cdot h[k]$$ (14.6)

With these considerations, digital low-pass filters can be designed that have an increasing slope of the filter edge with the increase of the filter order. The minimum attenuation of these filters in the stopband reaches 20 dB and is independent of the filter order. The ripple of the filter is reduced and the attenuation in the stopband is increased by weighting the filter coefficients with a window function $h'[k] = h[k] \cdot w[k]$. The following equation:

$$w[k] = a - b \cdot cos\left(\frac{2\pi \cdot k}{N}\right)$$ (14.7)

describes the *Hanning* (with $a = b = 0.5$) and *Hamming* (with $a = 0.54$ and $b = 0.46$) functions, two of the best-known window functions. The window method can be used to calculate the coefficients of an FIR filter with a given cutoff frequency, slope and attenuation in the stopband. In the literature [2, 3], the *optimal* method is described as another design method, which leads to a lower filter order.

Fig. 14.6 FIR filter

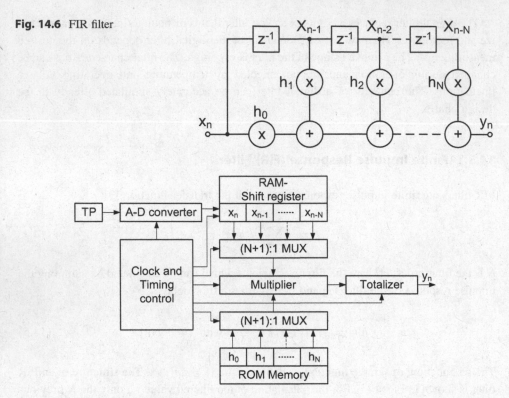

Fig. 14.7 Digital filter – basic structure

A digital filter that operates in real time can be implemented in terms of hardware – as shown in Fig. 14.7. The high clock frequency and the parallel processes allow a more complex filter structure. An application-specific integrated circuit (ASIC) can integrate a digital filter together with the antialiasing low-pass filter (TP) and the A/D converter. An FPGA[3] also needs an external filter and the less complex CPLD[4] s can process the digital values from an external A/D converter. The digital control can change the resolution of the A/D converter, or the cut-off frequency of the filter.

In software terms, the Eq. 14.6 can also be solved with a digital signal processor (DSP) or a microcontroller. This is shown in principle in Fig. 14.8. Pointer 1 points to the vector location where the next digital value is stored. After storing, the multiplications according to Eq. 14.6 between the filter coefficients and the digital values take place in the second step. The pointers 2 and 3 are incremented synchronously and point to the values to be multiplied. At the same time, the results of the multiplications are summed up. In the last

[3] FPGA – field programmable gate array.

[4] CPLD – complex programmable logic device.

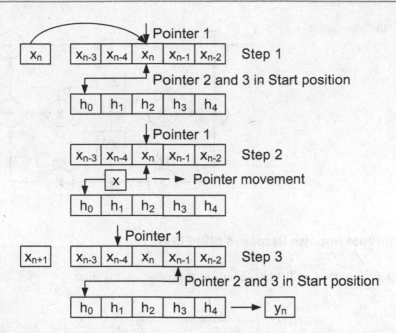

Fig. 14.8 FIR filter – software solution

step, the filtered value y_n is output and the pointers are reset. The duration of the presented sequence must be smaller than the sampling period.

Unlike the DSPs, which can calculate with floating point numbers, most 8-bit microcontrollers prefer to calculate with fixed point numbers or with integers. For this reason, the filter coefficients are quantized, which leads to a deviation from the ideal transfer function. Improperly selected variable types for the calculation of the digital filter can also lead to a non-ideal effect. The maximum intermediate result produced by the multiplications can be well estimated because both the bit resolution of the A/D converter and the magnitude of the coefficients are known. For FIR filters, overflow can usually occur when the intermediate results are summed. Variable types that span many more bits than the bus width of the microcontroller reduce the risk of overflow, but slow down the computation speed. To increase the computation speed while keeping the overflow risk low, the scaling method can be used. The filter coefficients are all divided by the same factor. If random overflow still cannot be avoided, the results can be limited (saturation method). FIR filters are easy to implement, are tolerant to quantization of the filter coefficients and have only zeros, therefore they are always stable. Because of the high order of these filters, many mathematical operations are required.

Fig. 14.9 IIR filter second order

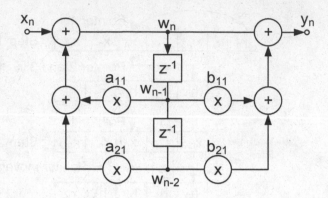

14.3.2 Infinite Impulse Response (IIR) Filter

The transfer function of an IIR filter emulates that of an analog filter:

$$H(z) = \frac{1 + b_1 \cdot z^{-1} + \ldots + b_N \cdot z^{-1}}{1 - a_1 \cdot z^{-1} - \ldots - a_M \cdot z^{-1}} \tag{14.8}$$

These filters have both pole and zero points and are recursive, resulting in an infinitely long impulse response. They can easily become unstable due to improper quantization of the filter coefficients and overflow. Overflow must be prevented by downscaling the coefficients and limiting (saturating) the intermediate results. However, these filters provide a high filter slope at a relatively low order of the filter, resulting in a reduction of mathematical operations per filtered value. The transfer function of a filter $(2 \cdot n + 1)$ -th order can be decomposed as follows:

$$H(z) = H_1 + \prod_{k=1}^{n} H_{2k}(z) \tag{14.9}$$

with $H_1(z)$ being the transfer function of a first order filter and $H_{2k}(z)$ the transfer function of a second order filter. This decomposition reduces the deviation from the ideal transfer function due to rounding. The structure of a second order digital IIR filter stage is shown in Fig. 14.9.

14.3.3 Filtering Using the Example of an FIR Filter

The following example shows the implementation of an FIR filter using an 8-bit microcontroller of the ATmega family in integer arithmetic. The microcontroller is to filter AC voltages with a peak value of up to 0.5 V. The analog signal is band-limited with an external low-pass filter and provided with a DC offset of +0.5 V. The microcontroller

Table 14.1 Filter coefficients

Calculated coefficients	Integer coefficients (char cH[k])
−0.011408912567595797	−1
−0.017514503197375211	−2
−0.017666528020283574	−2
−0.0013393806576736487	0
0.036486634909693227	5
0.092260634158284921	12
0.15297263055073201	20
0.20026952064209985	26
0.21814559786716198	28

samples at 5 kHz the resulting signal and quantizes it. The digital filter shall have a cutoff frequency of 340 Hz, a ripple of 1 dB in the passband, and an attenuation of 40 dB at 816 Hz. The specified conditions can be met with a 16th order equiripple filter. Because of coefficient symmetry, only the first nine are listed in Table 14.1. The coefficients calculated with Matlab are multiplied by 128 and then rounded. After calculation, the filtered values must be divided by 128, which means a shift to the right by seven digits so that the amplitude of the filtered signal is not distorted. The limitations due to the integer arithmetic must be investigated in a simulation phase so that the desired conditions are fulfilled by the digital filter.

The sampling period is set with a timer interrupt whose service routine (ISR) is listed below:

```
ISR (TIMER1_COMPA_vect)
{
    uiADCValue = ADC; //the converted value is stored temporarily
    ADCSRA |= (1 << ADSC); //a new AD conversion is started
    ucADCFlag = 1;
}
```

The filtering takes place in the endless loop of the main function:

```
#define n 17 //number of filter coefficients
//globally declared variables
volatile uint16_t uiADCValue; //cache for the quantized value
uint16_t uiFIFOin[n]; //input buffer for the n sampled values
int iAktPosFIFOin = n-1; //pointer to the current position in the
                         //input buffer (pointer 1)
uint8_t ucLvFIFOin; //control variable for the input buffer (pointer 2)
long lYout = 0; //filtered value
if(ucADCFlag)
```

```
{
    ucADCFlag = 0;
    lYout = 0; //the filtered value is initialized
    uiFIFOin[iAktPosFIFOin] = uiADCValue; //the sampled value is saved
    ucLvFIFOin = iAktPosFIFOin; //the control variable is recalculated
    for(uint8_t k = 0; k < n; k++) //k = pointer 3
    { //in the loop the new filtered value is calculated
        lYout = lYout + uiFIFOin[ucLvFIFOin] * cH[k];
        ucLvFIFOin++;
        if(ucLvFIFOin == n) ucLvFIFOin = 0;
    }
    //the current position of the pointer is decremented
    iAktPosFIFOin-;
    if(iActPosFIFOin < 0) iActPosFIFOin = n -1;
    if(lYout < 0)lYout = 0; //value limit to prevent the overflow
    else lYout = lYout / 128; //the coefficients were multiplied by 128
}
```

After filtering, the filtered values *lYout* can be processed further.

14.4   I/O Control Logic

The values measured by the digital sensor are made available in digital form to a microcontroller via a serial interface. The sensors are preconfigured as slaves, which means that they cannot initiate communication themselves. Some digital sensors have external outputs that can be used to trigger an interrupt on the microcontroller. This notifies the master that a new reading has been done or that the reading has exceeded a set limit. The sensors usually have a serial interface such as I^2C or SPI. The microcontroller as the master initiates the communication, reads the measured values and transmits the settings. According to these setting values, the I/O control logic drives the components of the sensor and parameterizes it. The I/O control logic can perform the following actions:

- Control of the serial interface;
- Start of a new measurement according to the measurement mode: single measurement mode or automatic measurement mode;
- Control of the measurement if several sensors are present;
- Storage of the measured values in registers or in a FIFO buffer;
- Storage of the received settings in the corresponding registers;
- Parameterization of the sensor according to the stored settings;
- Change of the measuring rate with corresponding adjustment of the sampling frequency and the digital low pass filter;

- If required, perform temperature compensation or DC offset correction of the measured values;
- Comparison of the measured values with the set limit values;
- Control of the external interrupt outputs.

14.5 Abstraction of the I/O Pins

In general, sensors have additional control inputs besides the serial data lines. To keep the software code as easily portable as possible, it is recommended to describe the wiring of these control inputs on the microcontroller in a single place (board abstraction layer in the software architecture). For this purpose, there is a data structure in the software module of the sensor in which these control connections are listed:

```
typedef struct
{
    tspiHandle SENSORspi;
    volatile uint8_t* Pin1_DDR;
    volatile uint8_t* Pin1_PORT;
    uint8_t Pin1_pin;
    uint8_t Pin_state;
    volatile uint8_t* Pin2_DDR;
    volatile uint8_t* Pin2_PORT;
    uint8_t Pin2_pin;
    uint8_t Pin2_state;
} SENSOR_pins;
```

In case of an SPI connection this data structure stores the structure tspiHandle (see Chap. 8) for the control of the chip select input and the addresses of the DDR and PORT registers for the control of Pin1 and Pin2 of the sensor. Thus the control of the sensors remains independent of the board configuration. The same SPI module can thus be used to control different sensors connected to the same bus, and one sensor software module can be used to control several sensors of the same type. If the sensor has no controllable pins, its SPI data structure only stores the data structure for controlling the chip select pin.

```
typedef struct
{
    tspiHandle SENSORspi;
} SENSOR_pins;
```

The placeholder SENSOR in the above examples must be chosen differently for each sensor, it is recommended to use the name of the sensor. For all sensors connected to the same microcontroller, one data structure is declared and initialized in the main file for each sensor.

14.6   Integer Arithmetic

An 8-bit microcontroller of the ATmega family implements arithmetic operations (addition, subtraction and multiplication) with integer and fixed-point data types. Fixed-point data types allow the mapping of rational numbers as multiples of 2^{-7} or 2^{-15}. This mapping is used to approximate the real numbers. Division is not provided for in hardware and is performed as a library function by shifting and applying shift, multiply and subtract.

In the programming language C, on the other hand, real numbers are approximated with the floating-point types *float* and *double*. For the reasons mentioned in Chap. 3, integer data types should be preferred in programming. These represent a bounded, closed interval of integers. This interval is called the value range.

14.6.1   Microcontroller Internal Number Formats

The working registers of an 8-bit microcontroller of the ATmega family and the memory cells that are individually addressable are all eight bits in size. To be able to address a 64 kByte memory area, register pairs are used. A distinction is made between *unsigned* and *signed* integer types. A signed integer type stores integers, an unsigned type stores only natural numbers and the 0. Table 14.2 represents the internal number representation of a hypothetical 3-bit microcontroller in binary format. When the most significant bit is 1, the same number content is interpreted differently. The interpretation depends on the selected integer type. Incrementing the highest value in the internal representation (all bits 1) causes a jump to the lowest value (all bits 0). Decrementing the lowest value causes a jump to the highest value. This overflow can also be observed with the *unsigned data type*. With the *signed data type,* a jump takes place when the most significant bit is changed, which is referred to as a two's complement overflow.

An 8-bit integer type takes values in the interval [0; 255] in the *unsigned representation* and values in the interval [−128; +127] in the *signed representation*. The microcontroller supports the handling of integer types wider than eight bits by the status register SREG:

- **Bit 5** – Half carry flag (H) – is set if after an arithmetic operation one of the byte halves of the result exceeds the value 9. This bit is used for the representation of binary coded decimal numbers.
- **Bit 4** – Sign bit (S) – is calculated as follows: $S = N \oplus V$, and is used when testing *signed data types*.

Table 14.2 Microcontroller internal number representation

Internal binary representation	111	000	001	010	011	100	101	110	111	000
Unsigned	7	0	1	2	3	4	5	6	7	0
Signed (two's complement representation)	−1	0	1	2	3	−4	−3	−2	−1	0
Signed (one's complement representation)	0	0	1	2	3	−3	−2	−1	0	0

- **Bit 3** – Two's complement overflow flag (V) – is set when the most significant bit of the result of an arithmetic or logic operation has changed. This bit indicates a possible two's complement overflow.
- **Bit 2** – Negative flag (N) – maps the most significant bit of the result of an arithmetic or logic operation.
- **Bit 1** – Zero flag (Z) – is set when the result of an arithmetic or logic operation is zero.
- **Bit 0** – Carry flag (C) – is set to signal a carry at the higher order location.

A detailed description of the status register can be found in the data sheet [6]. The complete list of operations affecting the individual bits can be found in [7].

14.6.2 Unsigned Integer Types

The value range of an unsigned integer type that is n-bit large is $[0; +2^n - 1]$. n can be 8, 16 or 32. With these data types it must be noted that:

$$(2^n - 1) + 1 = 0 \quad resp. for \ m < (2^n - 1) \quad (2^n - 1) + m = m - 1$$
$$0 - 1 = 2^n - 1 \qquad\qquad\qquad\qquad\qquad 0 - m = 2^n - m \qquad (14.10)$$

Operations (see Eq. 14.10) which return a number outside the range of values as a result lead to an overflow error. A careful analysis of the algorithm to be implemented and the choice of appropriate data types lead to a compact program that executes quickly and without errors. For the following considerations, two operands *ucNumber1* and *ucNumber2* and the result *ucResult* of type *unsigned char* are chosen. Similar considerations apply to other unsigned integer types. The following cases are to be considered when using unsigned integer types:

- **Assignment:** When a variable of type `unsigned char` is assigned a negative value (e.g. `ucNumber = −1;`), the variable stores the two's complement of the number amount (255 in this example) after the statement is executed.

- **Addition and multiplication: In** principle, the addition and multiplication of two variables `ucNumber1` and `ucNumber2` can lead to a range overrun. If this is not taken into account, the result will be as follows:
- `ucResult = (ucNumber1 + ucNumber2) % 256` resp.
- `ucResult = (ucNumber1 * ucNumber2) % 256`.
- **Division:** A rounding error can occur with integer division.
- **Comparison:** *Instruction1* from the following example is never executed because the condition is evaluated as false for all values of *number1* and *instruction2* is always executed because for all values of the variable *ucnumber2,* the condition is true:

```
if(ucnumber1 < 0) statement1;
if(ucnumber2 > -7) statement2;
```

- **Shift:** As long as the value range is not exceeded, shifting to the left by n bits corresponds to a multiplication by 2^n. Otherwise, the result is as follows:
`ucResult = (ucNumber1 < < n) % 256;`
Shifting to the right by n bits is equivalent to an integer division with 2^n.

14.6.3 Signed Integer Types

The signed integer data types can be 8, 16, or 32 bits in size, which corresponds to 1, 2, or 4 bytes. To represent negative numbers, the most significant bit of the data type is used as the sign. This halves the value range of the absolute value, since one bit is omitted for the absolute value representation (the value range with eight bits would then be $0 \ldots$ 127 and $-0 \ldots -127$). A disadvantage of this representation is that the zero is present twice and closed arithmetic is no longer possible.

In the alternative one's complement representation, the negative numbers are created by bitwise inversion of a positive number. In this representation, however, the value 0 is also present twice if all bits are 0 or all are 1.

Microcontrollers instead use the two's complement for the representation of negative numbers. In this representation, 0 is uniquely defined when all bits of a number are 0. Because the number of 2^n combinations of an n-bit number is even, there is no symmetry of positive and negative numbers with respect to zero. The range of values of an n-bit signed integer type is $[-2^{n-1}; +2^{n-1} - 1]$. In this representation, if $-x$ is a negative number and $+x$ is its absolute value, then the familiar equation holds:

$$x + (-x) = 0 \tag{14.11}$$

To calculate the representation of a negative number in two's complement, the positive number is inverted bit by bit and a 1 is added to the result. A similar procedure is used to calculate the absolute value of a negative number, as in the following example code:

```
uint8_t ucNumber;
signed char scNumber = -5; //-5₁₀ = 1111 1011₂
ucNumber = ~scNumber; //result: ucNumber = 0000 0100₂ = 4₁₀
ucNumber = ucNumber + 1; //result: ucNumber = 0000 0101₂ = +5₁₀
```

Equation 14.12 represents the two's complement overflow in mathematical form. A numerical example of such a two's complement overflow is illustrated in Table 14.3. The variable scNumber is of type signed char.

$$\begin{aligned}(2^{n-1}-1)+1 &= -2^{n-1}\\(-2^{n-1})-1 &= (2^{n-1}-1)\end{aligned} \tag{14.12}$$

Table 14.4 explains operations with signed, 8-bit variables that lead to a misinterpretation or to an unexpected result. The variable scNumber is of type signed char and n is a natural number.

A rounding error also occurs when dividing signed data types. Table 14.4 shows the difference between integer division and right shifting of negative values.

14.6.4 Detection and Prevention of Overflow

The errors caused by the overflow have more serious consequences than rounding errors, especially in recursive algorithms. An overflow that occurs when calculating an FIR filter will cause a single disturbance in the filtered signal in the form of a jump with the duration of a sampling period. An overflow occurring when computing an IIR filter will cause the output signal to oscillate. The decay of the oscillation is highly dependent on the filter order. The higher the filter order, the longer the oscillation will last.

The information provided by the status register can be used to detect an overflow. To detect an overflow, each result must be tested. This has a negative impact on program complexity and execution time. To minimize the effect of an overflow when it is detected, one of the limits of the value range is assigned to the result depending on the direction of the overflow. This method is called saturation.

To prevent an overflow in principle, either variables with a larger value range are selected or all input values are divided by the same constant. This second option is known as scaling.

Table 14.3 Numerical example for the two's complement overflow

Operation	Value before operation	Value after operation
scNumber++;	+127	−128
scNumber--;	−128	+127

Table 14.4 Operations with signed integers

Operation	Operands	Statement	Result
Allocation	scNumber +127 < n < 256	scNumber = n;	scNumber = n − 256 The internal representation of scNumber and n are the same
Division	scNumber =−(2n + 1)	scNumber = scNumber/2;	scNumber = −n
	scNumber =−(2n)		scNumber = −n
	scNumber = −1		scNumber = 0
Rightwrd shift	scNumber =−(2n + 1)	scNumber = scNumber >> 2;	scNumber = −(n + 1)
	scNumber =−(2n)		scNumber = −n
	scNumber = −1		scNumber = −1
Multiplication	scNumber = −128	scNumber = scNumber *(−1);	scNumber = −128
Comparison	scNumber +127 < n < +255	if(scNumber > n) statement;	The statement is never executed because the condition is always false.

In safety-relevant applications, it is essential to perform a plausibility and value range check after important operations. In fact, value range violations have already led to dramatic accidents, for example the crash of the Ariane rocket during the maiden flight 501 in 1996.[5]

References

1. Meroth, A., & Tolg, B. (2008). *Infotainmentsysteme im Kraftfahrzeug*. Friedr. Vieweg & Sohn Verlag & GWV Fachverlage GmbH.
2. von Grüningen, D. C. (2008). *Digitale Signalverarbeitung*. Hanser.
3. Roppel, C. (2006). *Grundlagen der digitalen Kommunikation*. Hanser.
4. Werner, M. (2006). *Nachrichten-Übertragungstechnik*. Friedr. Vieweg & Sohn Verlag & GWV Fachverlage GmbH.
5. Borucki, L. (1985). *Grundlagen der Digitaltechnik*. Teubner Verlag.
6. Microchip. (2021). 8-bit Atmel Microcontroller with 4/8/16k bytes in-system programmable flash. www.microchip.com. Accessed 2 Apr 2021.
7. 8-bit AVR® Instruction set. http://ww1.microchip.com/downloads/en/devicedoc/atmel-0856-avr-instruction-set-manual.pdf. Accessed 2 Apr 2021.

---

[5]Cf. http://esamultimedia.esa.int/docs/esa-x-1819eng.pdf and http://sunnyday.mit.edu/accidents/Ariane5accidentreport.html respectively.

Environmental Sensors

Abstract

In this chapter, four digital environmental sensors are presented, a digital air pressure sensor with temperature measurement, a humidity sensor with temperature measurement, a temperature sensor and a fine dust sensor.

Environmental sensors allow the acquisition of data from air or water. They are required in various applications, on the one hand in the acquisition of atmospheric data such as pressure, temperature and humidity with the aim of determining the weather or the ambient conditions, for example in the storage of products. Secondly, they provide information about the quality of the environment, for example the concentration of fine dust or the concentration of certain gases in the air. Thirdly, their data is needed for the indirect determination of variables, for example the dew point, the height above sea level or an air flow rate. In this chapter, four sensors are presented as examples: the digital pressure sensor MPL3115, the humidity sensor SI7021, the temperature sensor TMP75 and the fine dust sensor SDS011.

15.1 MPL3115 Digital Air Pressure Sensor

MPL3115 [1, 2] and [3] is a digital air pressure sensor that can be used for measuring absolute air pressure in the range 200 hPa ... 1100 hPa. The sensor is internally calibrated for measurements above 500 hPa (mbar). The 20-bit A/D conversion allows a resolution of 0.01 hPa (1 Pa). The measured air pressure can be converted internally into altitude, which converts the sensor into an altimeter. In addition to air pressure, it can also measure

A. Meroth, P. Sora, *Sensor networks in theory and practice*,
https://doi.org/10.1007/978-3-658-39576-6_15

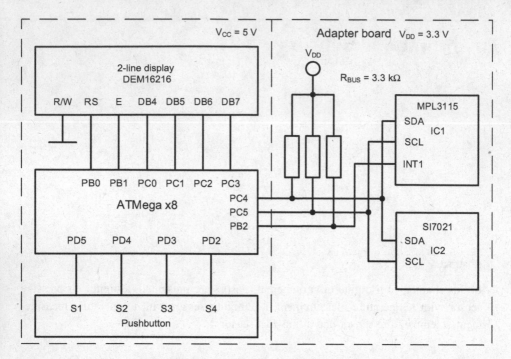

Fig. 15.1 Electronic weather station with MPL3115 and SI7021

temperature in the range −40 °C ... +85 °C. The sensor is preconfigured as an I^2C slave and can communicate with a master in Fast-mode. Via I^2C, the master transfers the desired settings and reads out the measured values. Events that have occurred can be signaled to the master via adjustable interrupts and controllable outputs.

Figure 15.1 shows the circuit of an electronic weather station consisting of an air pressure sensor MPL3115 and a humidity sensor SI7021. The humidity sensor is described in Sect. 15.2.

15.1.1 Functionality

Basically, sensors for detecting the absolute pressure work with a membrane which is applied to a chamber with stable walls in a vacuum or at a very low reference pressure and which seals it hermetically. As soon as the sensor is exposed to atmospheric pressure, the diaphragm experiences a force in the direction of the vacuum and is deflected due to its elasticity (see Fig. 15.2). The amount of deflection is related to pressure via elasticity, a fixed shape and material parameter. Thus, to determine the pressure, one must determine the deflection, which by its nature is very small. In the case of a micromechanical sensor (MEMS sensor), the values for the membrane thickness are a few µm, the distance between

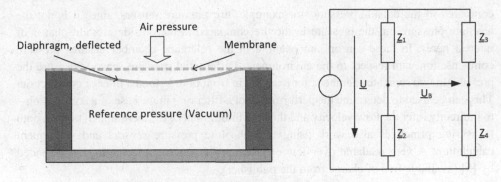

Fig. 15.2 Principle of a pressure sensor (left) and bridge circuit (right)

the membrane and the base plate a few tens to hundreds of μm and the diameter of the pressure chamber a few hundred μm.

In principle, there are two ways to measure the deflection:

- Resistive measurement exploits the effect that the resistance of a material changes due to its strain. This can be a metallic material, a metal oxide or a semiconductor, in the latter case the piezoresistive effect is mostly used. Since the resistance changes are extremely small, the measurement will be made with a bridge circuit (Fig. 15.2 right). The impedances (resistors in the case of resistive sensors) of the bridge form two voltage dividers and the bridge voltage U_B is 0 V in the balanced case; as soon as the bridge is detuned, the voltage changes. Depending on the measuring principle, one resistor (usually Z_2), two diagonally opposite resistors (Z_2 and Z_3) or all four can be changed, whereby it is usually ensured that two resistors change in opposite directions to the measured variable, whereby the sensitivity is considerably increased.

A very clear derivation and calculation basis can be found in Chap. 8 of [4]. Resistive sensors are often sensitive to temperature fluctuations and are compensated accordingly.

- Capacitive measurement exploits the effect that the capacitance of the diaphragm changes relative to the base plate when the diaphragm bends. The capacitance is determined with alternating current either in a bridge circuit or by detuning the resonant frequency of a resonant circuit (Chap. 9 in [4]).
- In piezoelectric measurement, the electrostatic field strength changes due to the deformation of a crystal as charges are displaced.
- Other measuring elements can be magnetic Hall sensors or plunger coils. These are not used in MEMS technology.

In addition to absolute pressure sensors, there are also relative pressure sensors and differential pressure sensors. Relative pressure sensors measure the pressure in a space compared to the outside pressure (for example, tire pressure sensors, since it is not the absolute pressure but the pressure in the tire compared to the outside pressure that is of interest here). In these sensors, an opening in the reference chamber ensures pressure compensation with respect to the environment. Differential pressure sensors measure the pressure in two separate chambers, for example in front of or behind a filter or constriction. This can be used to determine the performance of a filter or – in the case of a constriction – to indirectly infer the flow velocity and thus the throughput of the medium. Of course, both measuring principles also work using two absolute pressure sensors and subsequent calculation. A very readable overview of pressure measurement technology is provided by [5] (available free of charge from the publisher).

15.1.2 Structure of the MPL3115

A schematic block diagram of the device is shown in Fig. 15.3.

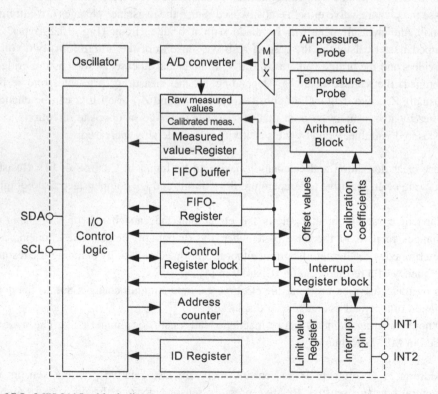

Fig. 15.3 MPL3115 – block diagram

15.1.2.1 Measuring Probe

The air pressure acts on the air pressure sensor through a small hole in the sensor housing. This sensor is built into a bridge circuit whose analog output voltage is proportional to the absolute air pressure. The absolute air pressure p measured at height H above sea level can be calculated using the barometric altitude formula [6]:

$$p = p_0 \cdot e^{-\frac{H}{8000}}, \tag{15.1}$$

where p_0 is the normal air pressure at sea level. With the measured value of the absolute air pressure p, if the altitude H is known, the air pressure at sea level p_0 can be calculated with Eq. 15.2 from [7]:

$$p_0 = \frac{p}{\left(1 - \frac{H}{44,\,330.77}\right)^{\frac{1}{0.1902632}}} \tag{15.2}$$

The coefficient 44,330.77 accounts for the dependence of the atmospheric pressure on the global mean temperature ($T_m = 15\,°C = 288.15\,K$) and the vertical temperature gradient ($\gamma = -0.65\,K/100\,m$) (see [6]):

$$44,330 \approx \left|\frac{T_m}{\gamma}\right| \tag{15.3}$$

The sensor calculates the altitude with the normal air pressure at sea level $p_0 = 1013.25$ hPa using Eq. 15.4 [1]. In order to be able to calculate a more accurate altitude, the current relative air pressure must be stored in the volatile registers BAR_IN_MSB_REG (address 0x14) and BAR_IN_LSB_REG (0x15) with a resolution of 2 Pa/LSB. A possible altitude correction is stored in the register OFF_H_REG (0x2D) in meters in two's complement. This correction is always taken into account when calculating the altitude.

$$H = 44,330.77 \cdot \left(1 - \left(\frac{p}{p_0}\right)^{0.1902632}\right) \tag{15.4}$$

The absolute air pressure is temperature-dependent. In order to compensate for the temperature influence, the module also has a temperature sensor. The analog voltages generated by the sensors are amplified and filtered with a low-pass filter. After limiting the frequency spectra, the measurement signals are quantized and further digitally processed.

15.1.2.2 Register

The settings and the measurement results of the MPL3115 device are stored in volatile, 8-bit registers. These registers occupy the memory area 0x00 ... 0x2D and are addressed individually via an address counter. The address counter is loaded with the desired address via I^{2}C. When a register is read or written, the address counter is incremented and thus another register is addressed. Exceptionally, when FIFO mode is disabled, the address counter is loaded with value 0x00 or 0x06 after accessing the register with address 0x05 or 0x0B. These exceptions allow continuous reading of the measured value registers and the status register. After power-up, the contents of most registers are set to 0. Refer to data sheet [1] for the operating modes that allow the register contents to be changed. The values are stored in the registers in big-endian format.

15.1.2.2.1 Control Register Block

The control register block consists of five registers that occupy the address range 0x26 ... 0x2A. These registers control the entire measurement process of the sensor. The first register CTRL_REG1 (0x26) stores the measurement settings and has the following configuration:

- **Bit 7** – ALT – by setting this bit the measured air pressure is converted to altitude and stored in the measuring register instead of the air pressure value;
- **Bit 6** – RAW – if this bit is set, the uncompensated measured values are stored by the A/D converter directly into the measured value registers; with this selection, the FIFO storage mode and the interrupts must be deactivated;
- **Bit 5:3** – OS[2:0] – a well-known method for reducing the measurement noise level is the averaging of several measurement results (oversampling). The oversampling rate (2OS) and the minimum measurement duration t_{OS} are determined via these three bits:

$$t_{OS} = 2^{OS} \cdot 4 + 2 \text{ in ms} \tag{15.5}$$

- **Bit 2** – RST – setting this bit resets the entire sensor and loads the registers with the preset values. The bit is then reset.
- **Bit 1** – OST – if this bit is set in standby mode, a single measurement, which takes into account the ALT, RAW and OST bits, is started. The measured values are stored and the internal control resets the bit afterwards.
- **Bit 0** – SBYB – the sensor is in standby mode when this bit is 0 or in active mode when the bit is 1.

The CTRL_REG2 register is used for alarm setting and determination of the sampling period in FIFO mode. The sampling period is set with bits 3:0 of register CTRL_REG2 in the range [1 ... 2^{15}] s. The CTRL_REG3 (0x28) register configures the two interrupt outputs as push-pull or open-drain and sets the polarity of the outputs. Register CTRL_REG4 (0x29) is used to enable the interrupts, and register CTRL_REG5 is used to switch the enabled interrupts to the INT1 or INT2 output. If several interrupts are switched to the same output, they are ORed.

15.1.2.2.2 Status Registers

If FIFO mode is deactivated, bits 7, 6 and 5 of the status register (0x00) signal that unread measured values (temperature and/or air pressure values) have been overwritten. Bits 2, 1 and 0 signal the saving of new measured values. Bits 0 and 5 relate to temperature values, bits 1 and 6 relate to air pressure values, and bits 2 and 7 relate to both. The register with address 0x06 maps the contents of the status register.

15.1.2.2.3 Measured Value Registers

15.1.2.2.3.1 Absolute Measured Values

The measured air pressure or the calculated altitude is stored in the registers OUT_P_MSB_REG (0x01), OUT_P_CSB_REG (0x02) and OUT_P_LSB_REG (0x03) left-justified. The corrected air pressure value is stored unsigned in Pascal (1 Pa = 0.01 mbar) on 20 bits, two of them for the decimal places. The stored value takes into account any offset value stored in two's complement in register OFF_P_REG (0x2B) with a resolution of 4 Pa/LSB. The MPL3115_Read_DataReg function stores the status register and the measurement registers into the vector ucDataByte. ucDataByte[0] stores the contents of the status register, ucDataByte [4] stores the contents of the OUT_P_MSB_REG register, etc. The following code calculates the absolute air pressure in Pa and stores it in the variable lPressureData.

```
lPressureData = ucDataByte[1];
lPressureData = (lPressureData << 8) | ucDataByte[2];
lPressureData = (lPressureData << 8) | ucDataByte[3];
lPressureData = lPressureData >> 6;
```

The corrected altitude is stored in meters in two's complement on 20 bits, four of them for the decimal places. By storing the current relative air pressure into the registers BAR_IN_MSB_REG (0x14) and BAR_IN_LSB_REG (0x15) with a resolution of 2 Pa/LSB, the local altitude is calculated more accurately. A possible correction can be stored into the register OFF_H_REG (0x2D) in meters, in two's complement.

The measured temperature is stored in the registers OUT_T_MSB_REG (0x04) and OUT_T_LSB_REG (0x05) in °C left-justified. The temperature value occupies 12 bits, four of them for the decimal places. The temperature correction can be stored in the register OFF_T_REG (0x2C) in two's complement with a resolution of 0.0625°C/LSB. The following program code calculates the positive temperature with a resolution of 0.1 °C from the values read with the function MPL3115_Read_DataReg and stores it in the variable uiTemperatureData.

```
uiTemperatureData = ucDataByte[4];
uiTemperatureData = ((uiTemperatureData << 8) | ucDataByte[5]) >> 4;
uiTemperatureData = (uiTemperatureData * 10) >> 8;
```

Table 15.1 Memory registers for the minima and maxima of the measured values

	Minima		Maxima	
	Register	Address	Register	Address
Air pressure	P_MIN_MSB_REG	0x1C	P_MAX_MSB_REG	0x21
	P_MIN_CSB_REG	0x1D	P_MAX_CSB_REG	0x22
	P_MIN_LSB_REG	0x1E	P_MAX_LSB_REG	0x23
Temperature	T_MIN_MSB_REG	0x1F	T_MAX_MSB_REG	0x24
	T_MIN_LSB_REG	0x20	T_MAX_LSB_REG	0x25

15.1.2.2.3.2 Relative Measured Values

After each measurement, the arithmetic unit of the sensor calculates the differences between the current and previous measured values and stores them in two's complement, similar to the absolute measured values. The air pressure difference/altitude difference is stored in the registers OUT_P_DELTA_MSB (0x07), OUT_P_DELTA_CSB (0x08) and OUT_P_DELTA_LSB (0x09), while the temperature difference is stored in the registers OUT_T_DELTA_MSB (0x0A) and OUT_T_DELTA_LSB (0x0B).

15.1.2.2.3.3 Extreme Values

The sensor stores the minima and maxima of the measured air pressure and temperature values in the same format as the absolute measured values. The memory registers are shown in Table 15.1. They can be set to 0x00 at any time to search for a new minimum or maximum.

15.1.2.2.3.4 FIFO Registers

The sensor provides a FIFO buffer for 32 measurement data sets with 5 bytes each for the intermediate storage of the measurement results, which facilitates the asynchronous reading of the measurement values. The byte configuration of a measurement data set is the same as that of the absolute measurement values. Via bits 7:6 of register F_SETUP_REG (0x0D), FIFO mode is deactivated at "00", FIFO mode is activated at "01" or "10". If the two bits have the combination "01", the oldest measurement data set is overwritten and at "10" the saving of the measurement values is stopped as soon as the buffer is full. The combination "11" is not implemented. When the number of unread data sets reaches the fill limit set by bits 5:0 of register F_SETUP_REG, bit 6 in register F_STATUS_REG (0x0D) is set and may cause an interrupt to be triggered. Another interrupt can be triggered in FIFO mode if the buffer is full and bit 7 of register F_STATUS_REG has been set as a result. Bits 5:0 of the last register store the number of unread measurement records from the buffer. The stored data sets are read from register F_DATA_REG (0x0E) in FIFO mode. Alternatively, the register OUT_P_MSB_REG can be used for reading. The other registers, which store the absolute measurement values, return the value 0x00 when read in FIFO mode. The function *MPL3115_Read_DataReg* can be used for reading the measurement results, which is listed below. When calling the function, the address of the first register from the register range (ucfirst_reg_address), the address of the variable that is to store the measured values (uctarget_address) and the number of registers to be read (ucnumber_of_reg) are passed as parameters.

```c
uint8_t MPL3115_Read_DataReg(uint8_t ucfirst_reg_address,
                             uint8_t* uctarget_address,
                             uint8_t ucnumber_of_reg)
{
    uint8_t ucDeviceAddress, ucI;
    //address of the air pressure sensor form
    ucDeviceAddress = MPL3115_DEVICE_TYP_ADDRESS << 1;
    ucDeviceAddress |= TWI_WRITE; //write mode
    TWI_Master_Start(); //start
    if((TWI_STATUS_REGISTER) != TWI_START) return TWI_ERROR;
    TWI_Master_Transmit(ucDeviceAddress); //send device address
    if((TWI_STATUS_REGISTER) != TWI_MT_SLA_ACK) return TWI_ERROR;
    //the start address of the register area is sent
    TWI_Master_Transmit(ucfirst_reg_address);
    if((TWI_STATUS_REGISTER) != TWI_MT_DATA_ACK) return TWI_ERROR;
    TWI_Master_Start(); //restart
    if((TWI_STATUS_REGISTER) != TWI_RESTART) return TWI_ERROR;
    ucDeviceAddress = (MPL3115_DEVICE_TYP_ADDRESS << 1) | TWI_READ;
    TWI_Master_Transmit(ucDeviceAddress); //send address in read mode
    if((TWI_STATUS_REGISTER) != TWI_MR_SLA_ACK) return TWI_ERROR;
    //the contents of the addressed registers are read in
    for(ucI = 0; ucI < (ucnumber_of_reg - 1); ucI++)
    {
        uctarget_address[ucI] = TWI_Master_Read_Ack();
        if((TWI_STATUS_REGISTER) != TWI_MR_DATA_ACK) return TWI_ERROR;
    }
    uctarget_address[ucnumber_of_reg - 1] = TWI_Master_Read_NAck();
    if((TWI_STATUS_REGISTER) != TWI_MR_DATA_NACK) return TWI_ERROR;
    TWI_Master_Stop(); //stop return TWI_OK;
}
```

15.1.2.2.3.5 Interrupt Registers

The events that can trigger an interrupt are defined by setting the corresponding bit in the INT_SOURCE_REG (0x12) register:

- **Bit 7** – signals a new measured value based on the settings from the PT_DATA_CFG_REG register;
- **Bit 6** – the FIFO buffer triggers an interrupt on overflow or when the number of unread records has reached the set fill limit;

- **Bit 5** – the measured air pressure/calculated altitude is in the range $p_g \pm \Delta p$; p_g is stored in registers P_TGT_MSB_REG (0x16) and P_TGT_LSB_REG (0x17) and Δp ($\neq 0$) is stored in registers P_WIND_MSB_REG (0x19) and P_WIND_LSB_REG (0x1A);
- **Bit 4** – the measured temperature is in the range $t_g \pm \Delta t$; t_g is stored in registers T_TGT_REG (0x18) and Δt ($\neq 0$) is stored in registers T_WIND_REG (0x1B);
- **Bit 3** – the measured air pressure/calculated altitude crosses one of the limits: Δp, $p_g - \Delta p$ or $p_g + \Delta p$;
- **Bit 2** – the measured temperature exceeds one of the limit values: Δt, $t_g - \Delta t$ or $t_g + \Delta t$;
- **Bit 1** – the air pressure change triggers an interrupt;
- **Bit 0** – the temperature change triggers an interrupt.

The desired interrupts must be enabled with the setting of register CTRL_REG4 and switched to 1 of the interrupt pins with CTRL_REG5.

The following function MPL3115_Write_DataReg can be used to change the contents of the registers that immediately follow each other, such as the interrupt registers or the registers that store the limit values. The new contents must be stored in an array, the address of which together with the address of the first register from the range and the number of setting values are passed as parameters when the function is called. Up to ten setting values can be transferred.

```
uint8_t MPL3115_Write_DataReg(uint8_t ucfirst_reg_address,
                        uint8_t* ucsource_address,
                        uint8_t ucnumber_of_reg)
{
    uint8_t ucDeviceAddress, ucI, ucArray[12];
    //form the address of the air pressure sensor
    ucDeviceAddress = MPL3115_DEVICE_TYP_ADDRESS << 1;
    ucDeviceAddress |= TWI_WRITE;//write mode
    ucArray[0] = ucDeviceAddress;
    ucArray[1] = ucfirst_reg_address;
    for(ucI = 0; ucI < ucnumber_of_reg; ucI++)
        ucArray[ucI + 2] = ucsource_address[ucI];
    TWI_Master_Start(); //start
    if((TWI_STATUS_REGISTER) != TWI_START) return TWI_ERROR;
    for(ucI = 0; ucI < (ucnumber_of_reg + 2); ucI++)
    {
        TWI_Master_Transmit(ucArray[ucI]);
        if((TWI_STATUS_REGISTER) != TWI_MT_DATA_ACK)
            return TWI_ERROR;
    }
    TWI_Master_Stop(); //stop
    return TWI_OK;
}
```

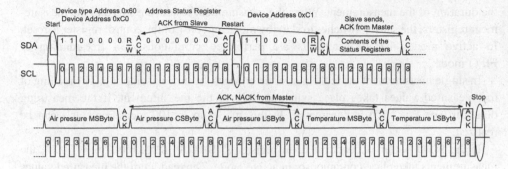

Fig. 15.4 MPL3115 – serial communication

15.1.2.2.3.6 Device Identification Registers

The device identification register WHO_I_AM_REG (0x0C) stores the value 0xC4 and can be used in a bus system to identify the device and to test the bus.

15.1.3 Serial Communication

The MPL3115 sensor implements the I^{2}C protocol in Fast-mode and enables communication with the master at a maximum clock frequency of 400 kHz. It responds to the 7-bit device type address 0x60, not also to the general call address 0x00. Figure 15.4 depicts the timing that occurs when the status register and measurement registers are read using the MPL3115_Read_DataReg function. At a clock frequency of 370 kHz, the entire process takes 258 μs.

15.1.4 Power Modes

The device is in standby mode after power up, with most registers set to 0x00. In this working mode settings can be changed, the analog part of the sensor is switched off to save energy and full access to the registers is guaranteed. By setting bit 0 of control register 1, the sensor switches to active mode. In this mode the measurements are performed continuously, but the contents of the registers can be changed only partially.

15.1.5 Measuring and Reading Modes

The complexity of the device allows different measurement and reading strategies. The sensor can measure air pressure and temperature in single measurement, free-running or FIFO mode. In standby mode, measurements can be made in single measurement mode. A measurement is started by setting bit 1 of control register 1. The sensor is in active mode for the duration of the measurement. When the measurement results are saved to the measurement registers, the sensor switches to standby mode and bit 1 of the control register is reset. In active mode, the measurements are performed continuously in free-running or FIFO mode.

In single measurement mode the master determines the time of sampling. The reading of the measured values takes place synchronously to the measurement, if the measuring duration (see Eq. 15.5) is considered. In this mode up to 100 measurements/s can be realized.

If the FIFO buffer is deactivated (in register F_SETUP_REG, bits 7:6 = "00"), the measurements take place continuously in active mode. The readout of the measured values can be time-controlled, synchronous or measurement-dependent. With time-controlled readout, the master determines the time of reading. However, this does not guarantee that all measured values are read exactly once. By configuring the PT_DATA_CFG_REG register, new measured values are signalled by appropriately set bits (for temperature and/or air pressure) in the status register. This makes it possible to read all measured values synchronously. The state of these bits can be tested in polling mode, but this leads to a blocking wait of the microcontroller. If additionally the interrupt registers are configured, the measurement of a new value is signaled to the master via an external interrupt. For this purpose, the configured interrupt output of the sensor must be connected to an interrupt-capable input of the master. If required, the measured values can also only be read when they have reached or exceeded a limit value.

If the FIFO mode is activated (bits 7:6 in register F_SETUP_REG are "01" or "10") and the sensor is in active mode, the measurements are started internally in a time-controlled manner and the measured values are stored in the FIFO buffer. The time period between two measurements is set with bits 3:0 of control register 2. With the appropriate setting of the interrupt registers, an interrupt is triggered when the buffer is full or has reached the set number of measurement records. The interrupt determines the time of reading and, despite asynchronous reading, no measured values are lost in this case. The communication scope is thus greatly reduced, which leads to a relief of the master and to energy saving.

15.1.6 Initialization of the MPL3115 Sensor

The adjustable registers of the MPL3115 are volatile and must therefore be set anew after each power-up, initialization or reset. The master must configure the control and interrupt registers to set the measurement and interrupt mode and load the offset values, the limit values and the current relative air pressure as a reference for measuring the altitude.

15.2   Humidity Sensor SI7021

The air humidity or air moisture is defined as the water vapour content of the air. A distinction is made between absolute and relative humidity. Absolute humidity indicates the mass of water vapour in a volume of air, is measured in g/m^3 or kg/m^3 and is independent of temperature. Saturation state is the state in which the maximum water vapour content is reached at a certain air temperature and air pressure. In this state, a relative humidity of 100% is reached. The capacity of the air to absorb water vapour increases with increasing temperature and decreases with increasing air pressure. The relative humidity RH^1 is the ratio between the current mass of water vapour and that in the saturation state (at the same air temperature and air pressure) and is measured as a percentage. The dew point temperature or dew point is the temperature at which the relative humidity reaches 100%. The water vapour condenses when the temperature falls below the dew point.

The measurement of relative humidity plays a major role in meteorology, transportation and storage of goods, and human health and comfort. Humidity can affect the accuracy and reliability of other measurement sensors, so it is specified as a parameter of many measurements. The influence on e.g. sound velocity or interferometric measurements must even be compensated.

There are materials whose conductivity or dielectric constant varies with the moisture absorbed from the air. These properties are used to build electronic sensors whose measurement methods for measuring relative humidity are based on measuring electrical resistance or electrical capacitance.

15.2.1 Structure of the SI7021

SI7021 [8, 9] is an electronic sensor for measuring relative humidity, whose sensing device is a capacitor. Moisture exchange between the dielectric of the capacitor and the air takes place through a hole in the housing of the device. To compensate for the temperature dependence of the relative humidity, the SI7021 also has a temperature sensor as shown in Fig. 15.5.

The sensor does not have to be initialized, approx. 15 ms must be waited before the first measurement after start-up. The analog output voltage of a sensor is converted via a D/A ·converter, the digital value is converted by means of the calibration coefficients in an arithmetic block and the result is stored in the measured value register. This is a 16-bit register which can only be read via the serial interface. The conversion linearizes the characteristic curves of both sensors and compensates the dependence of the humidity values with the temperature. In addition, the arithmetic block calculates the checksum of

---

[1] RH – relative humidity.

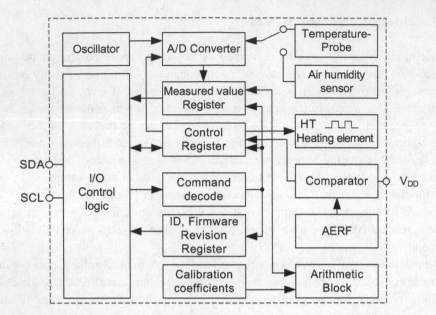

Fig. 15.5 SI7021 block diagram

the measured values, which can be transmitted if required. Several checksums are transmitted along with the identification number. The calibration coefficients are determined and stored at the factory for each sensor to enable the individual sensors to be replaced in a circuit without software changes.

The 8-bit read-write control register has the following configuration:

- **Bit 7, Bit 0** – RES1:0 – these two bits are used to set the bit resolution of the D/A converter as shown in Table 15.2.
- **Bit 6** – VDDS – is a bit that can only be read and is set by the hardware when the supply voltage is below the 1.9 V limit. Below 1.8 V the functionality of the sensor is no longer guaranteed.
- **Bit 5, Bit 4, Bit 3, Bit 1** – are reserved.
- **Bit 2** – HTRE – the internal heating element HT of the sensor is switched on at 1, otherwise off.

If the sensor is used at temperatures below the dew point, there is a risk that water vapor will condense on the sensor or dielectric of the sensing device, resulting in incorrect results. An offset of the humidity values can occur if the sensor has been exposed to high humidity for a long time. To avoid these possible distortions, the internal heating element can be switched on and off repeatedly.

The command decoder decodes the 1- or 2-byte commands received via I^{2}C and controls the internal assemblies of the sensor accordingly. These commands are placed in

Table 15.2 SI7021 Bit resolution and measuring duration of temperature and humidity

		Temperature		Humidity	
RES1	RES0	Bit resolution (bit)	Measuring time (ms)	Bit resolution (bit)	Measuring time (ms)
0	0	14	10.8	12	12
0	1	12	3.8	8	3.1
1	0	13	6.2	10	4.5
1	1	11	2.4	11	7

a message immediately after addressing in the write mode of the slave, as shown in Fig. 15.6.

The sensor stores an electronic identification number in eight bytes, which uniquely identifies it in a complex network, and the firmware version in one byte, which distinguishes it from the predecessor and successor versions. A master can read this information via I^2C.

15.2.2 Serial Communication

The sensor is configured as an I^2C slave and supports communication at up to 400 kBit/s. Its initial device type address is 0x80 (reset with the R/W bit). It can be supplied with voltages between 1.9 V and 3.6 V. The SDA and SCL inputs are not 5 V tolerant, which means that the master must either be supplied with the same voltage as the sensor or accept the low voltage level as a logical "1". Also, a voltage divider or level shifter must be installed to protect against electrical destruction. The slave responds to 1- or 2-byte commands. The 16-bit values are transmitted with the higher-order byte first.

In order to save energy, the device only works in single measurement mode. Each additional measurement must be started separately.

15.2.2.1 Access to the Control Register

Before the first measurement, the measurement resolution must be set via the control register. Four combinations of bit resolutions for the measurement of temperature and relative humidity can be selected via the bits RES1 : 0 (see Table 15.2). When selecting the bit resolution, the conversion time and the accuracy of the measurements (max. 3% for relative humidity between 0% and 80% and max. 0.4 °C for temperature) must be taken into account. To be able to measure the two quantities with resolutions whose combinations are not present in the table, such as 12 bits for both measurements, the bit resolution must be reset before each measurement.

Figure 15.6a shows the reading of the control register. The master addresses the sensor in write mode and then sends the command code 0xE7. After a restart sequence and addressing in read mode (R/W bit = 1), the slave sends the contents of the register. The

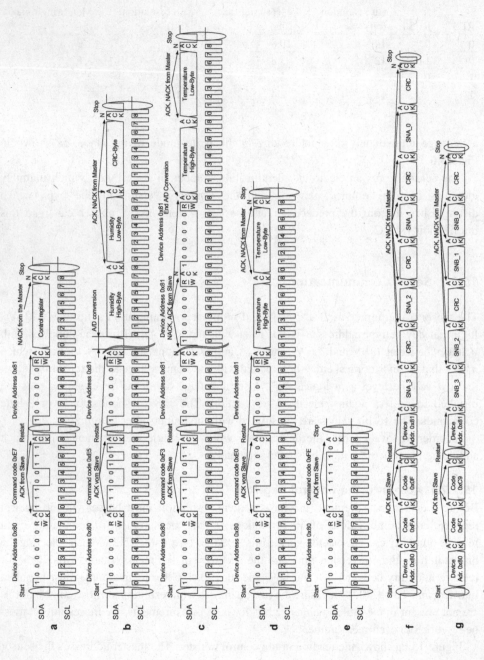

Fig. 15.6 SI7021 I²C communication

master acknowledges receipt of the byte with NACK and terminates communication with a stop sequence. Reading the control register can be used, for example, to monitor the supply voltage of the device.

To write to the control register, the master addresses the slave in write mode and sends another byte with the new contents of the register after command code 0xE6. The slave must acknowledge receipt of each byte with ACK. The master ends the communication with a stop sequence.

15.2.2.2 Measurement of Relative Humidity

A new measurement of the relative humidity is started with the code 0xE5 in Hold mode (see Fig. 15.6b) or with 0xF5 in No-Hold mode. After the slave has received one of the codes and confirmed it with ACK, the A/D conversion begins, the duration of which depends on the bit resolution and can be found in Table 15.2. After RESTART and re-addressing, the slave in hold mode stretches the SCL signal during the conversion and then, after releasing the SCL signal, it sends the 16-bit measurement value to the master. When the master confirms receipt of the low-order byte of the measured value with ACK, the slave sends another CRC[2] byte which is acknowledged by the master with NACK. Regardless of the bit resolution set for the measurement, bit 15 of the measured value is the most significant bit and bits 1:0 have the binary value "10". The measured value of relative humidity read by the master has been temperature compensated by the sensor by internal calculations and can be converted to % according to the formula presented in Sect. 15.2.3.

Clock stretching is an option provided by the I^2C protocol. This option allows a slave to slow down the clock frequency on bit level or to keep the clock signal low on byte level after receiving a byte as long as it is internally busy. This stops the communication but does not interrupt it. The microcontrollers of the ATmega family have implemented this option for the TWI interface.

The following function SI7021_Read_ValueHoldMode enables the start of a new measurement in Hold mode and implements the time history from Fig. 15.6b. The command code for a temperature or humidity measurement is passed as a parameter. Because the clock stretching is implemented on the data link layer of the protocol, it does not have to be considered in the software. The measured value is stored in the variable ucTempOrHumidValue [5] and the CRC byte in the variable ucChecksum. These variables, which are declared globally in the SI7021 module, can be accessed from the main program via interface functions. The function returns the value TWI_OK if the communication was error-free, otherwise TWI_ERROR.

---

[2]CRC – cyclic redundancy check.

```
uint8_t SI7021_Read_ValueHoldMode(uint8_t ucmeasure_cmd)
{
    uint8_t ucAddress;
    ucAddress = SI7021_DEVICE_TYPE_ADDRESS | TWI_WRITE; //write mode
    TWI_Master_Start(); //start
    TWI_Master_Transmit(ucAddress); //send device address
    if((TWI_STATUS_REGISTER) != TWI_MT_SLA_ACK) return TWI_ERROR;
    //send measurement type
    TWI_Master_Transmit(ucmeasure_cmd);
    if((TWI_STATUS_REGISTER) != TWI_MT_DATA_ACK) return TWI_ERROR;
    TWI_Master_Start(); //restart
    if((TWI_STATUS_REGISTER) != TWI_RESTART) return TWI_ERROR;
    ucAddress = SI7021_DEVICE_TYPE_ADDRESS | TWI_READ; //read mode
    TWI_Master_Transmit(ucAddress); //send device address
    if((TWI_STATUS_REGISTER) != TWI_MR_SLA_ACK) return TWI_ERROR;
    //most significant byte read
    ucTempOrHumidValue[0] = TWI_Master_Read_Ack();
    if((TWI_STATUS_REGISTER) != TWI_MR_DATA_ACK) return TWI_ERROR;
    //read least significant byte
    ucTempOrHumidValue[1] = TWI_Master_Read_Ack();
    if((TWI_STATUS_REGISTER) != TWI_MR_DATA_ACK) return TWI_ERROR;
    //read checksum
    ucChecksum = TWI_Master_Read_NAck();
    if((TWI_STATUS_REGISTER) != TWI_MR_DATA_NACK) return TWI_ERROR;
    TWI_Master_Stop(); //stop
    return TWI_OK;
}
```

In no-hold mode the slave answers the addressing in read mode after the RESTART sequence with NACK as long as the A/D conversion is running. At the end of the conversion, it confirms the address with ACK and sends the measured value. and, if required, the CRC byte.

15.2.2.3 Measurement of Temperature

A new temperature measurement in Hold mode is started with the code 0xE3 and in No-Hold mode with 0xF3 (Fig. 15.6c). The communication is similar to that described in Sect. 15.2.2.2. If required, the checksum of the temperature value can also be read.

The temperature is also measured before each humidity measurement to be able to compensate for the temperature influence. This temperature value remains stored until the next measurement and can be read with the command 0xE0 as shown in Fig. 15.6d. There is no checksum for this transmission. Regardless of the bit resolution selected, bit 15 for the measurement of the measured value is the most significant bit and bits 1:0 have the binary value "00". The measured value can be converted to °C using the formula in Sect. 15.2.3.

15.2.2.4 Reading the Electronic ID and the Firmware Revision

The electronic identification number of the sensor is read in two steps using 2-byte instructions, as shown in Fig. 15.6f. To secure the reading, a CRC byte is calculated and appended for each byte in the first step and for each 2-byte block in the second.

To read the firmware version, the master sends the command 0x84 and 0xB8 in an I^{2}C read message after the addressing phase. In the slave response, the byte with the firmware version is not followed by a CRC byte.

15.2.3 Calculation of Temperature and Relative Humidity

Based on the measured value read and stored in the variable ucTempOrHumidValue [5] in this example, the relative humidity is calculated using the formula from [8]:

$$\varphi = \frac{125 \cdot \varphi_{read}}{2^{16}} - 6 \; in\%$$

(15.6)

In order to be able to calculate the humidity with a resolution of 0.1%, Eq. 15.6 is multiplied by 10:

$$\varphi = \frac{1250 \cdot \varphi_{read}}{2^{16}} - 60 \; in \; 0,1\%$$

(15.7)

The SI7021_Get_Humidity function can be called after a relative humidity measurement has been completed and returns the measured value in 0.1% as an integer:

```
uint16_t SI7021_Get_Humidity(void)
{
    uint16_t uihumid = 0;
    long lhumid;
    //the 2 bytes read out are combined
    uihumid = (ucTempOrHumidValue[0] << 8) + ucTempOrHumidValue[1];
    //the formula is multiplied by 10 so that the resolution
    //of the result is 0.1 %
    lhumid = (long)uihumid * 1250;
    lhumid = (lhumid / 65536) - 60;
    if (lhumid < 0) lhumid = 0; //values less than 0 %, are set to 0
    //Values greater than 100 % are set to 100 %
    else if (lhumid > 1000) lhumid = 1000;
    uihumid = lhumid;
    return uihumid;
}
```

After a temperature measurement, the read value can be stored in the same variable as above with the high-order byte at location [0] and the air temperature calculated using the formula from [8]:

$$t = \frac{175{,}72 \cdot t_{read}}{2^{16}} - 46{,}85 \text{ in } °C \tag{15.8}$$

In the following, the implementation of Eq. 15.8 in the program code is presented. The first example function float SI7021_Get_Temperature(void) calculates the temperature as a fixed point number and returns this value.

```
float SI7021_Get_Temperature(void)
{
    float ftemp;
    uint16_t uitemp;
    //the two read bytes are combined
    uitemp = (ucTempOrHumidValue[0] << 8) + ucTempOrHumidValue[1];
    ftemp = (((float) uitemp * 1757.2) / ((float) 65356)) -468.5;
    return ftemp;
}
```

For the second example function, the coefficients of the equation were multiplied by a factor of 100, resulting in all operands and the result becoming integers. Because the multiplication increased the temperature resolution to 0.01 °C, the result can be divided by 10 as an integer.

```
int SI7021_Get_Temperature(void)
{
    long ltemp;
    //the temperature is calculated in integer arithmetic
    //the two bytes read are combined
    ltemp = (ucTempOrHumidValue[0] << 8) + ucTempOrHumidValue[1];
    //the multiplication by 100 of the coefficients produces integers
    ltemp = ltemp * 4393;
    //the fraction (temp*17572)/65536) is truncated by 4
    ltemp = ltemp / 16384;
    //division by 10 results in a resolution of 0.1°C
    ltemp = (ltemp - 4685) / 10;
    return (int)ltemp;
}
```

A test program using the integer arithmetic function is more than 800 bytes shorter after compilation with AVRStudio Ver.6.2 than the test program using the first example function, regardless of whether optimization by code length was turned on or turned off completely. This is related to the float library of the used Gnu-C compiler.

With the calculated values of relative humidity and temperature, the dew point temperature can be calculated using the Magnus formula presented in [9]:

$$\tau = \frac{B1 \cdot \left(ln\left(\frac{\varphi}{100}\right) + \frac{A1 \cdot t}{B1+t} \right)}{A1 - ln\left(\frac{\varphi}{100}\right) - \frac{A1 \cdot t}{B1+t}} \tag{15.9}$$

with τ = dew point temperature,
φ = relative humidity,
t = measured air temperature,
A1 = 17.625; B1 = 243.04, const.

15.2.4 Testability

The SI7021 offers mechanisms to determine whether the read measured value matches the desired measured variable or whether errors occurred during message transmission. This is advantageous in a complex network.

The AND of a humidity value with 0x0003 always results in 0x0002, that of a temperature value in 0x0000, thus the measured values can be distinguished.

The arithmetic block of the sensor calculates a checksum for each measured value immediately following a measurement start in the form of a CRC code that can be read by the master to check the transmission. Such checksums must be read when transmitting the electronic ID that uniquely identifies a sensor in a network as described in Sect. 15.2.2.4. In general, the calculation of CRC codes is based on polynomial division (see [7]). There are different methods to calculate a CRC code, for the SI7021 the CRC-8 method was implemented by the company Maxim Integrated [10], which is used for the 1-Wire devices. The procedure uses the generator polynomial $x^8 + x^5 + x^4 + 1$, the code is initialized with 0x00.

To determine the CRC code of a byte, all 256 possible values can be calculated in advance and stored in a lookup table (LUT) according to the order of the byte values. In this case, the byte whose CRC code is to be calculated becomes the address of the memory cell in which the code was stored. No mathematical or logical operations are required. For the 2-byte blocks, determining the CRC code using this method is difficult to implement because the table must be 64 kbytes in size. In such a case, the CRC code must be computed, for example with the following function `uint8_t ComputationCRC (uint16_t ucvalue)`, which implements the above procedure. The function calculates the CRC code of a 1- or 2-byte value and returns it.

```
uint8_t ComputationCRC(uint16_t ucvalue)
{
    unsigned long ulTwoBytevalue = 0, ulMask = 0x800000;
    unsigned long ulPolynomGenerator = 0x988000;
    uint8_t ucIndex = 0, ucCRC;
    //initialization of the code
    ulTwoBytevalue |= (unsigned long)ucvalue << 8;
    while(ucIndex < 16) //polynomial division
    {
        if(ulTwoBytevalue & ulMask)
        {
            ulTwoBytevalue = ulTwoBytevalue ^ ulPolynomGenerator;
        }
        ucIndex++;
        ulMask = ulMask >> 1;
        ulPolynomGenerator = ulPolynomGenerator >> 1;
    }
    //CRC code as remainder of polynomial division
    ucCRC = ulTwoBytevalue;
    return ucCRC;
}
```

15.3   Temperature Measurement with the TMP75

TMP75 is a digital temperature sensor that measures temperature in the specified range of
−40 °C to +125 °C with a typical accuracy of ±0.5 °C [11]. It can be configured and read
out as a slave via I^{2}C or SMBus in Fast- or High-speed-mode. Up to eight identical sensors
can be operated on the same bus thanks to the three address inputs. A rough block diagram
of the sensor is shown in Fig. 15.7. A diode is used as the temperature sensing device. The
analog voltage proportional to the temperature is digitized by an A/D converter with
adjustable bit resolution and stored in a 12-bit temperature register. This register can
only be read via the serial interface. The TMP75 can operate in free-running or single
measurement mode. In free-running mode, the temperature is measured continuously. In
order to save energy and thereby also reduce its own heating, which leads to measurement
deviations, the sensor can be switched to an energy saving mode. In this mode only the
serial interface remains active. Via the built-in thermostat function, the sensor can signal
the exceeding of a maximum or the falling below of a minimum temperature via an open-
drain output. This output requires an external pull-up resistor. The two temperature limits
are stored in two 16-bit registers and can be set at any time via the software. Another
register is used to configure the sensor and control the measurement. Access to the
individual registers is realized via the address register.

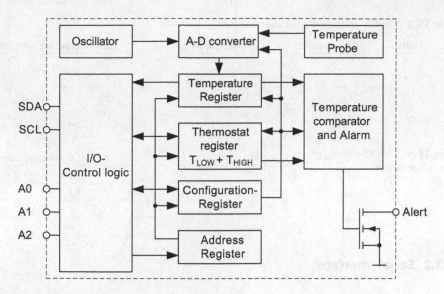

Fig. 15.7 TMP75 – block diagram of the sensor

15.3.1 Sensor Configuration

The 8-bit configuration register with address 0x01 can be read or written. The meaning of the individual bits is explained below:

- **Bit 7 – OS** – Setting this bit by the control software while the sensor is in power save mode starts a single temperature measurement. After the measurement is completed and the temperature value is saved, the sensor returns to power save mode. This bit is always read as 0.
- **Bit 6:5 – R1:R0** – These two bits are used to set the temperature resolution (Table 15.3).
- **Bit 4:3 – F1:F0** – These bits are used to set the number of consecutive overshoots and undershoots of the temperature limits to trigger a temperature alarm (Table 15.4).
- **Bit 2 – POL** – This bit determines the polarity of the alarm signal. If the bit is 0, the alarm signal is set to low in the event of an alarm.
- **Bit 1 – TM** – If this bit is 1, the thermostat operates in comparator mode, otherwise in interrupt mode.
- **Bit 0 – SD** – Setting this bit by the control software puts the sensor into energy-saving mode.

Table 15.3 TMP75 resolution and measuring time

R1	R0	Bit resolution	Temperature step (°C)	Max. measuring time (ms)
0	0	9	0.5	37.5
0	1	10	0.25	75
1	0	11	0.125	150
1	1	12	0.0625	300

Table 15.4 TMP75 Coding of the number of errors

F1	F0	Consecutive error
0	0	1
0	1	2
1	0	4
1	1	6

15.3.2 Serial Interface

The TMP75 implements the I^2C and SMBus protocols for communication with a master. The maximum transmission rate can reach 400 kBit/s in Fast-mode and 3.4 MBit/s in High-speed-mode. The minimum limit for the frequency of the bus clock is 1 kHz and thus it deviates slightly from the corresponding requirement of the I^2C specification (0 Hz). The three address inputs can be used to set the least significant bits of the address. These inputs can be connected either to GND ("0") or to Vcc ("1"), thus realizing up to eight different addresses. Via the three address inputs of the sensor TMP175, which is similar to the TMP75 and has the same characteristics, up to 27 different addresses can be realized. These inputs can be connected to Vcc or GND or not. The exact, setting addresses can be found in the data sheet [11].

The basic address of the TMP75, if all address inputs are connected to GND, is 0x90 and is the same as that of the A/D and D/A converter module PCF8591. In terms of circuitry, this must be taken into account when networking the two deyices on a bus so that all bus participants can be uniquely identified. For the example in Fig. 15.8 the address of IC1 is 0x9E (like the address of the real-time clock device MAX31629) and that of IC2 0x96.

The temperature values and the configuration byte are stored in volatile registers. The desired configuration and possibly the temperature limits must be loaded by the master during the initialization phase of the bus. To access the contents of a register, the master must send the address of the target register after initiating communication and addressing the slave. During this process, the read/write bit is reset. During a write operation, the master continues to send one or two bytes and then terminates communication. After each byte sent, the master checks whether the sensor has confirmed receipt with ACK. To read data, the master must generate a RESTART sequence after addressing the destination register and send the slave address with the read/write bit set. Furthermore, it generates the clock for the transmission and receives the data. The last received byte is acknowledged with NACK, all others with ACK.

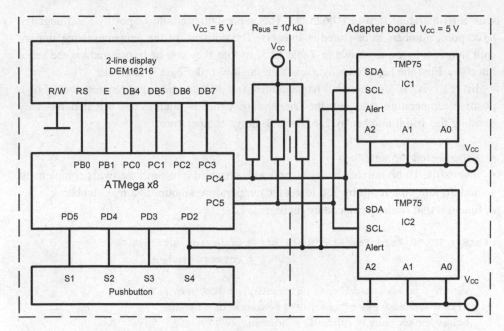

Fig. 15.8 TMP75 – wiring of two temperature sensors

The TMP75 and TMP175 sensors respond to the reserved address 0x00 (general call), acknowledge its reception and if the following byte is 0x06, they load the registers with the initial values after reset and store the state of the address inputs.

If one of the participants in an I²C bus "hangs up" and pulls one of the bus lines permanently low, the entire communication breaks down. Therefore, the sensor implements a timeout function that automatically intervenes if one of the bus lines remains low for longer than 54 ms between a START and a STOP sequence. In such a case, the function leads to a reset of the own serial interface.

15.3.3 Temperature Measurement

After the sensor has been started up, the bit resolution and the measurement mode must be set via the configuration register. If the focus is on temperature measurement and energy consumption, setting bit 0 in the configuration register selects the energy saving mode in which single measurement operation is possible. Each temperature measurement is started by the master by setting bit OS from the same register. The end of the measurement is not signalled, the maximum measurement duration must be taken into account. The bit resolution affects the measurement duration, but without affecting the accuracy. The temperature register of the sensor can be accessed via address 0x00, has a size of 12 bits (T11:0) and stores the measured temperature value in a two's complement representation with a temperature step of 0.0625 °C. This representation allows positive and negative fixed-point numbers to be stored as integers. The contents of the temperature register are split into two bytes, as shown in Table 15.5, so that they can be transferred via the serial interface. First the high-order byte and then the low-order byte is read out.

Bit T11 is "0" for positive temperatures and "1" for negative temperatures. For a positive temperature read out, the temperature value is obtained as a real number by dividing the 16-bit number by 256. For a negative temperature:

- remembered the sign, then
- inverts the 16-bit number bit by bit and adds a one to it to form the two's complement
- and dividing the result by 256 to get the temperature amount as a real number.

A function that reads a temperature register is:

```
uint8_t TMP75_Read_Temperature(uint8_t ucdevice_address,
                               uint8_t uctemp2read)
{
    uint8_t ucDeviceAddress, ucTempHigh, ucTempLow;
    //form theaddress of the TMP75 temperature sensor
    ucDeviceAddress = (ucdevice_address << 1) |
                    TMP75_DEVICE_TYPE_ADDRESS;
    ucDeviceAddress |= TWI_WRITE; //write mode
    TWI_Master_Start(); //start
    if((TWI_STATUS_REGISTER) != TWI_START) return TWI_ERROR;
    TWI_Master_Transmit(ucDeviceAddress); //send device address
    if((TWI_STATUS_REGISTER) != TWI_MT_SLA_ACK) return TWI_ERROR;
    /*the address of the desired temperature register is sent*/
    TWI_Master_Transmit(uctemp2read);
    if((TWI_STATUS_REGISTER) != TWI_MT_DATA_ACK) return TWI_ERROR;
    TWI_Master_Start(); //restart
    if((TWI_STATUS_REGISTER) != TWI_RESTART) return TWI_ERROR;
    ucDeviceAddress = (ucdevice_address << 1) |
                    TMP75_DEVICE_TYPE_ADDRESS;
    ucDeviceAddress |= TWI_READ;
    /*send device address in read mode*/
    TWI_Master_Transmit(ucDeviceAddress);
    if((TWI_STATUS_REGISTER) != TWI_MR_SLA_ACK) return TWI_ERROR;
    /*The contents of the addressed register are read*/
    ucTempHigh = TWI_Master_Read_Ack();
    if((TWI_STATUS_REGISTER) != TWI_MR_DATA_ACK) return TWI_ERROR;
    ucTempLow = TWI_Master_Read_NAck();
    if((TWI_STATUS_REGISTER) != TWI_MR_DATA_NACK) return TWI_ERROR;
    TWI_Master_Stop(); //stop
    iTemperature = (ucTempHigh << 8) + ucTempLow;
    return TWI_OK;
}
```

Table 15.5 TMP75 temperature value coding

Temperature [°C]	Highest byte				Low byte							
	T11	T10	T9	T8	T7	T6	T5	T4	T3	T2	T1	T0
+23.75	0	0	0	1	0	1	1	1	1	1	0	0
0	0	0	0	0	0	0	0	0	0	0	0	0
−17.25	1	1	1	0	1	1	1	0	1	1	0	0

When calling the function, the device chip address of the sensor `ucdevice_address` and the address of the register to be read `uctemp2read` are passed as parameters. The temperature value is stored in two's complement representation to the variable:

```
int iTemperature;
```

From the main program one accesses this variable declared in the software module `TMP75.c` via the following function:

```
int TMP75_Get_Temperature(void)
{
    return iTemperature;
}
```

15.3.4 Thermostat Function

The sensor compares the stored temperature value with the contents of two thermostat registers in the background. Both are read-write registers, have the same size as the temperature register and use the two's complement representation for storing the limit temperatures. To encode a given temperature value, the following must be observed:

- if the value is positive, it is multiplied by 256;
- if it is negative, the temperature amount is multiplied by 256, the 2-byte result is inverted bit by bit and a one is added to it (two's complement).

The thermostat register with address 0x02 stores the lower limit temperature, the register with address 0x03 the upper limit. Exceeding the upper limit or falling below the lower limit leads to an alarm condition, which is signalled at the Alert output (see Fig. 15.9). If the sensor is used as a two-point controller, thus the thermostat function is in the foreground and a fast reaction in case of an alarm is desired, the free-running measurement mode is selected by resetting bit SD. In this mode, the temperature measurement takes place continuously. The TM bit can be used to select one of the two thermostat modes and the POL bit can be used to select the polarity of the alert signal in the event of an alarm. In comparator mode (TM = 0), the alert output is switched high (for POL = 1) when the temperature value exceeds the upper limit temperature and switches it low the next time the temperature falls below the lower limit. In interrupt mode (TM = 1) with POL = 1, the alert output is switched high both when the temperature exceeds the upper limit and when it falls below the lower limit. After a measuring period or after calling an SMBus alert response function, the output is switched low again. Bits F1:0 can be used to set the number of successive overshoots/undershoots that trigger an alarm according to Table 15.4. In this

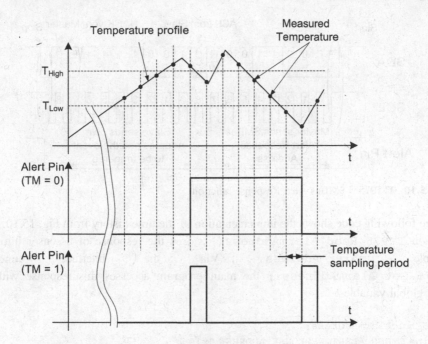

Fig. 15.9 The thermostat function of the sensorTMP75

way, the triggering of an alarm by temperature noise can be avoided. Figure 15.9 shows the thermostat function graphically for the settings POL = 1, F1 = 0 and F0 = 1.

In a circuit with several sensors, as shown in Fig. 15.8, the alert outputs are connected together to form an AND wiring. Because the alarm trigger cannot be identified from the resulting signal in comparator mode, the interrupt mode must be selected with POL = 0. Using an SMBus alert response function, the timing of which is shown in Fig. 15.10 for IC1 as the trigger, the master can identify the slave as well as the event that triggered the alert. With the previous settings, the alert pin will go low upon the occurrence of an event that triggers an alarm. The master responds and initiates an I^2C communication. It also sends the reserved address 000110 and sets the read/write bit. This alert response address, which is implemented in the SMBus protocol, is acknowledged by the TMP75 sensors with ACK. The slave that has triggered the alarm sends its own 7-bit address and with the eighth bit codes the alarm event as follows: with 0 the temperature falls below the lower limit and with 1 the temperature reaches or exceeds the upper limit. The master acknowledges the response of the slave with NACK, terminates the communication and the slave deactivates its alarm output. If two or more sensors trigger an alarm at the same time, they will also try to respond to the SMBus alert-response address at the same time. The slave with the smallest address wins the arbitration, places its address and the event to be triggered in coded form on the data line, and disables its alert output. However, the alert signal remains active and the master must repeat the procedure until all sensors that triggered the alert have identified themselves.

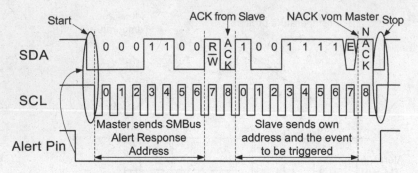

Fig. 15.10 TMP75 – SMBus Alert Response function

The following code shows the implementation of the time history from Fig. 15.10. The function `TMP75_Read_AlarmAddress()` stores the response of a sensor into the variable `ucAlarmAddress`. Via the interface function `TMP75_Get_AlarmAddress()` the main program accesses this response without using global variables.

```
uint8_t ucAlarmAddress;
#define SMBUS_ALERT_RESPONSE_ADDRESS 0x19
uint8_t TMP75_Read_AlarmAddress(void)
{
    TWI_Master_Start(); //start
    if((TWI_STATUS_REGISTER) != TWI_START) return TWI_ERROR;
    //send SMBus alert response address
    TWI_Master_Transmit(SMBUS_ALERT_RESPONSE_ADDRESS);
    if((TWI_STATUS_REGISTER) != TWI_MT_SLA_ACK) return TWI_ERROR;
    /*address of the sensor triggering the alarm is read in*/
    ucAlarmAddress = TWI_Master_Read_NAck();
    if((TWI_STATUS_REGISTER) != TWI_MR_DATA_NACK) return TWI_ERROR;
    TWI_Master_Stop(); //stop
    return TWI_OK;
}


uint8_t TMP75_Get_AlarmAddress(void)
{
    return ucAlarmAddress;
}
```

15.4   Fine Dust Sensor SDS011

Fine dust is a mixture of solid particles and fine droplets suspended in the air (aerosol). Fine dust can pose a significant health hazard due to its respirable nature and, in some cases, even resorption (absorption into the bloodstream). Therefore, fine dust sensors are increasingly used to measure air quality. As a rule, fine dust is not classified according to its material composition, but according to the size of the particles. The nomenclature PMx is used for this purpose, where x denotes the maximum particle diameter in the aerosol. Larger particles up to 10 μm mainly irritate the mucous membranes, while particles smaller than 2.5 μm can penetrate deep into the lungs and cause respiratory diseases [12].

The SDS011 fine dust sensor from Nova Fitness, a spinoff of the University of Jinan in China, has gained a certain notoriety, as it can be purchased for about 25€ and is used in numerous environmental monitoring projects. It measures particulate matter from 0.3 μm to 10 μm at concentrations up to 999.9 μg/m^3 and a nominal resolution of 0.3 μg/m^3 in the PM2.5 (particles up to a maximum of 2.5 μm in diameter) and PM10 (particles up to a maximum of 10 μm in diameter) fractions. These fractions are required by law for air quality measurements. Numerous publications can be found on this sensor, its accuracy is viewed critically, but it has a good dynamic range of a few seconds and is probably sufficiently accurate in the 3 μ range, while larger particles tend to be measured inaccurately, especially at high humidity [13]. However, if the dependencies of the deviations are known, they can be corrected. However, it can be used very well at least for the qualitative measurement of the development of the fine dust concentration.

15.4.1 Measuring Principle

The sensor is only very sparsely documented by the manufacturer. It uses the principle of laser scattering. The principle is based on the knowledge that a single particle in a plane laser beam behaves like an aperture (see Fig. 15.11). As soon as a beam of light hits a particle, it is partially absorbed, partially reflected (negligible for small particles), and partially scattered. This scattering results in an interference pattern, as with an aperture. If the particle circumference πd is put in relation to the wavelength λ, the following applies for

$$0.1 \leq \frac{\pi d}{\lambda} \leq 10 \tag{15.10}$$

the model of Mie scattering.[3] This puts the course of the scattering intensity at a detector in the optical far field of the particle (producible by lens systems in the measuring chamber)

---

[3] After Gustav Mie, German physicist 1868–1957.

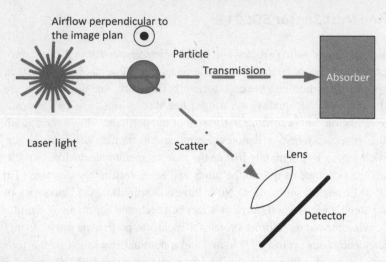

Fig. 15.11 Principle of laser scattering

into a complex relation to the particle diameter. For expensive sensors, detector arrays with many measuring channels are used to characterize the interference patterns. In the case of a particle passing by at a known speed, however, temporal progressions at a point-shaped detector also allow conclusions to be drawn about the diameter with sufficient accuracy. Therefore, a uniform particle flow is generated with the aid of a built-in fan [14, 15].

From the time-varying patterns, the microcontroller built into the sensor calculates the particle size distribution according to an algorithm not further described by the manufacturer.

According to the manufacturer, the laser diode has a service life of 8000 h.

15.4.2 Control and Serial Communication

The SDS011 sends out a data packet with the measurement data every second (exactly every 1004 ms) via a serial interface (9600 baud, 8 data bits, no parity, 1 stop bit), this looks as follows:

Byte number	Name	Content
0	Message header	0xAA
1	Command	0xC0
2	DATA 1	PM2.5 low byte
3	DATA 2	PM2.5 high byte
4	DATA 3	PM10 low byte
5	DATA 4	PM 10 high byte
6	DATA 5	ID Byte 1

(continued)

Byte number	Name	Content
7	DATA 6	ID Byte 2
8	Checksum	Checksum (arithm. sum of the data bytes 1 ... 6)
9	Message tail	0xAB

The 16-bit values indicate the values in 0.1 µg/m^3, i.e.

$$PMx = ((PMx \text{ high byte} \gg 8) + PMx \text{ low byte})/10$$

To read this information, a UART Receive Interrupt Service routine is defined in the module uart.h/uart.c, which describes the two global 8-bit variables RecFlag and ByteRec:

```
ISR(USART_RX_vect)
{
    RecFlag=1;
    ByteRec=UDR0;
}
```

With the two macros

```
#define UART_BYTE_RECEIVED 1
#define UART_IDLE 0
```

the following function UARTcheckRec() returns whether a new byte has arrived and UARTgetRec() returns the byte itself:

```
uint8_t UARTcheckRec()
{
    if (RecFlag)
    {
        RecFlag = 0;
        return UART_BYTE_RECEIVED;
    }
    else return UART_IDLE;
}

uint8_t UARTgetRec()
{
    return ByteRec;
}
```

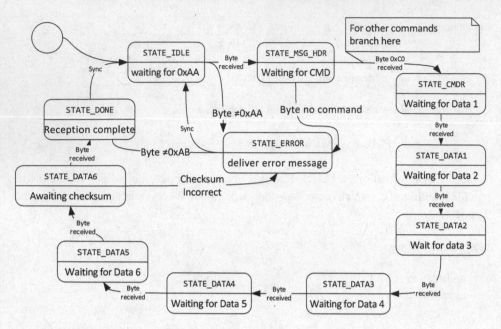

Fig. 15.12 State machine receiving data from the SDS011

This is more elegant than global variables and completely decouples the uart.c/.h module from the main program.

The actual receiving mechanism is a state machine that follows the scheme in Fig. 15.12.

In the main loop, the very lean state machine of the sensor is now constantly called (polling):

```
int main(void)
{
    uint8_t erg;
    Init();
    sei();
    while (1)
    {
        erg=SDS011StateMachine();
        ...
    }
}
```

This checks whether a byte was received and only if this is the case, the states are processed, these must be defined of course as macro as arbitrary different numbers, which we spare ourselves here:

```
uint8_t SDS011StateMachine()
{
    uint8_t Recv;
    if(UARTcheckRec())
    {
        Recv=UARTgetRec();
        switch (state)
        {
            case STATE_IDLE : if (Recv ==0xAA) state=STATE_MSG_HDR;
                            else state=STATE_ERROR; CSCalc=0; break;
            case STATE_MSG_HDR : if (Recv==0xC0) state = STATE_CMDR;
                            else state=STATE_ERROR; break;
            case STATE_CMDR : PM2_5=Recv;
                            state=STATE_DATA1; CSCalc+=Recv; break;
            case STATE_DATA1 : PM2_5|=((uint16_t)Recv)<<8;
                            state=STATE_DATA2; CSCalc+=Recv; break;
            case STATE_DATA2 : PM10=Recv;
                            state=STATE_DATA3; CSCalc+=Recv; break;
            case STATE_DATA3 : PM10|=((uint16_t)Recv)<<8;
                            state=STATE_DATA4; CSCalc+=Recv; break;
            case STATE_DATA4 : ID=Recv;
                            state=STATE_DATA5; CSCalc+=Recv; break;
            case STATE_DATA5 : ID|=((uint16_t)Recv)<<8;
                            state=STATE_DATA6; CSCalc+=Recv; break;
            case STATE_DATA6 : CSRec=Recv;
                            if (CSRec==CSCalc) state=STATE_CHECKSUM;
                            else state=STATE_ERROR; break;
            case STATE_CHECKSUM : if (Recv==0xAB) state = STATE_DONE;
                            else state=STATE_ERROR; break;
            case STATE_DONE: state=STATE_IDLE; return REC_COMPLETE;
            case STATE_ERROR : state=STATE_IDLE; return NotOK;
            default: break;
        }
    }
    if (state != STATE_ERROR) return OK;
    else return NotOK;
}
```

The checksum is the lower byte of the sum of the data bytes (without header, command byte and tail) for any communication with the sensor.

The global variables for the data (PM10, PM2_5 and ID) can still be equipped with suitable Get functions, so that the complete state machine can be encapsulated in a module SDS011.c/.h:

```
unsigned int SDS011GetID()
{
        return ID;
}
nsigned int SDS011GetPM2_5()
{
        return PM2_5;
}
unsigned int SDS011GetPM10()
{
        return PM10;
}
```

Since the sensor has a limited lifetime and one measurement per second is usually not needed, you can put it to sleep and wake it up by command (for example every minute). Alternatively, you can configure the sensor to work for 30 s at a time and then sleep for 1–30 min. To do this, send a byte sequence to the sensor (for example, using the functions shown in Sect. 7.1.6):

```
static const char SLEEPCMD[19] = {
    0xAA,// head
    0xB4,// command id
    0x06,// data byte 1
    0x01,// data byte 2 (set mode)
    0x00,// data byte 3 (sleep)
    0x00,// data byte 4
    0x00,// data byte 5
    0x00,// data byte 6
    0x00,// data byte 7
    0x00,// data byte 8
    0x00,// data byte 9
    0x00,// data byte 10
    0x00,// data byte 11
    0x00,// data byte 12
    0x00,// data byte 13
    0xFF,// data byte 14 (device id byte 1)
    0xFF,// data byte 15 (device id byte 2)
    0x05,// checksum
    0xAB// tail
};
```

If byte 3 is set to 0, the sensor is put to sleep, if it is set to 1, it is woken up again. The checksum must of course be adjusted to 0x06. Since the sensor responds to the command, the state machine must be adjusted. In this case, the second byte (after the header) is 0xC5 instead of 0xC0. In this case, the six following data bytes contain a response in the form of a system status instead of the dust concentrations. With the understanding built up here, it should not be a problem to build up the corresponding code using the protocol manual [16].

With the help of the manual, further commands can be sent to the sensor: You can connect several SDS011 in parallel, give them their own ID so that you can address individual sensors. It is also possible to query them in turn via their ID according to the master-slave principle. This is explained in detail in the manual and always follows the same scheme described here.

References

1. Freescale Semiconductor. *MPL3115A2, I^2C Precision Altimeter*. Data Sheet Rev 4.0, 09/2015.
2. Frescale Semiconductor. *Miguel Salhuana Application Note AN4519 – Data Manipulation and Basic Settings of the MPL3115A2 Command Line Interface Driver Code*. Rev 0.1 08/2012.
3. Freescale Semiconductor. *Miguel Salhuana Application Note AN4481 – Sensor I^2C Setup and FAQ*. Rev 0.1, 07/2012.
4. Mühl, T. (2020). *Elektrische Messtechnik, Grundlagen, Messverfahren, Anwendungen* (6. Aufl.). Springer.
5. Bräutigam, M. (2017). *Elektronische Druckmesstechnik – Technisches Knowhow für den Anwender, JUMO Fulda*. https://campus.jumo.de. Accessed 15 Feb 2021.
6. Roedel, W., & Wagner, T. (2011). *Physik unserer Umwelt: Die Atmosphäre*. Springer.
7. Werner, M. (2008). *Information und Codierung*. Vieweg + Teubner.
8. Silicon Labs. (2015). *SI7021 – I2C Humidity and temperature sensor*. www.silabs.com.
9. Silicon Labs. (2015). *AN607 – SI70xx – Humindity sensor designer's guide*. www.silabs.com.
10. Maxim Integrated. (2015). *AN27 – Understanding and using redundancy checks with Maxim 1-wire and iButton Products*. www.maximintegrated.com.
11. Texas Instruments. (2015). *TMP175 – Digital temperature sensor with two-wire interface*. www.ti.com.
12. Lattanzio, L. *Feinstauberfassung zur Messung der Luftqualität*. https://www.sensirion.com/de/ueber-uns/newsroom/sensirion-fachartikel/feinstauberfassung-zur-messung-der-luftqualitaet/. Accessed 27 Dez 2020.
13. Budde, M., & Schwarz, A. et al. *Potenzial und Grenzen des kostengünstigen SDS011Partikelsensors bei der Überwachung urbaner Luftqualität, 48*. Jahrestagung der Gesellschaft für Umweltsimulation. http://www.teco.edu/~budde/publications/GUS2019_budde.pdf. Accessed 6 Jun 2020.
14. Vogel, M. (2012). *Wächter der Luft*. Physik im Alltag, Ausgabe 1/2012, S. 44 f. https://www.pro-physik.de/restricted-files/94561. Accessed 7 Juni 2020.
15. Wang, S. (2009). *Partikelgrößenbestimmung mittels eines Laser-Optischen Partikelzählers mit zwei Empfangswinkelbereichen*. Dissertation Uni Cottbus.
16. Nova Fitness Co., Ltd. (2015). *Laser dust sensor control protocol V1.3*. http://www.inovafitness.com/en/a/chanpinzhongxin/95.html. Accessed 27 Dez 2020.

Accelerometers

Abstract

In this chapter, two linear micromechanical accelerometers for different value ranges are presented.

Micromechanical acceleration sensors have experienced an incredible boom in recent years. In the past, measurement methods existed for the position of an object in space (for example in aircraft, spacecraft or ships), as part of an position control system, so-called inertial sensors. The accelerations here thus correspond to the order of magnitude of fractions of the acceleration due to gravity. Other applications deal with linear accelerations or decelerations which occur, for example, when a vehicle is involved in an accident (crash sensor for the airbag) or in the case of rapidly accelerating vehicles and aircraft (see also Chap. 26 for rocket acceleration). Their value ranges are several times the acceleration due to gravity (10 ... 100 g). In various cases, integration of acceleration is a means of determining or verifying a velocity and a distance travelled. Finally, there are applications for rotational movements (angular rate sensors, see Chap. 17).

In the past, these sensors were implemented with mechanical gyroscopes. Micromechanical sensors are cheap and produced in huge masses, some applications are:

- Position sensors
- Measurement of delays for passive safety systems (airbag, belt tensioners, etc.)
- Vibration measurement
- Active suspensions in vehicles

© The Author(s), under exclusive license to Springer Fachmedien Wiesbaden GmbH, part of Springer Nature 2023
A. Meroth, P. Sora, *Sensor networks in theory and practice*,
https://doi.org/10.1007/978-3-658-39576-6_16

- Rate-of-turn sensors for active stabilization of cornering and braking (see Chap. 17)
- Alarm systems (tilt, fall)
- Sensors for crash tests
- Detection of damage events (e.g. during transport)
- Video games
- Determination of sleep phases
- Camber measurement for hard disks
- Accident data recorder

to name just a few. The working principle of the MEMS sensors is explained in the next section, here it should only be noted that there are also analog acceleration sensors on the market, which output a voltage proportional to the acceleration. We have decided to use digital sensors throughout.

16.1   Acceleration Sensor ADXL312

ADXL312 is a microelectromechanical sensor (MEMS) [2] that uses a capacitive measurement method to measure acceleration along three mutually perpendicular axes in the range of ± 12 g (1 g $\approx$ 9.81 m/s^2). The capacitive measurement method provides good linearity and a low temperature coefficient [1]. With a measurement rate of up to 3200 measurements/s, a bandwidth from 0 Hz to 1600 Hz is covered. With the measurement of the acceleration due to gravity, the device can also be used as a tilt sensor.

In principle, the sensing device of such a sensor for a measuring axis consists of a double-sprung seismic mass which can move between two stationary electrodes along a straight line. Together with the electrodes, the seismic mass forms two capacitors whose change in capacitance is proportional to the movement or acceleration of the device, as in Fig. 16.1a. To increase the sensitivity of the probe, a comb structure is used, as described in [1]. The probe is electrically integrated into a bridge circuit (see Fig. 16.1b), whose analog deflection voltage is sampled, digitized and stored after digital filtering. Depending on the measuring rate and the complexity of the networking, the measured values can be stored either in data registers or in intermediate buffers with 32 memory locations per axis. The stored values can then be read via a serial interface.

The sensor has a set of registers that can provide flexibility in measurement configuration, data storage, communication, and power management. A master connected to the device can access these registers to change settings, read measurement data, test the device, or identify it in a network. The power consumption is to be optimized depending on the concretely pursued goals, in order to be able to use the device also in battery-operated devices. With flexible interrupt management and two interrupt outputs, the device can be used in complex network structures.

Fig. 16.1 ADXL312 – Basic design of a MEMS capacitive acceleration sensor

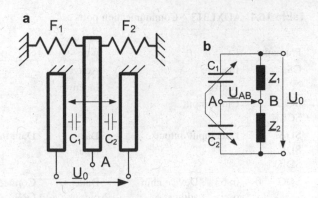

16.1.1 Networking of the ADXL312

Two serial interfaces are provided for networking the sensor, which is configured as a slave: SPI and I^2C. These interfaces share the SDI (Data In), CLK (clock signal), SDO (Data Out) and CS (Chip Select) pins (see Table 16.1). The Chip Select signal automatically activates one of the interfaces. A master can communicate with the sensor in 4-wire or 3-wire SPI mode at a maximum data rate of 5 Mbit/s in SPI mode 3. The most significant bit of a byte is sent first. By default, 4-wire mode is active. 3-wire mode can be enabled after power-up by setting bit 6 of the DATA_FORMAT register. The elements of the data structure that uniquely identify the sensor for use with a global SPI module have the following meaning:

- 1st element – the address of the DDR register to which the slave select line is connected;
- 2nd element – the address of the PORT register to which the slave select line is connected;
- 3rd element – the corresponding bit of the slave select line;
- 4th element – 1 if the line is controlled (0 if with permanently connected line).

After SPI communication has been initiated by setting the CS line low, a start byte is sent that contains the address of the desired register. The register addresses are in the range: 0x00 ... 0x3F. The most significant bit of the start byte determines the direction of the data flow:

- a 0 signals the overwriting of the addressed register (WRITE); the start byte is followed by the new content of the register.
- with a 1 a read command is coded; the master sends after the start byte any further byte (mostly 0x00 or 0xFF) to receive the response of the sensor.

Table 16.1 ADXL312 – Communication ports

Pin name		I^2C		SPI 4-wire	SPI 3-wire
CS	0	Inactive		Active	
	1	Active		Inactive	
SCL/ SCLK		Clock input			
SDA/ SDI/ SDIO		Data input/output		Data input	Data input/output
SDO	0	0×53	Device chip address	Data output	Connected directly to V_{DD} or via 10 kΩ at GND
	1	0×31			

If bit 6 of the start byte is set, the internal address counter is incremented with each register access, which enables reading or writing of several contiguous registers in one process. In this case, the start byte contains the address of the first register.

In the case of a point-to-point connection between a master and the sensor, the master can uniquely determine the communication interface. In the case of SPI networking (Fig. 16.2a), the SPI interface is selected by the master pulling the CS line low, and the sensor correctly evaluates the incoming bytes as an SPI message. When the master communicates with the slave1 (CS = High, CS_1 = Low), a data flow takes place at the SDI input of the sensor, which together with the clock signal can be interpreted by the sensor as an I^2C message. To avoid bus disturbances, it must be prevented that the signal at the SDI input becomes Low while CS is High [2]. An OR gate (see Fig. 16.2b) can solve the problem in the case of 4-wire transmission. If you want to use the high bandwidth of the sensor and read out the resulting large amount of data, the SPI interface must be used.

If the measurement frequency is low or only the acceleration due to gravity is measured, the sensor can be networked via I^2C, as shown in Fig. 16.3. In this case the ADXL312 is used to determine the horizontal position of the electronic compass HMC5883. The CS pin is hardwired to V_{DD} and the 7-bit I^2C address is fixed at 0x53. Furthermore, the communication with the sensor via I^2C is considered in more detail, which can take place at 400 kBit/s. The clock frequency of the interface is set via the TWBR register of the microcontroller. Depending on the clock frequency of the master (in Hz) and the interface, the content of the register is calculated:

```
#define TWI_SCL_FREQ 400000UL //TWI clock frequency in Hz for the master
#define TWI_MASTER_CLOCK (((F_CPU / TWI_SCL_FREQ) - 16 ) / 2 +1)
```

and saved by calling the function TWI_Master_Init(TWI_MASTER_CLOCK).

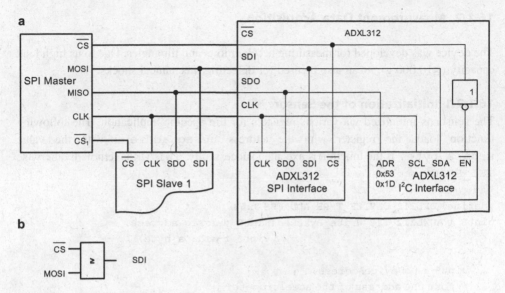

Fig. 16.2 ADXL312 – SPI networking

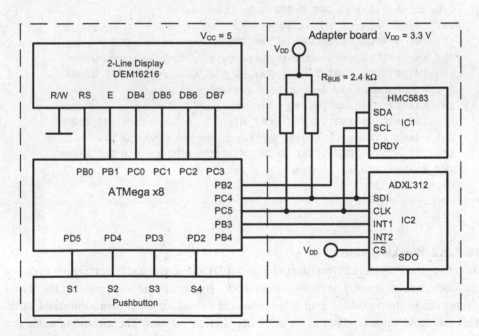

Fig. 16.3 ADXL312 – I²C networking

16.1.2 Measurement Data Acquisition

The device was developed for measuring accelerations and tilt angles. Due to its high load capacity ($\leq$10,000 g), it can also be used for detecting mechanical shocks.

16.1.2.1 Initialization of the Sensor

The sensor is initialized via various registers for the specific application. The following function loads the register with the address `ucreg_address` with the value `ucdata_byte`. If the function is executed successfully, `TWI_OK` is returned, otherwise `TWI_ERROR`.

```
#define ADXL312_DEVICE_TYPE_ADDRESS 0x53
uint8_t ADXL312_I2C_Write_ByteReg(uint8_t ucreg_address,
                                  uint8_t ucdata_byte)
{
    uint8_t ucDeviceAddress;
    //form the address of the accelerometer
    ucDeviceAddress = ADXL312_DEVICE_TYPE_ADDRESS << 1;
    ucDeviceAddress |= TWI_WRITE; //write mode
    TWI_Master_Start(); //start
    if((TWI_STATUS_REGISTER) != TWI_START) return TWI_ERROR;
    TWI_Master_Transmit(ucDeviceAddress); //send device address
    if((TWI_STATUS_REGISTER) != TWI_MT_SLA_ACK) return TWI_ERROR;
    /*the address of the destination register is sent*/
    TWI_Master_Transmit(ucreg_address);
    if((TWI_STATUS_REGISTER) != TWI_MT_DATA_ACK) return TWI_ERROR;
    TWI_Master_Transmit(ucdata_byte); //the data byte is transmitted
    if((TWI_STATUS_REGISTER) != TWI_MT_DATA_ACK) return TWI_ERROR;
    TWI_Master_Stop(); //stop
    return TWI_OK;
}
```

16.1.2.2 Working Mode

By setting the `POWER_CTL` register (address 0x0D), the sensor can be set directly to one of the modes: sleep, standby or measurement mode. By setting the same register, the sensor can remain in sleep mode as long as the measured values are lower than a threshold value. This threshold value is stored in the `THRESH_INACT` register. The measuring mode is a free-running measuring mode in which the analog values of the three measuring axes are sampled at a set frequency, digitized and then stored.

While the other bits are 0, bit 3 of the `POWER_CTL` register can be used to switch between standby mode (bit $3 = 0$) and measurement mode (bit $3 = 1$). After power-on, the device boots up in standby mode to save power.

16.1.2.3 Measurement Frequency Setting

Bits 3:0 of register BW_RATE (0x2C) code the measuring rate. The smallest adjustable measuring frequency is 6.25 Hz, which corresponds to code 0110. The measuring frequency doubles with incrementing the code value and reaches 3200 Hz at 1111. The possible bandwidth of the accelerations to be measured is half of the set measuring frequency. By setting bit 4 of the same register, measurements in the range 12.5 Hz ... 400 Hz are performed with reduced current consumption. In this power saving measuring mode the noise level may be higher.

16.1.2.4 Measurement Data Format

The entire measuring range of ± 12 g is divided into a further three ranges to achieve a higher measuring resolution. The A/D conversion is performed over the four measuring ranges either with a constant 10-bit resolution, or with a constant step size of approx. 2.9 mg. The settings which are possible with the DATA_FORMAT register (0x31) can be taken from the following description:

- **Bit 7** – SELF_TEST – by setting this bit an electrostatic force is generated internally, which leads to the displacement of the seismic mass. An offset is superimposed on all three axes in order to be able to test the sensor.
- **Bit 6** – SPI – the sensor powers up to 4-wire SPI mode. Setting this bit switches to 3-wire SPI mode.
- **Bit 5** – INT_INVERT – with 0 the interrupt outputs are high-active, with 1 they become low-active;
- **Bit 4** – = 0;
- **Bit 3** – FULL_RES – determines the step size of the A/D converter, as listed in Table 16.2. If this bit is set, the measured values are converted with the lowest step size, the bit resolution changes automatically with the selection of the measuring range;
- **Bit 2** – JUSTIFY- with 0 the measurement results are stored as a 16-bit number, right-justified in two's complement format in two 1-byte registers; with 1 they are stored left-justified;
- **Bit 1:0** – determine the measuring range according to Table 16.2.

16.1.2.5 Saving Measurement Data

The measured values from each measuring axis are stored as a 16-bit number either in two 1-byte registers or in a 32x2-byte FIFO memory buffer. Before saving, each acceleration value of all axes is added to the content of the corresponding offset register. This saves corresponding computing time of the master. The offset registers are 1-byte volatile registers and store the values in two's complement with a resolution of 11.6 mg. The contents of these registers are set to 0x00 at power-up. The sensor's storage modes are set with bits 7:6 of register FIFO_CTRL (0x38) as follows:

Table 16.2 ADXL312 – Setting the measuring range, bit resolution and step size

DATA_FORMAT	Bit3, Bit1:Bit0							
	000	001	010	011	100	101	110	111
Measuring range/g	±1.5	±3	±6	±12	±1.5	±3	±6	±12
Bit resolution/Bit	10	10	10	10	10	11	12	13
Step size/mg	2.9	5.8	11.6	23.2	2.9	2.9	2.9	2.9

Table 16.3 ADXL312 – Data and offset registers

Measuring axis	Data register		Offset register
	Higher byte	Low byte	
X	DATA_X_MSB 0x33	DATA_X_LSB 0x32	OFS_X 0x1E
Y	DATA_Y_MSB 0x35	DATA_Y_LSB 0x34	OFS_Y 0x1F
Z	DATA_Z_MSB 0x37	DATA_Z_LSB 0x36	OFS_Z 0x20

- **Bit 7:6 = 00**; in bypass mode the measured values are stored directly into the data registers in little-endian format (see Table 16.3). In this mode, the sensor allows the unread data set, or the unread data registers, to be overwritten. By selecting this mode, all memory cells of the FIFO buffer are set to 0x00.
- **Bit 7:6 = 01**; in FIFO mode a new data set is only stored in the buffer as long as free memory locations are available. Reading a data set in this mode frees up a memory location.
- **Bit 7:6 = 10**; in stream mode the new data sets are continuously stored in the buffer. When the buffer is full, the oldest data record is overwritten by the newest.
- **Bit 7:6 = 11**; in trigger mode, the last "n" measured values remain stored before an interrupt is triggered. Further $(32 - n)$ data sets fill the buffer fully and then an interrupt is triggered at pin INT1 if bit 5 of register FIFO_CTRL is 0 or at INT2 if bit is 1. In this mode, the events that trigger an interrupt can be examined. New data sets can only be stored in this mode after clearing the buffer as shown in the following code section:

```
#define FIFO_CTRL_REG 0x38 //address of the register
//switch over to bypass
ADXL312_I2C_Write_ByteReg(FIFO_CTRL_REG, 0x00);
//switch to trigger, store 15 measured values before event,
// interrupt at INT1 0xDF = 1101 1111
ADXL312_I2C_Write_ByteReg(FIFO_CTRL_REG, 0xDF);
```

In bits 4:0 of register FIFO_CTRL the number "n" is stored in trigger mode, in FIFO and stream mode the fill limit of the buffer. When the unread records from the FIFO buffer reach this fill limit, an interrupt is triggered. The current number of unread data sets from the buffer is stored in bits 5:0 of the FIFO_STATUS register (0x39).

16.1.2.6 Reading the Measured Values

In bypass mode, the last stored data set in the data registers is available for reading. In this mode, only individual data registers can also be read, but there is no protection mechanism that prevents an unread data set from being overwritten. In the memory modes that use the FIFO buffer, the oldest, unread data set is in the data registers. After reading the six data registers, with the jump of the internal address counter to address 0x38, the next data set is shifted into the data registers within 5 µs. This delay time must be observed when reading data from the FIFO buffer in order to ensure that a consistent data set is read. By calling the following function a complete data set is read out via I²C and the measured values are stored in a vector of type unsigned int.

```
uint8_t ADXL312_I2C_Read_DataReg(uint16_t* uidata_word)
{
    uint8_t ucDeviceAddress, ucDataLSByte, ucDataMSByte, ucI;
    //form the address of the accelerometer sensor
    ucDeviceAddress = ADXL312_DEVICE_TYPE_ADDRESS << 1;
    ucDeviceAddress |= TWI_WRITE; //write mode
    TWI_Master_Start(); //start
    if((TWI_STATUS_REGISTER) != TWI_START) return TWI_ERROR;
    TWI_Master_Transmit(ucDeviceAddress); //send device address
    if((TWI_STATUS_REGISTER) != TWI_MT_SLA_ACK) return TWI_ERROR;
    //the start address of the register area is sent
    TWI_Master_Transmit(DATA_X_MSB_REG);
    if((TWI_STATUS_REGISTER) != TWI_MT_DATA_ACK) return TWI_ERROR;
    TWI_Master_Start(); //restart
    if((TWI_STATUS_REGISTER) != TWI_RESTART) return TWI_ERROR;
    ucDeviceAddress = (ADXL312_DEVICE_TYPE_ADDRESS << 1) | TWI_READ;
    //send device address in read mode
    TWI_Master_Transmit(ucDeviceAddress);
    if((TWI_STATUS_REGISTER) != TWI_MR_SLA_ACK) return TWI_ERROR;
    //the contents of the addressed registers are read in
    for(ucI = 0; ucI < 3; ucI++)
    {
        //the lower order byte is read
        ucDataLSByte = TWI_Master_Read_Ack();
        if((TWI_STATUS_REGISTER) != TWI_MR_DATA_ACK) return TWI_ERROR;
        if(ucI < 2)
        {
            //higher order byte is read
            ucDataMSByte = TWI_Master_Read_Ack();
            if((TWI_STATUS_REGISTER) != TWI_MR_DATA_ACK)
```

```
                  return TWI_ERROR;
            uidata_word[ucI] = (ucDataMSByte << 8) + ucDataLSByte;
      }
      else
      {
            //last register is read
            ucDataMSByte = TWI_Master_Read_NAck();
            if((TWI_STATUS_REGISTER) != TWI_MR_DATA_NACK)
                  return TWI_ERROR;
            uidata_word[ucI] = (ucDataMSByte << 8) + ucDataLSByte;
      }
   }
   TWI_Master_Stop(); //stop
   return TWI_OK;
}
```

16.1.3 Offset Determination

If the absolute values of the acceleration are required, such as for the calculation of the tilt angle, the offset values must be taken into account. To determine the offset, the master reads the measured accelerations and calculates the arithmetic mean over at least two or more measured values in order to filter out any peaks (outliers) (moving average). With a rotational movement around each axis, the maxima (positive values) iValueMax_i and the minima (negative values) iValueMin_i are recorded and stored. The offset value for the i. axis is calculated with the equation:

$$iOffset\_i = \frac{iValueMax\_i + iValueMin\_i}{2} \qquad (16.1)$$

according to the selected measuring range. The calculated offset values must be corrected by a factor depending on the selected step width (see Table 16.3) before they are stored in the offset registers:

$$iOffsetKorr\_i = -\left(iOffset\_i \cdot \frac{OffsetReg\_i\ Resolution}{Measuring\ range\ Resolution} \right) \qquad (16.2)$$

For a selected measuring range of ± 1.5 g and a resolution of the offset register of 11.6 mg, the corrected offset value of the X axis can be calculated and stored with the following program section:

```
#define OFFSET_X_REG 0x1E //address of the offset register
int iValueMax_X, iValueMin_X, iOffset_X; //declaration of the variables
iOffset_X = (iValueMax_X + iValueMin_X) / 8; //divide by (11.6/2.9) x 2
iOffset_X = ~iOffset_X + 1; //multiplication with (-1)
ADXL312_I2C_Write_ByteReg(OFFSET_X_REG, iOffset_X); //save the offset
```

The measured values are added internally with the contents of the offset registers before they are stored in the data registers. Values around 0 for the X- and Y-axis show the horizontal position of the sensor.

16.1.4 Interrupt Mode

ADXL312 has a complex interrupt structure that is intended to relieve the master. The freely configurable interrupts control two push-pull outputs when triggered. The fulfillment of a condition triggering an interrupt is signaled by the internal logic by setting the corresponding bit in the INT_SOURCE (0x30) register, which is described below.

- **Bit 7** – DATA_READY – if this bit is set, then it indicates that a new data set is present. This interrupt source can be used in bypass mode to read all measured values without blocking waiting.
- **Bit 6, 5, 2** – not used
- **Bit 4** – ACTIVITY – this bit is set when a measured value exceeds the threshold value stored in the TRESH_ACT register (0x24).
- **BIT 3** – INACTIVITY – if the measured values fall below the value stored in the TRESH_INACT (0x25) register over the TIME_INACT (0x26) time, this bit is set. The time is stored in the register in seconds.
- **Bit 1** – WATERMARK – the bit can be set in FIFO or stream mode if the number of unread records from the FIFO buffer is equal to the fill limit.
- **Bit 0** – OVERFLOW – the set bit indicates that unread records have been overwritten.

The TRESH_ACT and TRESH_INACT registers are 8-bit registers that store the unsigned threshold values with a resolution of 46.4 mg. The INT_ENABLE and INT_MAP registers have the same bit assignment as the INT_SOURCE register. The INT_ENABLE (0x2E) register enables each interrupt source by setting the appropriate bit. The enabled interrupts can be assigned to output INT1 or INT2 through the INT_MAP (0x2F) register. If a bit is set in this register, the corresponding enabled interrupt is assigned to output INT2; otherwise to INT1. Several interrupts can be assigned to the same output via an OR operation as in Fig. 16.4. Further details regarding interrupts can be found in the data sheet of the sensor [2].

The following program section is used to set the DATA_READY interrupt in addition to the existing interrupts (see Fig. 16.4).

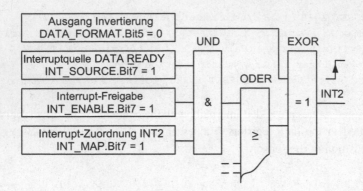

Fig. 16.4 ADXL312 – Interrupt configuration and pin assignment

```
uint8_t ucRegByte;
ADXL312_I2C_Read_ByteReg(INT_SOURCE_REG, &ucRegByte);
ucRegByte |= 0x80; //the bit 7 is set
//interrupt source is selected
ADXL312_I2C_Write_ByteReg(INT_SOURCE_REG, ucRegByte);
ADXL312_I2C_Read_ByteReg(INT_MAP_REG, &ucRegByte);
ucRegByte |= 0x80;
//interrupt allocation
ADXL312_I2C_Write_ByteReg(INT_MAP_REG, ucRegByte);
ADXL312_I2C_Read_ByteReg(INT_ENABLE_REG, &ucRegByte);
ucRegByte |= 0x80;
ADXL312_I2C_Write_ByteReg(INT_ENABLE_REG, ucRegByte); //interrupt enable
```

16.1.5 ADXL312 as Inclination Sensor

The following is an example of a way to use the ADXL312 as a tilt sensor. Figure 16.5a shows the axis arrangement of the sensor, which is wired as in Fig. 16.3. When the angle between the positive direction of an axis and the vector of gravitational acceleration is 180°, +1 g is measured at that axis. Depending on the sensor position, the acceleration due to gravity $\overline{g}$ is divided into the components g_x, g_y and g_z measured by the sensor, where

$$g^2 = g_x^2 + g_y^2 + g_z^2 \tag{16.3}$$

16.1.5.1 Theoretical Approach

The aim is to calculate the tilt angle β (see Fig. 16.5b) between the vertical and the sensor's own Z-axis and the direction of tilt via the angle α. The tilt angle takes values between 0°

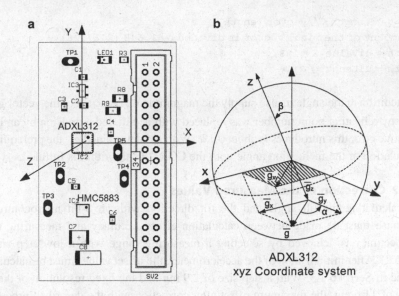

Fig. 16.5 ADXL312 as inclination sensor

and 180° and can be calculated using the CORDIC algorithm if the components g_z and g_{xy} are known. The component g_z is measured directly and g_{xy} can be calculated using the Pythagorean theorem:

$$g_{xy} = \sqrt{g_x^2 + g_y^2}, \tag{16.4}$$

but this is time-consuming for an 8-bit microcontroller. The tilt direction is determined by the angle α between the vector $\overline{g_{xy}}$ and the Y-axis. For very small tilt angles, noise affects the accuracy of the direction calculation because the X and Y values are very small. This angle takes values from 0° to 360° and can be calculated using the CORDIC algorithm presented in Chap. 18 (Eq. 18.5). At the end of the iterative procedure for the angle calculation, the final vector $\overline{g'_{xy}}$ is almost on the X-axis and provides a measure of the magnitude. According to [4], the vector magnitudes g_{xy} and g'_{xy} have the following relationship:

$$g'_{xy} = k \cdot g_{xy}, k > 1$$
$$k = \prod_{i=0}^{n-1} \sqrt{1 + 2^{-2 \cdot i}} \tag{16.5}$$

where n is the number of iterations. For $n = 8$, $k \approx 1.646760258$. Dividing the final vector length by the factor k gives the desired magnitude g_{xy}. The angle calculation function presented in Chap. 18 with the complement:

```
uint16_t uiLength; //vector length
/*the amount of the final vector is divided by 1.648 (211/128)*/
uiLength = uiValue_X * 128;
uiLength = uiLength / 211;
```

can, in addition to the angle α, also supply the magnitude uiLength of the vector $\overline{g_{xy}}$. The division by a floating point number was replaced with a shift and a division by an integer. The relative error this introduces is about 0.1%. With the determined g_{xy} the prerequisite for the calculation of the inclination angle with the CORDIC algorithm is fulfilled.

16.1.5.2 Correction of the Measured Values

When calculating the tilt angle and the tilt direction with an 8-bit microcontroller, a compromise must be made between calculation effort, accuracy and measuring rate. A higher accuracy is achieved by selecting a measuring range with a low step size (see Table 16.2). After initialization of the accelerometer, the offset values must be calculated as described in Sect. 16.1.3. With a step size of 2.9 mg and the fixed resolution of the offset registers of 11.6 mg, the maximum offset after correction with the described procedure is $\pm 7 \cdot 2.9$ mg if:

$$iValueMax\_i + iValueMin\_i = n \cdot 8 - 1 \ n \in N \tag{16.6}$$

which affects the accuracy of the calculations. Better results can be obtained if the offset values calculated with Eq. 16.1 are added directly to the read-out measured values. With an additional three additions, a maximum offset of ± 1 is obtained.

The correction of the measured values is aimed at calculating the correction factors by which the measured values are multiplied so that Eq. 16.7 is fulfilled.

$$g_{x\_korr}^2 + g_{y\_korr}^2 + g_{z\_korr}^2 = const. \tag{16.7}$$

- The maximum positive values for all three axes are calculated with consideration of the offset: $iValueMax\_i = iValueMax\_i + iOffset\_i$
- the highest value of all three values is determined: $iValueMax = \max (iValueMax\_i)$
- the correction factors for the three axes are calculated: $iValueKorr\_i = iValueMax/ iValueMax\_i$.

Before the measured acceleration values are used to calculate the tilt angle and direction, the corresponding offset values are added to them and the results are multiplied by the correction factors as in the following code snippet for the X axis:

```
iValue_X = iValue_X + iOffset_X;
iValue_X = (iValue_X * iValueMax) / iValueMax_X;
```

16.1.5.3 Calculation of the Angle of Inclination and the Direction of Inclination

With the corrected measured values of the X- and Y-axis the angle α and the amount g_{xy} are calculated as described. By calling the function ADXL312_Get_TiltAngle() with the amount uixy_value and the corrected value of the Z-axis uiz_value as parameters, the tilt angle is also calculated in tenths of a degree.

```
uint16_t ADXL312_Get_TiltAngle(uint16_t uixy_value,
                               uint16_t uiz_value)
{
    uint8_t ucQuadrant = 0, ucI;
    uint16_t uiXY_Value, uiZ_Value, uiAngle = 0;
    //the angles as support points in tenths of a degree
    uint16_t uiPhi[8] = {450, 265, 140, 71, 36, 18, 9, 4};
    if(uixy_value & 0x8000) uixy_value = ~uixy_value + 1; //X is negative
    if(uiz_value & 0x8000) //if true, the Y-value is negative
    {
        uiz_value = ~uiz_value + 1;
        ucQuadrant |= 0x01;
    }
    //approximation of the angle in the 1st quadrant in eight steps
    for(ucI = 0; ucI < 8; ucI++)
    {
        if(uiz_value & 0x8000)
        {
            uiAngle -= uiPhi[ucI];
            uiXY_Value = uixy_value - ((int)uiz_value >> ucI);
            uiZ_Value = uiz_value + (uixy_value >> ucI);
        }
        else
        {
            uiAngle += uiPhi[ucI];
            uiXY_Value = uixy_value + ((int)uiz_value >> ucI);
            uiZ_Value = uiz_value - (uixy_value >> ucI);
        }
        uixy_value = uiXY_Value;
        uiz_value = uiZ_Value;
    }
    switch(ucQuadrant) //calculation of the real angle
    {
        case 0: uiAngle = 900 - uiAngle; break;
        case 1: uiAngle = 900 + uiAngle; break;
    }
    return uiAngle;
}
```

16.2   MMA6525

MMA6525 is a 2-axis accelerometer capable of measuring accelerations in the range ±105.5 g with a resolution of 0.055 g [3]. It was developed for equipping airbag control units in the automotive sector, but can also be used for other applications in which high accelerations are to be expected.

16.2.1 Sensor Structure

The block diagram of the sensor is shown in Fig. 16.6. To meet the high safety requirements of an airbag sensor, a DSP is provided for the control and data processing of each axis. The accelerations are measured continuously in free-running mode. The sensing device provide an analog voltage proportional to the acceleration. This voltage is amplified, digitized by a 12-bit A/D converter, and then digitally filtered. The selectable cut-off frequency of the filter is in the range 50 Hz ... 1000 Hz. The digital acceleration values can be read out with or without sign. The measuring range of the sensor is covered with unsigned values in the range +128 ... +3968 and with signed values in the range −1920 ... +1920. The middle of a range (2048 or 0) represents the value 0 g. Each DSP controls the entire digital information flow of a

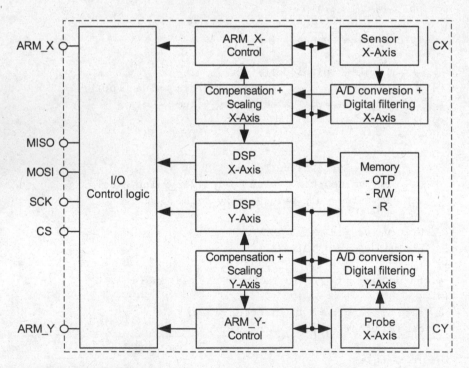

Fig. 16.6 MMA6525 – Block diagram

measurement axis, exchanges data with the internal memory and is connected to the host
microcontroller via a 4-wire SPI interface. The internal DSP can put the measurement sensor
into a test mode by exploiting the electrostatic measurement principle. Exceeding an adjustable
acceleration range can be signaled asynchronously via the ARM_X and ARM_Y pins in time-
critical applications, for example to unlock an airbag. These pins can of course also trigger an
interrupt.

16.2.2 Register Block

The user can access the registers in the address range 0x00 ... 0x1D.

16.2.2.1 OTP[1]-Register

The registers in the address range 0x00 ... 0x08 are programmed once by the manufacturer
and cannot be reprogrammed. Address 0x08 stores the value 0xE1 for identification of the
measuring range ±105.5 g. Each device is assigned a unique 32-bit identification number,
which is stored at addresses 0x00 ... 0x03. At the manufacturer, the acceleration value is
read in test mode and stored at address 0x04 for the X-axis and at address 0x05 for the
Y-axis. When repeating the measurements in test mode, the measured values must not
deviate from the stored values by more than ± 10%. A checksum has been calculated and
stored for these nine memory cells. In normal operation, the checksum is recalculated and
compared with the stored one. If there is a discrepancy, a persistent error is signalled.

16.2.2.2 Read-Write Registers

The stored values in the address range 0x0A ... 0x13 are used to configure the axes and can
be written and read. The registers with addresses from 0x0E to 0x13 are used to set the
interrupt-capable outputs ARM_X and ARM_Y and to store the positive and negative
acceleration limits for each axis. The sensor can be reset via the control register DEVCTL
(address 0x0A) and the test mode for the corresponding axis can be activated and the
cut-off frequency of the digital filter selected via the configuration registers DEVCFG_X
(0x0C) and DEVCFG_Y (0x0D). The general configuration of the sensor is done via the
register DEVCFG (0x0B) which is described in the following:

- **Bit 7** – OC – with the reset of this bit the sensor provides dynamic acceleration values;
- **Bit 6** – reserved;
- **Bit 5** – ENDINIT – the master sets this bit to 1 to end the initialization phase and start the
  normal operating mode; this disables writing of the memory area except for register
  DEVCTL and a checksum of this register area is calculated and stored.
- **Bit 4** – SD – if this bit is 1, the measured values are output unsigned;

---

[1] OTP – one time programmable.

- **Bit 3** – OFMON – if set, the sensor logic monitors the exceeding of the stored acceleration limits;
- **Bit 2 . . . 0** – with these three bits set to 0 the ARM outputs are disabled.

16.2.2.3 Read-Only Registers

The register values from the address range 0x14 to 0x17 are updated internally and can only be read. The offset values for the X- and Y-axis determined by the sensor logic are stored in the register OFFCORR_X (0x16) or OFFCORR_Y (0x17). The current status of the sensor is mapped by the DEVSTAT register (0x14).

- **Bit 7** – reserved;
- **Bit 6** – IDE – is set if the checksums are inconsistent;
- **Bit 5** – reserved;
- **Bit 4** – DEVINIT – is only set during the internal initialization phase;
- **Bit 3** – MISOERR – if set, signals an SPI transmission error;
- **Bit 2 . . . 1** – OFF_X, OFF_Y – the bits are set when the offset limits have been reached;
- **Bit 0** – DEVRES – is set during initialization as a result of a reset.

The error bits set by the sensor (except DEVINIT) can be cleared by reading the status register.

16.2.3 SPI Communication

The sensor MMA6525 is configured as SPI slave in mode 0. The data structure of a sensor whose chip select input is connected to the PB0 pin of an ATmega168 microcontroller looks as follows (Chap. 14):

```
MMA65XX_pins MMA65XX_1 = {{/*CS_DDR*/ &DDRB,
                           /*CS_PORT*/ &PORTB,
                           /*CS_pin*/ PB0,
                           /*CS_state*/ ON}}; //ON = 1
```

An SPI message to the sensor always consists of two bytes, the more significant bit is transmitted first. While receiving a command, the sensor sends a response to the previous message. Each command is secured with a parity bit, which is checked by the sensor. If there is a discrepancy, an SPI error is signaled. The feedback sent by the sensor is also secured with a parity bit, which can be checked by the microcontroller. It is calculated with odd parity.

16.2.3.1 Initialization of the Sensor

The initialization of the sensor should take place at the earliest 10 ms after it has been supplied. During initialization, the sensor is configured and its error status is checked. In addition, its identification number can be checked and the sensor and measuring chain can be tested in test mode. Any errors that occur must be intercepted and handled by the software. The error-free initialization is terminated by setting the ENDINIT bit in the DEVCFG register, thus starting the normal working mode of the sensor. In the following program code, analogous to the I/O pin abstraction presented in Chap. 14, a data structure MMA65XX_pins is defined first, which contains the pin definitions including the SPI slave select input. In the case of the MMA65XX this may look exaggerated, since the chip does not contain any additional pins, but in the sense of uniformity such an approach makes sense.

```
typedef struct {
    tspiHandle MMA65XXspi;
} MMA65XX_pins;
```

This is handled in the same way for all the following components with SPI interface, without going into more detail.

```
uint8_t MMA65XX_Init (MMA65XX_pins sdevice_pins)
{
    #define NO_ERROR 0
    #define DEVINIT 0x20
    #define ERROR_MASK 0x5F
    volatile uint8_t ucAnswer[4];
    uint8_t ucReg, ucError;
    //the slave select connection of the sensor is initialized
    //(microcontroller output set to high)
    SPI_MasterInit_CS(sdevice_pins.MMA65XXspi);
    //the error flags will be reset
    MMA65XX_Read_Reg(sdevice_pins, DEVICE_STATUS_REG, ucAnswer);
    //are the error flags reset?
    MMA65XX_Read_Reg(sdevice_pins, DEVICE_STATUS_REG, ucAnswer);
    ucError = ucAnswer[3] & ERROR_MASK;
    if(ucError) return ucError;
    //saving the Device Config Reg. in ucReg
    MMA65XX_Read_Reg(sdevice_pins, DEVICE_CONFIG_REG, ucAnswer);
    ucReg = ucAnswer[3];
    ucReg |= DEVINIT;
    MMA65XX_Write_Reg(sdevice_pins, DEVICE_CONFIG_REG, ucReg, ucAnswer);
    return NO_ERROR;
}
```

The function MMA65XX_Init() performs a simplified initialization of the sensor. The first read of the status register clears the error bits set during power-up. The second read of the register checks whether all error bits have been cleared, otherwise the function is acknowledged with an error code. If no persistent errors are present, the ENDINIT bit is set and the function returns the value NO_ERROR.

16.2.3.2 Reading a Register

The composition of a read command and the response of the sensor, which follows with the sending of another message, is shown in Table 16.4. The parity bit P depends on the address or the content of the register.

The content of a register can be read with the following example function. When calling the function MMA65XX_Read_Reg() the SPI data structure of the sensor sdevice_pins, the address of the register to be read ucaddress and the address of a 4-byte vector ucarray2read must be passed as parameters. In the function, the parity bit that accompanies the message is calculated. In the first two bytes of the vector the response of the sensor to the previous message is stored and in the last two the content of the register according to Table 16.4, or an error message.

```
void MMA65XX_Read_Reg(MMA65XX_pins sdevice_pins,
                      uint8_t ucaddress,
                      volatile uint8_t* ucarray2read)
{
    uint8_t ucDummy = 0x00, ucI, ucMask = 0x01, ucCnt = 0x00;
    for(ucI = 0; ucI < 8; ucI++)
    {
        if(ucaddress & ucMask) ucCnt++;
        ucMask = ucMask << 1;
    }
    if(!(ucCnt % 2)) ucaddress |= 0x80;
    //the chip select line is set to low
    SPI_Master_Start(sdevice_pins.MMA65XXspi);
    //the address of the register is transferred
    ucarray2read[0] = SPI_Master_Write(ucaddress);
    ucarray2read[1] = SPI_Master_Write(ucDummy);
    SPI_Master_Stop(sdevice_pins.MMA65XXspi);
    //the response of the sensor to the read command is read out
    SPI_Master_Start(sdevice_pins.MMA65XXspi);
    ucarray2read[2] = SPI_Master_Write(ucDummy);
    ucarray2read[3] = SPI_Master_Write(ucDummy);
    SPI_Master_Stop(sdevice_pins.MMA65XXspi);
}
```

16.2.3.3 Writing a Register

The write instruction of a register and the expected response is summarized in Table 16.5 according to [3]. Bit 13 of the instruction encodes a register operation, while bit 14 encodes the write. The least significant byte of the instruction maps the new contents of the register.

Table 16.4 MMA6525 – Read command of a register and response

	MSB	Bit													LSB
	15	14	13	12	11	10	9	8	7	6	5	4	3	2	1	0
Command	P	0	0	A4	A3	A2	A1	A0	0	0	0	0	0	0	0	0
				Address of the register												
Reply	0	1	0	P	1	1	1	0	D7	D6	D5	D4	D3	D2	D1	D0
									Contents of the register							

Table 16.5 MMA6525 – Write command of a register and response

	MSB	Bit													LSB
	15	14	13	12	11	10	9	8	7	6	5	4	3	2	1	0
Command	P	1	0	A4	A3	A2	A1	A0	D7	D6	D5	D4	D3	D2	D1	D0
				Address of the register					New content of the register							
Reply	0	0	1	P	1	1	1	0	D7	D6	D5	D4	D3	D2	D1	D0
									New content of the register							

16.2.3.4 Reading Out the Acceleration Values

The query of an acceleration value stores a new value into the measuring register of the corresponding axis on the rising edge of the chip select signal. However, this value can only be read with the next message (query). The query consists of a 2-byte command that is transmitted to the sensor via SPI. Bit 13 on 1 is used to code the read command of a measured value. The following options can be selected:

- **Bit 14** – AX – codes the axis: 0 for the X-axis, 1 for the Y-axis;
- **Bit 12** – OC – at 0 the dynamic acceleration is measured, otherwise the static acceleration;
- **Bit 2** – SD – with 0 the values are output signed, with 1 unsigned;
- **Bit 1** – ARM – with 0 the ARM function is deactivated.

Bits 11 and 10 S1:S0 from the response indicate the following:

- 00 – the query is made in the initialization phase;
- 01 – the query is made during normal operation;
- 10 – acceleration was measured in test mode;
- 11 – an error has occurred.

Bits 9:0, 15 and 14 from the response form the measured acceleration value. The measured acceleration can be read out with the function MMA65XX_Get_Acceleration. The parameter ucop_code codes the read command according to Table 16.6. The acceleration value is stored at the address uiaccel_data. To measure and read unsigned the dynamic value of the acceleration in the X direction with the ARM function disabled, the command code is 0x200C (b0010 0000 0000 1100). With the call: MMA65XX_Get_Acceleration(MMA65XX_1, 0x200C, & uiAccelData), the function is executed and if executed without errors, the summarized value of the acceleration is stored in the variable uiAccelData.

```
uint8_t MMA65XX_Get_Acceleration(MMA65XX_pins sdevice_pins,
                                 uint16_t uiop_code,
                                 uint16_t *uiaccel_data)
{
    #define ERROR_CODE 0x0C00
    #define NO_ERROR 0
    uint16_t uiData, uiMask = 0x01;
    uint8_t ucI, ucCnt = 0x00, ucDummy = 0x00;
    uint8_t ucCodeHigh, ucCodeLow, ucData;
```

```
ucCodeLow = uiop_code;
ucCodeHigh = uiop_code >> 8;
SPI_Master_Start(sdevice_pins.MMA65XXspi);
//the read command is transferred
SPI_Master_Write(ucCodeHigh);
SPI_Master_Write(ucCodeLow);
SPI_Master_Stop(sdevice_pins.MMA65XXspi);
//the reply to the previous message is read
SPI_Master_Start(sdevice_pins.MMA65XXspi);
ucData = SPI_Master_Write(ucDummy);
uiData = (uint16_t) ucData << 8;
uiData |= SPI_Master_Write(ucDummy);
SPI_Master_Stop(sdevice_pins.MMA65XXspi);
if((uiData & ERROR_CODE) == ERROR_CODE) return 1; //internal error
for(ucI = 0; ucI < 16; ucI++) //parity check
{
    if(uiData & uiMask) ucCnt++;
    uiMask = uiMask << 1;
}
if(!(ucCnt % 2)) return 2; //parity error
uiData = uiData << 2;
uiData |= ucData >> 6;
*uiaccel_data = uiData; //unsigned 12-bit acceleration value
return NO_ERROR;
}
```

In the first step, the command code is transmitted via SPI; in the second, the response of the sensor to the previous request is read with the transmission of two dummy bytes and summarized in the variable uiData. The function then tests whether bits 10 and 11 of this variable are both 1 and if so, it returns the value 1 to signal an internal error of the sensor. The parity of the response is then checked. If this is not true, the function returns a value of 2, which should mean parity error. If the response is error free, the acceleration value is summarized and stored to address uiaccel_data.

Table 16.6 Read command of the measured values and response of the sensor

	MSB	Bit															LSB
	15	14	13	12	11	10	9	8	7	6	5	4	3	2	1	0
Command	0	AX	1	OC	0	0	0	0	0	0	0	0	1	SD	ARM	P
Reply	D1	D0	AX	P	S1	S0	D11	D10	D9	D8	D7	D6	D5	D4	D3	D2

References

1. Glück, M. (2005). *MEMS in der Mikrosystemtechnik*. Springer Fachmedien Wiesbaden.
2. Analog Devices Inc. (2017). ADXL312 – Digital Accelerometer. https://www.analog.com/media/en/technical-documentation/data-sheets/ADXL312.pdf. Accessed 3 Apr 2021.
3. NXP Semiconductors. MMA65xx, dual-axis SPI inertial Sensor. www.nxp.com. Accessed 3 Apr 2021.
4. Volder, J. (1959). The CORDIC computing technique. *IRE Transactions on Electronic Computers, 8*(3), 330–334.

Angular Rate Sensors

17

Abstract

This chapter describes the L3GD20 angular rate sensor.

Angular rate sensors (or gyroscopes) are inertial sensors that measure the rotational speed of a body. They are mostly used for position stabilization and reached great popularity with the introduction of the Electronic Stability Program (ESP) in vehicles at the end of the last century. Some sensors, such as the Bosch BMI160 are referred to as 6-axis sensors and include linear accelerometers (Chap. 16) in the three spatial axes as well as the angular rate sensors about the three spatial axes. In this chapter the sensor L3GD20 from ST Microelectronics is described as a pure gyroscope.

When a ship, aircraft or spacecraft rotates about its longitudinal axis, it is called rolling, with the longitudinal axis pointing forward and the reference point always being the center of mass. A vehicle is said to roll. If the rotation is about the normal axis (axis pointing downwards), it is called yaw. And a rotation around the transversal axis is called pitching in nautics, aircraft and vehicles. The easiest way to remember the underlying Euler angles is to use the right-hand rule: If the thumb points in the direction of the respective axis, then the curved fingers point in the positive direction of rotation. Figure 17.1 shows the respective movements.

Fig. 17.1 Axes of rotation in an aircraft

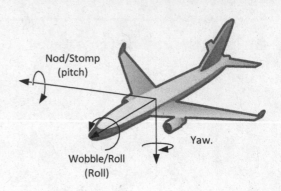

Nod/Stomp (pitch)

Yaw.

Wobble/Roll (Roll)

17.1   Gyroscope

To measure rates of rotation, one can make use of either the Coriolis force or the Sagnac effect. The Coriolis force is a force that acts on a body as it moves relative to a spinning reference system. The Sagnac effect is related to the invariability of the speed of light. A phase shift occurs between two beams of coherent light directed clockwise and counter-clockwise through mirrors on the same path in a circle as the entire structure, including the light source, rotates. Highly sensitive laser gyroscopes used for attitude stabilization of spaceships or airplanes are based on this effect. MEMS sensors are used in consumer technology. These contain micromechanical inertial masses which are spring-mounted and excited to vibrate by an electrical pulse. The Coriolis force that occurs causes them to be minimally deflected in a rotating reference system, resulting in a change in capacitance, which in turn can be measured via the shift of a resonance frequency.

A digital gyroscope is an integrated circuit that provides a digital signal proportional to the angular velocity about one or more axes. The L3GD20 sensor ([1, 2]) measures angular velocity about three axes with the positive measurement direction counterclockwise, as shown in Fig. 17.2. The sensor offers three measurement ranges: ± 250, ± 500, and ± 2000 deg/s. An integrated temperature sensor measures the temperature every second with a resolution of 1 K and can be used to determine the temperature development between two measuring times. The total angle of rotation can be calculated as an integral of the angular velocity. A configurable high-pass filter can be used to remove the DC component of the signal and thus determine the angular acceleration. This high-pass filter can also be used to reduce the influence of high-frequency vibrations. The configuration of the circuit and the reading of the measured values can be done via SPI or via I^2C. An elaborate interrupt concept with two adjustable outputs can lead to a relief of the host microcontroller.

Fig. 17.2 Axis arrangement of
the L3GD20

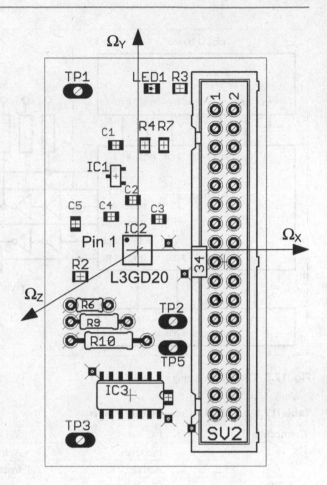

17.1.1 Wiring of the L3GD20

Figure 17.3 shows an example of the wiring of a sensor L3GD20 with a microcontroller of the ATmegax8 family via SPI in 4-wire mode. The two modules are supplied with different voltages and the level adjustment is realized with Schmitt triggers. The pull-up resistors are used in the initialization phase to prevent an I^2C communication from being initiated by mistake.

17.1.2 Communication Interfaces

For communication with the outside world, the sensor provides two interrupt outputs and four interface connections, the meaning of which can be found in Table 17.1.

The device has two serial interfaces: I^2C and SPI, which use the same pins. An I^2C communication with up to 400 kBit/s can take place via the pins SDA and SCL if the pin

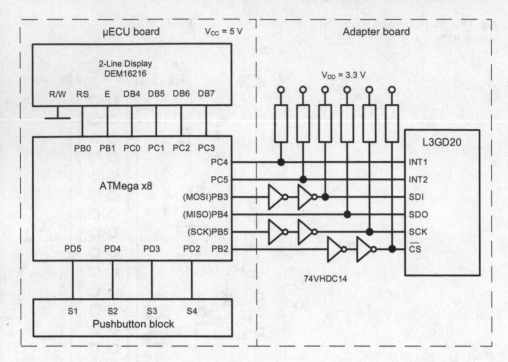

Fig. 17.3 L3GD20 – wiring example

Table 17.1 L3GD20 – Communication ports

Connection		I^2C	SPI 4-wire	SPI 3-wire
CS	0	Inactive	Active	
	1	Active	Inactive	
SCL/SPC		Clock input		
SDA/SDI/SDO		Data connection	Data input	Data input/output
SDO		Device chip address	Data output	Not used

CS is switched high. The SDO connector acts as an address input during this transfer and allows two sensors to be connected to the same I^2C bus. This interface is disabled during an SPI transfer.

In SPI mode the communication takes place with up to 10 MBit/s while the CS input is switched low. In SPI-3-wire mode, a single pin is used for bidirectional data transmission as described in Sect. 10.4. This mode is activated by setting bit 0 SIM in register CTRL_REG4 .

In the following we deal with the control of the sensor only in the classic SPI 4-wire mode with separate connections for the data input and output. Communication with the sensor takes place in SPI mode 3 and the most significant bit is transmitted first, see Fig. 17.5.

Table 17.2 L3GD20 – Configuration of the start byte during SPI transmission

Bit7	Bit6	Bit5	Bit4	Bit3	Bit2	Bit1	Bit0
R/W	MS	Register address					

For the example circuit shown in Fig. 17.3, the data structure for driving the pins, including the chip select input (Sect. 14.5), is as follows:

```
typedef struct
{
    tspiHandle L3GD20spi;
} L3GD20_pins;
```

This structure must be filled in the main module so that it can be used from anywhere.

```
L3GD20_pins L3GD20_1 = {{/*CS_DDR*/ &DDRB,
                         /*CS_PORT*/ &PORTB,
                         /*CS_pin*/ PB2,
                         /*CS_state*/ ON}}; //ON = 1
```

The four values occupied here correspond exactly to the data structure tspiHandle, which is in the first place in the structure L3GD20_pins. All registers of the block have a size of eight bits and are individually addressable. To read or write the content of a register, the master sends a start byte with the configuration from Table 17.2 after switching the slave select line low.

The most significant bit of the start byte determines the direction of the data transfer. If this bit is 0, a write operation is started, otherwise a read operation. After the start byte, the microcontroller sends a data byte for writing or any dummy byte for reading. In an SPI sequence, the start byte can be followed by several data bytes or dummies. If bit 6 (MS) is 0, a new data byte overwrites the old one, with each dummy byte the content of the addressed register is read again. When the MS bit is set, the internal address counter is incremented after each transfer of a byte, allowing the reading or writing of an entire range of contiguous registers in a single operation. Bits 5:0 from the start byte store the address of the first register from the range. An example of such a read function is listed here:

```
void L3GD20_Read_RegBlock(L3GD20_pins sdevice_pins,
                          volatile uint8_t* ucarray2read,
                          uint8_t ucfirst_address,
                          uint8_t ucbyte_number)
{
    uint8_t ucAddress = 0, ucDummy = 0x80, ucI;
    //ucfirst_address the first address of the register area to be read
    //READ = 0x80 means a read command
```

```
//INKR_ON = 0x40 leads to incrementing of the internal address
    //counter after each read register.
ucAddress = ucfirst_address | READ | INKR_ON;
/*the slave select line is set low*/
SPI_Master_Start(sdevice_pins.L3GD20spi);
/*the address of the register transmitted in read mode*/
SPI_Master_Write(ucAddress);
for(ucI = 0; ucI < ucbyte_number; ucI++)
{
    //the read byte is stored in the array ucarray2read
    ucarray2read[ucI] = SPI_Master_Write(ucDummy);
}
//the slave select line is set to high and thus
//the SPI transmission finished
SPI_Master_Stop(sdevice_pins.L3GD20spi);
}
```

When calling the function, the data structure (`sdevice_pins`), the address of the array in which the data is to be stored (`ucarray2read`), the address of the first register from the area to be read (`ucfirst_address`) and the number of registers to be read (`ucbyte_number`) must be passed as parameters.

17.1.3 Working Modes

About 10 ms after switching on, the sensor is in power-down mode (Fig. 17.4), in which the current consumption is reduced to approx. 5 µA. Only the communication interfaces are

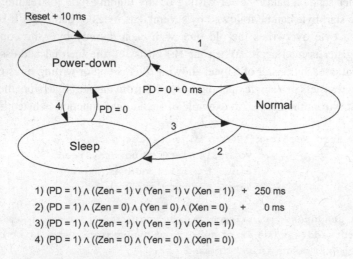

1) $(PD = 1) \wedge ((Zen = 1) \vee (Yen = 1) \vee (Xen = 1))$ + 250 ms
2) $(PD = 1) \wedge (Zen = 0) \wedge (Yen = 0) \wedge (Xen = 0)$ +   0 ms
3) $(PD = 1) \wedge ((Zen = 1) \vee (Yen = 1) \vee (Xen = 1))$
4) $(PD = 1) \wedge ((Zen = 0) \wedge (Yen = 0) \wedge (Xen = 0))$

Fig. 17.4 L3GD20 – Operating modes State transition diagram

Table 17.3 Cut-off frequencies of the high-pass filter (HPF) and the low-pass filters (LPF1 and LPF2)

Cut-off frequency LPF2/Hz		DR1:0			
		00	01	10	11
BW1:0	00	12.5	12.5	20	30
	01	25	25	25	35
	10	25	50	50	50
	11	25	70	100	100
Cut-off frequency LPF1/Hz		32	54	78	93
Sampling frequency/Hz		95	190	380	760
Cut-off frequency HPF/Hz					
HPCF3:0	0000	7.2	13.5	27	51.4
	0001	3.5	7.2	13.5	27
	0010	1.8	3.5	7.2	13.5
	0011	0.9	1.8	3.5	7.2
	0100	0.45	0.9	1.8	3.5
	0101	0.18	0.45	0.9	1.8
	0110	0.09	0.18	0.45	0.9
	0111	0.045	0.09	0.18	0.45
	1000	0.018	0.045	0.09	0.18
	1001	0.009	0.018	0.045	0.09

supplied in this mode. In sleep mode, the sensor consumes about 2 mA, much more than in power-down mode, but the switch to normal mode occurs much faster. By setting the Power_down bit from the CTRL_Reg1 register and activating at least one measurement axis, the sensor is switched to Normal mode. In this mode, the angular velocities are measured continuously with a previously set measuring frequency (see Table 17.3). The switching times which are not entered in Fig. 17.4 are six sampling periods or one, depending on whether an internal low-pass filter LPF2 is activated or not.

17.1.3.1 Angular Velocity Measurement

The sensor's sensing elements supply analog voltages which are amplified and whose frequency spectra are limited by a low-pass filter LPF1 so that no alias effect occurs. With the selection of the measuring frequency, the cut-off frequency of the LPF1 filter is automatically set. It is also possible to filter the measured values with a second low-pass filter LPF2 and/or a high-pass filter HPF before they are stored internally. In this way, high-frequency and/or low-frequency components of the signals can be filtered out.

The control register CTRL_REG1 with address 0x20 is used to select the operating mode of the sensor, the sampling rate and the cut-off frequencies of the low-pass filters LPF1 and LPF2:

Table 17.4 High-pass modes

HPM1	HPM0	High-pass mode
x	0	Normal mode; the DC component of the measured values is set to 0 when the REFERENCE register (address 0x25) is read out.
0	1	Reference mode; the contents of the REFERENCE register are subtracted from the measured value before it is stored in the output register.
1	1	Autoreset mode; the DC component of the measured values is set to 0 when an interrupt is triggered.

- **Bit 7:6** – DR 1:0 – selection of the sampling frequency according to Table 17.3
- **Bit 5:4** – BW 1:0 – selection of the cut-off frequency of the low-pass filter LFP2 according to Table 17.3
- **Bit 3** – PD = 0 – the sensor is in power-down mode
- **Bit 2** – Zen = 1 – enable Z-axis
- **Bit 1** – Yen = 1 – enable Y-axis
- **Bit 0** – Xen = 1 – enable the X-axis

If the sensor operates in normal mode, a value is recorded after each sampling period for each enabled axis, which is quantized with a 16-bit A/D converter and the measured values are stored in two's complement. The control register CTRL_REG2 (0x21) determines the operating mode of the high-pass filter and its cut-off frequency (see Table 17.3) depending on the selected sampling frequency:

- **Bit 7:6** – must both be set to 0
- **Bit 5:4** – HPM 1:0 – determine the working mode of the high-pass filter according to Table 17.4
- **Bit 3:0** – HPCF 3:0 – Cut-off frequency configuration of the high-pass filter according to Table 17.3

17.1.3.2 Intermediate Storage of the Measured Values

After recording, the measured values for each axis can be additionally filtered before saving. Bit 4 (HPEN) and bits 1:0, (Out_Sel 1:0) from control register CTRL_REG5 (0x24) can be used to select the type of filtering before saving:

- OUT1_SEL1:0 = 00 – no filtering
- OUT1_SEL1:0 = 01 – Filtering with the high pass HPF
- HPEN = 0 ∧ OUT1_SEL1 = 1 – Filtering with the low pass LPF2
- HPEN = 1 ∧ OUT1_SEL1 = 1 – Filtering with a bandpass consisting of the lowpass LPF2 and the highpass HPF; the cutoff frequency of the lowpass must be higher than that of the highpass.

Measuring axis	Higher byte	Low byte
X	OUT_X_H 0x29	OUT_X_L 0x28
Y	OUT_Y_H 0x2B	OUT_Y_L 0x2A
Z	OUT_Z_H 0x2D	OUT_Z_L 0x2C

Table 17.5 L3GD20 – Output registers with their addresses

The measured value for an axis can be stored either in a 16-bit output register or in one of the 32x16-bit FIFO[1] data buffers.

17.1.3.2.1 Direct Saving

In this mode the three FIFO buffers are deactivated. The measured values are stored in the output registers for the duration of one sampling period, after which they are overwritten. The configuration of these registers in little-endian format[2] is listed in Table 17.5. In this storage format, the least significant byte of the measured value is stored at the starting address and then the most significant byte is stored. The storage format of the registers can be changed to big-endian by setting bit 6, BLE of the control register CTRL_REG4 (0x23). These output registers are part of the FIFO buffers and store the oldest unread measurement value.

17.1.3.2.2 Buffered Storage

Setting bit 6, FIFO_EN in control register CTRL_REG5 (0x24) activates the storage of the measured values in the buffers. In order to optimize the readout of the buffers and to make it flexible, several storage modes can be configured via the register FIFO_CTRL_REG (0x2E):

- **Bit 7:5** – FM 2:0 – determine the FIFO mode
- **Bit 4:0** – WTM 4:0 – these bits store the fill limit of the data buffers [0, 30]; an interrupt can be triggered when this fill limit is exceeded.

The information about the fill level of the data buffers is stored by the sensor in the read-only register FIFO_SRC_REG (0x2F):

- **Bit 7** – WTM – if set, this bit indicates that the number of unread measured values from the buffer has exceeded the set limit (WTM 4:0).
- **Bit 6** – OVR – is set when the data buffer is full.
- **Bit 5** – EMPTY – is set if there are no unread measured values in the buffers.
- **Bit 4:0** – FSS 4:0 – this counter counts the unread measured values from the data buffer

---

[1] FIFO – first in first out.

[2] Little-endian format – in a two-byte word, the low-order byte is stored at the first address; in big-endian format, the high-order byte is stored first.

17.1.3.2.2.1 *Bypass Mode (FIFO_EN = 1 and FM2:0 = 000)*
If the FIFO_EN bit is set and the FM 2 : 0 bits are reset, the bypass mode is activated and the entire contents of the three data buffers are cleared. The measured values are stored in the output registers for the duration of one sampling period, as with direct storage. This mode is used to switch between the other FIFO modes.

17.1.3.2.2.2 *FIFO Mode (FIFO_EN = 1 and FM2:0 = 001)*
In this mode the measured values fill the data buffer. When the set filling limit is exceeded or at the latest when the buffers are full, an interrupt can be triggered to signal the master to start reading out. The oldest stored measured values are available for readout in the output registers. In order to be able to store further measured values in this mode, the bypass mode must be activated first and the FIFO mode immediately afterwards.

17.1.3.2.2.3 *Stream Mode (FIFO_EN = 1 and FM2:0 = 010)*
In this mode, the counter FSS 4 : 0 is incremented with each new measured value stored in the buffer. When the output register is read out, the oldest unread measured value is transferred and the FSS 4 : 0 counter is decremented. When the capacity of the buffer is reached and no more memory is available, the oldest measured value is overwritten.

In the Stream-to-FIFO (FIFO_EN = 1 and FM 2 : 0 = 011) and Bypass-to-Stream (FIFO_EN = 1 and FM 2 : 0 = 100) modes, the sensor can switch between Stream and FIFO mode, or between Bypass and Stream mode, in order to obtain more detailed information about the course of an interrupt.

17.1.3.3 Reading Out the Measured Values
The stored measured values are always read out from the output registers (Table 17.5). The readout can take place serially either via SPI or I²C. When selecting the interface and the bus clock, the selected measuring frequency must be taken into account.

17.1.3.3.1 Reading Out the Directly Stored Measured Values
If all three measuring axes are enabled, three measured values of 2 bytes each must be transmitted within one sampling period (1/ODR), i.e. a total of 48 data bits, plus the control bits. To determine whether a new data set is available, bit 3 (ZYXDA) of register STATUS_REG (address 0x27) can be polled, i.e. regularly at short intervals. The host microcontroller is relieved if the sensor is configured in such a way that a fully completed measurement process is signaled at the INT2 output. If bit I2_DRDY is set in register CTRL_REG3, output INT2 is set high as soon as a new data set is available, and reset after a measured value has been completely read out. The new measured values can be read for each axis individually as in Fig. 17.5a or for all three axes in one read operation as in Fig. 17.5b. Figure 17.5b shows the timing of read operations of several registers via SPI. Set bit 7 of the control byte encodes the read operation, set bit 6 causes the address to increment after each byte read, and bits 5:0 form the starting address of the register block.

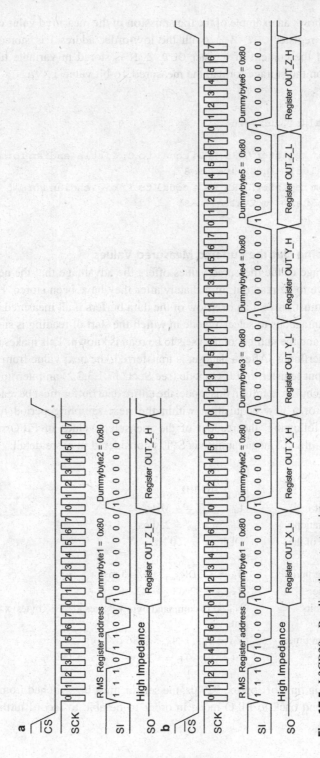

Fig. 17.5 L3GD20 – Reading the output registers

Figure 17.5a shows an example of the transmission of the measured value of the Z-axis. If the content of register OUT_Z_L (with the low-order address) is stored in variable ucZValue1 and the content of register OUT_Z_H is stored in variable ucZValue2, then, depending on the storage format, the measured 16-bit value iZValue is calculated as follows:

```
//uint8_t ucZValue1, ucZValue2;
//int iZValue;
//bit6, BLE from register CTRL_REG4 equal to 0, little-endian format
iZValue = ucZValue1 + (ucZValue2 << 8);
//bit6, BLE from register CTRL_REG4 equal to 1, big-endian format
iZValue = ucZValue2 + (ucZValue1 << 8);
```

17.1.3.3.2   Reading Out the Buffered Measured Values

The buffered storage of the measured values offers the advantage that the new measured values do not have to be read out immediately after they have been stored. However, the selected strategy must prevent an overflow of the data buffers if all measured values are to be read. It is advantageous to select a mode in which the start of reading is signaled via an external interrupt and the number of data sets to be read is known. This makes the query for new data sets superfluous. After a data set is transferred, the next value from the buffer is moved to the output register. In FIFO mode (see Sect. 17.1.3.2.2) an interrupt is triggered as soon as the data buffer is full. In this mode, the entire data buffer must be read out and the sensor prepared for a new acquisition within the next sampling period following the interrupt. In the following, the readout of the measured values in FIFO mode with a microcontroller of the ATmega family via SPI is considered in more detail:

• Microcontroller clock frequency:	$f_q = 16$ MHz
• Maximum SPI bit rate:	$f_{Bit} = f_q/4 = 4$ MHz
• Minimum bit duration:	$T_{Bit} = 1/f_{Bit} = 250$ ns
• Highest data rate of the sensor:	$ODR_{max} = 760$ Hz
• Shortest sampling period:	$T_A = 1/ODR_{max} \approx 1.315$ ms
• FIFO storage time:	$T_{FIFO} = 32 \times T_A = 42$ ms
• Total number of bits:	$N_{Bit} = (1$ command byte $+ 3$ axes $\times 2 \times 32$ bytes$) \times 8$ bits $= 1544$ bits
• Total transmission time:	$T_D = T_{BIT} \times N_{Bit} = 386$ µs
• Remaining time:	$\Delta t = T_A - T_D = 929$ µs

In the remaining time of approx. 929 µs, the sensor must be switched from FIFO mode to bypass mode and back to FIFO mode in order to be able to record further measured values. In the time until T_{FIFO} is reached, the microcontroller can process the data and perform further tasks.

In stream mode an interrupt can be triggered when the set fill limit is reached. The advantages of this mode are that more time is available for the transfer of the unread data sets and the recording of the data and saving in the data buffers runs continuously.

The buffered measured values can be read according to the example in Fig. 17.5. The first address from the output register range (0x28) is selected as the start address; for each measured value read, the master must transmit any two dummy bytes. After each byte read, the internal address counter of the sensor is incremented; after the last address from this range (0x2D), the address counter automatically jumps to the first.

17.1.3.4 Interrupt Control

Thanks to a complex interrupt structure of the sensor, a microcontroller can read out large amounts of data as a master without blocking waiting and monitor the angular velocity or angular acceleration in real time in order to solve control tasks. The device has two pins that can be configured as interrupt outputs. Interrupt 1 occurs as a result of monitoring the measured values, while interrupt 2 is triggered depending on the fill level of the FIFO buffer, or when a new measured value is present. The interrupt modes are managed by the control register CTRL_REG3 (0x22):

- **Bit 7** – I1_Int1 – setting this bit enables interrupt 1;
- **Bit 6** – I1_Boot – if this bit is set, pin INT1 indicates the booting of the device;
- **Bit 5** – H_L_active – configures the level of the INT1 output when an interrupt is triggered;
- **Bit 4** – PP_OD = 0/1 – output INT1 is configured as push-pull or open-drain;
- **Bit 3** – I2_DRDY = 1 – an interrupt 2 is triggered when a new measured value is present;
- **Bit 2** – I2_WTM = 1 – an interrupt 2 is triggered when the fill limit of the FIFO is exceeded;
- **Bit 1** – Ovrn = 1 – an interrupt 2 is triggered when the capacity of the FIFO is reached;
- **Bit 0** – I2_Empty = 1 – an interrupt 2 is triggered if there are no unread measured values in the FIFO.

If several bits of bit 2:0 are set simultaneously, the ORing of the FIFO events selected by this will trigger an interrupt. Via the register INT1_CFG (0x30) (see Table 17.6) the interrupt events and their logical operations which trigger interrupt 1 are selected:

Table 17.6 Interrupt 1 – Configuration registers

AND/OR	LIR	ZHIE	ZLIE	YHIE	YLIE	XHIE	XLIE

Table 17.7 INT1 – SRC registers

0	IA	ZH	ZL	YH	YL	XH	XL

Table 17.8 L3G20 – Threshold value register

Axis	Higher Byte	Low Byte
X	INT1_THS_XH 0x32	INT1_THS_XL 0x33
Y	INT1_THS_YH 0x34	INT1_THS_YL 0x35
Z	INT1_THS_ZH 0x36	INT1_THS_ZL 0x37

- **Bit 7** – AND/OR = 1 – the selected events are ANDed, otherwise ORed;
- **Bit 6** – LIR = 0 – the interrupt 1 state is automatically cleared when the event that caused it disappears. If LIR = 1, this state is retained until the next access to the INT1_SRC register, which stores the events triggering interrupt 1.
- **Bit 5:0** – if a bit is set, an interrupt 1 can be triggered if the measured values of an axis (X, Y or Z) have exceeded ("H") or fallen below ("L") the set value limit.

The event that triggers an interrupt 1 (Table 17.7) is stored in the INT1_SRC register (0x31). This register can only be read and has the following configuration:

- **Bit 7** – is always 0;
- **Bit 6** – IA – is set when an interrupt 1 has been triggered;
- **Bit 5:0** – are set internally when the measured values of an axis (X, Y, or Z) exceed or fall below the set limit values.

17.1.3.4.1 Threshold Setting for Interrupt 1

To ensure monitoring of the measured values, there is a 16-bit register for each axis in which the amount of the limit value is stored. The internal A/D converter has a 16-bit resolution, but measurements are made in both directions of rotation, so the amount can be no more than 15 bits. Table 17.8 lists the limit value register pairs and their addresses for all axes.

17.1.3.4.2 Setting the Pulse Duration at Pin INT1

The measured values are continuously compared with the threshold values. The comparison can take place with unfiltered or filtered measured values and can lead to the triggering of an interrupt 1 immediately or only after an adjustable delay. Bit 4, HPEN and bit 3:2, INT1_SEL1:0 from the control register CTRL_REG5 (0x24) are used to determine the type of filtering before the comparison as follows:

- INT1_SEL1:0 = 00 – no filtering;
- INT1_SEL1:0 = 01 – Filtering with the high-pass HPF;
- HPEN = 0 ∧ INT1_SEL1 = 1 – Filtering with the low pass LPF2;
- HPEN = 1 ∧ INT1_SEL1 = 1 – Filtering with a bandpass consisting of the lowpass LPF2 and the highpass HPF.

The INT1_DURATION register (0x38) can be used to set the delay with bits 6:0 in multiples of the sampling period and to set whether the delay occurs only when the event triggering the interrupt occurs or also when it disappears. If bit 7 WAIT of the register is set, the switching on and off of the output pulse is delayed. If this bit is reset, the delay takes place only on power-up, as shown in Fig. 17.6.

17.1.3.5 Temperature Measurement

The temperature sensor of the gyroscope measures the ambient temperature and stores it every second with a resolution of 1 °C into the register OUT_TEMP (0x26) in two's complement format. The stored value is actually neither accurate enough nor resolved enough for absolute temperature measurements, since the sensor is calibrated for relative

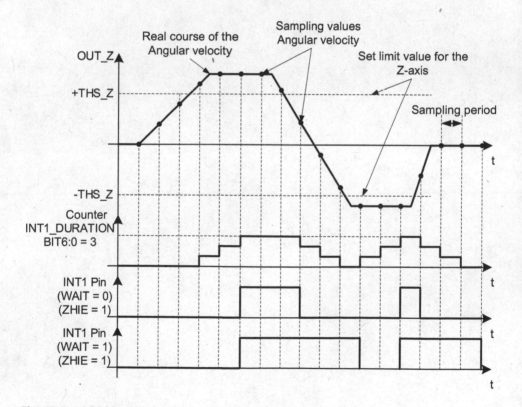

Fig. 17.6 L3GD20 – Interrupt 1 example

measurements and is used for temperature compensation. The temperature difference between two measurements stored in the signed variables cTemp1 and cTemp2 is calculated as follows:

$$cDeltaTemp = (-1) \cdot (cTemp2 - cTemp1)$$

A positive result means an increase in temperature.

References

1. STMicroelectronics. L3GD20 – MEMS motion sensor: Three-axis digital output gyroscope. www.st.com
2. STMicroelectronics. AN4505 Application Note. L3GD20: 3-axis digital output gyroscop. www.st.com

Magnetic Field Sensors

18

Abstract

In this chapter the magnetic field sensor HMC5883 is introduced.

Magnetic field measurements are used in a wide variety of fields, for example in the measurement of the earth's magnetic field (compass), for the non-reactive measurement of electric current or as a proximity sensor. In this chapter we present a magnetic field sensor which is particularly suitable as a compass.

Magnetometers are used to measure the magnetic flux density $\vec{B}$. This is measured in the SI[1] unit Tesla (T). In the SI system, a tesla is defined as

$$1 \ T = 1 \frac{V \cdot s}{m^2} = 1 \frac{kg}{A \cdot s^2} \tag{18.1}$$

and measuring ranges of magnetometers are in the order of magnitude of approx. 10^{-15} T to 10 T. Magnetic flux densities are sometimes also given in Gauss, since the physicist and mathematician Carl Friedrich Gauss developed the first magnetometer in 1832.

$$1 \ Gs = 10^{-4} T \tag{18.2}$$

The magnetic flux density is linked to the magnetic field strength $\vec{H}$ via the material equations of electrodynamics:

---

[1] SI – International System of Units.

© The Author(s), under exclusive license to Springer Fachmedien Wiesbaden GmbH, part of Springer Nature 2023
A. Meroth, P. Sora, *Sensor networks in theory and practice*,
https://doi.org/10.1007/978-3-658-39576-6_18

$$\vec{B} = \mu \vec{H} \tag{18.3}$$

The permeability number μ is $\mu = \mu_0 \mu_r$ and the magnetic field constant is

$$\mu_0 = 1.25666 \ldots 10^{-6} \frac{N}{A^2} \approx 4\pi \cdot 10^{-7} \frac{N}{A^2} \tag{18.4}$$

The dimensionless relative permeability number μ_r is 1 in vacuum (and approximately in air).

From Ampère's Law

$$\oint \vec{H} d\vec{s} = I \tag{18.5}$$

the magnetic flux density can be derived for different geometrical arrangements and is calculated for example around a long cylindrical semiconductor:

$$|\vec{B}| = \frac{I}{2\pi r} \mu \tag{18.6}$$

In numbers: At a distance of 5 cm from the axis of a straight conductor carrying a current of 50 A, the magnetic flux density is

$$|\vec{B}| = \frac{50\ A}{2\pi \cdot 0.05\ \text{m}} \cdot 4\pi \cdot 10^{-7} \frac{N}{A^2} = 2 \cdot 10^{-4} T \tag{18.7}$$

There are many different ways to measure magnetic fields. On the one hand, induction effects can be used, which occur when a conductor (coil) moves in a magnetic field. On the other hand, there are various galvanomagnetic effects ([1]), such as the anisotropic magnetoresistive effect (AMR) or the Hall effect and many others.

The AMR effect is based on the observation that the electrical resistance of some ferromagnetic (i.e. magnetized) materials changes depending on the strength and direction of the magnetic field. Many consumer electronics magnetic field sensors are based on this effect, including the one we describe.

The Hall effect is based on the fact that charge carriers in a semiconductor are deflected from their direction of flow by the Lorenz force under the influence of a magnetic field. As a result, a voltage can be measured perpendicular to the direction of the current on a semiconductor through which a current is flowing if, in turn, a magnetic field acts perpendicular to it.

18.1   HMC5883 Magnetic Field Sensor

The HMC5883 device ([2]) enables the measurement of the flux density of magnetic fields in the strength of the geomagnetic field. The measurement method is based on the anisotropic magnetoresistive effect (AMR) already briefly mentioned above. In order for magnetic field sensors to achieve higher sensitivity, the magnetoresistive elements are integrated in a Wheatstone bridge. When the field lines are perpendicular to the measurement bridge, the measurement voltage is zero, or reaches a maximum when they are parallel. In order to determine the exact direction of the magnetic field, a second measuring bridge is required, the orientation of which is rotated by 90° with respect to the first. An additional magnetic field is generated via a current-carrying coil, which is used for offset compensation. A permanent magnetization of the magnetic sensors can occur under the influence of strong magnetic pulses, or by a constant magnetic field acting over a longer period of time. To avoid this effect, the sensor must be demagnetized regularly. In the case of the HMC5883, this is done by means of direction-changing, strong and very short magnetic pulses that are generated internally via an additional coil.

The natural geomagnetic field in Europe has a magnetic flux density of approx. 40 ... 50 microtesla (μT) [3]. Large ferromagnetic bodies as well as deposits can lead to local magnetic anomalies of several μT. The HMC5883 can measure magnetic flux density along three perpendicular axes, ranging from 1 milligauss (mG) to about eight gauss, which corresponds to 0.1 μT ... 800 μT in SI. Thus, the device can be used to measure magnetic flux densities, as a compass or to detect local anomalies of the geomagnetic field.

18.1.1   Structure of the HMC5883

A block diagram of the HMC5883 magnetic field sensor is shown in Fig. 18.1.

18.1.1.1 Sensing Elements
The sensor block contains three Wheatstone bridges, each consisting of four anisotropic magnetoresistive resistors. In addition, it contains coils for offset compensation and demagnetization. In each measurement cycle, the three measurement voltages are successively digitized by a 12-bit A/D converter and stored as 16-bit integers. The converted voltages cover the discrete range $-$ 2048 ... +2047, or 0xF800 ... 0x7FF. If the range is exceeded, the value $-$4096 (0xF000) is stored for the affected axis. The measuring voltages are linearly dependent on the temperature, but no temperature compensation takes place internally. Compensation can be implemented by software as described in Sect. 18.1.3.

18.1.1.2 Control Logic
The control logic is responsible for the initialization of the device after switching on the supply voltage and for the control of the entire measuring sequences, the register addressing and the serial interface. The control of the interface has priority over the internal

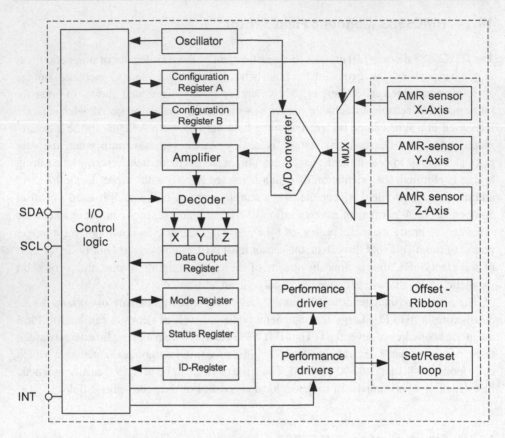

Fig. 18.1 Block diagram of the sensor HMC5883

processes. The control logic controls the status bits according to the state of the measurement sequence and switches the DRDY pin to signal the master to store a new measurement value. An external pull-up resistor is not necessary for this pin.

18.1.1.3 Serial Communication
The device is equipped with a serial interface that implements the I^2C protocol in standard and Fast-mode and enables data transmission at up to 400 kBit/s. The 7-bit address of the slave is 0x1E.

18.1.1.4 Register Block
The register block consists of thirteen 8-bit registers, which are individually addressable via an address counter and whose addresses occupy the range 0x00 ... 0x0C. The content of a register is accessible if it has been addressed before. With random access, the address counter is loaded with the desired value via the serial interface. In the sequence shown in Fig. 18.2, the pointer is set to address 0x03.

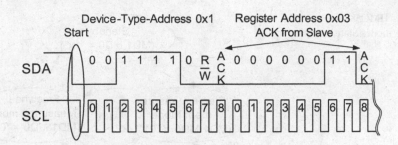

Fig. 18.2 HMC5883 – random register addressing

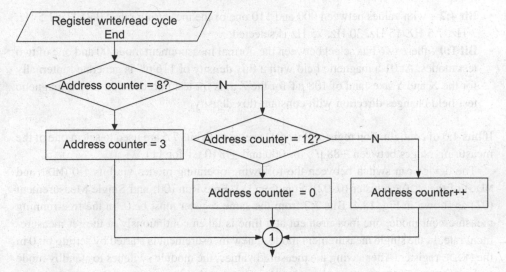

Fig. 18.3 HMC5883 – Control of the internal address counter

The master addresses the device via its I^2C address in write mode and then sends the address of the selected register. After each write or read of a register, the address counter is changed as shown in the flow chart Fig. 18.3. The control logic controls the register addressing and reading of the measured value registers. This speeds up the software configuration of the device.

18.1.1.4.1 Configuration Registers
Three registers are used for the overall configuration of the device. Configuration register A (address 0x00) is used to determine the measuring modes, the measuring rate and the number of averaged measured values for each measuring cycle.

- **Bit 7** – is always 0;
- **Bit 6:5** – defines the number of averaged measured values/the measuring cycle: 00–1, 01–2, 10–4 and 11–8 measured values;

Fig. 18.4 HMC5883 –
Operating modes State transition
diagram

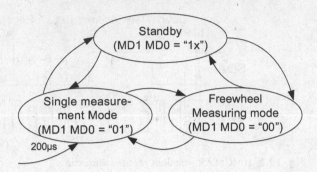

- **Bit 4:2** – with values between 000 and 110 one of the measuring rates: 0.75 Hz, 1.5 Hz, 3 Hz, 7.5 Hz, 15 Hz, 30 Hz, 75 Hz is selected;
- **Bit 1:0** – these two bits select between the normal measurement mode 00 and one of two test modes. At 01 a magnetic field with a flux density of 116 µT is generated internally for the X and Y axes and of 108 µT for the Z axis for test purposes. At 10 the magnetic test field changes direction with constant flux density.

If bits 4:0 of configuration register B (0x01) are set to 0, bits 7:5 are used to select one of the measuring ranges between ±88 µT for 000 and ± 810 µT for 111.

The device can switch between the following operating modes via bits 1:0 (MD1 and MD0) of the MODE register (0x02): Standby (1x), Free Run (00) and Single Measurement (01) as shown in Fig. 18.4. Bits 7:2 from the same register must be 0. In the free-running measurement mode, one measurement at a time is taken continuously at the set measurement rate. In the single measurement mode, a new measurement is started by setting bit 0 in the MODE register. After saving the measured values, the module switches to standby mode and bits MD1 and MD0 are set.

```
uint8_t HMC5883_Set_Config(uint8_t ucsamples_avg,
                           uint8_t ucout_rate,
                           uint8_t ucmeas_mode,
                           uint8_t ucmeas_range,
                           uint8_t ucop_mode)
{
    uint8_t ucConfigReg[3], ucDeviceAddress, ucI;
    ucConfigReg[0] = ucsamples_avg | ucout_rate | ucmeas_mode;
    ucConfigReg[1] = ucMeasGain[ucmeas_range];
    ucConfigReg[2] = ucop_mode;
    //form the address of the electronic compass
    ucDeviceAddress = HMC5883_DEVICE_TYPE_ADDRESS << 1;
    ucDeviceAddress |= TWI_WRITE; //write mode
    TWI_Master_Start(); //start
    if((TWI_STATUS_REGISTER) != TWI_START) return TWI_ERROR;
    //send device address
```

```
    TWI_Master_Transmit(ucDeviceAddress);
    if((TWI_STATUS_REGISTER) != TWI_MT_SLA_ACK) return TWI_ERROR;
    //the address of the 1st configuration register is sent
    TWI_Master_Transmit(0x00);
    if((TWI_STATUS_REGISTER) != TWI_MT_DATA_ACK) return TWI_ERROR;
    //the data bytes are sent
    for(ucI = 0; ucI < 3; ucI++)
    {
        TWI_Master_Transmit(ucConfigReg[ucI]);
        if((TWI_STATUS_REGISTER) != TWI_MT_DATA_ACK) return TWI_ERROR;
    }
    TWI_Master_Stop(); //stop
    return TWI_OK;
}
```

The function HMC5883_Set_Config() can be used to reconfigure the sensor. An example for calling the function is explained below.

```
#define SAMPLES_AVG8 0x60 // mean value of eight measurements
#define OUT_RATE_1 0x00 //0.75 Hz measuring rate
#define MEAS_MODE_NORM 0x00 //the normal measuring mode is selected
#define MEAS_GAIN_5 0x04 //±400 µT measuring range
#define OP_MODE_SINGLE 0x01 //single measurement mode
HMC5883_Set_Config(SAMPLES_AVG8, OUT_RATE_1, MEAS_MODE_NORM,
                   MEAS_GAIN_5, OP_MODE_SINGLE);
```

18.1.1.4.2   Measured Value Registers
A new measured value is calculated as the average of two recordings. A demagnetizing field with alternating direction is generated before each recording. This avoids a possible permanent magnetization of the magnetoresistive resistors. The measured values of the magnetic flux density are stored for each axis in two registers in big-endian format as shown in Table 18.1.

18.1.1.4.3   Status Register (Address 0x09)
The status register provides information about the status of the measurement process and data readout.

- **Bit 7:5, 3:2** – the bits are always 0;
- **Bit 4** – is not documented; on the tested specimens, this bit is set when unread measured values have been overwritten. It is reset after reading a complete set of measured values.
- **Bit 1** – this bit is set when the MODE register is read or with the start of the data readout. It is reset: when saving the configuration register A, or the MODE register, or when a

Table 18.1   HMC5883 – Data registers

Axis	X		Z		Y	
Register	MSByte	LSByte	MSByte	LSByte	MSByte	LSByte
Address	0×03	0×04	0×05	0×06	0×07	0×08

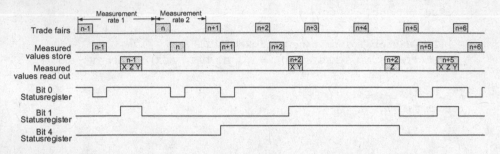

Fig. 18.5   HMC5883 – Free-running measuring mode – Schematic representation with asynchronous data reading

complete set of measured values has been read out. As long as this bit is set, no new measured values are stored in the data registers.

- **Bit 0** – is reset for the duration of the storage of the measured values in the data registers (at least 250 μs).

The change of the individual bits of the status register during free-running measurement operation with asynchronous data read is shown in Fig. 18.5.

18.1.1.4.4   Identification Registers

The HMC5883 has three read-only registers with addresses 0x0A, 0x0B and 0x0C which are used to identify the device and to test the serial communication. The stored contents of the registers are: 0x48, 0x34 and 0x33 which are the ASCII codes of the characters "H", "4" and "3".

18.1.2 Reading HMC5883 Measured Values

According to [3], the duration of a measurement with averaging of eight recordings takes 6 ms. The reading of the six data registers can be done for example with the transmission of nine bytes:

- 1 byte for addressing the sensor in write mode;
- 1 byte – the address of the first data register (0x03);
- 1 byte for addressing the sensor in read mode;
- 6 bytes for the transfer of the data registers.

In the ideal case, the transmission of the 9 x 9 bits, at a maximum bit rate of 400 kBit/s, would take 202.5 μs.

18.1.2.1 Reading Measured Values in Single Measurement Mode

In single measurement mode, each measurement is started by setting bit 0 in the MODE register. Since the measurement duration is considerably longer than the transmission duration, the readout of the *nth* measured value can be carried out immediately after the start of the $(n + 1)$ measurement, as shown in the following code excerpt:

```
HMC5883_Write_ByteReg(MODE_REG, 0x01); //start of a new measurement
HMC5883_Read_DataReg(uiData); //read the six data registers
```

In this measuring mode, bit 1 in the status register is set after reading one or more data registers. However, this bit is reset with each new start, so that only individual registers can be read in this mode. By means of a blocking wait, a master can use the query of the state of bit 0 from the status register to determine the correct time at which the data registers can be read out immediately after a measurement process, as can be seen in the following program excerpt.

```
uint8_t ucStatusReg = 0x01;
//the function HMC5883_Read_ByteReg reads the content
//of the register MODE and stores it into the variable ucStatusReg
//it waits until bit 0 in the status register goes low
while(ucStatusReg & 0x01)
{
    HMC5883_Read_ByteReg(STATUS_REG, ucStatusReg);
}
//if bit 0 of the status register goes high again, end of storage
while(!ucStatusReg) HMC5883_Read_ByteReg(STATUS_REG, ucStatusReg);
```

18.1.2.2 Reading Measured Values in Free-Running Measuring Mode

A special feature of the sensor is that in this measuring mode, new measured values are recorded when all data registers are read incompletely, but cannot be stored in the data registers. In free-running mode, the measured values can be read in asynchronous or synchronous mode.

18.1.2.2.1 Asynchronous Reading in Free-Running Mode

As described in Sect. 18.1.1.4.2, overwriting of unread measured values is signalled but not blocked. Over the duration of the entire reading of the six data registers, the saving of new measured values is blocked. This allows asynchronous reading at any time without the danger that the contents of the data registers originate from different measurements.

18.1.2.2.2   Synchronous Reading in Free-Running Mode

If all measured values have to be read out, this is referred to as synchronous reading. A first possibility to read the data synchronously would be to determine the read time by reading bit 0 from the status register. Especially at higher data rates, this method is not recommended because of the blocking wait.

In order to avoid uneconomical waiting, the module has the DRDY connection which is controlled by the control logic at the same time as bit 0 from the status register. By switching this connection to low, the master can be signalled that a new measured value is ready for reading, if the connection is suitable. If the pin connected to the DRDY is interrupt-capable by the master, the corresponding interrupt service routine can instruct the readout of the measured values in good time.

18.1.3   Calibration of the Sensor

The aim of calibration is a more accurate conversion of the measured values into units of magnetic flux density. Bits 7:5 of configuration register B can be used to select one of the eight measuring ranges. For each measuring range, the manufacturer specifies a gain factor which corresponds to the digital value when measuring a flux density of 1 Gs ($100\,\mu T$). Bits 1:0 of configuration register A can be used to activate test modes in order to generate internal magnetic fields of known magnitude and thus calibrate the sensor.

For example, for the fifth measuring range (bit 7:5 = 100), the measurement of a maximum flux density of $\pm 400\,\mu T$ is recommended and a gain factor of 440 is specified. Under the influence of the test fields, digital values of $\pm 1.16 \times 440 = \pm 510$ for the X and Y axes and $\pm\,1.08 \times 440 = \pm 475$ for the Z axis can be expected. The calibration is in each case related to a measuring range and is temperature-dependent. The following steps are necessary for the calibration:

- the measuring range is selected;
- test mode 1 is activated and a single measurement is started;
- the measured values are read out;
- test mode 2 is activated and a single measurement is started;
- the measured values are read out;
- from the positive and negative measured values, the offset and correction value of the amplification factor are calculated for each axis.

When calibrating a sensor in the fifth measuring range, the values from Table 18.2 have been read out.

The offset and gain factors as integers are calculated as follows using axis X as an example and the results for all three axes are stored in Table 18.3.

Table 18.2 Example calibration: measured values

+X_Meas	−X_Meas	+Y_Meas	−Y_Meas	+Z_Meas	−Z_Meas
+569	−550	+525	−525	+499	−480

Table 18.3 Example calibration: calculated offset values and gain factors

X_Offset	Y_Offset	Z_Offset	V_X_Korr	V_Y_Korr	V_Z_Korr
+9	0	+9	510/560	1	475/490

Example

- $X\_Offset = (+X\_Meas + (-X\_Meas))/2 = (569-550)/2 = +9$
- $V\_X\_Korr = X\_Test/(+X\_Meas - X\_Offset) = 510/560$ ◀

To calculate the calibrated, digital values of the flux density for this measuring range, the corresponding offset values are first subtracted from the read-out values and then the results are multiplied by the corrected gain factors. By dividing the corrected values by 4.4 (100/440), the measurement results are expressed in µT. If the measuring range or the ambient temperature changes, the calibration procedure should be repeated.

18.1.4 HMC5883 as Electronic Compass

The sensor can be used as an electronic compass because of its high sensitivity. The geomagnetic field runs parallel to the earth's surface in our latitudes and as a vectorial quantity it is always directed towards the magnetic north pole [4, 5]. Therefore, the orientation of the sensor in the magnetic field can be determined from the measured values of the X (X_Meas) and Y (Y_Meas) axes. The calculation of the angle between the Y-axis of the sensor and the north pole is explained in the following example. At 0° the direction of the positive Y-axis should point to the magnetic north pole. As a prerequisite for the angle determination, the points P_i (X_Meas_i/Y_Meas_i) must be located on a circle, which is usually not the case. The radius of the circle is meaningless, therefore for the calibration of the sensor as an electronic compass a simplified procedure can be used, the basis of which is presented in [4]. This procedure leads to a lower computational cost. After the selection of the desired measuring range, the following steps are carried out:

1. In a magnetically undisturbed environment, the horizontally placed sensor is rotated around its Z-axis. The smallest (negative) and the largest (positive) measured value of the axes X (X_Min and X_Max) and Y (Y_Min and Y_Max) are determined and stored during the rotation.
2. For the two axes, the offset values are calculated as integers, stored and further subtracted from the measured values.

$$\begin{cases} X\_Offset = (X\_Max + X\_Min)/2 \\ Y\_Offset = (Y\_Max + Y\_Min)/2 \end{cases} \tag{18.8}$$

Considering the offset values, step 1 is repeated and it is checked that $X\_Max \approx X\_Min$ and $Y\_Max \approx Y\_Min$. The current maxima X_Max and Y_Max are stored. If, for example, $X\_Max > Y\_Max$, then the measured values continue to be corrected as follows:

$$\begin{cases} X\_Korr = X\_Meas - X\_Offset \\ Y\_Korr = ((Y\_Meas - Y\_Offset) \cdot X\_Max)/Y\_Max \end{cases} \tag{18.9}$$

With the corrected values, taking into account the quadrant in which these values are located, the sought angle between the Y-axis and the north pole direction can be calculated using the trigonometric function arctangent.

$$\beta = arctan \left(\frac{X\_Korr}{Y\_Korr} \right) \tag{18.10}$$

18.1.5 Angle Calculation with the CORDIC Algorithm

To calculate the angle from Eq. 18.9 with a microcontroller, a division would first have to be performed and then a function called to calculate the arctangent of the quotient. An alternative would be to calculate the possible angle values in advance with an acceptable resolution corresponding to the quotient X_Korr/Y_Korr as an integer and store these values in a table (LUT).[2] The stored values would then be accessed using the quotient, which is used as the table index. The first solution is time consuming, the second is extremely space consuming. As an alternative, the application of the CORDIC algorithm for the calculation of the angle is presented [6, 7].

CORDIC is the acronym of COordinate Rotation Digital Computer and refers to a mathematical algorithm that can be used to calculate approximately trigonometric functions. A vector $\overrightarrow{V_0}$ in the xy-plane, whose components are X_0 and Y_0, leads in the sense of the algorithm by clockwise rotation with the angle α_0 to a vector $\overrightarrow{V_1}$ with the components X_1 and Y_1 (see Fig. 18.6).

$$\begin{cases} X_1 = X_0 + Y_0 \cdot \tan \alpha_0 \\ Y_1 = Y_0 - X_0 \cdot \tan \alpha_0 \end{cases} \tag{18.11}$$

---

[2]Lookup table.

Fig. 18.6 HMC5883 – electronic compass

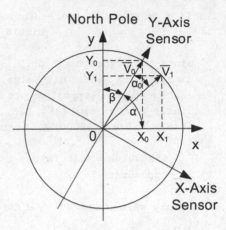

Table 18.4 Support values for the CORDIC algorithm

$\tan\alpha_i$	2^0	2^{-1}	2^{-2}	2^{-3}	2^{-4}	2^{-5}	2^{-6}	2^{-7}
α_i	450	265	140	71	36	18	9	4

In an iterative process, the vector $\overrightarrow{V_1}$ is further rotated in discrete steps until the ordinate of the resulting vector is approximately zero.

$$\begin{cases} X_{i+1} = X_i + Y_i \cdot \sigma_i \cdot \tan\alpha_i \\ Y_{i+1} = Y_i - X_i \cdot \sigma_i \cdot \tan\alpha_i \end{cases} \text{ with} \tag{18.12}$$

$$\sigma_i = \begin{cases} +1, \ wenn \ Y_i > 0 \\ -1, \ wenn \ Y_i < 0 \end{cases} \tag{18.13}$$

With $\tan\alpha_i = 2^{-i}$ the component of the rotating vector can only be calculated by the basic mathematical functions of a microcontroller "addition" and "digit shift to the right", which leads to an acceleration of the angle calculation. Table 18.4 lists the corresponding angles α_i for eight iteration steps as integers in tenth degree resolution.

If the abort criterion of the algorithm is met or the maximum number n of steps is reached, the total rotation angle is calculated

$$\alpha = \sum_{i=0}^{n-1} \sigma_i \cdot \alpha_i \tag{18.14}$$

and this gives the angle we are looking for $\beta = 90^\circ - \alpha$. The following function returns the angle between the Y-axis and the north pole when called with the corrected measured values X_Korr and Y_Korr.

```c
uint16_t HMC5883_Get_Angle(uint16_t uix_value,
                           uint16_t uiy_value)
{
    uint8_t ucQuadrant = 0, ucI;
    uint16_t uiX_Value, uiY_Value, uiAngle = 0;
    //the angles as support points in tenths of a degree
    uint16_t uiPhi[8] = {450, 265, 140, 71, 36, 18, 9, 4};
    if(uix_value & 0x8000) //if true, the X-value is negative
    {
        uix_value = ~uix_value + 1;
        ucQuadrant |= 0x01;
    }
    if(uiy_value & 0x8000) //if true, the Y-value is negative
    {
        uiy_value = ~uiy_value + 1;
        ucQuadrant |= 0x02;
    }
    //approximation of the angle in the 1st quadrant in eight steps
    for(ucI = 0; ucI < 8; ucI++)
    {
        if(uiy_value & 0x8000)
        {
            uiAngle -= uiPhi[ucI];
            uiX_Value = uix_value - ((int)uiy_value >> ucI);
            uiY_Value = uiy_value + (uix_value >> ucI);
        }
        else
        {
            uiAngle += uiPhi[ucI];
            uiX_Value = uix_value + ((int)uiy_value >> ucI);
            uiY_Value = uiy_value - (uix_value >> ucI);
        }
        uix_value = uiX_Value;
        uiy_value = uiY_Value;
    }
    switch(ucQuadrant) //calculate of the real angle
    { //dependent on the quadrant
        case 0: uiAngle = 2700 + uiAngle; break;
        case 1: uiAngle = 900 - uiAngle; break;
        case 2: uiAngle = 2700 - uiAngle; break;
        case 3: uiAngle = 900 + uiAngle; break;
    }
    return uiAngle;
}
```

References

1. Tränkler, H.-R., & Reindl, L. M. (Eds.). (2014). *Sensortechnik*. Springer Vieweg.
2. Honeywell International Inc. (2017). HMC5883L – Three-Axis digital compass IC. https://aerospace.honeywell.com.
3. Knödel, K., Krummel, H., & Lange, G. (2005). *Geophysik*. Springer.
4. Honeywell International Inc. (2015). Michael Caruso application of Magnetoresistive sensors in navigation systems. www.honeywell.com
5. Honeywell International Inc. (2015). AN203 – Compass heading using magnetometers. www.honeywell.com
6. Volder, J. (1959). The CORDIC computing technique. *IRE Transactions on Electronic Computers, 8*(3), 330–334.
7. Kuhlmann, M., Parhi, K. K., & P-CORDIC. (2002). A precomputation based rotation CORDIC algorithm. *EURASIP Journal on Applied Signal Processing, 9*, 936–943.

Proximity Sensors

Abstract

In this chapter, an ultrasonic sensor and an optical proximity sensor are presented.

Proximity sensors are sensors that can detect an object without direct contact. Depending on the measuring principle, a distinction is made between ultrasonic, capacitive, inductive and optical detectors. Depending on the measuring principle, proximity sensors show different sensitivities and can therefore partly only be used for the digital detection of a proximity (proximity switch) or for distance measurement. The two sensors presented in this chapter are suitable for distance measurement at a distance of a few centimeters to a few meters.

19.1 Ultrasonic Proximity Sensors

Ultrasonic proximity sensors detect objects without contact according to the pulse-echo principle. The frequency range of ultrasound is above the human hearing range (>20 kHz). To convert the electrical pulses into mechanical pulses and vice versa, the sensors use ultrasonic transducers, which are mostly based on the piezoelectric effect. A piezoelectric crystal deforms under the influence of an electric field. Conversely, when an externally caused elastic deformation occurs on its surface, an electric voltage proportional to the pressure is generated. Such a sensor emits a sonic pulse sequence that propagates as a

A. Meroth, P. Sora, *Sensor networks in theory and practice*,
https://doi.org/10.1007/978-3-658-39576-6_19

longitudinal wave.[1] When hitting an object, the sound pulses are reflected as an echo. By measuring the propagation time of the transmitted sound pulses and the received echo, if the propagation speed of the sound is known, the distance between the sensor and the reflecting object can be calculated. The speed of sound in air is approximately 340 m/s [1] and depends on temperature, relative humidity and pressure. Equation (19.1) [2] describes the temperature dependence of the speed of sound c on the air temperature T in K:

$$c = c_0 \cdot \sqrt{1 + \frac{T}{273.15}} \qquad (19.1)$$

where c_o is the propagation speed of the sound at 0 °C. The speed of sound increases at 5 m/s with an increase in air pressure of 30 hPa [2] and constant air temperature. A change in relative humidity from 0% to 100% leads to an increase in the speed of sound with about 1 m/s at an air temperature of 20 °C [2]. The sound intensity at constant frequency decreases quadratically with distance. Sound attenuation is directly proportional to the square of the sound frequency. Higher frequencies can therefore only be used for measuring shorter distances, but allow a higher measurement rate.

19.1.1 Measuring Principle

An ultrasonic sensor measures the time Δt between the emission of a sound pulse sequence and the reception of the reflected sound wave. The distance d from the sensor to the object is calculated as follows:

$$d = c \cdot \frac{\Delta t}{2} \qquad (19.2)$$

According to the number of ultrasonic transducers used, a distinction is made between single-head and dual-head systems. A single-head system has a transducer that can both transmit and receive sound. The aperture angle of the sound lobe of such a transducer is about 5° [3]. In the transmitting phase, the transducer is excited to oscillate by the controlling electronics and generates a sound pulse sequence with a total duration that depends on the sound frequency. The transducer cannot receive immediately after transmission because it must swing off after excitation. During this intermediate phase, the sound waves reflected from nearby objects are not perceived. Depending on the decay time, the size of the blind zone can be calculated. The decay time is many times longer than the duration of the pulse sequence. After the settling time, the transducer is switched to receive

---

[1] Longitudinal wave is a wave in which the direction of pressure change is the same as the direction of wave propagation [26].

Fig. 19.1 Ultrasonic two-head
system

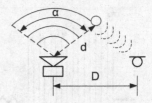

and the electronics perform the measurements within an adjustable time window. It must be checked whether the received oscillations are correlated with the transmitted ones in order to exclude measuring errors. To counteract the sound attenuation and to increase the range of the measurement, the received signals are amplified differently. The signals received later are amplified more. Such a system is simple and compact to implement, but cannot detect nearby objects. The time interval between two measurements must be large enough for the amplitude of the echoes caused by the previous measurement to decay.

A two-head system has one transducer for transmitting and a second for receiving. Such a sensor can receive and evaluate incoming signals immediately after the transmission phase. The waiting time due to the decay time and the resulting blind zone are eliminated. A blind zone also arises in this system because of the limitation of the aperture angle of the sound lobe α, but it is much smaller than in the single-head system. The smallest distance d to an object that can be detected with this system is (see Fig. 19.1):

$$d = \frac{D/2}{\sin \frac{\alpha}{2}} \tag{19.3}$$

where D is the distance between two transducers. This system allows the detection of multiple objects because of the larger aperture angle.

The amplitude of the reflected sound pulses depends on the distance to the detected object, the size and the surface of the object. A large object can reflect the sound waves better than a small one, a smooth surface better than a rough one. There are materials that reflect the sound wave well, others that absorb it well.

Ultrasonic sensors are used for distance measurement, for pattern and object detection, as motion detectors or proximity sensors. If several ultrasonic sensors are active in one room, they can operate in synchronous or multiplex mode. In synchronous operation, all sensors transmit their sound pulses simultaneously, provided they do not interfere with each other. In multiplex operation, the sensors are activated and evaluated one after the other.

19.1.2 SRF08: Ultrasonic Measuring Module

SRF08 is a discretely constructed, microcontroller-controlled ultrasonic sensor ([4]). It uses transducers with an aperture angle of approx. 60° and generates sound pulses with a

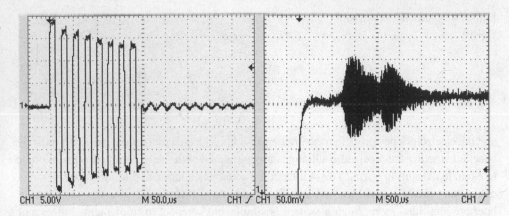

Fig. 19.2 SRF08 transmitted pulses (left) and received pulses (right)

frequency of 40 kHz for detection and measurement (Fig. 19.2 right). The distance to objects up to 6 m away is measured with a resolution of 1 cm and the measurement results can be converted into centimeters, inches or microseconds. Configuration of the sensor and readout of the measured values are done via the I^2C-bus.

19.1.2.1 Structure
A block diagram of the SRF08 ultrasonic sensor is shown in Fig. 19.3. The microcontroller excites the sound transmitter to oscillate via a level converter and an amplifier with an electrical pulse sequence (see Fig. 19.2 left). The received sound pulses are converted into electrical pulses by the sound receiver (Fig. 19.2a), which can be amplified to improve the range before digitization.

Due to the large opening angle and the long range, the sensor can detect several objects. The measured runtimes can be converted to lengths using Eq. (19.2) before storing in the measured value registers. The sensor also has a photoresistor that can be used to estimate light brightness. The distance measurements are configured using the range and gain registers. By storing a value from the range [0, 31] into the gain register (address 0x01), the analog signal is amplified by the sound receiver by a factor from the range [94, 1025] to achieve optimum measurement results. The echoes are evaluated within an adjustable time window, which corresponds to a multiple of 256 µs. By storing values from the range [0, 140] into the range register (0x02), the range is limited to values between 43 mm and approx. 6 m in a grid of 43 mm. The codes stored in the command register 0x00 lead to the start of a new measuring process, or to a change of the I^2C address.

19.1.2.2 Serial Communication
The configuration of the sensor, the start of new measurements, the readout of the measured values, and the change of address are carried out via the I^2C bus. The sensor is configured as a slave with the I^2C-7 bit address 0x70, which can be set by the user to one of a further

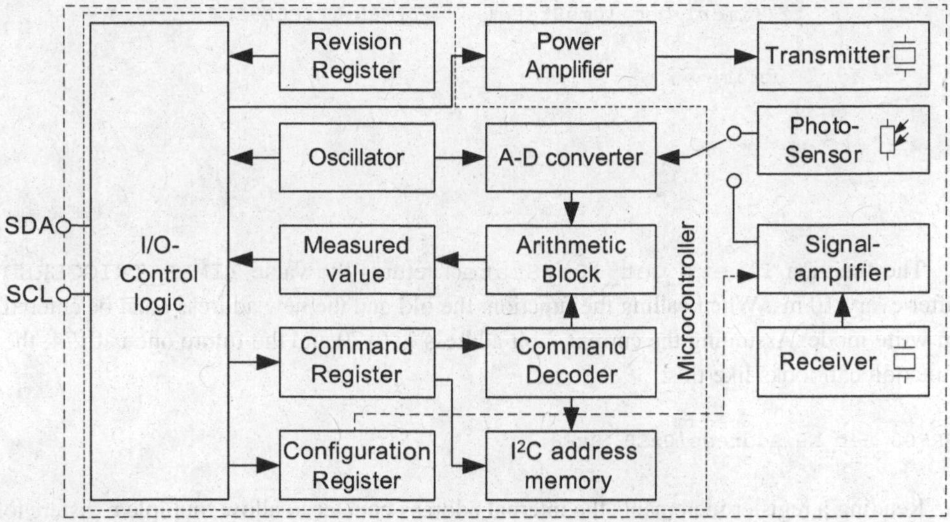

Fig. 19.3 SRF08 – Block diagram

15 addresses from the range [0x71, 0x7F] to enable multiplex operation via a bus. All sensors connected to a bus can be addressed simultaneously via address 0x00. This makes it possible, for example, to start all measurements at the same time. Writing a value to a register starts with a START sequence and takes place in a single I^{2}C message. The master transmits the sensor address followed by the register address and the value to be stored. Each byte must be acknowledged by the slave with ACK, the master ends the transmission with a STOP sequence. To change the address of the slave, the master stores the bytes 0xA0, 0xAA and 0xA5 followed by the new address into the command register. A pause of at least 50 ms must be observed between two transmissions:

```
void SRF08_Set_NewAddress(uint8_t ucold_address,
                          uint8_t ucnew_address)
{
    #define COMMAND_REG0
    uint8_t ucTime, ucI;
    uint8_t ucAddress_Sequence[4] = {0xA0, 0xAA, 0xA5, 0x00};
    ucAddress_Sequence[3] = ucnew_address;
    for(ucI = 0; ucI < 4; ucI++)
    {
        SRF08_Write_ByteReg(ucold_address, COMMAND_REG,
                            ucAddress_Sequence[ucI]);
        ucTime = 0;
        while(ucTime < 6)
        { //after each byte sent there must be a 50ms pause to be observed
```

```
if(Timer1_get_10msState() == TIMER_TRIGGERED)
{
  ucTime++;
}
    }
  }
}
```

The function `Timer1_get_10msState()` returns the value `TIMER_TRIGGERED` after every 10 ms. When calling the function, the old and the new address must be entered in write mode. Assuming the current 7-bit address is 0x70 and the future one is 0x74, the function call looks like this:

`SRF08_Set_NewAddress(0xE0, 0xE8);`

Reading a register increments the internal address counter to allow multiple registers to be read in an I^{2}C transfer.

19.1.2.3 Measurement Value Acquisition

The SRF08 sensor operates in single measurement mode; each measurement is started by storing a corresponding command code into the command register and is performed within the set time window. Reading out the measured values can be done at the end of this time window. A timer interrupt, appropriately configured and once started with the measurement, can specify the time for the readout without loading the microcontroller. Alternatively, the `SOFTWARE_REVISION` register (address 0x00) can be read. This register stores the value 0xFF during the measurement, otherwise a value corresponding to the software revision. However, this method also keeps the microcontroller busy during the measurement.

19.1.2.3.1 Distance Measurement Mode

In the distance measurement mode, either the runtime (command code 0x52) or the distance is measured in inches (0x50) or in centimeters (0x51). The measurements in this mode start with the smallest gain factor from the named range. This is increased every 70 μs until it reaches the set value. By testing, these parameters can be optimized depending on the objectives pursued. Up to 17 echo signals can be evaluated in one measurement process. The 16-bit results are stored in increasing order in the measured value registers in the address range [0x02, 0x23]. The measurement of the next object is stored at addresses 0x02 and 0x03 in big-endian format. The measured values can be read individually or as a whole as follows. The module address in write mode `ucdevice_address` and the vector address `uidata_word`, at which the measured values are stored, are passed as parameters.

```
uint8_t SRF08_Read_DataWordReg(uint8_t ucdevice_address,
                               uint16_t* uidata_word)
{
    uint8_t ucDeviceAddress, ucDataLSByte, ucDataMSByte, ucI;
    TWI_Master_Start(); //start
    if((TWI_STATUS_REGISTER) != TWI_START) return TWI_ERROR;
    TWI_Master_Transmit(ucdevice_address); //send module address
    if((TWI_STATUS_REGISTER) != TWI_MT_SLA_ACK) return TWI_ERROR;
    //the start address of the register area is sent
    TWI_Master_Transmit(DATA1_H_ADDR);
    if((TWI_STATUS_REGISTER) != TWI_MT_DATA_ACK) return TWI_ERROR;
    TWI_Master_Start(); //restart
    if((TWI_STATUS_REGISTER) != TWI_RESTART) return TWI_ERROR;
    ucDeviceAddress = ucdevice_address | TWI_READ; //read mode
    //send module address in read mode
    TWI_Master_Transmit(ucDeviceAddress);
    if((TWI_STATUS_REGISTER) != TWI_MR_SLA_ACK) return TWI_ERROR;
    //the contents of the registers are read in
    for(ucI = 0; ucI < 17; ucI++)
    {
        ucDataMSByte = TWI_Master_Read_Ack();
        if((TWI_STATUS_REGISTER) != TWI_MR_DATA_ACK) return TWI_ERROR;
        if(ucI < 16)
        {
            ucDataLSByte = TWI_Master_Read_Ack();
            if((TWI_STATUS_REGISTER) != TWI_MR_DATA_ACK)
                return TWI_ERROR;
            uidata_word[ucI] = (ucDataMSByte << 8) + ucDataLSByte;
        }
        else
        {
            ucDataMSByte = TWI_Master_Read_NAck();
            uidata_word[ucI] = (ucDataMSByte << 8) + ucDataLSByte;
            if((TWI_STATUS_REGISTER) != TWI_MR_DATA_NACK)
                return TWI_ERROR;
        }
    }
    TWI_Master_Stop(); //stop
    return TWI_OK;
}
```

19.1.2.3.2   Artificial Neural Network[2]: Measurement Mode

The measurements in Artificial Neural Network mode are started with one of the command codes 0x53, 0x54 or 0x55. The measurements take place with the highest gain factor regardless of the settings. The entire measurement time window is divided into 32×2048 µs time intervals, each assigned to a register starting at address 0x04. If no reflected signal was received or detected in a time interval, a 0 is stored in the corresponding register, otherwise a value unequal to 0. This results in a pattern that partially maps the real distance to the reflecting objects at close range. A more accurate mapping can be achieved using additional ultrasonic sensors. By evaluating this pattern with neural networks, object outlines or movement of objects can be detected. With the command 0x53 the distance to the next object in inches, with 0x54 the distance in centimeters and with 0x55 the runtime in microseconds is measured simultaneously with the above mentioned measurements and the result is stored in the registers with the addresses 0x02 and 0x03. For reading out the results the function listed above can be used.

19.1.2.3.3   Light Brightness Measurement

With each distance measurement, the sensor also measures the light brightness and stores the result in the register with address 0x01. This register can be read out like all other 8-bit registers with the following function:

```
uint8_t SRF08_Read_ByteReg(uint8_t ucdevice_address,
                           uint8_t ucreg_address,
                           uint8_t* ucdata_byte)
{
    uint8_t ucDeviceAddress;
    TWI_Master_Start(); //start
    if((TWI_STATUS_REGISTER) != TWI_START) return TWI_ERROR;
    /*send module address in write mode*/
    TWI_Master_Transmit(ucdevice_address);
    if((TWI_STATUS_REGISTER) != TWI_MT_SLA_ACK) return TWI_ERROR;
    //the address of the desired register is sent
    TWI_Master_Transmit(ucreg_address);
    if((TWI_STATUS_REGISTER) != TWI_MT_DATA_ACK) return TWI_ERROR;
    TWI_Master_Start(); //restart
    if((TWI_STATUS_REGISTER) != TWI_RESTART) return TWI_ERROR;
    ucDeviceAddress = ucdevice_address | TWI_READ; //read mode
    /*send module address in read mode*/
    TWI_Master_Transmit(ucDeviceAddress);
    if((TWI_STATUS_REGISTER) != TWI_MR_SLA_ACK) return TWI_ERROR;
    /*the addressed register is read in*/
    *ucdata_byte = TWI_Master_Read_NAck();
```

---

[2] **Artificial Neural Network – artificial neural networks.**

```
    if((TWI_STATUS_REGISTER) != TWI_MR_DATA_NACK) return TWI_ERROR;
    TWI_Master_Stop(); //stop
    return TWI_OK;
}
```

19.2   SI114x: Optical Proximity Sensor

The SI1141, SI1142 and SI1143 devices have one photosensor for visible light and two for infrared light [5–7]. They enable proximity sensing (PS) to an object up to 50 cm away from the sensor and direct ambient light sensing (ALS). Depending on the type, a device from this family controls up to three infrared light-emitting diodes with voltage pulses and measures the reflected rays in order to detect an object. The illuminance of the white and infrared light can be measured to minimize the influence of ambient light on the proximity measurement.

19.2.1 SI114x: Operating Modes

The sensors of the SI114x series do not have a reset pin. An internal comparator ensures that the entire device is switched off at supply voltages below the $V_{DD\text{-}OFF}$ level (approx. 1 V). As soon as this limit is exceeded, hardware initialization takes place, which is completed after approx. 25 ms. The sensor is then in standby mode. In this mode the communication interface is active and commands can be received and executed. A RESET command (code 0x01) enables a complete reinitialization of the device, as shown in Fig. 19.4.

To complete the initialization, the controlling master must first store the value 0x17 into the register HW-KEY (address 0x07). The standby mode is exited to perform measurements in single measurement mode, or in automatic measurement mode. After the selected measurement set has been completed and the measurement values have been stored, the sensor returns to standby mode to save power. One of the measurement modes can be selected via the MEAS_RATE register (0x08). After the initialization phase, this register is set to 0x00 and thus the single measurement mode is selected. In this mode single proximity measurements can be started with the command PS_FORCE (0x05), ambient light measurements with the command ALS_FORCE (0x06) or a complete measurement with the command PSALS_FORCE (0x07).

The sensor is in automatic measurement mode after a non-zero compressed[3] value has been stored in the 8-bit register MEAS_RATE (address 0x08) and one of the measurement

---

[3] Compression will be explained later.

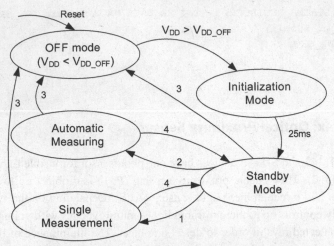

1) (MEAS_RATE = 0) & (PS_FORCE | ALS_FORCE | PSALS_FORCE)
2) (MEAS_RATE = 1) & (PS_AUTO | ALS_AUTO | PSALS_AUTO)
3) RESET command
4) End of a measurement

Fig. 19.4 SI114x – state transition diagram

types has been started. The start is done for the proximity measurements with the command PS_AUTO (code 0x0D), for the ambient light measurement with the command ALS_AUTO (0x0E). All measurements take place automatically after the command PSALS_AUTO (0x0F). An automatic measurement mode can be stopped with one of the commands: PS_PAUSE (0x09), ALS_PAUSE (0x0A) or PSALS_PAUSE (0x0B) to stop without leaving the automatic measurement mode.

The value stored in the MEAS_RATE register sets the sampling rate at which the sensor wakes up from standby to start measurements when needed. The uncompressed value k multiplied by 31.25 µs (1/32 kHz) specifies the time T_{mess} between two possible measurements. The sampling period for the proximity measurement $n * T_{mess}$ and for the ambient light measurement $m * T_{mess}$ are multiples of the basic sampling period T_{mess}, where n and m are 16-bit integers. If n or m is zero, the measurement of the corresponding quantity is not performed in the automatic measurement mode. Because the PS_RATE (0x0A) register for storing n and ALS_RATE (0x09) register for storing m are 8-bit in size, the values must be compressed. Figure 19.5 shows the calculation path to be able to calculate the uncompressed value from a compressed value (stored in the variable ucCompressed).

The following function calculates and returns the compressed value of a positive, 16-bit number passed as a parameter when the function is called.

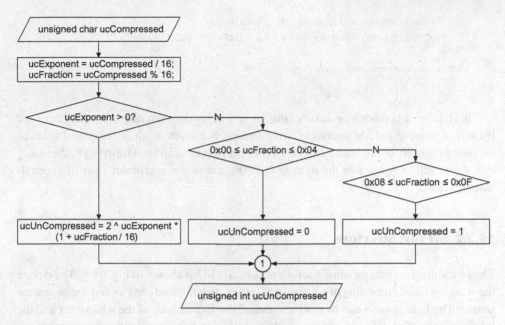

Fig. 19.5 SI114x – calculation of uncompressed values

```
uint8_t SI114x_Set_CompressedValue(uint16_t uiuncompressed)
{
    uint8_t ucExponent = 0xFF, ucFraction, ucCompressed;
    uint16_t uiUnCompressed = uiuncompressed;
    //the uncompressed value 0 is assigned 0x04 and the 1 is assigned 0x08
    if(!uiuncompressed) ucCompressed = 0x04;
    else if (uiuncompressed == 1) ucCompressed = 0x08;
    else
    {
        //the leading 1 in the binary representation of the number is
        //searched and the higher-order nibble of the compressed
        //number is calculated
        while(uiUnCompressed)
        {
            uiUnCompressed = uiUnCompressed >> 1;
            ucExponent++;
        }
        //the least significant nibble of the compressed number
        //is calculated
        if(ucExponent >= 4) ucFraction =
            uiuncomponent >> (ucExponent - 4);
        else ucFraction = uiuncompressed << (4 - ucExponent);
        ucFraction &= 0x0F;
```

```
        //the compressed number is calculated
        ucCompressed = ucExponent * 16 + ucFraction;
    }
    return ucCompressed;
}
```

This type of data compression saves storage space, but does not map bijectively between the sets of compressed and uncompressed numbers. For example, the compressed value of all binary numbers of the form *1111 1xxx xxxx xxxx* is in hexadecimal form 0xFF. Decoding this value will only recover the number 0xF800, causing a maximum error of approximately 3.2%.

19.2.2 SI114x: Structure

The block diagram of a proximity sensor of type SI1143 is shown in Fig. 19.6. The device has a sensor block consisting of a white light sensor, two infrared sensors and a temperature sensor. The light sensors can be used to measure the illuminance of the white light and the infrared light. The infrared sensor is additionally used for proximity measurement, or detection of objects. The temperature sensor is recommended for measuring the

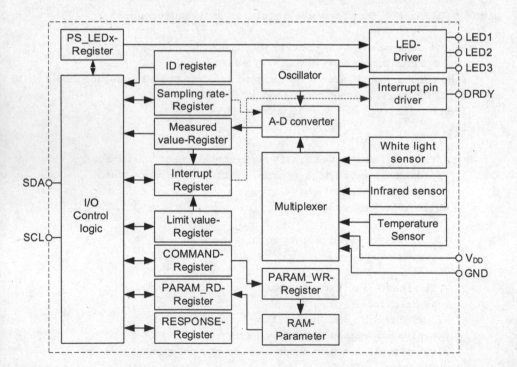

Fig. 19.6 SI1143 – block diagram

temperature difference between two measurements. The analog voltages at the outputs of the sensors and additionally the supply and ground voltages can be digitized with a 17-bit A/D converter.

For proximity measurement, the SI1143 module has an LED driver that can be used to control up to three infrared emitting diodes. This is done by voltage pulses whose duration and amplitude are adjustable. For each emitting diode, one proximity measurement can be selected and executed. This triple measurement and evaluation of the measured values can even be used to determine the spatial position of a detected object relative to the module.

The device has an extensive register block with 8-bit registers that can be addressed individually. Incrementing of the internal register address counter is switched off if bit 6 of a register address is set during a write or read access. Thus, it is possible to repeatedly read or write the contents of a register in a single I^2C transfer (between a START and STOP sequence) without repeating the addressing of the device and the register. The 16-bit measurement results are stored in two registers each in little-endian format, as listed in Table 19.1. Furthermore, the sensor has a register group for measurement and interrupt settings, one with thresholds for interrupt triggering, identification and status registers. All registers are directly addressed via the serial interface.

Settings of the measuring ranges, the amplification factors, the DC offset, the enable of the measuring mode or the limit hysteresis can be stored in the volatile parameter area. The parameter area is accessed by transmitting commands. The coded commands accepted by the device are listed in [5] and are loaded into the COMMAND register (0x18) for execution. To read a parameter through the serial port, the PARAM_QUERY command is used, whose code is 100a aaaa. The five least significant bits of the code form the address of the parameter to be read. After executing the command, the desired parameter is available for reading in the PARAM_RD (0x2E) register. To change a parameter, first the new value is loaded into the PARAM_WR (0x17) register and then the PARAM_SET (*101a aaaa*) command code is loaded into the COMMAND register.

Table 19.1 SI1143 – measured value registers

Name	Address	Description
ALS_VIS_DATA0	0x22	Memory register for the result of a visible light measurement
ALS_VIS_DATA1	0x23	
ALS_IR_DATA0	0x24	Memory register for the result of a measurement of infrared light
ALS_IR_DATA1	0x25	
PS1_DATA0	0x26	Memory register for a proximity measurement with the first transmit diode
PS1_DATA1	0x27	
PS2_DATA0	0x28	Memory register for a proximity measurement with the second transmit diode
PS2_DATA1	0x29	
PS3_DATA0	0x2A	Memory register for a proximity measurement with the third transmit diode
PS3_DATA1	0x2B	
AUX_DATA0	0x2C	Memory registers for the temperature, the supply voltage or the ground voltage measurement
AUX_DATA1	0x2D	

An internally successfully executed command is signaled by incrementing a counter in the RESPONSE register (0x20), or by setting the counter to 0x01 if the RESPONSE register was previously set to 0 with the command NOP (0x00). The execution time of the commands stored in the COMMAND register varies. Successful execution of an instruction can be determined by repeatedly evaluating the RESPONSE register. This procedure is illustrated in the following program code. If the procedure takes longer than 25 ms, it should be aborted and the command repeated.

```
#define COMMAND_REG 0x18 //address of the register COMMAND
#define RESPONSE_REG 0x20 //address of the register RESPONSE
#define NOP 0x00 //NOP instruction; sets the RESPONSE register to 0x00
uint8_t ucResponse; //globally defined variable in module SI114x
uint8_t SI114x_Write_CommandReg(uint8_t ucdata_byte)
{
    uint8_t ucErrorFlag = 0, ucRepeatCnt = 0;
    ucResponse = 0x01;
    //the response register is reset
    SI114x_Write_ByteReg(COMMAND_REG, NOP);
    //the response register is read
    SI114x_Read_ByteReg(RESPONSE_REG, &ucResponse);
    //if the reset did not work, the procedure is repeated
    while (ucResponse != 0)
        SI114x_Read_ByteReg(RESPONSE_REG, &ucResponse);
    //ucdatabyte is written to the Command Register;
    //a command should be executed
    SI114x_Write_ByteReg(COMMAND_REG, ucdata_byte);
    //the response of the sensor to the command is stored
    //in the variable ucResponse
    SI114x_Read_ByteReg(RESPONSE_REG, &ucResponse);
    while (!ucResponse && !ucErrorFlag)
    {
        //the query is repeated until when the sensor responds
        //to the command or 25 ms have elapsed
        SI114x_Read_ByteReg(RESPONSE_REG, &ucResponse);
        ucRepeatCnt++;
        if (ucRepeatCnt == 100)
        {
            ucErrorFlag = 1;
            return TWI_ERROR;
        }
    }
    return ucResponse;
}
```

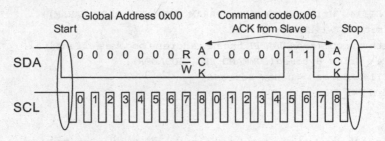

Fig. 19.7 I^2C – global reset command

19.2.3 Serial Communication

The SI114x series proximity sensors are built as I^2C slaves and can be addressed by a master via the 7-bit address 0x5A. They also respond to the global address 0x00 and to the global I^2C reset command 0x06, but do not allow 10-bit addressing. The global reset command causes the device to initialize, similar to what happens after the RESET command. Figure 19.7 shows the timing of this command. The serial interface is also active in standby mode and allows a maximum transfer rate of 3.4 MBit/s. The I^2C address of these devices can be changed by software. First the new address is stored as I^2C-ADDR parameter (address 0x00). After executing the BUSADDR command (code 0x02), the new address becomes effective and is retained until the next initialization of the device.

The SI114x_Set_NewAddress() function listed below is used to save the passed ucnew_address call parameter as a new slave address. The function returns the value 0x00 if the change was successful, otherwise 0x01.

```
//register addresses
#define PARAM_WR_REG 0x17
#define COMMAND_REG 0x18
//commands
#define PARAM_SET 0xA0
#define BUSADDR 0x02
//parameter addresses
#define I2C_ADDR_PARAM 0x00
//variables
uint8_t ucSI114xAddress = 0x5A;
//function start
uint8_t SI114x_Set_NewAddress(uint8_t ucnew_address)
{
    //ucnew_address is stored in the PARAM_WR register,
    //which is to be transmitted at the parameter address I2C_ADDR
    SI114x_Write_ByteReg(PARAM_WR_REG, ucnew_address);
    //the address of the parameter is entered in the command register
    //write mode is loaded and thus the transfer of the new
    //address into RAM at address 0x00 executed
```

```
   if(!SI114x_Write_CommandReg(PARAM_SET | I2C_ADDR_PARAM))
      return TWI_ERROR;
   //with the command BUSADDR the new address becomes effective
   SI114x_Write_ByteReg(COMMAND_REG, BUSADDR);
   ucSI114xAddress = ucnew_address;
   return TWI_OK;
}
```

To change the content of a register, the master must initiate communication with the slave with a START sequence, address the slave via the current address in write mode, send the address of the desired register and finally transmit the new value. When the slave has acknowledged each received byte with ACK, the write operation was successful and the master terminates the communication with a STOP sequence.

19.2.4 Measurements with the SI114x

A proximity sensor of the SI114x series enables the measurement of proximity and/or ambient light. The proximity measurements take place on up to three measurement channels *PS1, PS2* and *PS3*. With the ambient light measurement selected, both the light intensities of the visible and infrared light, as well as the temperature, the supply voltage or the ground voltage can be measured. Each measurement channel can be selected and deselected, resulting in high flexibility of the measurement process. The number of selected measurements influences the total measurement time and the total energy consumption of the device. The selected measurements are performed one after the other and the measurement results of each measurement channel are stored in a pair of registers (see Table 19.1).

Bits 3:0 of the RESPONSE register form a counter that is incremented with each successfully executed instruction. This register stores an error code if an overflow has occurred during a measurement. In such a case, the settings must be adjusted to the ambient conditions and the measurements repeated. Some results are obtained by subtracting two measured values, which can lead to a negative value that would be falsely signaled as an overflow. To avoid this, a calculated offset is applied to all measured values. This adjustable global offset is stored in compressed form in the ADC_OFFSET register (0x1A) and must be taken into account when evaluating the results.

Proximity measurement (detection of an object) can also be used for gesture recognition [7]. The devices have 1 . . . 3 measuring channels *PS1, PS2* and *PS3*. These can be used to control the connected infrared diodes with voltage pulses, to measure the light intensity of the reflected beams and to store the measured values in the corresponding registers.

Each device has a small and a large infrared sensor. The small one can be used when the outdoor lighting is high (usually in sunlight), the large one is used under normal lighting conditions to achieve a higher sensitivity. The influence of ambient light is compensated by a twofold measurement. First, the infrared ambient light is measured with the transmitter diode turned off, and then the proximity measurement is taken. The difference of the two measurements is stored.

In order to achieve more accurate measurements, to increase the distance to the detected objects and to minimize the influence of ambient light (light intensity and pulsating ambient light), there are numerous setting options. As an example, the proximity measurement in the single measurement mode of the SI1141 is examined in more detail, which only has the measurement channel *PS1* and only controls one light emitting diode.

19.2.4.1 Measuring Mode: Setting and Selection of the Measuring Channel

By setting the MEAS_RATE register to 0x00, only the individual measurements that were previously enabled in the CHLIST parameter (0x01) can be started. The configuration of the parameter is as follows:

- **Bits 7, 3** – are reserved;
- **Bit 6** – EN_AUX = 1 enables the measurement of the additional channel (temperature/supply voltage/ground voltage measurement);
- **Bit 5** – EN_ALS_IR – Enable infrared light measurement;
- **Bit 4** – EN_ALS_VIS – Enable white light measurement;
- **Bit 2** – EN_PS3 – Enable proximity measurement at channel PS3;
- **Bit 1** – EN_PS2 – Enable proximity measurement at channel PS2;
- **Bit 0** – EN_PS1 – Enable proximity measurement at channel PS1;

Further parameters PSLED12_SELECT (0x02) and PSLED3_SELECT (0x03) are used to assign the measurement channels to the connected infrared diodes. With PSLED12_SELECT = 0x01 and PSLED3_SELECT = 0x00 the proximity measurement takes place at channel *PS1* and the measured values are stored in the register pair PS1_DATA0 / PS2_DATA1.

19.2.4.2 Setting the Infrared Light Pulses

The infrared light pulses sent to detect objects can reach a higher light intensity if the current intensity through the transmitting diode is larger. The current through LED1 is set by bits 3:0 (LED1_I) of register PS_LED21 (0x0F) and can be calculated using Eq. (19.4). If LED1_I = 0000, the current through the LED is switched off.

$$
I/mA = \begin{cases} LED1\_I \cdot 5.6 & |0 < LED1\_I < 3 \\ (LED1\_I - 2) \cdot 22.4 & |2 < LED1\_I < 13 \\ (LED1\_I - 12) \cdot 22.4 & | \quad LED1\_I > 12 \end{cases} \qquad (19.4)
$$

The duration of the light pulses t_P is set via parameter PS_ADC_GAIN (0x0B) and is given by Eq. (19.5). The setting range of the parameter is 0 ... 7.

$$
t_P = 25.6 \ \mu s \cdot 2^{PS\_ADC\_GAIN} \qquad (19.5)
$$

The period of the ADC clock is also increased by the factor $2^{PS\_ADC\_GAIN}$.

Table 19.2 SI1141 –
sensitivity selection and
adjustment to the ambient
illumination

	PS_ADC_MISC Ambient lighting	
0×04		0×24
Normal		Strong
	PS1_ADCMUX Sensitivity	
0×00		0×03
Low		High

19.2.4.3 Selection of the Measuring Sensor and the Measuring Settings

Two infrared sensors of different sizes allow proximity measurements with different sensitivities, which can be selected via parameter PS1_ADCMUX (0x07) according to Table 19.2. A correction of the light pulse duration for measurements under direct sunlight can be done via parameter PS_ADC_MISC (0x0C).

In order to reduce the measurement noise during quantization, a recovery time t_R is set for the A/D converter before each proximity measurement. This recovery time, which is to be set with bits 6:4 (PS_ADC_REC) of parameter PS_ADC_COUNTERS (0x0A) with Eq. (19.6), should be large enough to minimize the measurement noise. For proximity measurements that take place under pulsating ambient light, care must be taken that the two successive measurements happen close in time to keep the measurement inaccuracies small, which means small recovery times. The choice of recovery time must also take into account the manufacturer's recommendation that PS_ADC_REC maps the ones complement of PS_ADC_GAIN.

$$t_R/ns = \begin{cases} 50 \cdot 2^{PS\_ADC\_GAIN} & |PS\_ADC\_REC = 0 \\ 50 \cdot 2^{PS\_ADC\_GAIN} \cdot \left(2^{2+PS\_ADC\_REC} - 1\right) & |PS\_ADC\_REC > 0 \end{cases} \qquad (19.6)$$

19.2.4.4 Initialization of the Sensor

The following program code represents the initialization of a sensor of type SI1141 for proximity measurement in single measurement mode. The current through the transmitting diode is set to 11.2 mA and the duration of the light pulses is multiplied by a factor of 16 (PS_ADC_GAIN = 4). From this, PS_ADC_REC = 0x03 is used to calculate the recovery time. The sensor should work under normal ambient light and the measurements should take place with high sensitivity.

```
uint8_t SI114x_Init(void)
{
    //initialization is completed
    if(SI114x_Write_ByteReg(HW_KEY_REG, HARDWARE_INIT))
        return TWI_ERROR;
    //single measurement mode
    if(SI114x_Write_ByteReg(MEAS_RATE_REG, 0x00)) return TWI_ERROR;
    //enable measuring channel PS1
    if(!SI114x_Write_Param(CHLIST_PARAM, 0x01)) return TWI_ERROR;
    //the measuring channel PS1 is assigned LED1, LED2 is deactivated
    if(!SI114x_Write_Param(PSLED12_SEL_PARAM, 0x01)) return TWI_ERROR;
    //LED3 is deactivated
    if(!SI114x_Write_Param(PSLED3_SEL_PARAM, 0x00)) return TWI_ERROR;
    //the current through LED1 is set to 11.2mA, switched off by LED2
    if(SI114x_Write_ByteReg(PS_LED21_REG, 0x02)) return TWI_ERROR;
    //the current through LED3 is switched off
    if(SI114x_Write_ByteReg(PS_LED3_REG, 0x00)) return TWI_ERROR;
    //the pulse width is increased by a factor of 16
    if(!SI114x_Write_Param(PS_ADC_GAIN_PARAM, 0x04)) return TWI_ERROR;
    //the measurements take place under normal lighting
    if(!SI114x_Write_Param(PS_ADC_MISC_PARAM, 0x04)) return TWI_ERROR;
    //the large infrared sensor is selected
    if(!SI114x_Write_Param(PS1_ADCMUX_PARAM, 0x03)) return TWI_ERROR;
    //setting the recovery time
    if(!SI114x_Write_Param(PS_ADC_COUNTER_PARAM, 0x30))
        return TWI_ERROR;
    return TWI_OK;
}
```

The SI114x_Init () function returns the value TWI_OK (0) if it was executed without errors. The settings must be adapted to the specific application.

19.2.4.5 Starting the Measurement
After the desired settings have been saved, a proximity measurement can be started with the command PS_FORCE. According to [5], the duration of a measurement is 155 µs for the default settings. The end of a measurement can be signaled via an interrupt.

19.2.4.6 Reading the Measured Values
The measurement result is stored in the registers PS1_DATA0 and PS1_DATA1 and can be read out with the function SI114x_Read_WordReg () . This function, which is listed below, reads the contents of two registers whose address follow each other, combines the contents to a 16-bit value and stores this value in a variable.

```
uint8_t SI114x_Read_WordReg(uint8_t ucreg_address,
                            uint16_t* uidata_word)
.{
    uint8_t ucDeviceAddress, ucDataLSByte, ucDataMSByte;
    //address of the Proximity Sensor
    ucDeviceAddress = ucSI114xAddress << 1;
    ucDeviceAddress |= TWI_WRITE; //write mode
    TWI_Master_Start(); //start
    if((TWI_STATUS_REGISTER) != TWI_START) return TWI_ERROR;
    TWI_Master_Transmit(ucDeviceAddress); //send device address
    if((TWI_STATUS_REGISTER) != TWI_MT_SLA_ACK) return TWI_ERROR;
    //the address of the desired register is sent
    TWI_Master_Transmit(ucreg_address);
    if((TWI_STATUS_REGISTER) != TWI_MT_DATA_ACK) return TWI_ERROR;
    TWI_Master_Start(); //restart
    if((TWI_STATUS_REGISTER) != TWI_RESTART) return TWI_ERROR;
    ucDeviceAddress = (ucSI114xAddress << 1) | TWI_READ; //read mode
    TWI_Master_Transmit(ucDeviceAddress); //send device address
    if((TWI_STATUS_REGISTER) != TWI_MR_SLA_ACK) return TWI_ERROR;
    //the content of the addressed register are read
    ucDataLSByte = TWI_Master_Read_Ack();
    if((TWI_STATUS_REGISTER) != TWI_MR_DATA_ACK) return TWI_ERROR;
    ucDataMSByte = TWI_Master_Read_NAck();
    *uidata_word = (ucDataMSByte << 8) + ucDataLSByte;
    if((TWI_STATUS_REGISTER) != TWI_MR_DATA_NACK) return TWI_ERROR;
    TWI_Master_Stop(); //stop
    return TWI_OK;
}
```

The address of the variable in which the measurement result is stored together with the address of register PS1_DATA0 are passed as parameters when the function is called. After reading the first register, the internal address counter is incremented and the second register can be read in the same I^2C session.

```
uint16_t uiPSValue;
SI114x_Read_WordReg(PS1_DATA0_REG, &uiPSValue);
```

19.2.5 Interrupts

An SI114x has an open-drain output that can serve as an interrupt trigger for a master. A pull-up resistor must be connected between this output and the supply voltage of the module. It is activated by storing 0x01 into the configuration register INT_CFG (0x03).

The measurement channels that can trigger an interrupt are enabled in register
IRQ_ENABLE (0x04). After a proximity measurement via channel *PS1,* an interrupt can
be triggered when a measurement is completed or when the measurement result crosses a
set threshold value, i.e. exceeds it from above or below. In order for an interrupt to be
triggered after each completed *PS1* measurement, the value 0x04 must be stored in the
IRQ_ENABLE register and the value 0x00 in the IRQ_MODE1 (0x05) register.

When an interrupt-triggering condition is fulfilled, the corresponding bit (bit 2 for
channel *PS1*) in the IRQ_STATUS status register (0x21) is set and the INT output is
switched low. This state remains stored until it is reset by the master by overwriting with
1 the set bits in the IRQ_STATUS register.

The interrupt initialization listed below can be added to the SI114x_Init()
function.

```
//interrupt settings
//the interrupts via channel PS1 are enabled
if(SI114x_Write_ByteReg(IRQ_ENABLE_REG, 0x04)) return TWI_ERROR;
//a completed measurement at channel PS1 triggers an interrupt
if(SI114x_Write_ByteReg(IRQ_MODE1_REG, 0x00)) return TWI_ERROR;
//INT output of the device is activated
if(SI114x_Write_ByteReg(INT_CFG_REG, 0x01)) return TWI_ERROR;
//the interrupt status register is read out
if(SI114x_Read_ByteReg(IRQ_STATUS_REG, &ucResponse)) return TWI_ERROR;
//overwriting the register with the same value
//resets all interrupts
if(SI114x_Write_ByteReg(IRQ_STATUS_REG,ucResponse)) return TWI_ERROR;
```

19.2.6 SI114x: Network Identification

The devices of the SI114x series can be identified in a network via the read-only registers
PART_ID (0x00), REV_ID (0x01) and SEQ_ID (0x02). The PART_ID register stores the
value 0x41 for a SI1141 type device, and the value 0x42 for SI1142 or 0x43 for SI1143.
Reading this register provides the device type so that the appropriate initialization and
correct control can be performed.

The revision number of the internal sequence control stored in register SEQ_ID can
play an important role. For example, the 16-bit theshold value, the exceeding of which can
lead to an interrupt when measuring in automatic measurement mode with channel *PS1,* is
stored in the register pair PS1_TH0 (0x11) and PS1_TH1 (0x12) starting with revision
A11 (memory code 0x09), and up to this revision (memory code smaller than 0x09) it is
stored in compressed form in register PS1_TH0 . Further differences can be found in the
data sheet [5]. 0x00 is always stored in the REV_ID register.

References

1. Meroth, A., & Tolg, B. (2008). *Infotainmentsysteme im Kraftfahrzeug*. Friedr. Vieweg & Sohn.
2. Hering, E., & Schönfelder, G. (Eds.). (2012). *Sensoren in Wissenschaft und Technik*. Vieweg +Teubner.
3. Tränkler, H.-R., & Reindl, L. M. (Eds.). (2014). *Sensortechnik*. Springer Vieweg.
4. Ultraschall-Modul SRF08. (2021). Datenblatt. http://www.roboter-teile.de/datasheets/srf08.pdf. Accessed 2021, Febuary 23.
5. Silicon Labs. (2015). SI1141/42/43 Proximity/Ambient light sensor IC with I^2C interface. www.silabs.com
6. Silicon Labs. (2015). AN498 – SI114x Designer's guide. www.silabs.com
7. Silicon Labs. (2015). AN580 – INFRARED GESTURE SENSING. www.silabs.com

Digital-to-Analog and Analog-to-Digital Converters

Abstract

This chapter describes D/A and A/D converter modules.

Already in the third chapter the possibilities of a microcontroller were described, to read in analog signals with on-board means or to generate them via an integrating circuit with PWM. These integrated possibilities are often limited, e.g. the Atmega8x offers only one A/D-converter, which can at best be multiplexed and whose reliable resolution is only eight bits (at ten bits nominal resolution). On the other hand, the generation of analog signals via PWM quickly reaches its limits, since at a resolution of eight bits, for example, only a 256th of the clock frequency is available. External converters, which can be controlled via SPI or I^2C, can help here. The functionality of the devices is explained in the respective section.

20.1 MCP48XX SPI-Driven Digital-to-Analog Converters

The devices of the MCP48XX series [1, 2] are 1- or 2-channel, serial, unipolar D/A converters with a bit resolution of 8, 10 or 12 bits (see Table 20.1). They feature simple external circuitry, thanks to a single supply voltage V_{DD} between 2.7 V and 5.5 V and the internal precise reference voltage of 2.048 V. The D/A converters are controlled via a serial, unidirectional (SDO line missing) SPI interface that can be clocked at a frequency of up to 20 MHz and have a settling time of 4.5 µs.

The devices include a dual data register, a D/A converter, and an analog amplifier as shown in Fig. 20.1.

A. Meroth, P. Sora, *Sensor networks in theory and practice*,
https://doi.org/10.1007/978-3-658-39576-6_20

Table 20.1 MCP48xx – SPI driven D/A converters

Module MCP	Bit resolution	Channels	Gain factor	
			1	2
			Step size /mV / U_{out}/mV	Step size /mV / U_{out}/mV
4801	8	1	8 / 0 … 2040	16 / 0 … 4080
4802	8	2	8 / 0 … 2040	16 / 0 … 4080
4811	10	1	2 / 0 … 2046	4 / 0 … 4092
4812	10	2	2 / 0 … 2046	4 / 0 … 4092
4821	12	1	0.5 / 0 … 2047.5	1 / 0 … 4095
4822	12	2	0.5 / 0 … 2047.5	1 / 0 … 4095

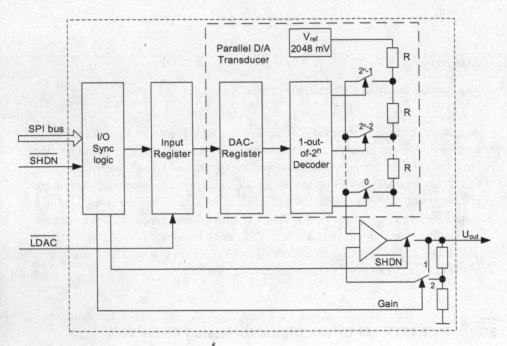

Fig. 20.1 MCP48x1 – block diagram

20.1.1 The SPI Interface

An SPI-compliant protocol is implemented in the I/O control logic to provide communication with a microcontroller. The device is preconfigured as a slave and a master must send the bytes in SPI mode 0 or 3, with the most significant bit first. This is a unidirectional transfer because the device does not have a transmit pin (there is no SDO pin). Data transmission begins with the activation of the chip select (or slave select) line, followed by the transmission of two bytes needed to produce an analog value. The first byte transmitted will occupy bits 15:8 and the second byte will occupy bits 7:0 from the input register. On the rising edge from the chip select, the two bytes are stored into the input register.

To be able to design a software module for the MCP48XX in general, a data structure is defined in the main file and initialized for each device connected to the SPI bus, depending on the concrete wiring. For each controllable pin (in this case, slave select, latch DAC, and shutdown), the corresponding DDR and PORT registers, as well as the port pin, must be specified, noting whether they are connected in the circuit. A data structure of the following form is declared in the software module of the device (Sect. 6.1.5):

```
typedef struct
{
    tspiHandle MCP48XXspi;
    volatile uint8_t* LDAC_DDR;
    volatile uint8_t* LDAC_PORT;
    uint8_t LDAC_pin;
    uint8_t LDAC_state;
    volatile uint8_t* SHDN_DDR;
    volatile uint8_t* SHDN_PORT;
    uint8_t SHDN_pin;
    uint8_t SHDN_state;
} MCP48XX_pins;
```

20.1.2 The Input Register

The 16-bit input register contains control bits as well as data bits and is configured as follows:

- **Bit 15** – is always 0 for MCP48x1, while for MCP48x2 0 is for channel A and 1 is for channel B;
- **Bit 14** – reserved;
- **Bit 13** – Gain – if this bit has the value 0, the analog signal is amplified by a factor of two;
- **Bit 12** – Shutdown – at 1 the analog voltage is switched on to the output; the 1-channel converters have an additional shutdown input;
- **Bit 11:0** – Data bits for MCP482x (Bit 11 = MSB, Bit 0 = LSB);
- **Bit 11:2** – Data bits for MCP481x (Bit 11 = MSB, Bit 2 = LSB, Bit 1:0 irrelevant);
- **Bit 11:4** – Data bits for MCP480x (Bit 11 = MSB, Bit 4 = LSB, Bit 3:0 irrelevant);

20.1.3 The D/A Converter

The D/A converter mainly consists of a voltage divider with 2^n equal resistors, where n is the bit resolution. The voltage divider is connected to the internal reference voltage and to the ground of the device. An analog electronic switch is connected to each common connection between two resistors and to the ground. By connecting the second connection of all 2^n switches, the output of the D/A converter is formed. The number stored in the DAC register drives a single switch through the 1-of-2^n decoder. The input voltage of this switch is passed through to the common output. When the control input LDAC is switched low, then the data bits from the input register are stored into the DAC register, thus starting the D/A conversion, which is completed after the settling time.

If U_{REF} is the reference voltage of the converter, then define the step size U_{LSB} as the analog voltage corresponding to one LSB:

$$U_{LSB} = \frac{U_{REF}}{2^n} \tag{20.1}$$

Equation 20.1 can be used to calculate the voltage across the i-th resistor:

$$U_i = \left(2^i - 1\right) \cdot U_{LSB}. \tag{20.2}$$

The internal D/A converter converts positive binary numbers in the range $0 \ldots 2^n - 1$ into a positive voltage, theoretically between 0 V $\ldots (U_{REF} - U_{LSB})$ V. The relative difference between the resistance values is small and thus a good linearity of the output voltage is ensured over the entire temperature range.

20.1.4 The Analog Output Amplifier

In order to keep the power consumption of the device small, the resistance of the voltage divider must be large, but this means that at different loads the real output voltage will differ from the calculated one. The output voltage becomes less sensitive to the external load if an analog amplifier with low output resistance is used. With a gain factor of 2, which can be set via the bit gain of the input register, output voltages between 0 V $\ldots 2*(U_{REF} - U_{LSB})$ V can be realized, but not greater than V_{DD}. This also doubles the step size. Thanks to rail-to-rail technology, the analog output voltage can practically assume values in the range 10 mV $\ldots (V_{DD} - 40$ mV$)$.

Bit 12 (Shutdown) of the input register, or the shutdown pin for the 1-channel D/A converters, can be used to decouple the output pin from the amplifier output and switch it to high impedance.

In applications where the analog voltage output is not time critical (for example, generating a variable reference voltage or offset voltage, or calibrating a sensor), the LDAC line can remain low all the time. On the rising edge of the chip select signal, the data bytes are stored into the input register and at the same time the data bits are stored into the DAC register and the conversion is started.

20.1.5 Synchronous Control of Two D/A Converters

If one wants to generate a PAM signal[1] with fixed sampling rate or two analog signals with defined phase shift, the analog voltages must be output time controlled. In this case the time of the conversion start has to be determined via the LDAC signal by switching the control signal from high to low at the end of the SPI transmission. The timing of the SPI transmission for generating synchronous voltages with the circuit in Fig. 20.2b is shown in Fig. 20.3. The order in which one addresses the D/A converters is not important. After the data bytes have been sent to the two D/A converters, the master synchronously starts the conversions by switching the LDAC signal.

A sample code for the control shown in Fig. 20.3 for the circuitry shown in Fig. 20.2 is given below. With:

```
#define ON 1
#define OFF 0
```

the SPI data structures of the two devices are initialized in the main file:

```
//declaration for IC1
MCP48XX_pins MCP48XX_1 = {{&DDRD, &PORTD, PD6, ON},
                          &DDRB, &PORTB, PB2, ON,
                          OFF, OFF, OFF};
//declaration for IC2
MCP48XX_pins MCP48XX_2 = {{&DDRC, &PORTC, PC4, ON},
                          &DDRB, &PORTB, PB2, ON,
                          OFF, OFF, OFF};
```

In the first step the previously calculated bytes are transferred to the first D/A converter,

```
//excerpt from the main file
//the slave select line of IC1 is set to low
SPI_Master_Start(MCP48XX_1.MCP48XXspi);
//the most significant byte for IC1 is transmitted
SPI_Master_Write(ucMSByte_IC1);
//the least significant byte for IC1 is transmitted
SPI_Master_Write(ucLSByte_IC1);
//the slave select line is set to high and thus
//the SPI transfer is finished
SPI_Master_Stop(MCP48XX_1.MCP48XXspi);
```

and then to the second:

---

[1] PAM signal – Pulse Amplitude Modulated signal.

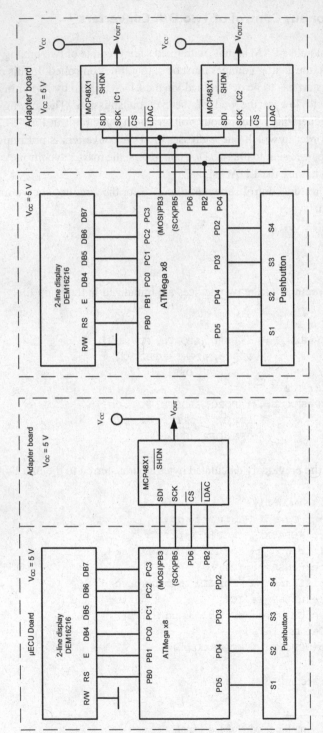

Fig. 20.2 Connection of an SPI-controlled D/A converter from the MCP48x1 series to a microcontroller (**a**) and networking of two D/A converters (**b**)

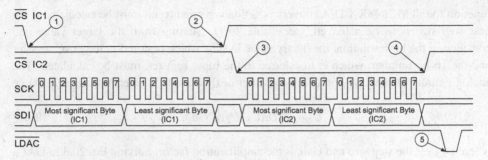

1 – Start of SPI communication with IC1;
2 – End of SPI communication with IC1. The 2 bytes are stored in the input register of IC1;
3 – Start of SPI communication with IC2;
4 – End of SPI communication with IC2. The 2 bytes are stored in the input register of IC2;
5 – The data bits of the two interfaces are transferred simultaneously into the corresponding DAC register. Start of the D/A conversion which is finished after the settling time.

Fig. 20.3 Time domain representation of the SPI transmission for the generation of an analog value pair with two A/D converters of the MCP48x1 series

```
//the slave select line of IC2 is set to low
SPI_Master_Start(MCP48XX_2.MCP48XXspi);
//the most significant byte for IC2 is transmitted
SPI_Master_Write(ucMSByte_IC2);
//the least significant byte for IC2 is transmitted
SPI_Master_Write(ucLSByte_IC2);
//the slave select line is set to high and thus
//the SPI transfer is finished
SPI_Master_Stop(MCP48XX_2.MCP48XXspi);
```

and finally the conversion is started by switching the LDAC signal:

```
//the common LDAC line of the D/A converters is set low and then high
//(the converted values are output simultaneously)
MCP48XX_Set_LDACLow(MCP48XX_1);
MCP48XX_Set_LDACHigh(MCP48XX_1);
```

20.1.6 Software Example

When choosing a specific D/A converter, the achievable step size and the output voltage swing play a decisive role. The presented MCP48XX series offers a better step size (Eq. 20.1) at output voltages smaller than 2048 V. The output amplifier can be used to extend the output voltage swing, but this worsens the step size. The analog target value is often calculated with the aid of a conversion function. For the design of a general drive function for all MCP48XX-D/A converters, a floating point result must be calculated if the best step size is to be achieved. (see Table 20.1). Starting from the target value and considering the bit resolution, the binary value DAC_n, which is stored in the DAC register and the 16-bit number, which is transferred to the input register, must be calculated. The analog output voltage U_{OUT} of a D/A converter of this family is determined by Eq. 20.3:

$$U_{OUT} = DAC_n{}^* U_{LSB}{}^* Gain \tag{20.3}$$

where U_{LSB} is the step size and Gain is the amplification-factor. Solving Eq. 20.3 to DAC_n gives:

$$DAC_n = \frac{V_{out}}{U_{LSB}{}^* Gain} \tag{20.4}$$

The best step size can be achieved by selecting the amplification factor Gain $= 1$ for output voltages smaller than the reference voltage.

In the following a function is presented which can be used to control any D/A converter from the series. Based on the desired output voltage, $fvalue$ (in mV with a resolution of 0.5 mV), and the device used, $ucdevice$ (0 for MCP480X, 1 for MCP481X and 2 for MCP482X), the function calculates the binary value DAC_n and the gain factor $ucGain$. This binary value together with the selected channel $ucchannel$ and the output voltage enable $ucout$ form the variable $uiInputRegister$, which is transferred to the input register of the D/A converter. With a logical "0" the first channel is selected and with a logical "1" the second channel is selected. The output voltage is enabled with a logical "1". The concrete wiring of the device is taken into account in the SPI data structure. The function checks the analog output voltage entered as a parameter and limits it if the range is exceeded. The calculated value for the input register is transferred via SPI while the LDAC signal remains low. The conversion is started with the rising edge of the chip select signal.

```
void MCP48XX_Set_OutputVoltage(MCP48XX_pins sdevice_pins,
                               uint8_t ucdevice,
                               uint8_t ucchannel,
                               uint8_t ucout,
                               float fvalue)
{
    uint8_t ucStepSize, ucLSByte, ucMSByte, ucGain = 2;
    uint16_t uiVoltageOut, uiInputRegister = 0x0000;
    uint16_t uiCommandBits = 0x0000;
    //voltage limitation to the maximum achievable value
    if(fvalue >= (VOLTAGE_REFERENCE * 2))
        fvalue = (VOLTAGE_REFERENCE * 2) - 1;
    //if channel 1 was selected, then bit 15 is set
    if(ucchannel) uiCommandBits = 0x8000;
```

```
    //if the output voltage is lower than the reference voltage,
    //bit 13 is set
    if(fvalue < VOLTAGE_REFERENCE)
    {
        uiCommandBits |= 0x2000;
        ucGain = 1;
    }
    //if output enable, then bit 12 is set
    if(ucout) uiCommandBits |= 0x1000;
    switch(ucdevice)
    {
        case MCP480X:
            uiVoltageOut = fvalue; //output voltage as integer number
            ucStepSize = (VOLTAGE_REFERENCE * ucGain) / 256;
            //calculation of the step size
            uiInputRegister = (uiVoltageOut / ucStepSize) << 4;
            //binary value is determined
        break;
        case MCP481X:
            uiVoltageOut = fvalue;
            ucStepSize = (VOLTAGE_REFERENCE * ucGain) / 1024;
            uiInputRegister = (uiVoltageOut / ucStepSize) << 2;
        break;
        case MCP482X:
            uiVoltageOut = fvalue * 2;
            ucStepSize = (2 * VOLTAGE_REFERENCE * ucGain) / 4096;
            uiInputRegister = uiVoltageOut / ucStepSize;
        break;
    }
    //calculation of the input register
    uiInputRegister |= uiCommandBits;
    //the least significant byte of the input register is determined
    ucLSByte = uiInputRegister;
    //the most significant byte of the input register is determined
    ucMSByte = uiInputRegister >> 8;
    //the slave select line is set to low
    SPI_Master_Start(sdevice_pins.MCP48XXspi);
    //the MSByte of the input register is transferred
    SPI_Master_Write(ucMSByte);
    //the LSByte of the input register is transferred
    SPI_Master_Write(ucLSByte);
    //the slave select line is set to high and
    //thus the SPI transmission is terminated
    SPI_Master_Stop(sdevice_pins.MCP48XXspi);
}
```

20.2   PCF8591 I²C-Controlled D/A and A/D Converter

The PCF8591 [3] device integrates both a D/A and an A/D converter on a single silicon chip and is powered by a single voltage, as shown in Fig. 20.4. The conversion rate is the same for both directions and is determined by the bus frequency. The device has one analog output and four analog inputs and is controlled via an I²C-bus. Thanks to the three address inputs, up to eight identical devices can be connected to a single two-wire bus. Thus, the analog periphery of a microcontroller could be extended with eight analog outputs or with up to 32 analog unbalanced inputs.

20.2.1 I²C Communication

Communication with a microcontroller configured as a master takes place at a transmission rate of max. 100 kBit/s. The transmission rate plays an active role for the conversion processes, because new conversions are started with the transmission. The master starts the communication with the transmission of the device address of the component. The R/W bit, as the least significant bit of this address, controls the internal data flow of the device. If this

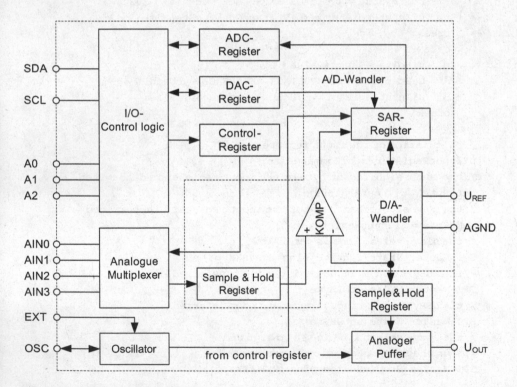

Fig. 20.4 PCF8591 – block diagram

bit is 0 (write command), then the next byte is stored in the control register in preparation for the following conversion. The next 1 to n bytes are stored in the DAC register and output as an analog value at the end of a byte (after nine clocks). This burst mode allows the highest conversion rate: 100 kBit/s / 9 Bit/Sample = 11.1 kS/s.[2]

If the R/W bit is 1 (read command), the following happens:

- the slave starts a new A/D conversion with the confirmation of its own address with ACK. The conversion takes the existing settings from the control register (input configuration and input channel) into account.
- with the following eight clocks the slave shifts bitwise the content of the ADC register to the data line (the result of the previous conversion). If the master responds with ACK, a new conversion is started either from the same input channel or from the next one, if the auto increment bit in the control register is set. With a NACK the master signals the end of the communication to the slave.

After power-on, the microcontroller reads the value 0x80 as the first byte from the ADC register.

20.2.2 The D/A Converter

The device's 8-bit D/A converter operates in the same parallel mode as described in Sect. 20.1.3. The internal voltage divider, which takes care of the D/A conversion, is connected to a reference voltage and to the internal analog ground, which is galvanically isolated from the digital one. Thus, the output voltage range of the converter $U_{REF} - U_{AGND}$, the step size U_{LSB}

$$U_{LSB} = \frac{U_{REF} - U_{AGND}}{256} \tag{20.5}$$

and the output voltage

$$U_{OUT} = U_{AGND} + U_{LSB} \cdot OutputByte \tag{20.6}$$

are calculated. When the analog ground is connected to the digital one, the dynamic range becomes equal to U_{REF}, which means that an analog positive voltage smaller than U_{REF} can be converted to a binary value smaller than 256. The settling time is 90 μs and the conversion rate can reach 11.1 kS/s. After conversion, the analog value is stored in a sample-and-hold circuit and output via an analog buffer that can be switched off at pin U_{OUT}. The control register controls this analog buffer. In the active state, its value is

---

[2]kS/s – kilo Samples/Second.

retained until a new conversion takes place. The analog output can reach 100% of the supply voltage in the unloaded state, and only 90% when loaded with 10 kΩ. With the following definitions:

```
#define PCF8591_DEVICE_TYPE_ADDRESS 0x90
#define REFERENCE_VOLTAGE 5000 //mV
#define ANALOG_GROUND_VOLTAGE 0 //mV
#define OUTPUT_STEP_SIZE
    ((REFERENCE_VOLTAGE - ANALOG_GROUND_VOLTAGE) / 256)
#define D_A_CONVERSION_OPCODE 0x40
```

the program section for the conversion of a voltage value iout_value entered in millivolts could look as follows:

```
uint8_t ucDeviceAddress, ucOut_Byte;
//the value required to drive the D/A converter is calculated taking into
//account the reference voltage and the ground voltage
ucOut_Byte = (iout_value - ANALOG_GROUND_VOLTAGE) / OUTPUT_STEP_SIZE;
//form address of the PCF8591
ucDeviceAddress = (ucdevice_address << 1) |
                    PCF8591_DEVICE_TYPE_ADDRESS;
ucDeviceAddress |= TWI_WRITE; //write mode
//the master initiates the I²C communication
TWI_Master_Start();
if((TWI_STATUS_REGISTER) != TWI_START) return TWI_ERROR;
//send device address
TWI_Master_Transmit(ucDeviceAddress);
if((TWI_STATUS_REGISTER) != TWI_MT_SLA_ACK) return TWI_ERROR;
TWI_Master_Transmit(D_A_CONVERSION_OPCODE);
if((TWI_STATUS_REGISTER) != TWI_MT_DATA_ACK) return TWI_ERROR;
//the data byte is sent
TWI_Master_Transmit(ucOut_Byte);
if((TWI_STATUS_REGISTER) != TWI_MT_DATA_ACK) return TWI_ERROR;
//the master terminates the communication
TWI_Master_Stop();
return TWI_OK;
```

The time-dependent course of the communication can be seen in Fig. 20.5a. If one wants to generate an analog value and the reference and offset voltage of the analog ground are known, then Eq. 20.6 can be resolved according to the binary value and one obtains:

$$OutputByte = \frac{U_{OUT} - U_{AGND}}{U_{LSB}}. \tag{20.7}$$

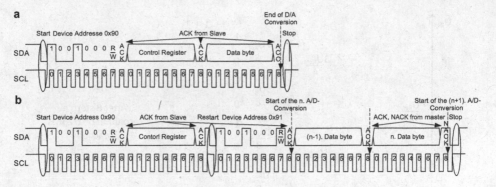

Fig. 20.5 PCF8591 – I^{2}C communication for controlling the D/A and A/D converter

To realize a single D/A conversion, the master starts the communication and then sends the address of the slave with the R/W bit set to 0 (write command), another byte stored in the control register and finally the byte to be converted. The master ends the communication with a stop command. The slave must acknowledge each received byte with ACK, otherwise the master interrupts the communication. As soon as the slave has acknowledged the data byte with ACK, the analog voltage is stored in the Sample&Hold circuit and if bit 6 of the control register is 1, the analog output buffer is activated.

20.2.3 The A/D Converter

The four analog inputs can be wired as shown in Fig. 20.7 and the resulting channels can be routed to the input of the A/D converter via the analog multiplexer. The wiring of the inputs and the channel selection are controlled by the control register. The analog channels can be either unbalanced (the voltage is measured against analog ground) or balanced (the difference of the voltages applied to the balanced inputs is measured). When converting an unbalanced channel, the lower drive limit is U_{AGND} and the upper drive limit is U_{REF} and the result is stored as a positive integer in the ADC register. When converting a balanced channel, the lower drive limit is $-(U_{REF} - U_{AGND})/2$ and the upper limit is $+(U_{REF} - U_{AGND})/2$ and the result is signed stored.

The A/D converter works according to the principle of *successive aproximation*[3] and consists of the D/A converter described above, which is temporarily used for the A/D conversion, an analog comparator and a SAR[4] register. In this A/D converter, the number of steps is equal to the bit resolution. Step by step, each bit starting with bit 7 is tested whether it is 1 or 0. In the first step, bit 7 (MSB) of the SAR register is set to 1, resulting in

---

[3] See also Chap. 3, Sect. 3.8.3.

[4] SAR – Successive Aproximation Register.

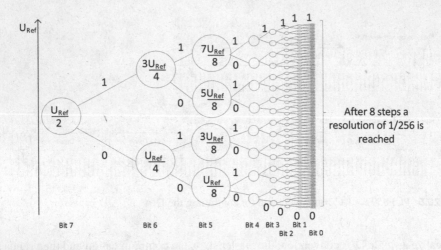

Fig. 20.6 Principle of successive approximation

an analog voltage at the output of the D/A converter equal to $V_{REF}/2$. The comparator compares the voltage to be converted with that at the output of the D/A converter. If it is greater, bit 7 remains at 1, otherwise it is set to 0. In the second step, bit 6 is tested and set to 1. The voltage at the output of the D/A converter is now *Bit* 7 $\cdot$ *UREF*/2 + *UREF*/4. After comparison with the input voltage, bit 6 remains set or it is reset. With each further step the voltage is successively approached, after the eighth step the approximation is smaller than U_{LSB} and the SAR register contains the binary value which is stored in the ADC register. Figure 20.6 illustrates this situation (see also Sect. 3.8.3).

The following code excerpt presents an example of a read function that enables the start of a new A/D conversion and the reading of the corresponding binary value via a parameterized input of the input configuration (ucinput_mode), the input channel (ucchannel) and the state of the analog output buffer (ucanalog_out). Additionally, the device chip address (*ucdevice_address*) is passed, as well as the address of the variable that is to store the converted value (**ucdigital_out*). The chronological sequence when calling this function is shown in Fig. 20.5b).

In the variables ucControlByte and ucDeviceAddress the parameters are put together to form the new configuration of the control register and the total address.

```
uint8_t ucDeviceAddress, ucControlByte;
ucControlByte = ucanalog_out | ucchannel | ucinput_mode;
ucDeviceAddress = (ucdevice_address << 1) | PCF8591_DEVICE_TYPE_ADDRESS;
ucDeviceAddress |= TWI_WRITE;
```

After the master has initiated the communication, he must check whether the bus is free, if not, the function is aborted with an error code.

```
TWI_Master_Start();
if((TWI_STATUS_REGISTER) != TWI_START) return TWI_ERROR;
```

If the bus is free, the master sends the address of the device with the R/W bit set to 0 in order to be able to change the content of the control register. If the slave confirms the reception of the address with ACK, the master also sends the new content of the control register. The configuration of the inputs and a new input channel can thus be changed with each call of the function.

```
TWI_Master_Transmit(ucDeviceAddress);
if((TWI_STATUS_REGISTER) != TWI_MT_SLA_ACK) return TWI_ERROR;
TWI_Master_Transmit(ucControlByte);
if((TWI_STATUS_REGISTER) != TWI_MT_DATA_ACK) return TWI_ERROR;
```

After a new start of the communication, the master sends the address of the device again, this time with the R/W bit set to 1, in order to be able to read out the contents of the ADC register. On the falling edge of the ACK bit, which the slave uses to acknowledge the reception of the address, a new A/D conversion is started. During the next eight clock pulses, the conversion takes place and the result of the previous conversion, which remained stored in the ADC register, is transferred. However, this byte is discarded by the function because it comes from a different input channel or input configuration. The master responds with ACK to receive the desired result and a new conversion is started.

```
TWI_Master_Start();
if((TWI_STATUS_REGISTER) != TWI_RESTART) return TWI_ERROR;
ucDeviceAddress = (ucdevice_address << 1) | PCF8591_DEVICE_TYPE_ADDRESS;
ucDeviceAddress |= TWI_READ; //read mode
TWI_Master_Transmit(ucDeviceAddress);
if((TWI_STATUS_REGISTER) != TWI_MR_SLA_ACK) return TWI_ERROR;
TWI_Master_Read_Ack();
if((TWI_STATUS_REGISTER) != TWI_MR_DATA_ACK) return TWI_ERROR;
```

The result is stored at the address of the variable ucanalog_in and thus it is available in the main file. The master is caused to respond with NACK and then to terminate the communication. The function reports an error-free output.

```
*ucanalog_in = TWI_Master_Read_NAck();
if((TWI_STATUS_REGISTER) != TWI_MR_DATA_NACK) return TWI_ERROR;
TWI_Master_Stop();
return TWI_OK;
```

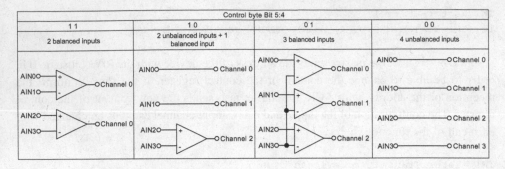

Fig. 20.7   PCF8591 – wiring of analog inputs

20.2.4 The Control Register

The control register controls the wiring of the analog inputs, the analog multiplexer and the analog buffer and has the following configuration:

- **Bit 7** – is reserved for further developments and should always be set to 0.
- **Bit 6** – switches the analog output to the active state with a logical "1"; a "0" switches this output off, thus saving current
- **Bit 5:4** – are control bits for the wiring of the analog inputs (see Fig. 20.7)
- **Bit 3** – like bit 7
- **Bit 2** – Auto increment bit; if bit 2 = 1, the analog channels are switched automatically one after the other with the transmission of each byte. The voltage converted first is that of channel 0. For the operation to work without errors, bit 6 must be set to 1.
- **Bit 1:0** – are control bits for the analog multiplexer: the bits encode the selected channel (for example 00 means channel 0 etc.)

20.2.5 The Oscillator

The device needs a clock for the A/D conversion. If pin EXT is connected to ground, this clock is generated by the internal oscillator and is externally accessible at pin OSC. If pin EXT is connected to V_{CC}, then an external oscillator must provide the required clock at pin OSC.

In the example circuit Fig. 20.8 all three external address pins are connected to ground, thus the total address of the device becomes 0x90. The internal oscillator is enabled via the EXT pin and the D/A converter output range is equal to V_{CC}. The reference voltage pin is connected to V_{CC} and the analog ground pin AGND is connected to digital ground.

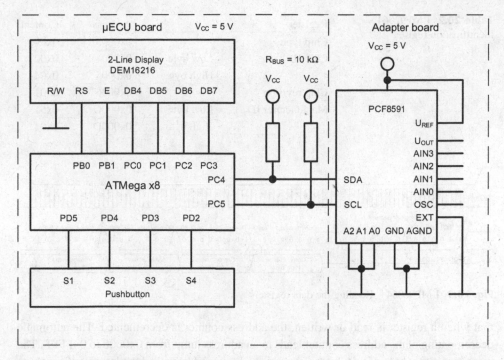

Fig. 20.8 Connecting a PCF8591 to a microcontroller

20.3   Current Measurement with the LMP92064

LMP92064 is an SPI-networkable device designed to measure a DC current and a DC voltage simultaneously [4]. The device has two 12-bit A/D converters that are used to measure the voltages across external resistors. With a conversion rate of 125 kS/s, fast changes in electrical quantities can be measured. The manufacturer specifies a bandwidth of 70 kHz for the current measurement and 100 kHz for the voltage measurement. Because the two quantities are sampled simultaneously, it is also possible to calculate the electrical power and, by integrating, the electrical energy [5]. The device converts the values in free-running mode, an internal logic prevents overwriting of an old data set during data transmission, which can reach up to 20 Mbit/s.

20.3.1   LMP92064 Structure

For controlling the A/D converters, configuring the serial interface, identifying the device in a network and storing the measured values, LM92064 has a register set whose 8-bit registers are addressable with a 16-bit address counter. A special feature of the device is

Table 20.2 LMP92064 – Identification register

Register		Address	Content
Chip type		0x0003	0x07
Chip ID	Low byte	0x0004	0x00
	High byte	0x0005	0x04
Chip revision		0x0006	0x01
Manufacturer ID	Low byte	0x000C	0x51
	High byte	0x000D	0x04

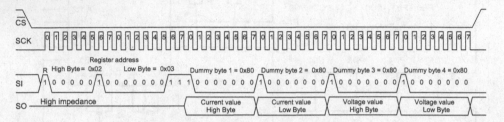

Fig. 20.9 LMP92064 – Reading the data registers

that when a register is read or written, the address counter is decremented. The automatic decrementing of the address counter leads to an address jump from 0x0000 to 0xFFFF. The more significant byte of a 16-bit register has the higher address, and the less significant byte has the lower address. The device can be reset by hardware by setting the RESET input high. The same can be achieved by software by setting bit 7 in configuration register A with address 0x0000 to 1. After power-up, this bit is cleared automatically. The other bits of the configuration registers can only be read. Therefore, they are not explicitly described here. There are several registers for identifying the device and testing the communication, which are listed in Table 20.2.

An internal reference voltage of 2.048 V for both measurement channels simplifies the wiring of the device. Two operational amplifiers with an input impedance of 100 GΩ ensure that the voltages to be measured are fed to the respective A/D converter unloaded. An impedance converter with a gain factor of 1 is used to measure the voltage. A positive voltage with respect to the ground terminal of the device up to the level of the reference voltage can be measured with a step size of $2048 \text{ mV}/2^{12} = 0.5 \text{ mV}$. The voltage measurement range is extended with a voltage divider up to U_{max} (see Fig. 20.9) which also leads to an increase of the step size by the factor $(R_1 + R_2)/R_2$:

$$U_{max} = U_{ref} \cdot \frac{R_1 + R_2}{R_2} \tag{20.8}$$

The voltage proportional to the current is accessed at an external shunt resistor and amplified by a factor of 25 (typical value) before it is digitized. This voltage can be a maximum of $2048 \text{ mV}/25 = 81.92 \text{ mV}$, which corresponds to a step size of $81.92 \text{ mV}/2^{12} = 20 \text{ μV}$. The digital value range of the A/D converter for current measurement is

Table 20.3 LMP92064 – measurement data register

Data register		Address
Voltage	Low byte	0x0200
	High byte	0x0201
Current	Low byte	0x0202
	High byte	0x0203

limited to 0 ... 3840 by the internal logic of the device, resulting in a maximum measurement voltage of 76.8 mV. Higher input voltages within the permissible voltage range result in the display of the digital value 3840 (or 0x0F00). The offset voltage of this amplifier is ±15 μV. The digitized values of the measuring voltages are stored in the 16-bit data registers whose addresses can be found in Table 20.3.

An incomplete reading of the data registers should prevent the overwriting of the old measured values. In the tested specimens it was found that the measured values are updated even if only the voltage values are read without the current values.

20.3.2 Serial Communication

LMP92064 is configured as a slave and has a four-wire SPI interface that enables bidirectional data transmission in mode 0 or 3 with the most significant bit of a byte first. The simple transmission protocol and a data rate of up to 20 Mbit/s enable the high conversion rate of the device to be exploited. In order to control several measuring units in a circuit with the same software module, it is useful to identify the individual devices with a definition array each. For the measuring sensor from Fig. 20.10 the following SPI data structure is declared in the main file (Chap. 14).

```
LMP92064_pins LMP92064_1 = {{/*CS_DDR*/ &DDRB,
                /*CS_PORT*/ &PORTB,
                /*CS_pin*/ PB0,
                /*CS_state*/ ON}}; //ON = 1
```

After the dedicated slave select pin of the SPI interface has been declared to output, the function is called:

```
SPI_Master_Init(SPI_INTERRUPT_DISABLE, SPI_MSB_FIRST,
           SPI_MODE_3, SPI_FOSC_DIV_4);
```

the microcontroller is set as master and configured for communication with the LMP92064. With an SPI message the master can address single registers or a register block. A message starts with setting the slave select line low followed by transmitting the address of the desired register, or the first register from the register block. By overwriting the most significant bit of the address to be sent, the operation to be performed can be coded:

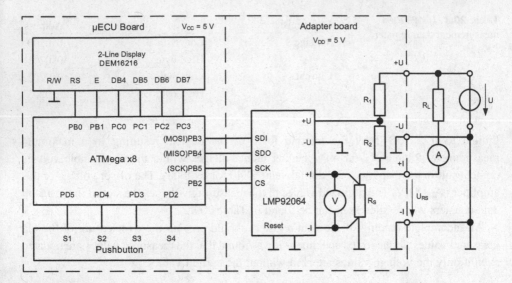

Fig. 20.10 LMP92064 – Wiring

0 means write, 1 means read. Figure 20.9 shows the time-dependent sequences for reading out the measured values. The highest address from the register range (0x0203) is transmitted. For each byte to be read out, the master transmits a dummy byte.

20.3.3 Measuring with the LMP92064

Thanks to the high input impedance of the measurement inputs, the low offset voltage of the operational amplifiers, the high resolution of the A/D converters and the high sampling rate, simultaneous, precise measurements of voltage and current in DC voltage networks can be realized. These measurements can be made for monitoring or for calculating power and energy. Figure 20.10 shows an example of the connection of an LMP92064 to a microcontroller, as well as the measurement and current calibration circuit.

20.3.3.1 Voltage Measurement
For the measurement of voltages greater than the reference voltage, a voltage divider is required, which can be dimensioned taking the following steps into account:

- a maximum current through the voltage divider I_D is selected which minimizes the losses and does not load the circuit to be measured
- starting from the maximum measurement voltage U_{max}, which can occur, the total resistance $R_1 + R_2$ is calculated such that:

$$I = \frac{U_{max}}{R_1 + R_2} \leq I_D \tag{20.9}$$

- From Eq. 20.8, the ratio R_1/R_2 is calculated. With the sum and the ratio of the resistors R_1 and R_2 are calculated. The accuracy of the resistors should be better than 1% to be able to realize precise voltage measurements.

The design of the following function is in accordance with these considerations, with which the best voltage resolution is achieved. The function called with the resistance values for R_1 (lr1) and R_2 (lr2) in Ω, reads the measurement register DATA_V_OUT_REG, calculates and returns the voltage value in mV. With a microcontroller clock frequency of 18.432 MHz, the calculation time of the voltage value is about 41 μs.

```
uint16_t LMP92064_Get_VoltageValue(LMP92064_pins sdevice_pins,
                                   long lr2, long lr1)
{
    unsigned long ulVoltage;
    ulVoltage = LMP92064_Read_WordReg(sdevice_pins, DATA_V_OUT_REG);
    //0.5mV = resolution of the A/D converter
    ulVoltage = (ulVoltage * (lr1 + lr2)) / 2;
    ulVoltage = ulVoltage / lr2; //lr1 + lr2 total resistance
    return ulVoltage;
}
```

If the ratio R_1/R_2 is chosen so that:

$$\frac{R_1}{R_2} = \begin{cases} (2 \cdot n) - 1 \ or \\ 2^n - 1 \end{cases} \ with \ n \in N \tag{20.10}$$

then the voltage value in mV can be calculated as follows:

```
ulVoltage = ulVoltage * n; //for R₁/R₂ = 2n-1
ulVoltage = ulVoltage << 2^n-1 ; //for R₁/R₂ = 2^n -1
```

which leads to a computation time below 2.5 μs.

20.3.3.2 Current Measurement

The current to be measured causes a voltage drop at the current measuring resistor, which is amplified and digitized as the basis for calculating the current intensity. In order to reduce the influence of the contact resistors R_{K1} and R_{K2} on the current measurement, the four-wire measuring method is used as in Fig. 20.11a. These contact resistances are created when the measurement resistor is soldered. The layout in the case of an SMD shunt used for current measurement using the four-wire method is shown in Fig. 20.11b.

A precise measurement is only guaranteed if the entire measuring chain is calibrated. In the circuit shown in Fig. 20.10, a 150 mΩ SMD shunt was installed. The voltage drops at the shunt are measured according to the current values in Table 20.4. Apart from the

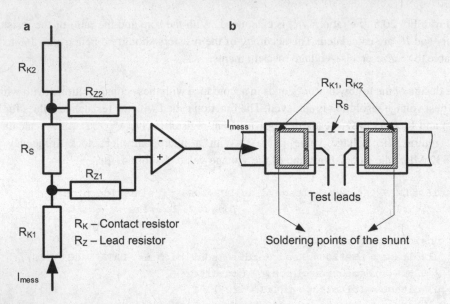

Fig. 20.11 Four-wire measuring method

Table 20.4 Measurement of the current measuring resistor

I_{mess}/mA	25.5	50.3	100	200	300	400
U_{RS}/mV	4	8	16	32.1	48	64.1

Table 20.5 Characteristic curve measuring current

I_{soll} [mA]	10	25	50	100	200	300	400	500
I_{ist} [mA]	9.5	24.5	49.5	99.3	199	298.6	398	498

measurement inaccuracies, this series of measurements results in a value of approx. 160 mΩ, which continues to be used for calculations.

According to a set current of 200 mA, the digital value 1504 is read from the data register of the device. With this value, a current step size of approx. 0.133 mA is calculated; the maximum current that can be measured with this shunt is:

$$I_{mess\ max} = 0.133\ mA \cdot 3840 \approx 510\ mA.$$

A current value is calculated from the digital value ulCurrent with the instruction:

```
ulCurrent = (ulCurrent * 133) / 1000; //the current step size is 0.133 mA
```

in approx. 35 µs is calculated. For comparison, Table 20.5 compares a series of set current values and the mean values of over 100 measurements.

References

1. Microchip Technology Inc. (2015). *MCP4801/4811/4821 1 Kanal SPI-D/A-Wandler*. www. microchip.com. Accessed 4 April 2021.
2. Microchip Technology Inc. (2015). *MCP4802/4812/4822 2 Kanal SPI-D/A-Wandler*. www. microchip.com. Accessed 4 April 2021.
3. NXP Semiconductors. (2013). *PCF8591 – 8-Bit A/D and D/A converter*. www.nxp.com. https:// www.nxp.com/docs/en/data-sheet/PCF8591.pdf. Accessed 3 April 2021.
4. Texas Instruments. (2014). *LMP92064 precision low-side, 125-kSps simultaneous sampling, current sensor and voltage monitor with SPI*. www.ti.com. Accessed 3 April 2021.
5. Parthier, R. (2008). *Messtechnik. Friedr.* Vieweg & Sohn.

Serial EEPROMs

Abstract

This chapter is dedicated to serial EEPROMs controlled by I^2C or SPI.

Read-only memories are non-volatile memories into which data is written at runtime and which remain persistently stored after the system has been switched off or reset. A distinction is also made here (Chap. 3) between EEPROMs and flash memories. Especially if you want to store larger amounts of data at runtime, for example in offline data loggers, read-only memories are needed as an extension of the limited memory space of a microcontroller. This chapter first describes typical EEPROM memories that are controlled with I^2C and SPI.

EEPROMs and flash memories usually use floating-gate field-effect transistors (or floating-gate FETs), as indicated in Fig. 21.1 [1]. A field-effect transistor is a voltage-controlled switching element. In an n-channel MOSFET, as sketched in the figure, two electrodes (drain and source) contact two heavily n-doped zones, which are regions where the silicon lattice (four outer electrons) is contaminated with impurity atoms that have five outer electrons. Here, therefore, there is an excess of charge carriers. The substrate of the semiconductor itself is weakly p-doped, so it has impurities with three outer electrons, thus an electron deficiency, usually called *hole excess*. A pn junction is formed at the boundary between the regions, which leads to a junction by recombination of the electrons with the holes. Therefore, no current can flow between the drain and source. This type of field effect transistor is called self-blocking or enhancement type (normal blocking).

If a positive voltage is now applied to another electrode (gate) insulated by a SiO_2 layer with respect to the substrate (bulk), which in turn can be connected to the source terminal,

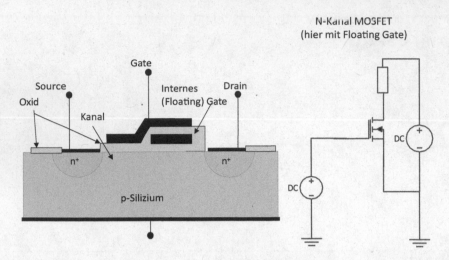

Fig. 21.1 MOSFET with floating gate

electrons are drawn from the voltage source into the substrate, which initially recombine with the holes in the substrate, but at a higher voltage[1] ($>V_{th}$) form a conductive channel enriched with electrons, which is formed between drain and source. From this moment, a current can flow. Since the gate-bulk path is itself a capacitance, no current flows into the gate, so the transistor is voltage controlled.

A floating-gate FET has an additional electrode between the gate and the substrate, which is hermetically insulated and embedded in the oxide layer. Due to the tunnel effect **(Fowler-Nordheim tunnel),** at a certain voltage some electrons succeed in overcoming the barrier of the insulator so that the insulated gate becomes negatively charged. They can no longer escape through the insulation and the FET now remains permanently blocked despite the positive gate. Only by removing the electrons with the help of UV light (EPROM) or a negative voltage (EEPROM) the floating gate is discharged again and the transistor switches normally again.

In other words:

- The gate electrode serves as a reading electrode: If a voltage $>V_{th}$ is present and the floating gate is uncharged, the transistor conducts and ground potential is present at drain (0 V = logical "0"). If the floating gate is charged, however, the FET blocks at V_{th} and battery voltage (=logical "1") is still present. Since EEPROMs are equipped with an inverting output driver, the signals are inverted, i.e. erasing means "1" at the output and writing "0".
- The gate electrode serves as a write electrode: If the voltage is increased significantly above the threshold voltage (typically 10 V), the floating gate is charged and the

---

[1] V_{th} stands for Threshold Voltage, typically 1.5–3.3 V.

memory cell is thus set to 1. To clear it again, the gate is connected to ground and the drain terminal is briefly connected to a high voltage (again >10 V). Then the electrons can tunnel back out of the floating gate via drain and the floating gate is neutral again. Technically, EEPROM cells are designed with two transistors [1], one of which switches the write voltage to the memory transistor.

Since complete tunneling takes a certain amount of time, writing such a cell takes a few milliseconds, which is significantly longer than reading.

Pure EEPROM cells also require a relatively large amount of space and energy. On the other hand, their long-term stability is usually high. Their writeability is typically specified with 10^6 write operations, the stability of the data with 10 years. For this reason, EEPROMs are generally used as low-capacity data memories (for example, for operating hours counters, error memories or for parameterizing control units).

21.1 Parallel Read-Only Memories

A parallel controllable read-only memory is characterized by a high data rate and simple control. The wiring of these memories is complex and the large number of pins makes them expensive and hardly applicable for microcontroller projects (see Fig. 21.2). In this example, an entire data byte (eight data bits) is transferred from the microcontroller to the memory or vice versa within one clock cycle. In addition to the eight data lines, ten address lines are required for addressing a memory cell and up to four control lines are required for driving the device. When doubling the memory capacity, the number of address lines increases by one.

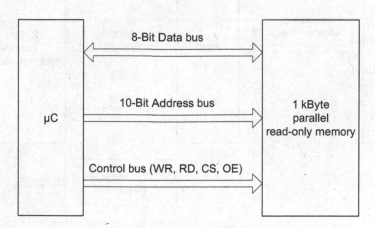

Fig. 21.2 External parallel memory connected to a microcontroller

21.2   Serial EEPROM Memory

The serial read-only memories are internally similar to the parallel memories, but externally the entire flow of information (data, addressing, and control) is through a serial interface, as shown in Fig. 21.3. The serial interfaces most commonly used for controlling serial memories are: SPI, I^2C, Microwire and UNI/O (1-wire interface from Microchip). The data exchange between microcontroller and a serial memory takes place bit by bit. For this reason, data rates comparable to parallel memories can only be achieved at much higher clock frequencies. The low number of pins required to drive such a memory results in a smaller size, lower power consumption and lower price. Therefore, they are preferred as external read-only memories for microcontroller applications.

21.2.1   M24C64: I^2C-Controlled EEPROM

As an example of I^2C-driven EEPROMs, the 24xx series is presented below and the M24C64-R device (see [2]) is described in more detail. This memory series is produced by most memory manufacturers with different memory capacities. The devices are controlled via an I^2C-Bus with clock frequencies of 100 kHz, 400 kHz or 1 MHz (depending on the manufacturer). A device of type 24xx00 has a memory capacity of 128 bits (16 bytes), while one of type 24xx102 can store 1 Mbit (128 kByte). From 1 kBit, the memories are organized in so-called pages (memory pages), which can be 8, 16 or 32 bytes

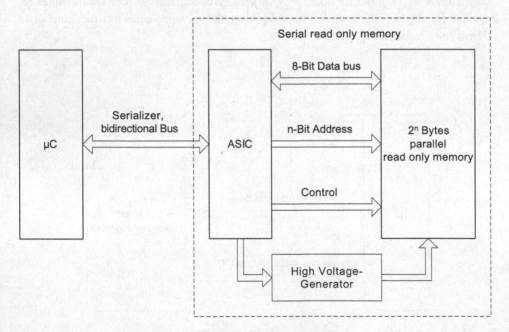

Fig. 21.3 External serial memory connected to a microcontroller

in size, depending on the size of the respective memory. A memory page is a contiguous memory area whose starting address is divisible by the size of the memory page. The devices in this family can be powered by voltages between 1.7 V and 5.5 V, and data retention is at least 100 years, according to the manufacturer.

The M24C64-R device has a capacity of 64 kBit (8 kByte) and has an internal 8-bit data bus. It requires an internal 13-bit address bus for addressing the individual memory cells (bytes) and has digital logic that controls the memory processes and implements the I^2C protocol. The supply voltage can vary between 2.5 V and 5.5 V in the temperature range from -40 °C to 85 °C, and between 1.7 V and 5.5 V for the -F and -DF versions. The M24C64-D version has an additional memory page (32 bytes) whose contents can be permanently write-protected.

Figure 21.4 shows a possible wiring of such memory devices; in this example, the EEPROMs provide additional non-volatile memory to the microcontroller to store settings, compensation values or extensive look-up tables (LUT). Because of the high number of erase/program cycles (>4 million), these memories can also be used in data loggers at relatively low sampling rates. In this example circuit, both the microcontroller and the memory devices are supplied with +5 V, so no level adjustment is needed for the I^2C-bus signals. The write control input is permanently switched low so that write protection is disabled.

21.2.1.1 I^2C Communication

The memory device is configured as a slave and waits in power saving mode for communication with a master. Communication can take place in standard mode (100 kHz), Fast-mode (400 kHz) or Fast–mode plus (1 MHz). The 7-bit device type address used to address the 24C64 series memories is 0x50. For control, the device has a pin for bidirectional data transfer (SDA), a clock input (SCL), a write-control input that can write-protect the entire memory area when switched high, and three device-chip address inputs (E0, E1, and E2). These three address inputs can be used to connect up to eight memories of this type to the same bus. The microcontroller distinguishes two or more memories connected to the same bus by their device chip address. While IC1 (Fig. 21.4) has all address inputs switched low and thus has the address 0x50, input E0 from IC2 is switched high and this results in address 0x51. The large clock frequency range and the address selection allow easy networking with other I^2C devices.

External initialization of the device is not provided. Before the master initiates communication with the slave, the device chip address inputs of the memory must be switched to the desired level if they are not hardwired.

The microcontroller starts the communication with an I^2C-START sequence, followed by the addressing phase. In this phase, the address of the device is transmitted. If the received address matches its own, the slave responds with ACK on the ninth clock pulse, otherwise the device switches to standby mode and only becomes active again after another START sequence. The addressing phase is followed by the communication phase, in which data is transmitted in one direction. The master ends the communication with an I^2C-STOP sequence, releases the bus again and the slave switches to power saving mode.

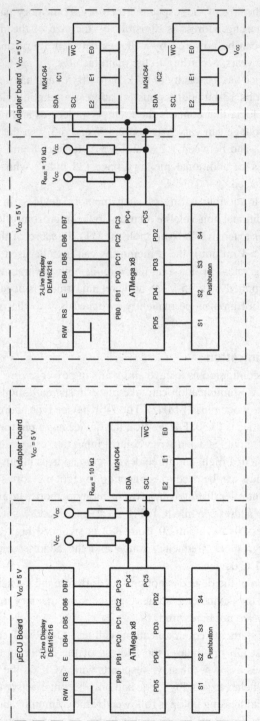

Fig. 21.4 Memory expansion of a microcontroller with serial EEPROMs

21.2.1.2 Reading

The internal address counter is reset after power-on. Before a read/write operation, it is set to the desired start address and incremented with each byte read/stored. The memory organization allows direct (random) or sequential access to the memory cells. The read functions can be executed even if the write control input is switched high. The following read functions are implemented:

21.2.1.2.1   Sequential Reading from the Current Address

To read one or more bytes from the current address, the following steps are performed:

- Step 1: the master initiates I^2C communication with a START sequence;
- Step 2: When the bus is free, the master sends the slave address with the R/W bit set to 1;
- Step 3: the slave confirms the receipt of its own address with ACK;
- Step 4: the slave sends the currently addressed byte with the next eight clocks and increments the address counter;
- Step 5: if the master confirms receipt of the byte with ACK, the slave will send another byte and step 4 is executed. If the master does not request any further bytes, it responds with NACK and step 6 is executed.
- Step 6: the master terminates the communication with a STOP sequence.

21.2.1.2.2   Sequential Reading with Direct Access

One or more bytes can also be read from a freely selected address. To do this, the master must start communication and, if the bus is free, send the slave address with the R/W bit set to "Write" (0). When the slave confirms receipt of the address with ACK, the master sends the desired 16-bit address, with the most significant byte first. The two bytes acknowledged by the slave with ACK are stored in the address counter and the read operation can begin. With a START sequence a restart of the communication takes place. From now on, reading with steps 2 to 6 from "Read from current address".

Sequential reading allows any number of bytes to be read. When the address counter reaches the end of memory, it jumps to the first address after incrementing.

21.2.1.3 Write

The write control input can be used to protect the memory against unintentional writing when it is switched high. If this input is not hardwired, it must be switched low before a write operation. Up to 32 bytes can be stored in one write cycle, provided they are all on the same memory page. The start address can be freely selected. Saving is performed in several steps:

- Step 1: the master initiates communication with the slave with a START sequence;
- Step 2: when the bus is free, the master sends the slave address with the R/W bit set low;
- Step 3: if an internal storage process is still running, the slave responds with NACK; otherwise it acknowledges receipt of the address with ACK;

- Step 4: the master sends the high-order address byte followed by the low-order address byte; the slave acknowledges the two bytes with ACK and stores them in the address counter;
- Step 5: the master sends the byte to be stored;
- Step 6: if the write control input is switched to high, the slave answers with NACK, otherwise with ACK and increments the address counter of the memory page; if the end of the memory page has been reached, this address counter jumps to the first address of the page;
- Step 7: if another byte is to be stored, then continue with step 5 otherwise with step 8;
- Step 8: the master ends the process with a STOP sequence.

The received bytes are stored in a 32-byte volatile memory and from there, after the I^2C-STOP sequence, in the EEPROM main memory. During the internal time-controlled write process, which takes up to 5 ms regardless of the number of bytes to be stored, the device switches off from the bus and an addressing attempt is acknowledged with NACK.

A flowchart of the above storage process is shown in Fig. 21.5 and the function code is listed below.

The function shall transfer a number ucbyte_number of bytes from a RAM array whose address is ucbyte_array to the 24C64 memory device identified by the device chip address ucdevice_address and store it starting with the uibyte_address address. To store an entire page, the starting address of the page must be calculated and transferred as uibyte_address. The function returns:

- with error-free execution: TWI_OK (= 0x00);
- in case of faulty data transmission: _24C64_WRITE_PROTECTED (= 0x80);
- in case of faulty transmission of the address or the start address of the memory area: TWI_ERROR (= 0x01).

It must be reacted according to the return value of the function.

```
uint8_t _24C64_Write_Page(uint8_t ucdevice_address,
                          uint16_t uibyte_address,
                          volatile uint8_t* ucbyte_array,
                          uint8_t ucbyte_number)
{
    uint8_t ucDeviceAddress, ucAddressByteHigh, ucAddressByteLow, ucI;
    //the byte address is decomposed in its high and low byte
    ucAddressByteLow = uibyte_address;
    ucAddressByteHigh = uibyte_address >> 8;
    //the address of the 24C64 memory device is formed
    ucDeviceAddress = (ucdevice_address << 1) | _24C64_ADDRESS;
    ucDeviceAddress |= TWI_WRITE; //write mode
    TWI_Master_Start(); //start
```

```
if((TWI_STATUS_REGISTER) != TWI_START) return TWI_ERROR;
//send device address
TWI_Master_Transmit(ucDeviceAddress);
if((TWI_STATUS_REGISTER) != TWI_MT_SLA_ACK) return TWI_ERROR;
//the higher byte of the byte address is sent
TWI_Master_Transmit(ucAddressByteHigh);
if((TWI_STATUS_REGISTER) != TWI_MT_DATA_ACK) return TWI_ERROR;
//the least significant byte of the byte address is sent
TWI_Master_Transmit(ucAddressByteLow);
if((TWI_STATUS_REGISTER) != TWI_MT_DATA_ACK) return TWI_ERROR;
for(ucI = 0; ucI < ucbyte_number; ucI++)
{
    //a data byte is sent
    TWI_Master_Transmit(ucbyte_array[ucI]);
    if(TWI_STATUS_REGISTER == TWI_MT_DATA_NACK)
        return _24C64_WRITE_PROTECTED;
}
TWI_Master_Stop(); //stop
return TWI_OK;
}
```

The following function call writes 32 bytes from the array ucBuffer[32] on the 32nd memory page (byte 1024:1055) of IC1 (see Fig. 21.4):

```
_24C64_Write_Page(0x00, 1024, ucBuffer, 32);
```

21.2.1.4 Testability

The electrical connection between the microcontroller and the memory chip is already tested in the addressing phase. If the device does not respond with acknowledge when receiving its address, then either the address is wrong, or the electrical connection is not OK, or it is currently writing data in EEPROM. The result of a write operation can be tested by reading the memory and comparing the sent data with the read data.

A write cycle lasts a maximum of 5 ms. During this time the device is not addressable. To avoid a blocking wait after a write cycle, the next write attempt must be delayed by the maximum memory time or the memory progress must be checked by polling. A flowchart of a function that checks the memory progress is shown in Fig. 21.6 and the code is listed below.

To check whether the last write process has been completed, the master initiates the I^{2}C communication with a START sequence and sends the address of the memory to be tested. If the address is acknowledged with ACK, another write process can be started, otherwise the master releases the bus again with a START-STOP sequence.

Fig. 21.5 Flow diagram of the
PageWrite function for the
M24C64 memory chip

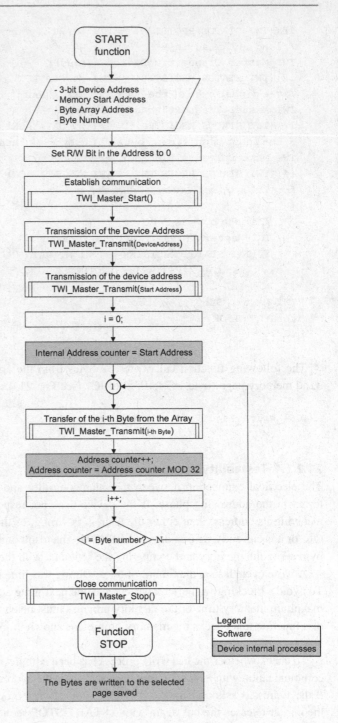

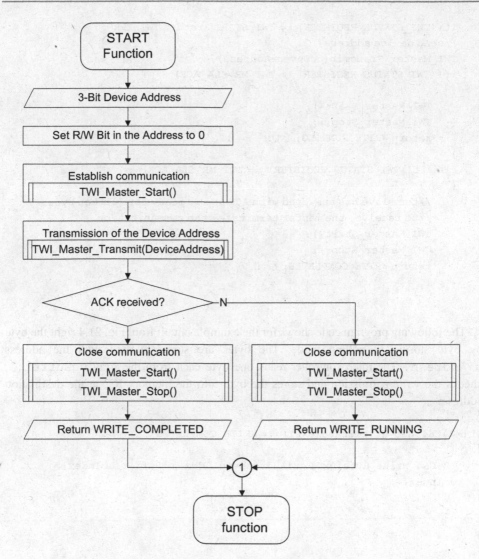

Fig. 21.6 Flow diagram – Write state query for the M24C64 memory device

```
uint8_t M24C64_Get_WriteState(uint8_t ucdevice_address)
{
    uint8_t ucDeviceAddress;
    //form the address of the 24C64 memory device
    ucDeviceAddress = (ucdevice_address << 1) | _24C64_ADDRESS;
    ucDeviceAddress |= TWI_WRITE; //write mode
    //start
    TWI_Master_Start();
```

```
    if((TWI_STATUS_REGISTER) != TWI_START) return TWI_ERROR;
    //send device address
    TWI_Master_Transmit(ucDeviceAddress);
    if((TWI_STATUS_REGISTER) != TWI_MR_SLA_ACK)
    {
        TWI_Master_Start();
        TWI_Master_Stop();
        return WRITE_RUNNING;
    }
    else if((TWI_STATUS_REGISTER) == TWI_MR_SLA_ACK)
    {
        //the slave has answered with ACK and is ready for further bytes
        //to receive, the master terminates the communication
        TWI_Master_Start();
        TWI_Master_Stop();
        return WRITE_COMPLETED;
    }
}
```

The following program code shows for the example circuit from Fig. 21.4 right the byte by byte storing in polling mode. The bytes are stored starting with the address uiAddress. The microcontroller reads one byte each from the array ucBuffer[], checks the write permission and stores the byte into the memory IC2 at the destination address:

```
if(M24C64_Get_WriteState(0x01) == WRITE_COMPLETED)
{
    _24C64_Write_Byte(0x01, uiAddress+ucIndex, ucBuffer[ucIndex])
    ucIndex++;
}
```

21.2.2 25LC256: SPI Controlled EEPROMs

SPI-controlled EEPROMs are offered by most manufacturers under the 25xx (or 95xx [3]) series with memory capacities from 1 kBit (128 bytes) to 1 Mbit (128 kBytes). The block diagram of such a memory can be seen in Fig. 21.7. As an example, the memory device 25LC256 [4] is presented below as the successor to the AT25256 [5]. The device, which has a memory capacity of 265 kBit (32 kByte), is organized in 64-byte pages and can be supplied with voltages between 2.5 V and 5.5 V (in some versions, such as 25AA256, from 1.8 V). Serial communication is via an SPI bus, which is driven in mode 0 or 3. The bus can be clocked at up to 10 MHz (or 20 MHz [3]). The most significant bit of a byte is

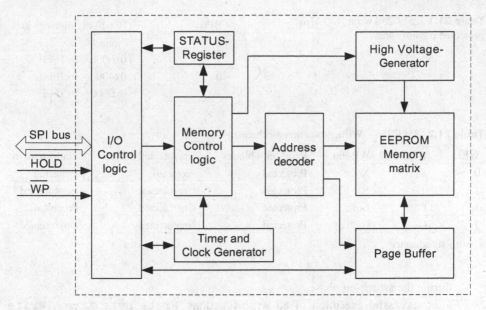

Fig. 21.7 Block diagram of the 25LC256 SPI memory

transmitted first. The storage process of a byte or an entire page takes a maximum of 5 ms, and the data retention is 100–200 years. Endurance is 1,000,000 erase/write cycles (up to 4,000,000 for [3]).

21.2.2.1 Write Protection of the Memory

Via the non-volatile status register of the memory you can:

- change the write protection of different memory areas,
- read out the write protection configuration,
- determine whether a write operation is in progress.

The configuration of the status register is as follows:

- **Bit 7** – WPEN (Write Protect Enable) – together with the Write Protect input is used for write protection of the status register (see Table 21.2);
- **Bit 3:2** – BP1:BP0 (Block Protection) – these 2 bits indicate which memory area is currently write protected (see Table 21.1);
- **Bit 1** – WEL (Write Enable Latch) – as long as this bit is Low, writing is disabled; writing is enabled when the Write Latch Enable function is called. This bit is always set to low:

Table 21.1 25LC256 write protected memory areas

BP1	BP0	Read-only memory area
0	0	None
0	1	0x6000 – 0x7FFF
1	0	0x4000 – 0x7FFF
1	1	0x0000 – 0x7FFF

Table 21.2 25LC256 – Write protection mechanisms

WEL	WPEN	WP-Pin	Read-only blocks	Unprotected blocks	Status register
0	X	X	Protected	Protected	Protected
1	0	X	Protected	Unprotected	Unprotected
1	1	Low	Protected	Unprotected	Protected
1	1	High	Protected	Unprotected	Unprotected

x = no significance

- during the switch-on phase,
- after successful execution of all write functions: Write Byte Array, Write Latch Disable or Write Status Register. This bit must be set high by software before each write operation to enable writing.
- **Bit 0** – WIP (Write-In-Proces) – this bit is set high by the internal logic as long as an internal memory cycle is active;

Bits 4, 5 and 6 are not assigned.

The internal logic of the device together with the write-protect pin implement different write-protect mechanisms against unintended modification of the memory contents, as shown in Table 21.2.

If the WEL bit in the Status Register is set low, or the WEL and WPEN bits are set high and the Write Protect pin is set low, then the entire Status Register is write-protected and no further changes can be made. While in the first case the write protection can be removed by software by calling the Write Latch Enable function, in the second case this is only possible by switching the WP pin high.

21.2.2.2 Read-Write Functions

The internal logic of the device implements the following functions (see Fig. 21.8):

- **Read Byte Array:** Read any number of bytes starting from a random address; after reading a byte, the internal address counter is incremented. When the last byte has been read (address 0x7FFF), the address counter jumps to the first address (0x0000).
- **Write Byte Array:** Writes a byte array starting from a random address; the number of bytes is limited to the page size. After each stored byte, the internal address counter is

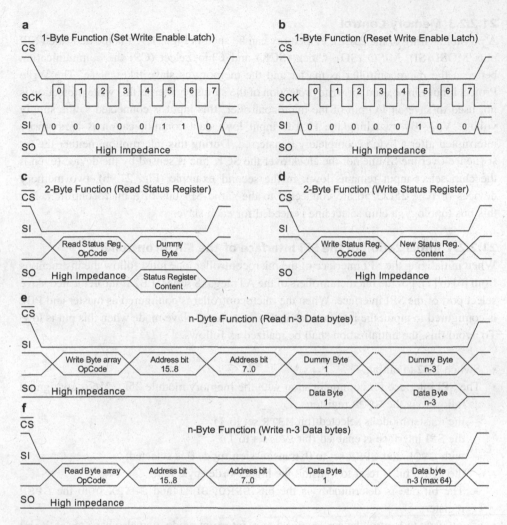

Fig. 21.8 Overview of the functions of the 25LC256

incremented. When the top address of the page is reached, the remaining bytes are stored starting from the first address of the page.

- **Write Latch Enable:** the Write Enable latch is set to high; this enables writing (see Table 21.2).
- **Write Latch Disable:** Resets the write enable latch; writing is disabled as a result.
- **Read Status Register:** The status register is read.
- **Write Status Register:** The content of the status register is changed (not the WIP and WEL bits).

21.2.2.3 Memory Control

A possible wiring of the 25LC256 memory can be seen in Fig. 21.9. Via the dedicated SPI pins MOSI (SI), MISO (SO), Clock (SCK) and Chip Select (CS) the communication between the microcontroller as master and the memory as slave takes place. The Write Protect input can be used for write protection of the status register. If this write protection is not used to save an I/O pin of the microcontroller, this input is connected to the supply voltage V_{CC}. By switching the HOLD input low, data communication is temporarily interrupted after a byte is completely transferred. During this interruption, neither the bit sequence over the SI line nor the clock over the SCK line is sensed by the device (even if the chip select input remains low). In the second example (Fig. 21.9b), two memory devices of type 25LC256 are connected to the same SPI bus of a microcontroller. For this bus topology, a chip select line is needed for each slave.

21.2.2.4 Initialization of the SPI Interface of the Microcontroller

When initializing the SPI interface of the microcontroller, one must follow the instructions from [6] to [7]. For the microcontrollers of the ATmegax8 series, PB2 is the dedicated slave select port of the SPI interface. When the microcontroller is configured as master and PB2 is configured to input, the interface internally switches to slave mode when this pin is low. To avoid this, the initialization shall be realized as follows:

- Switch PB2 to output.
- The SPI interface for communication with the memory module 25LC256 is configured as follows via the SPCR register:
  - the master mode is selected (bit MSTR set to 1);
  - the SPI interface is enabled (bit SPE set to 1);
  - with CPOL and CPHA set to 0, transmission mode 0 is selected;
  - the DORD bit is reset to transmit the higher-order byte first
  - The bit rate is determined via the bits SPR0, SPR1 and SPI2X from the SPSR register;
- if you want to control the communication in interrupt mode, you also have to set the bit SPIE in register SPCR to 1 before clearing the interrupt flag (readout of register SPCR followed by readout of register SPDR).

The design of a program according to the polling method is simpler, but leads to a blocking wait of the microcontroller. For the control of the devices according to the interrupt method, state machines must be implemented for all functions. The program for controlling the memory devices shown in Fig. 21.9 could be implemented with a mixture of the two methods. Depending on the concrete application, one determines which functions are called according to which procedure.

A flowchart for the design of the endless loop according to both methods can be seen in Fig. 21.10. The management of the interrupt functions is implemented using a state machine in the interrupt service routine of the SPI interface.

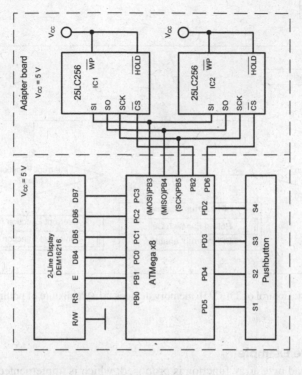

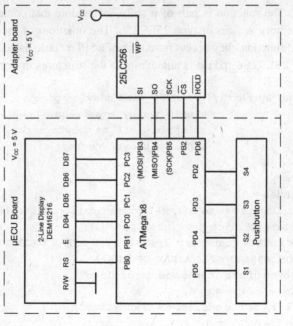

Fig. 21.9 Wiring of a 25LC256 memory

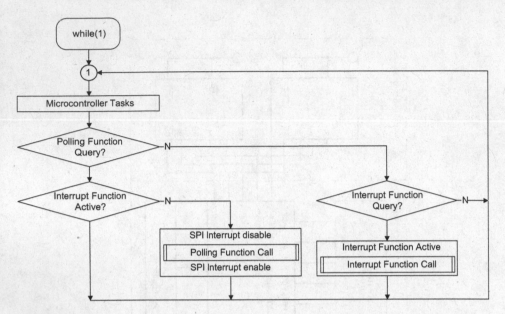

Fig. 21.10 Flowchart: Control of 25LC256 memory devices using a mixture of polling and interrupt methods

21.2.2.5 Software Example

As an example, a read byte array function is proposed, which is implemented according to the polling method. The function is part of a software module that contains the control functions for the memory devices of type 25LC256. The functions of the module can be called from the *main* function, they access functions of a SPI module listed in Chap. 8. The SPI data structure _25LC256_pins is analogous to the structures in Sect. 14.5.

```
void _25LC256_Read_ByteArray(_25LC256_pins sdevice_pins,
                             uint16_t uibyte_address,
                             volatile uint8_t* ucbyte_array,
                             uint16_t uibyte_number)
{
    uint8_t ucDummy;
    uint16_t uiIndex = 0;
    //chip select line is set to low, SPI communication is started
    SPI_Master_Start(sdevice_pins._25LC256spi);
    //the command code is transferred: READ_BYTE_ARRAY_OPCODE = 0x03
    SPI_Master_Write(READ_BYTE_ARRAY_OPCODE);
    //the higher byte of the address is determined
    ucDummy = uibyte_address >> 8;
    //the higher byte of the address is transmitted
    SPI_Master_Write(ucDummy);
```

```
//the least significant byte of the address is determined
ucDummy = uibyte_address;
//the least significant byte of the address is transmitted
SPI_Master_Write(ucDummy);
//uibyte_number Bytes from the addressed memory
// are transferred to the target buffer ucbyte_array
for(uiIndex = 0; uiIndex < uibyte_number; uiIndex++)
{
    ucbyte_array[uiIndex] = SPI_Master_Write(ucDummy);
}
//SPI communication is terminated, chip select line is set to high
SPI_Master_Stop(sdevice_pins._25LC256spi);
}
```

The design of the module allows its unchanged use (porting) with different connection configurations of the devices or with the control with different microcontrollers. For this to work correctly, the data structure _25LC256_pins (Sect. 14.5) must be adapted with the pin configuration in the *main* file. The data structures of the memory devices shown in Fig. 21.9 are explained below:

```
//this structure is defined in the header file of the module
typedef struct
{
    tspiHandle _25LC256spi;
    volatile uint8_t* WP_DDR;
    volatile uint8_t* WP_PORT;
    uint8_t WP_pin;
    uint8_t WP_state;
    volatile uint8_t* HOLD_DDR;
    volatile uint8_t* HOLD_PORT;
    uint8_t HOLD_pin;
    uint8_t HOLD_state;
} _25LC256_pins;
#define ON 1
#define OFF 0
//for IC1 on the right in the figure, this data structure is as follows:
_25LC256_pins _25LC256_1 = {{&DDRB, &PORTB, PB2, ON},
                            OFF, OFF, OFF,
                            OFF, OFF, OFF};
//and for the IC2 on the right side of the figure:
_25LC256_pins _25LC256_2 = {{&DDRD, &PORTD, PD6, ON},
                            OFF, OFF, OFF,
                            OFF, OFF, OFF};
```

References

1. Stiny, L. (2018). *Aktive elektronische Bauelemente – Aufbau, Struktur, Wirkungsweise, Eigenschaften und praktischer Einsatzdiskreter und integrierter Halbleiter-Bauteile* (4. Aufl.). Springer Vieweg Wiesbaden.
2. STMicroelectronics. (2015). *M24C24R – 64 kBit I²C-Bus serieller EEPROM*. www.st.com
3. STMicroelectronics. (2015). *M95256-xx SPI – serielle EEPROMs*. www.st.com
4. Microchip Technology Inc. (2015). *25LC256 – 256 kBit SPI-Bus serieller EEPROM*. www. microchip.com
5. Microchip Technology Inc. (2015). *AT25256 – 256 kBit serieller SPI-BUS EEPROM*. www. microchip.com
6. Microchip Technology Inc. *ATmega88 – 8 bit microcontroller*. www.microchip.com
7. Microchip Technology Inc. (2015). *AVR107 – Interfacing AVR serial memories*. www. microchip.com

Serial Flash Memory

22

Abstract

This chapter describes two typical flash memories with serial communication.

In contrast to the EEPROM devices described in Chap. 21, the flash devices presented here are characterized by high storage density and fast write operations. They are therefore used to store larger amounts of data. NOR flash memories, which are mainly used as program memory in the microcontroller, are usually controlled in parallel via an address bus and are therefore suitable for XIP (Execute-In-Place). They are the memory type of choice for program code, as they can be erased block by block due to their design, which speeds up the programming process. NAND flashes are cheaper and smaller per bit and are suitable for high memory densities, such as those found in memory cards. The serial flash devices presented here are not suitable as program memory due to data access via an SPI interface. Nevertheless, larger amounts of data occasionally occur, for example in data loggers or in memories for characteristic diagrams. The authors have also used them to store sampled voice messages.

EEPROMs are usually constructed in such a way that each individual cell can be erased or written. To speed up the process of erasing and to save space, larger groups of cells are placed together, as indicated in Fig. 22.1. The figure shows a NOR flash in NMOS technology with five memory cells of one bit each, where *flash* stands for the abrupt erasure of all blocks. Readout occurs when the read voltage V_{th} is applied to one of the word lines, and the data is then available on the bit line. The 0–1 transition (write) is performed by applying the write voltage to the word line to be written using hot-carrier

A. Meroth, P. Sora, *Sensor networks in theory and practice*,
https://doi.org/10.1007/978-3-658-39576-6_22

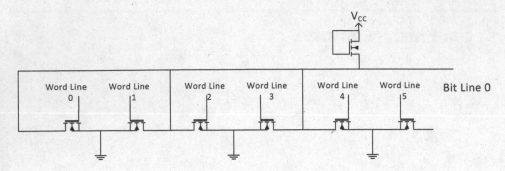

Fig. 22.1 NOR flash

injection. In this process, the charge carriers are accelerated to such an extent that their energy overcomes the insulation barrier. Erasing is done by applying the write voltage to the bit line while all word lines are grounded. Chains constructed in this manner can span several kilobytes. The term NOR is derived from the behavior at the word lines: The parallel connection of all open-drain transistors behaves like a logic NOR gate: if one of the transistors is connected through, the output is at logic "0"; if no transistor is connected through, it is at logic "1". If you want to increase the word width to, for example, eight bits, eight cells each are connected in parallel at the word lines, so that eight parallel bit lines output the entire word.

NOR Flash memory technology is characterized by very short read times and slow write times and is suitable where fast reading is required, for example in XIP (execution in place), i.e. in program memories of microcontrollers.

Flash memories with high storage density are constructed using NAND flash technology. In this process, self-locking transistors are not used as memory cells, as described in Chap. 21, but so-called depletion types, which are initially self-conducting. They have a continuously doped N-channel between drain and source. When a positive voltage is applied to the gate, this channel is interrupted by the charge carriers being displaced from the channel into the p-zone, and the transistor is turned off.

In NAND Flash, the memory cells are connected in series, as shown in Fig. 22.2. The series can be connected to ground and to the supply voltage by two self-blocking FETs, which are driven in push-pull. Thus, either transistor VL or transistor /VL conducts. When reading, first all gates of the memory cells are connected to ground (driven with "0"). This makes them all conductive. The gate of transistor VL is high, so the voltage on the bit line is also high, since transistor /VL is blocking at the same time. The capacitor is charging. To read, VL is now disabled, /VL is opened, and the cell transistor to be read is driven via the word line. If it is not charged, it blocks and a "1" is at the output. However, if it is charged, it continues to conduct and the voltage on the bit line goes to 0, see [1]. NAND memory cells save lots of connections between each other, so they are much smaller than NOR

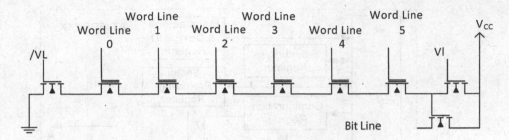

Fig. 22.2 NAND flash according to [1]

memory cells. However, due to the series connection, the charge carrier flow takes much more time to build up, so the read speed is smaller than that of NOR cells.

NAND flash devices are used wherever write speed and high storage density are more important than read speed, for example in SD cards or memory sticks. The memory density can be further increased by so-called multi-level cells (MLC), here the transistors are not merely switched on and off, but a certain voltage level is applied via the charge quantity stored on the floating gate, so that the output voltage can assume one of n levels. Thus log_2 n bits can be stored per cell.

Due to the technology, NAND flash memories in particular have many cells that are already defective during manufacture. Therefore, in addition to the data, a CRC checksum is also stored, which is suitable for forward error correction (FEC, see Chap. 6). An elaborate logic and a control mechanism for reading, writing, erasing and error correction, as well as usually a RAM buffer, make these devices relatively complex. Two essential parameters are used for characterization:

- The retention time is the time after which the data can still be read safely. It is about 20 years for NOR flash and about half that for NAND flash [2].
- Endurance indicates the number of write and erase cycles. This is now around 100,000 to one million cycles and is generally somewhat higher for NAND flash.

22.1   AT45DB161 Serial Flash Memory

The company Adesto Technologies manufactures flash memories with a memory capacity from 2 Mbit to 64 Mbit. These memory chips are controlled via a serial SPI bus at clock frequencies of up to 104 MHz and can be used in applications with low supply voltage, where space must be saved and attention must be paid to energy consumption. They have a word width of 8 bits (1 byte) and are supplied with voltages from 2.3 V to 3.6 V. From this voltage, a so-called "charge pump" is used to generate the high voltage with which the memory can be programmed or erased. As an example of this series of memories, we will take a closer look at an AT45DB161 ([3]), whose block diagram can be seen in Fig. 22.3.

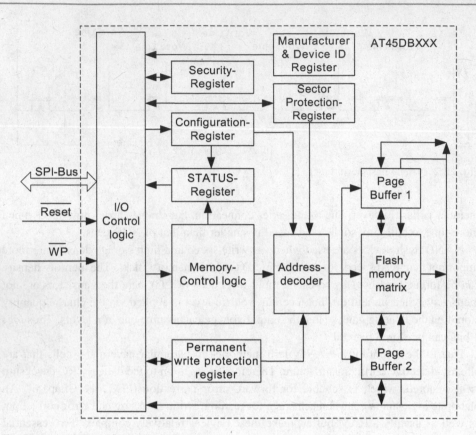

Fig. 22.3 AT45DBXXX – Block diagram

The memory is organized by the factory into 4096 memory pages with a size of 528 bytes and has two 528-byte SRAM caches. The page and buffer size can be changed to 512 ($=2^9$) bytes via the configuration register. There are different read functions of the main and intermediate memory at low (50 MHz) and high (85 ... 104 MHz) clock frequencies, respectively. In the following only the functions at low clock frequencies for a page size of 512 bytes are mentioned.

22.1.1 SPI Communication

The microcontroller must be configured as master, the bytes are transferred in mode 0 or 3 with the most significant bit first. Because the master is supplied with 5 V and the memory with 3.3 V, the level of the control signals must be adjusted. In this example, the level adjustment is implemented with voltage dividers, which allows reliable transmission at a bus frequency of about 4.5 MHz. No level adjustment is required for the data input from the microcontroller.

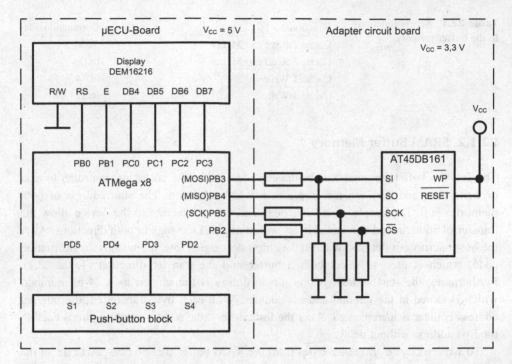

Fig. 22.4 AT45DB161 – Circuitry

The SPI data structure of the AT45DB161_pins device codes the Chip-Select, Write-Protect and Reset control connections. It is composed analogously to the data structure from the previous software example for the 25LC256 and is defined as follows for the circuitry in Fig. 22.4 (Sect. 14.5):

```
AT45DB161_pins AT45DB161_1 ={/*CS_DDR*/ &DDRB,
                             /*CS_PORT*/ &PORTB,
                             /*CS_pin*/ PB2,
                             /*CS_state*/ ON},
                             /*WP_DDR*/ OFF,
                             /*WP_PORT*/ OFF,
                             /*WP_pin*/ OFF,
                             /*WP_state*/ OFF,
                             /*RESET_DDR*/ OFF,
                             /*RESET_PORT*/ OFF,
                             /*RESET_pin*/ OFF,
                             /*RESET_state*/ OFF};
```

Table 22.1 Access functions
to the buffer memory

Function	Command code
Cache 1 Read (50 MHz)	0xD1
Cache 2 Read (50 MHz)	0xD3
Cache 1 Write	0x84
Cache 2 Write	0x87

22.1.2 SRAM Buffer Memory

The device's buffer memories can be directly accessed by the host microcontroller to read or write data, and they can also serve as a RAM extension. The start address of both memories is 0. The simple access functions and the architecture of the device allow the transfer of an unlimited number of bytes within one SPI message in both directions. After the host microcontroller has switched the chip select signal low, it transmits the instruction code, which is used to select both a buffer and the transfer direction (Table 22.1). Furthermore, the start address of the memory area is transmitted as a 24-bit number, which is stored in the internal address counter. With each byte transferred, the internal address counter is incremented. After the last address, the address counter jumps back to the first address without delay.

To write `uibyte_number` bytes from the RAM buffer `ucsource_buffer` of the microcontroller into a SRAM buffer starting at the address `uibuffer_address`, the following function can be used:

```
void AT45DB161_Write_Buffer(AT45DB161_pins sdevice_pins,
                            uint8_t ucop_code,
                            uint16_t uibuffer_address,
                            uint16_t uibyte_number,
                            uint8_t* ucsource_buffer)
{
    uint8_t ucAddressHighByte = 0x00, ucAddressMiddleByte;
    uint8_t ucAddressLowByte;
    //the 24-bit address is divided into single bytes
    ucAddressLowByte = uibuffer_address;
    uibuffer_address = uibuffer_address >> 8;
    ucAddressMiddleByte = uibuffer_address;
    //the communication is started
    SPI_Master_Start(sdevice_pins.AT45DB161spi);
    SPI_Master_Write(ucop_code); //the command code is transmitted
    //the 24-bit address is transmitted
    SPI_Master_Write(ucAddressHighByte);
    SPI_Master_Write(ucAddressMiddleByte);
    SPI_Master_Write(ucAddressLowByte);
    //uibyte_number Bytes are transferred
```

```
    for(uiI = 0; uiI < uibyte_number; uiI++)
    {
        SPI_Master_Write(ucsource_buffer[uiI]);
    }
    SPI_Master_Stop(sdevice_pins.AT45DB161spi);
}
```

With the call:

```
AT45DB161_Write_Buffer(AT45DB161_1, 0x84, 0, 10, ucBuffer);
```

for the example circuit shown in Fig. 22.4, 10 bytes would be stored from the variable ucBuffer [] into the first SRAM buffer starting at address 0.

22.1.3 Flash Main Memory

The AT45DB161 device is a 16-Mbit (2-Mbyte) memory with 20-year data retention and 100,000 write/erase cycles for each page. The complex memory architecture of the device is shown in Fig. 22.5. Eight pages form a block, and 32 blocks form a sector. There are 17 sectors in total because sector 0 is divided into 0a and 0b. When using a page size of 528 bytes, 16 additional bytes/page are available, but the calculation of the logical address for some functions becomes more complex. Switching to the binary $2^9 = 512$ bytes size is done by transmitting the command code *0x3D 2A 80 A6*. Restoring the factory setting (528 bytes/page) is realized with the transmission of the code *0x3D 2A 80 A7*.

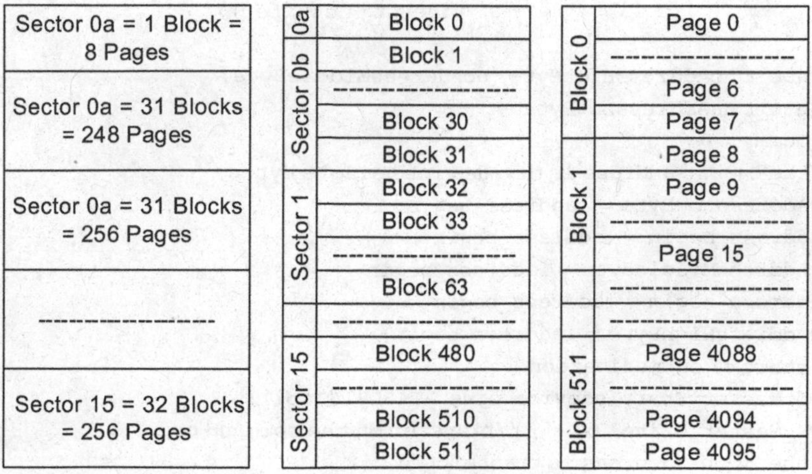

Sector 0a = 1 Block = 8 Pages	Sector 0b	0a	Block 0	Block 0	Page 0
			Block 1		-----------------
		0b	-----------------		Page 6
Sector 0a = 31 Blocks = 248 Pages			Block 30		Page 7
			Block 31		Page 8
	Sector 1		Block 32	Block 1	Page 9
Sector 0a = 31 Blocks = 256 Pages			Block 33		-----------------
			-----------------		Page 15
			Block 63		-----------------
-----------------			-----------------		-----------------
	Sector 15		Block 480	Block 511	Page 4088
Sector 15 = 32 Blocks = 256 Pages			-----------------		-----------------
			Block 510		Page 4094
			Block 511		Page 4095

Fig. 22.5 AT45DB161 – Memory architecture

22.1.4 Reading

The microcontroller as master can read data from the memory area directly (without buffering) or indirectly (with buffering).

22.1.4.1 Reading from the Entire Memory Area

The master starts the communication with the slave by switching the chip select signal from high to low. By transmitting the command code 0x03 followed by a valid 24-bit address from the address range of the device, the internal address counter of the main memory is loaded with the address. This function can take place at a clock frequency of up to 50 MHz. With the next eight clocks, the slave shifts out the contents of the addressed memory cell via the data output. The address counter is incremented with each byte read. As long as the master continues to generate the clock, the device will output the stored data. When the address counter reaches the end of a memory page, it jumps to the first address of the next page without delay. When the last address of the entire memory is reached, the address counter will jump to the first address. When the chip select signal is switched high, reading is terminated. The contents of the two buffer memories are not changed during the read operation.

Command code 0x01 (read from entire memory area in low power mode) reads data at a bus clock rate of up to 15 MHz, similar to the previous function.

The following function can be used to read bytes from a memory device starting at the address laddress_begin, uibyte_number in low power mode and store them in the array ucbuffer in the RAM memory of the microcontroller.

```
void AT45DB161_Read_MemoryArray(AT45DB161_pins sdevice_pins,
                                long laddress_begin,
                                uint16_t uibyte_number,
                                uint8_t* ucbuffer)
{
    uint8_t ucAddressHighByte, ucAddressMiddleByte;
    uint8_t ucAddressLowByte,
    ucDummy = 0;
    //the 24-bit address is divided into single bytes
    ucAddressLowByte = laddress_begin;
    laddress_begin = laddress_begin >> 8;
    ucAddressMiddleByte = laddress_begin;
    laddress_begin = laddress_begin >> 8;
    ucAddressHighByte = laddress_begin;
    //start of communication
    SPI_Master_Start(sdevice_pins.AT45DB161spi);
    SPI_Master_Write(0x03); //transfer of the command code
    //the 24-bit address is transmitted
    SPI_Master_Write(ucAddressHighByte);
```

```
SPI_Master_Write(ucAddressMiddleByte);
SPI_Master_Write(ucAddressLowByte);
//uibyte_number Bytes are read out
for((long)lI = 0; lI < uibyte_number; lI++)
{
    ucbuffer[lI] = SPI_Master_Write(ucDummy);
}
//the communication is terminated
SPI_Master_Stop(sdevice_pins.AT45DB161spi);
}
```

The following call would read a total of 256 bytes from the memory in the example circuit starting at address 1024 and store them in the array ucBuffer[].

```
AT45DB161_Read_MemoryArray(AT45DB161_1, 1024, 0x100, ucBuffer);
```

When a byte is transferred via the SPI interface, a byte exchange takes place between the master and the slave. This special feature allows the read and write functions to be structured similarly.

22.1.4.2 Reading Within a Memory Page

Similar to reading from the entire memory area, the master must start the communication. With the transmission of the command code 0xD2, followed by a valid 24-bit address, the internal address counter is loaded; four following dummy bytes provide the initialization of the process and from this sequence on the data is output as long as the master generates the clock. With each transmitted byte the address counter is incremented, but within the addressed page. When the end of the addressed page is reached, the address counter jumps to the first address of the same page. When the chip select signal is switched high by the master, the transfer is terminated. The contents of the buffers are not changed by this.

22.1.4.3 Loading a Memory Page Into a Buffer

This operation allows an entire memory page to be loaded into the buffer selected by the instruction code. The microcontroller can access this data by reading the buffer (indirect read). After starting communication, the master sends the command code 0x53 for the first buffer or 0x55 for the second, followed by the 24-bit starting address of the desired page. When the chip select signal is switched high, SPI communication is terminated and the device-internal timed transfer from the memory page to the buffer begins, which is completed after 200 µs. The microcontroller can only access the stored data indirectly. In the first step, the buffer is loaded with the data from the desired page and in the second it is read from it using the buffer read function. During the internal transfer, the microcontroller can store data in the other buffer.

534

Table 22.2 AT45DB161 – Storage times

Memory function	Storage period
Write buffer in flash page with erase	<25 ms
Write buffer in flash page without erase	<4 ms
Write data via buffer in flash page with erase	<25 ms
Write data via buffer in flash page without erase	8 µs/byte

22.1.5 Write

Error-free writing of data can only take place in previously erased areas. In the AT45DB161, only data from the buffers can be stored in the flash. Writing is an internally timed process with alternating duration (Table 22.2), during which the RDY/BUSY bit from the STATUS register is switched to 0. During writing, the other buffer can receive data from the microcontroller. If errors occur during writing, the EPE bit in the STATUS register is set to 1. Different write operations are implemented, which are briefly explained below.

22.1.5.1 Write Buffer in Flash Page with Erase

This operation writes the entire contents of a buffer into a memory page. The data to be stored must be transferred to the buffer previously. The host microcontroller starts the SPI communication and transmits the command code 0x83 to write the contents of buffer 1 (0x86 for buffer 2), followed by the 24-bit starting address of the desired page. When the communication is terminated, the internal process starts, the addressed page is erased first and then the data from the buffer is stored.

22.1.5.2 Write Buffer in Flash Page Without Erase

This process is similar to the one described above, buffer 1 is addressed via code 0x88 and buffer 2 via 0x89. In this case the memory page must have been erased previously. A function that writes the contents of a buffer into a memory page is presented below. The buffer is selected via the command code and the memory page is addressed via the uipage_address parameter. Depending on the selected command code, the memory page should be erased previously. A valid page address is a number between 0 and 4095.

```
void AT45DB161_Write_MemoryPageFromBuffer(
                          AT45DB161_pins sdevice_pins,
                          uint8_t ucop_code,
                          uint16_t uipage_address)
{
    uint8_t ucAddressHighByte, ucAddressMiddleByte;
    uint8_t ucAddressLowByte = 0x00;
    //the 24-bit address is divided into individual bytes
    uipage_address = uipage_address << 1;
    ucAddressMiddleByte = uipage_address;
```

```
    uipage_address = uipage_address >> 8;
    ucAddressHighByte = uipage_address;
    //SPI communication is started
    SPI_Master_Start(sdevice_pins.AT45DB161spi);
    SPI_Master_Write(ucop_code); //the command code is transferred
    //the start address of the desired memory area will be transferred
    SPI_Master_Write(ucAddressHighByte);
    SPI_Master_Write(ucAddressMiddleByte);
    SPI_Master_Write(ucAddressLowByte);
    //the SPI communication is stopped
    SPI_Master_Stop(sdevice_pins.AT45DB161spi);
}
```

For the example circuit, the following function would be called to store the contents of buffer 1 in page 5 (memory area 2048 … 2559) without erasing:

```
AT45DB161_Write_MemoryPageFromBuffer(AT45DB161_1, 0x88, 5);
```

22.1.5.3 Comparison Between Stored Page and Source Buffer
In order to check whether the saving of an entire buffer in the flash has taken place without errors, the two contents can be compared after the process has been completed. If the two contents match, then bit COMP in the status register is set to 0. To do this, the master must send the command code 0x60 followed by a 24-bit start address of the desired page via the SPI bus in order to compare the contents of this page with buffer 1 (buffer 2 is addressed via code 0x61).

22.1.5.4 Write Data Via Buffer in Flash Page with Erase
This operation is a combination between the write operation of a buffer and the write function "Write buffer in flash page with erase". The microcontroller starts the SPI communication and selects one of the buffers by transmitting a command code (0x82 for buffer 1 or 0x85 for buffer 2). Then it sends a 24-bit address, which is structured as follows: bits 23 to 21 are dummy bits, the following twelve are used to address a page, and the last nine are used to address the buffer from which the following data bytes are to be stored. If the end of the buffer is reached and further data are transmitted, they are stored starting at address 0 of the buffer. When the chip select signal is switched high, the microcontroller terminates the communication and starts the internal write process. First the addressed page is erased and then the entire contents of the selected buffer are stored in the page.

22.1.5.5 Memory Write with Direct Access Via Buffer 1
By sending the command code 0x02 followed by a 24-bit address, bits 20:9 address the page in which the data is stored and bits 8:0 set the start address of the buffer (see Fig. 22.6). The memory device waits for up to 512 more bytes to be stored into the buffer

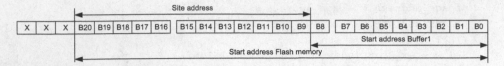

Fig. 22.6 24-bit address composition for direct access flash memories

Table 22.3 AT45DB161 – Erasing times

Memory area	Erasing time	Command code
Page	12 … 35 ms	0x81
Block	45 … 100 ms	0x50
Sector	1.4 … 2 s	0x7C
Total memory	22 … 40 s	0xC7 94 80 84

Delete a page (page)(0..4095)

| 0x81 | X | X | X | P11 | P10 | P9 | P8 | P7 | | P6 | P5 | P4 | P3 | P2 | P1 | P0 | X | | X | X | X | X | X | X | X | X |

Delete a block (0..511)

| 0x50 | X | X | X | B8 | B7 | B6 | B5 | B4 | | B3 | B2 | B1 | B0 | X | X | X | X | | X | X | X | X | X | X | X | X |

Delete sector 0a

| 0x7C | X | X | X | 0 | 0 | 0 | 0 | 0 | | 0 | 0 | 0 | 0 | X | X | X | X | | X | X | X | X | X | X | X | X |

Delete sector 0b

| 0x7C | X | X | X | 0 | 0 | 0 | 0 | 0 | | 0 | 0 | 0 | 1 | X | X | X | X | | X | X | X | X | X | X | X | X |

Deleting a sector (1..15)

| 0x7C | X | X | X | S3 | S2 | S1 | S0 | X | | X | X | X | X | X | X | X | X | | X | X | X | X | X | X | X | X |

Fig. 22.7 AT45DB161 – Deletion sequences

starting at the address 0x00. All bytes that have been transferred are stored in the previously erased page after communication ends. If less than 512 bytes have been transferred, the remaining bytes of the page remain unchanged.

22.1.6 Erase

As with any flash memory, programming is only possible in the erased state (all bytes 0xFF). Erasing the device can only be done area by area. To make the work with the device flexible, there is the possibility to erase single pages, blocks, sectors or the whole chip at once. The erase times are summarized in Table 22.3.

To erase one of the first three memory areas, the master sends over the SPI bus the corresponding command code followed by a 24-bit coded address of the memory area, as shown in Fig. 22.7. To erase the entire memory, only the 4-byte command code is transmitted. At the end of the SPI-transfer, the master turns the chip select signal high, which starts the internal erase process, meanwhile the RDY/BUSY bit from the STATUS register is set to 0. This means that the memory is busy. During this time, reading of the STATUS and Device ID registers and writing to the temporary memories are allowed.

22.1.7 Memory Protection

Several write protection mechanisms are implemented for the write protection of different memory areas.

22.1.7.1 Temporary Write Protection (Sector Protection)

The device has a non-volatile, 16-byte write-protect register in which you can specify in a coded manner which of the 17 sectors are to be protected against random writing or erasing. This register can be read or rewritten with a configuration of the protected sectors after previous erasure. The number of write/erase cycles of this register is limited to approximately 10,000. The activation/deactivation of the protection can be implemented either via the hardware or the software. With the Write Protect pin of the device switched low, the write protection of the selected sectors is enabled and the register itself becomes write protected. With the WP pin set to high, the write protection is enabled with the command code 0x3D 2A 7F A9 and disabled with 0x3D 2A 7F 9A.

22.1.7.2 Permanent Write Protection (Sector Lockdown)

It is possible to permanently protect an entire sector against writing/erasing. To do this, a microcontroller as master must transmit via SPI the command code 0x3D 2A 7F 30 followed by a 24-bit address of a byte within the sector to be protected. The operation cannot be reversed. The configuration of the permanently protected sectors is updated in a non-volatile, 16-byte read-only register.

22.1.8 Testability

If the device is installed in complex circuits, its complexity and memory size require extensive test capabilities that provide information about the state and identity of the device.

22.1.8.1 Status Registers

The 2-byte, read-only register can be used to check the result of some functions or the status of internal operations. The register is structured as follows:

- **Byte 1:**
  - **Bit 7** – RDY/BUSY – is 0 as long as an internal process is not completed.
  - **Bit 6** – COMP – is 0 if the data match when comparing the buffer and the selected page.
  - **Bit 5:2** – DENSITY – encodes the memory size of the device (1011 for 16 Mbit)
  - **Bit 1** – PROTECT – at 1 indicates that the sector write protection is active
  - **Bit0** – PAGE SIZE – indicates the page size; a 1 means 512 bytes

- **Byte 2:**
  - **Bit 7** – RDY/BUSY – like bit 7 of byte 1
  - **Bit 6** – reserved
  - **Bit 5** – EPE – indicates at 0 if an erase or write operation was successful.
  - **Bit 4** – reserved
  - **Bit 3** – SLE – at 0 further sectors can no longer be permanently write protected
  - **Bit 2** – PS2 – indicates at 1 that a write operation via buffer 2 has been temporarily suspended.
  - **Bit 1** – PS1 – indicates at 1 that a write operation via buffer 1 has been temporarily suspended.
  - **Bit 0** – ES – indicates that the erasing of a sector has been temporarily suspended.

22.1.8.2 Security Register (Security Register)

The safety register is a 128-byte register whose second half (byte 64:127) is provided by the manufacturer with an identification number that uniquely identifies each device. This part can neither be erased nor reprogrammed. The first part (byte 0:63) can be programmed once by the user and then only read.

22.1.8.3 Manufacturer and Chip ID Registers

This read-only register stores information about the manufacturer and the device in the JEDEC standard, which can be used in the system to identify the memory.

22.2   SST25WF0808 Serial Flash Memory

The SST25WF0808 device is a serial flash memory of the SST25XXYY family from Microchip Technology [4]. The memories of this family have a storage capacity of 512 kBit (64 kByte) to 64 Mbit (8 Mbyte) and are characterized by low power consumption and simple control. A block diagram of the 8-Mbit SST25WF0808 device is shown in Fig. 22.8. Each memory cell is addressable and the contents can be read or modified. The entire memory is divided into 256x4 kByte memory sectors, or 16x64 kByte memory blocks. These memory units can be erased individually. To prevent unauthorized modification of memory areas, they can be protected. The device is controlled via a 4-wire SPI interface, which can be clocked at up to 40 MHz. To double the bit rate during reading, a special serial two-line transmission can be used.

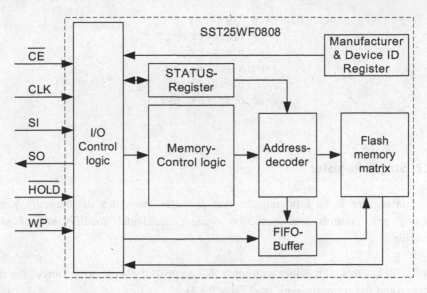

Fig. 22.8 SST25WF0808 – Block diagram

22.2.1 SPI Communication

The device is permanently set as an SPI slave. Communication can take place in mode 0 or 3 with the transmission of the more significant bit first. A host microcontroller is initialized as master with the call:

```
SPI_Master_Init(SPI_INTERRUPT_DISABLE, SPI_MSB_FIRST,
                SPI_MODE_0, SPI_FOSC_DIV_16);
```

The transfer of a byte starts after the CE (Chip Enable) input is set low while the HOLD input is high. The microcontroller can interrupt a data transfer to the memory by setting input HOLD low and establish SPI communication with another slave. Setting input HOLD back high resumes communication with the memory where it was interrupted. During the interruption, input CE must remain low. The SPI data structure of a memory SST25WF0808, whose inputs WP and HOLD are inactive (permanently connected to the supply voltage) and whose input CE is connected to pin PC1 of a microcontroller ATmega88 (Chap. 14, Sect. 14.5), looks as follows:

```
SST25PFXX_pins SST25PFXX_1 ={/*CS_DDR*/ &DDRC,
                  /*CS_PORT*/ &PORTC,
                  /*CS_pin*/ PC1,
                  /*CS_state*/ ON},
                  /*WP_DDR*/ OFF,
                  /*WP_PORT*/ OFF,
```

```
/*WP_pin*/ OFF,
/*WP_state*/ OFF,
/*HOLD_DDR*/ OFF,
/*HOLD_PORT*/ OFF,
/*HOLD_pin*/ OFF,
/*HOLD_state*/ OFF};
```

22.2.2 Status Register

The status register is an 8-bit register that provides the status of internally running operations and controls the protection against accidental modification of stored information.

- **Bit 7** – BPL – this bit together with input WP controls the write protection of the status register and the entire memory (see Table 22.4);

- **Bit 6** – reserved;
- **Bit 5:2** – these bits can be changed via software and determine the memory areas to be write protected (for details, see [4]); if bit 2 = bit 3 = bit 4 = 0, the entire memory area is writable;
- **Bit 1** – WEL – is controlled internally and when it is 1 indicates that contents of the memory can be changed;
- **Bit 0** – BUSY – the bit indicates, when it is 1, that an internal write operation has not yet been completed.

When repeating write and/or erase operations, pay attention to the duration of these operations (Table 22.5). A timer can be set with the maximum duration of a change

Table 22.4 SST25WF0808 – Write protection

WP-Pin	BPL bit	Status register	Memory
Low	1	R	R
Low	0	R/W	Protection-free areas are R/W
High	x	R/W	Protection-free areas are R/W

Table 22.5 SST25WF0808 – Duration of the internal write cycles

Memory area	Operation	Duration
Status register	Writing	<10 ms
Storage sector	Erase	40 ... 150 ms
Memory block	Erase	80 ... 250 ms
Total memory	Erase	0.5 ... 6 s
n-byte	Writing	$<(0.2 + n * 0.8/256)$ ms

operation to signal the end of the operation via an interrupt. Another way to determine the end of an operation is to periodically check the value of the BUSY bit in the status register in the endless loop of the main function, as shown in the following code excerpt:

```
#define BUSY 7
while(1)
{
    if(!(SST25_Read_StatusRegister(SST25PFXX_1) & (1 << BUSY)))
    {
        //write cycle is finished
    }
}
```

22.2.3 Read Functions

Regardless of the state of pin WP and bit BPL, all memory and the status register can be read. To read the status register, the microcontroller sends the code 0x05. The memory logic decodes the command and places the contents of the register in the transmit buffer. By sending another byte, the microcontroller reads this transmit buffer.

```
uint8_t SSTPFXX_Read_StatusRegister(SST25PFXX_pins sdevice_pins)
{
    //sdevice_pins data structure with the definition
    //of the controllable pins
    uint8_t ucDataByte;
    SPI_Master_Start(sdevice_pins.SST25PFXXspi); //CS on low
    //the read command of the status register is transferred
    SPI_Master_Write(0x05);
    ucDataByte = SPI_Master_Write(0xFF); //it becomes a dummy byte
        //to be able to read out the contents of the register
    SPI_Master_Stop(sdevice_pins.SST25PFXXspi); //CS on high
    return ucDataByte;
}
```

The following example represents a function for reading a stored byte. After sending the read code 0x03 (read at maximum clock frequency of 30 MHz), the 24-bit address of the desired memory cell and a dummy byte follow in order to read its contents. After reading a byte, the internal address counter is incremented. Thus it is possible to read a whole contiguous memory area in one SPI operation if the address of the first memory cell is given. After reading the memory cell with the highest address the address counter is loaded with the value 0x000000.

```
uint8_t SST25_Read_Byte(SST25PFXX_pins sdevice_pins,
                        unsigned long* uladdress)
{
    //sdevice_pins data structure with the definition
    //of the controllable pins
    uint8_t ucAddressHigh, ucAddressMiddle, ucAddressLow;
    uint8_t ucDataByte;
    SPI_Master_Start(sdevice_pins.SST25PFXXspi); //CS on low
    SPI_Master_Write(0x03); //operation code is transferred
    ucAddressLow = uladdress;
    ucAddressMiddle = uladdress >> 8;
    ucAddressHigh = uladdress >> 16;
    //high byte of the address is transmitted
    SPI_Master_Write(ucAddressHigh);
    //middle byte of the address is transmitted
    SPI_Master_Write(ucAddressMiddle);
    //low byte of the address is transmitted
    SPI_Master_Write(ucAddressLow);
    ucDataByte = SPI_Master_Write(0xFF); //a dummy byte is
        //transmitted to be able to read out a byte
    SPI_Master_Stop(sdevice_pins.SST25PFXXspi); //CS on high
    return ucDataByte;
}
```

22.2.4 Erase Functions

After erasing, all bits of the selected memory area are set to 1. A memory area can only be erased if it is not write-protected. Before erasing, writing must be enabled. This is done by transmitting the code 0x06, which sets bit WEL in the status register to 1. To erase a memory block, the command code 0xD8 must be transmitted and then any address from the target block. After setting the CE line high, the SPI communication is terminated and the internally controlled erase process begins. When the process is finished, the BUSY and WEL bits are set to 0. The following code example can be used to erase the entire contents of the memory.

```
void SST25_ChipErase(SST25PFXX_pins sdevice_pins)
{
    //sdevice_pins data structure with the definition
    //of the controllable pins
    SST25_WriteEnable(sdevice_pins); //write enable
    SPI_Master_Start(sdevice_pins.SST25PFXXspi); //CS on low
    //the entire content of the memory will be erased
```

```
        SPI_Master_Write(0x60);
        SPI_Master_Stop(sdevice_pins.SST25PFXXspi); //CS on high
}
```

22.2.5 Write Functions

The programming process only allows the bits to be switched from 1 to 0, so the memory area must be erased previously. As with erasing, the memory must not be protected and writing must be enabled. The master sends the command code 0x02 followed by the address of the first byte to be changed. A further n data bytes are transmitted and temporarily stored in a 256 byte FIFO buffer on the memory device. Of the n data bytes, only the last 256 are programmed. When the CE terminal is switched high, the SPI transfer is terminated and programming begins. At the end of programming the BUSY and WEL bits are set to 0. The write operation is optimized for 256 byte pages. The programming of a single byte can take 203.125 µs (see Table 22.5), while the programming of a whole page takes about 1 ms, which corresponds to a duration of just under 3.9 µs/byte.

The start address of the memory pages is a multiple of 256. After saving a byte, the internal address counter is incremented within the selected memory page. When the address counter reaches the last address of the page, it jumps to the first address of this page, which must be taken into account when storing data sets larger than 1 byte. An example of a function to store 1 byte is shown in the following program code.

```
void SST25_Write_Byte(SST25PFXX_pins sdevice_pins,
                      unsigned long* uladdress,
                      uint8_t ucbytetowrite)
{
    uint8_t ucAddressHigh, ucAddressMiddle, ucAddressLow;
    uint8_t ucDataByte;
    SST25_WriteEnable(sdevice_pins); //write enable
    SPI_Master_Start(sdevice_pins.SST25PFXXspi); //CS on low
    SPI_Master_Write(0x02); //operation code is transferred
    ucAddressLow = uladdress;
    ucAddressMiddle = uladdress >> 8;
    ucAddressLow = uladdress >> 16;
    //high byte of the address is transmitted
    SPI_Master_Write(ucAddress_High);
    //middle byte of the address is transferred
    SPI_Master_Write(ucAddress_Middle);
    //low byte of the address is transmitted
    SPI_Master_Write(ucAddress_Low);
    //the byte to be written is transferred
```

```
    SPI_Master_Write(ucbytetowrite);
    SPI_Master_Stop(sdevice_pins.SST25PFXXspi); //CS on high
}
```

22.2.6 2-Line Serial Interface

This interface uses the same connections as the SPI interface, but allows parallel transmission of 2 data bits and a bit rate of 80 MBit/s at a clock frequency of 40 MHz. The master sets the line CE low and transmits the code 0x3B followed by the address in SPI mode for reading. During the transmission of another dummy byte, the command is decoded and the address counter is loaded. The memory logic now also switches the DI pin to output and pushes out one bit at each clock pulse via the DO and DI pins. The master must compose a byte from the received bits every four clocks. Due to the high computational effort, transmission in this mode cannot be realized at higher bit rates with microcontrollers of the ATmega family. The 2-wire interface, on the other hand, can be implemented very well in programmable logic devices (CPLD and FPGA).

References

1. Stiny, L. (2018). *Aktive elektronische Bauelemente – Aufbau, Struktur, Wirkungsweise, Eigenschaften und praktischer Einsatzdiskreter und integrierter Halbleiter-Bauteile* (4. Aufl.). Springer Vieweg Wiesbaden.
2. Aravindan, A. (2021). *Flash 101: NAND Flash vs NOR Flash, embedded 2018*. https://www.embedded.com/flash-101-nand-flash-vs-nor-flash/. Accessed 1 Mar 2021.
3. Microchip Technology Inc. *AT45DB161–16 Mbit SPI Serial Flash Memory*. http://ww1.microchip.com/downloads/en/devicedoc/doc2224.pdf. Accessed 4 Apr 2021.
4. Microchip Technology Inc. *SST25WF080B – 8 Mbit 1,8 V SPI Serial Flash*. www.microchip.com. Accessed 2 Apr 2021.

Integrated Circuits for Audio Technology

23

Abstract

This chapter describes two components which together form a radio with amplifier. Both are controlled via I^{2}C.

In the field of consumer electronics, microcontrollers are often used for operating of equipment. In addition, the use of active boxes is popular, where the amplifier is built into the box. In these cases, it makes sense to use devices that can be controlled via a bus system. A distinction is made between devices in which the media signals are also transmitted digitally and those in which only the control signals (command & control) are transmitted digitally, while the media channel remains analog. For this book, we have decided to present only two devices of the latter category, since the former are generally too complex for an introduction in this context.

23.1 SI4840 Radio IC

The SI4840 ([2–4]) from Silicon Laboratories is an AM[1]/FM[2] receiver with a stereo decoder as shown in Fig. 23.1. It belongs to a series of ATDD[3] devices which have an analog tuner, which means that the frequency tuning is realized by an external analog voltage and the information about the settings is transmitted in digital form by the device.

---

[1] AM – Amplitude modulation.

[2] FM – Frequency Modulation.

[3] ATDD – Analog Tune Digital Display.

© The Author(s), under exclusive license to Springer Fachmedien Wiesbaden GmbH, part of Springer Nature 2023
A. Meroth, P. Sora, *Sensor networks in theory and practice*,
https://doi.org/10.1007/978-3-658-39576-6_23

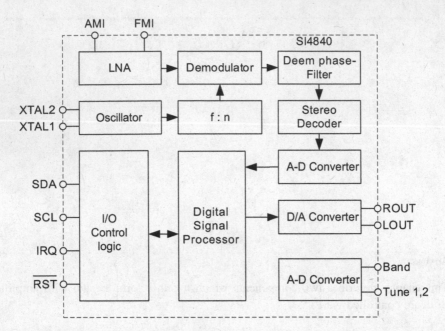

Fig. 23.1 SI4840 block diagram. (According to [2])

The device integrates all necessary components for the reception of the medium wave and ultra short wave frequency bands (see Table 23.1) with a very simple external circuitry, but must be controlled by a microcontroller. The device provides indication of stereo reception, reception of a valid station, and digital adjustment of pitch control and volume. To ensure good audio quality under different reception conditions, the device automatically switches between monophony and stereophony depending on the reception quality. Three reception quality parameters are monitored for this purpose:

- the reception field strength – RSSI[4];
- the signal-to-noise ratio (or signal-to-interference ratio) – SNR[5];
- Intersymbol interference due to multipath reception.

As soon as one of these three parameters falls below a digitally adjustable limit, the system switches to mono operation.

---

[4] RSSI – Received Signal Strength Indication.
[5] SNR – Signal to Noise Ratio.

Table 23.1 SI4840 received frequency bands

Modulation type	Frequency range
FM	87–108 MHz
	86.5–109 MHz
	87.3–108.25 MHz
	76–90 MHz
	64–87 MHz
AM	520–1710 kHz
	522–1620 kHz
	504–1665 kHz
	520–1730 kHz
	510–1750 kHz

23.1.1 Module Description

The broadcast signal is received via a suitable antenna (see [4]) connected to pin FMI (for FM)[6] or AMI (for MW)[7] and amplified with a low-noise amplifier LNA.[8] The demodulator requires a clock signal with a frequency of 32,768 Hz. This clock signal can be generated using a clock crystal with an accuracy of ±100 ppm,[9] or generated using an external oscillator. The frequency of the external oscillator is set to the frequency 32,768 ± 5% Hz using the internal adjustable frequency divider (:1 ... 4095). When transmitting the modulated audio signal, the higher frequency components are more disturbed, which leads to a deterioration of the signal-to-noise ratio. An improvement is achieved by boosting the high-frequency components of the signal with a so-called pre-emphasis filter (1st order high-pass filter) on the transmitter side before modulation. Because only the useful signal is amplified, while the noise power from the transmission path remains constant, the signal-to-noise ratio increases. To achieve a natural sound at the receiver end, the signal must be restored with a deemphasis filter (1st order low pass filter Fig. 23.2), resulting in the lowering of the higher frequency components. In Europe, a low-pass filter with a − 3 dB cut-off frequency of 3.18 kHz is used.

$$f_{-3\mathrm{dB}} = \frac{1}{2\pi RC} \tag{23.1}$$

and a time constant of

---

[6] VHF ultra-shortwave.

[7] MW – Medium Wave.

[8] LNA – Low Noise Amplifier.

[9] Ppm – parts per million; 1 ppm = 0.000001.

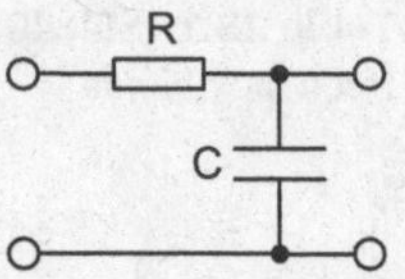

Fig. 23.2 SI4840 Deemphase
filter

$$\tau = RC = 50\mu s. \tag{23.2}$$

With this noise reduction method, an improvement of the noise-to-signal ratio of up to 7.8 dB is possible [1]. In the USA, a filter with a time constant of 75 μs is used, so attention must be paid to the choice of time constant.

An analog voltage is generated at the TUNE1 connection, which is used to select the frequency band or frequency tuning. A headphone or an audio amplifier can be connected to the ROUT and LOUT pins. The RST connection is used to safely reset the device. The SDA, SCL, and IRQ pins are available for communication with a microcontroller.

23.1.2 Selection of the Frequency Band and Frequency Tuning

The device can detect the frequency band itself. It generates the analog voltage needed for frequency tuning at pin TUNE1. To realize this, a precise voltage divider with a total resistance of 500 kOhm must be connected between pin TUNE1 and ground. A switch connected to the BAND pin can be used to select a frequency band. Examples of voltage dividers can be found in [4]. A potentiometer connected between TUNE1 and ground, with a wiper connected to pin TUNE2, provides the frequency tuning. To facilitate the process, the device provides information about the tuned frequency and quality of reception in digital form.

Figure 23.3 presents a circuit for frequency band selection and frequency tuning that is simple to implement in terms of circuitry. When the BAND pin of the SI4840 device is connected high, it expects the microcontroller to select the frequency band and transmit it serially. With the LNA_EN pin directly high (without a pull-up resistor), the limits of the frequency bands can be freely selected, otherwise the default values from Table 23.1 apply. The analog voltage needed for frequency tuning is generated by a D/A converter of type MCP4911, which uses the voltage at pin TUNE1 as reference voltage. The MCP4911 is similar in design to the D/A converters of the MCP48XX series (see Chap. 20) and is pin-compatible with an MCP4811 except for the VREF pin. The same functions as for the MCP48XX devices can be used for its control. When selecting another D/A converter, care must be taken to limit its output voltage to the voltage level from the TUNE1 pin.

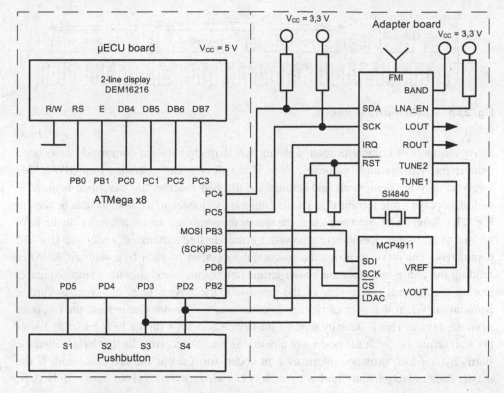

Fig. 23.3 SI4840 frequency band selection and frequency tuning via microcontroller

23.1.3 Initializing the Device

After the supply voltage of the device has reached its stable state, the device must first be initialized. To do this, the host microcontroller must set pin RST of the device to low and hold it at this level for at least 100 μs. In response, the device switches the IRQ signal to high and then back to low after 2 ms at the latest. After the falling edge of the IRQ signal, the device is in standby mode and accepts the Get_Status and Power_Up commands. In this phase, after transmitting a command, the microcontroller must wait another 2 ms for the response of the device or test in polling when bit CTS in the status byte is 1.

23.1.4 Communication with the Device

The device can be used to build a complete broadcast receiver with control from a microcontroller. The device implements an I²C- (with the SCL and SDA pins) and an SMB- (SCL, SDA and IRQ) compliant protocol and is preconfigured as a slave with the device type address 0x22. The serial interface can be clocked at up to 400 kHz.

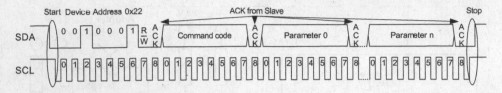

Fig. 23.4 SI4840 command structure

As elements of communication, a distinction is made between commands, associated parameters, settings and responses. Even if the device can operate as an MW (AM) receiver, only the commands and settings for the FM receiver are described here, MW no longer plays a role technically. The transmission sequence of a command can be seen in Fig. 23.4. After an I^2C start sequence, the master transmits the device address with the R/W bit set to 0 (write operation), followed by the desired command code and 0 ... n parameters. The slave confirms the successful reception of each byte with ACK. After sending the entire message, the master terminates communication with a stop sequence. During the internal conversion of the command, the device does not react to further commands. When the internal process is complete, the master can request the response from the device. This is done by sending the device address with the R/W bit set to 1 after the start sequence. With the next eight clocks, the master receives the first byte, called the status byte, which provides information in coded form about the last command. If the master does not require any further bytes, it acknowledges the received byte with NACK, otherwise with ACK. After the last received byte of the response, the master terminates communication with a stop sequence. The response of the receiver to a command consists of at least one byte. Bit 7 of the status byte always has the same meaning, CTS,[10] and indicates with 1 that the module is ready to receive a new command.

The following function reads the status byte to evaluate the CTS bit:

```
uint8_t SI4840_Get_StatusByte(void)
{
    uint8_t ucAddress = 0x23, ucStatusByte;
    TWI_Master_Start();
    if((TWI_STATUS_REGISTER) != TWI_START) return TWI_ERROR;
    TWI_Master_Transmit(ucAddress); //send device address
    if((TWI_STATUS_REGISTER) != TWI_MR_SLA_ACK) return TWI_ERROR;
    ucStatusByte = TWI_Master_Read_NAck();
    if((TWI_STATUS_REGISTER) != TWI_MR_DATA_NACK) return TWI_ERROR;
    TWI_Master_Stop();
    return ucStatusByte;
}
```

---

[10]CTS – Clear to Send (ready to receive).

After the call, the function returns either an error code or the status byte of the receiver. By repeatedly calling the function and evaluating the CTS bit, it is possible to determine when the block is ready to execute a new command.

23.1.4.1 Power_Up Command

A Power_Up function can be called to switch the device from standby to active state or to change the frequency band or its properties. The command is only accepted by the device after a successful reset sequence. The master can send up to six parameters after the device address and the Power_Up command code 0x91, which have the following meaning:

- Parameter 1
  - **Bit 7** – set to 1 if the receiver uses a 32.768 kHz crystal
  - **Bit 6** – codes the waiting time for the stabilization of the oscillations of the quartz (0 for 600 ms)
  - **Bit 5:0** – Frequency band index; for FM this is a number between 0 to 19 that encodes the frequency band, the time constant of the deemphasis filter, the frequency spacing between audio channels and the threshold of the received field strength.
- Parameter 2:3 together correspond to the lower frequency limit of the frequency band in 10 kHz
- Parameter 4:5 together correspond to the upper frequency limit of the frequency band in 10 kHz
- Parameter 6 encodes the frequency channel spacing; for FM, parameter 6 = 10

The Power_Up command could be implemented as follows:

```
uint8_t SI4840_Set_ATDDPowerUp(uint8_t ucband_index,
                               uint16_t uibottom_freq,
                               uint16_t uitop_freq,
                               uint8_t ucchannel_spacing)
{
    uint8_t ucAddress, ucDataToSend[7];
    ucDataToSend[0] = SI4840_ATDD_POWER_UP; //0xE1
    ucDataToSend[1] = ucband_index; //frequency band index
    //high byte of the lower frequency limit
    ucDataToSend[2] = uibottom_freq >> 8;
    //lowbyte of the lower frequency limit
    ucDataToSend[3] = uibottom_freq;
    //high byte of the upper frequency limit
    ucDataToSend[4] = uitop_freq >> 8;
    ucDataToSend[5] = uitop_freq; //lowbyte of the lower frequency limit
    ucDataToSend[6] = ucchannel_spacing; //frequency channel spacing
    ucAddress = SI4840_DEVICE_TYPE_ADDRESS | TWI_WRITE; //write mode
    TWI_Master_Start();
```

```
    TWI_Master_Transmit(ucAddress); //send device address
    if((TWI_STATUS_REGISTER) != TWI_MT_SLA_ACK) return TWI_ERROR;
    //send command code, frequency band, lower and upper frequency
    //and channel spacing
    for(uint8_t ucI = 0; ucI < 7; ucI++)
    {
        TWI_Master_Transmit(ucDataToSend[ucI]);
        if((TWI_STATUS_REGISTER) != TWI_MT_DATA_ACK) return TWI_ERROR;
    }
    TWI_Master_Stop();
    return TWI_OK;
}
```

With the call:

```
SI4840_Set_ATDDPowerUp(0x03, 8800, 10800, 10);
```

the device is set to the frequency band 88 ... 108 MHz with 50 µs time constant, 12 dB audio channel spacing and 28 dB receive field strength threshold.

The module only responds to the Power_Up command with the status byte. If bit 6 (ERR) of the status byte is set, an error has occurred when executing the command. Bits 3: 0 store the error code.

23.1.4.2 Power_Down Command

The Power-Down command (command code 0x11) puts the device into standby mode. The current consumption in this mode is approx. 10 µA. No parameters are required for this command. While in standby mode, the device accepts only the Power_Up and Get_Status commands. The device controls bits 7 (CTS) and 6 (ERR) in the status register in response to this command.

23.1.4.3 Get_Status Command

The master uses the Get_Status command (command code 0xE0) to request information from the device about the general status and configuration of the receiver. No additional parameters are required. The response to this command consists of four bytes that provide the following information:

- Status byte (byte 1)
  - **Bit 7** – CTS
  - **Bit 6** – HOSTRST – if it is 1, the device must be reset
  - **Bit 5** – HOSTPWRUP – if it is 1, the Power_Up command must be executed
  - **Bit 4** – INFORDY – indicates at 1 that information about the tuner is available
  - **Bit 3** – STATION – if it is 1 then a stable station is received
  - **Bit 2** – STEREO – indicates at 1 that a stereo broadcast is received.

- **Bit 1** – BCFG1 – indicates at 1 that the device accepts settings of the frequency band deviating from the default values.
- **Bit 0** – BCFG2 – indicates at 0 that the device detects the frequency band
- Byte 2
  - **Bit 7:6** – BANDMODE – for FM = 0, for AM = 1;
  - **Bit 5:0** – these bits encode the detected frequency band index.
- Byte 3:4 – these 2 bytes encode the frequency channel as a 4-digit number in BCD format; a code sequence for converting the frequency from the BCD code is presented in Sect. 23.1.5.

23.1.4.4 Audio Mode Command

This command can be used to change or query basic audio settings. The master must transmit a parameter after the command code 0xE2, the bits of which have the following meaning:

- **Bit 7** – if this bit is 1, the settings are queried and the following bits have no meaning
- **Bit 6:5** – are reserved
- **Bit 4** – at 0 the reception of a stereo transmission is indicated
- **Bit 3** – at 0 the volume around the transmitter frequency is attenuated by 2 dB
- **Bit 2** – at 0 the stereo mode is selected
- **Bit 1:0** – encode the audio mode: **0** – allows digital volume adjustment between 0 and 63, without pitch control; **1** – volume level is fixed at 59, pitch control is allowed; **2** – volume adjustment from 0 to 59 and pitch control are allowed; **3** – volume adjustment from 0 to 63 and pitch control are allowed.

The response of the device to this command consists of a status byte with bit 7 (CTS), bit 6 (ERR) and bits 4:0 with the same meaning as described above. Bit 5 is reserved.

23.1.4.5 Set_Property Command

Properties of the device such as clock frequency, frequency divider, volume, high and low frequency control and switching on and off of an audio channel can be changed using the Set_Property command. To implement this, the master sends a dummy byte equal to 0 after the 0x12 command code, followed by the 2-byte property code and the 2-byte property value. The higher-order byte is always transmitted first. The module does not provide a response to this command.

A detailed description of all properties can be found in [3].

23.1.4.6 Get_Property Command

The properties changed with the Set_Property command can be queried using the Get_Property command. To do this, the master must send the command code 0x13, followed by a dummy byte equal to 0 and the 2-byte property code. The block responds to this command with the status byte (CTS and ERR active), followed by a byte that is always 0 and the other 2 bytes encode the requested property value.

23.1.4.7 Get_Rev Command

No additional parameters are required when calling the function Get_Rev (command code 0x10). The response of the device to the command consists of the status byte with the active bits CTS and ERR as well as a further 8 bytes that provide information about the device. This information can be used to identify it in a complex circuit.

23.1.5 Searching for Broadcast Stations with the SI4840

The SI4840 module provides information about the frequency reached, the station or stereo reception. Due to these facts, it is obvious to conveniently switch from frequency channel to frequency channel by pressing a key, to search for the next transmitter, the next stereo broadcast or specifically for a frequency. As an example, the search for a frequency channel whose frequency is passed as a multiple of 100 kHz as a parameter when calling a function is presented.

The device supplies a voltage of 1.26 V for the frequency band of 88 . . . 108 MHz at pin TUNE1 for the potentiometric adjustment of the analog tuning voltage at pin TUNE2. As shown in Fig. 23.3, the microcontroller is now to control the tuning voltage for the radio device via a 10-bit D/A converter of type MCP4811. Using a voltage divider consisting of R1 = 1 kΩ and R2 = 1.6 kΩ, the tuning voltage reaches a value of 1.26 V at 2.048 V at the output of the D/A converter. The SPI data structure of the D/A converter (see Chap. 20) is:

```
MCP48XX_pins MCP4811_1 = {{ /*CS_DDR*/ &DDRD,
                            /*CS_PORT*/ &PORTD,
                            /*CS_pin*/ PD6,
                            /*CS_state*/ ON},
                            /*LDAC_DDR*/ &DDRB,
                            /*LDAC_PORT*/ & PORTB,
                            /*LDAC_pin*/ PB2,
                            /*LDAC_state*/ ON,
                            /*SHDN_DDR*/ OFF,
                            /*SHDN_PORT*/ OFF,
                            /*SHDN_pin*/ OFF,
                            /*SHDN_state*/ OFF};
```

The function MCP48XX_Set_Output_Voltage with the prototype:

```
void MCP48XX_Set_OutputVoltage(MCP48XX_pins sdevice_pins,
                            uint8_t ucdevice,
                            uint8_t ucchannel,
                            uint8_t ucout, float fvalue)
```

sets the voltage value at the output of the D/A converter, which was passed as parameter fvalue (in mV) when calling the function. This function is intended for the control of all devices of this family. When calling the function, the following parameters must be passed: the SPI data structure, the device type (e.g. MCP481X for a 10-bit D/A converter), the channel (CHANNEL0 for a single-channel D/A converter), the output enable (OUTPUT_ENABLE) and the voltage value to be set in mV.

After changing the tuning voltage, experience shows that the radio chip needs about 200 ms to switch the new frequency. In the following an algorithm is presented to search for a new frequency. The search starts at a tuning voltage of 1024 mV, half the output voltage of the D/A converter for the upper limit of the frequency band. An auxiliary variable uiVoltMax stores this value.

- Step 1: – The tuning voltage is set to the value uiVolt2Search = uiVoltMax;
- Step 2: – 200 ms are waited;
- Step 3: – The radio chip reports the frequency reached on request. If this frequency is the one you are looking for, then the search is successfully completed. If not, then continue with step 4.
- Step 4: – the value of the variable uiVoltMax is halved. If the new value is 0, then the search is unsuccessful.
- Step 5: – If the set frequency is greater than the searched frequency, then: uiVolt2Search -= uiVoltMax, otherwise uiVolt2Search += uiVoltMax. Continue with step 1. A result is obtained after 10 repetitions at the latest.

The presented algorithm is implemented using the SI4840_Seach_Freq frequency. The function returns the search result in a data structure:

```
typedef struct
{
    uint8_t ucStatus;
    uint8_t ucFBand;
    uint8_t ucFreq1;
    uint8_t ucFreq2;
} SI4840_response;
```

By evaluating the return of this function, it can be additionally determined whether the reception is of good quality at the frequency searched for, or whether the transmission is stereo.

```
uint8_t SI4840_Search_Freq(MCP48XX_pins sdevice_pins,
                    SI4840_response *sresponse,
                    uint16_t uifrequency)
```

```
{
    uint16_t uiFreq2Comp, uiFreq;
    //limiting the frequency to 108 MHz
    if(uifrequency < 1081) uiFreq = uifrequency;
    else uiFreq = 1080;
    if(ucFreqIndex == 0)
    {
        // the adjustment voltage is first set to 1024 mV
        uiVolt2Search = uiVoltMax;
        MCP48XX_Set_OutputVoltage(sdevice_pins, MCP481X, CHANNEL0,
                                  OUTPUT_ENABLE, uiVolt2Search);
        ucFreqIndex ++ ;
    }
    else
    {
        //the status of the radiochip is requested during the next runs
        if(SI4840_Get_ATDDStatus())
        {
            TWI_Master_Stop();
        }
        //and read his answer
        if(SI4840_Get_ATDDResponse(ucStatusArray,
                                   STATUS_RESPONSE_BYTE))
        {
            TWI_Master_Stop();
        }
        //the frequency reached is calculated
        uiFreq2Comp = (uint16_t)((ucStatusArray[2] >> 4) * 1000) +
                      (uint16_t)((ucStatusArray[2] & 0x0F) * 100) +
                      (ucStatusArray[3] >> 4) * 10 +
                      (ucStatusArray[3] & 0x0F);
        if(uiFreq2Comp == uiFreq)
        {
            //if the searched frequency is equal to the reached
            //frequency, a new search is initialized
            uiVoltMax = 1024;
            ucFreqIndex = 0;
            //response of the radiochip is stored in the response frame
            sresponse->ucStatus = ucStatusArray[0];
            sresponse->ucFBand = ucStatusArray[1];
            sresponse->ucFreq1 = ucStatusArray[2];
            sresponse->ucFreq2 = ucStatusArray[3];
            //frequency reached is returned
            return FREQUENCY_REACHED;
        }
```

```
    else
    {
        //if the two frequencies are different,
        //the voltage uiVoltMax is halved
        uiVoltMax = uiVoltMax >> 1;
        if (uiVoltMax == 0)
        {
            //when the voltage reaches zero, the search is finished
            //without reaching the desired frequency; the variables
            //for a new search are initialized .
            uiVoltMax = 1024;
            ucFreqIndex = 0;
            //frequency not reached is returned
            return FREQUENCY_NOT_REACHED;
        }
    }
    else
    { //the new voltage value is calculated
        if (uiFreq2Comp > uiFreq) uiVolt2Search -= uiVoltMax;
        else if (uiFreq2Comp < uiFreq)
            uiVolt2Search += uiVoltMax;
        MCP48XX_Set_OutputVoltage(sdevice_pins, MCP481X,
                                  CHANNEL0, OUTPUT_ENABLE,
                                  uiVolt2Search);
        //frequency is searched is reported back
        return FREQUENCY_SEARCH;
    }
  }
}
//this statement is only used to avoid a warning from the compiler
return FREQUENCY_SEARCH;
}
```

To avoid blocking waits, the frequency search in the main function is performed in a 200 ms task. The variable ucSearchFlag is set at the beginning of the search and reset when the searched frequency is reached.

```
if (Timer1_get_200msState() == TIMER_TRIGGERED)
{
        if (ucSearchFlag)
        {
                ucResponseFlag = SI4840_Search_Freq(MCP48XX_1, &
                                 sResponse_Frame, uiFreq2Search);
                if (ucResponseFlag == FREQUENCY_REACHED)
                {
                        //the result of the search is evaluated and
```

```
                    ucSearchFlag = 0;
              }
        }
}
```

23.2   LM48100Q Amplifier Module

The **LM48100** is an analog balanced mono amplifier (AB bridge amplifier) from Texas
Instruments, which is suitable, for example, for audio output stages in cars [5]. It has a
number of protection features and provides an output power of 1.3 W on an 8 Ω load at a
supply voltage of 3–5.5 V. You can't sound halls with it, but it is suitable for small active
boxes controlled by a microcontroller. The device has two inputs that can be mixed, with
each input changing volume in 32 steps. Mixer, volume control and the configuration of the
device are controlled via an I²C interface. Communication takes place at up to 400 kBit/s.

Power amplifier faults, such as short circuits on a loudspeaker line to ground or supply
voltage, line break, short circuits between the loudspeaker lines, overcurrent and

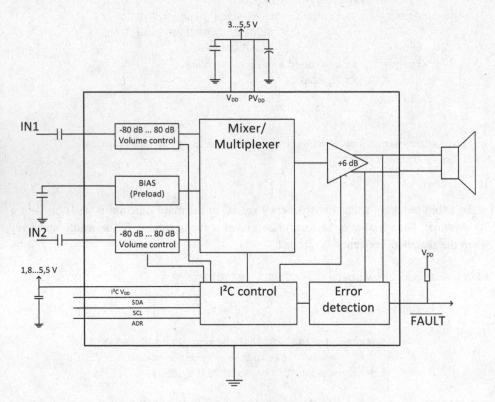

Fig. 23.5 Block diagram of the LM48100Q. (According to [5])

overheating can be detected continuously. In the event of a fault, a pin is connected to ground so that the fault can be identified with the microcontroller via a pin input. The diagnosis can be controlled in such a way that errors can be narrowed down.

Similarly, the device can be put to sleep via I^2C to reduce power consumption A block diagram of the device is shown in Fig. 23.5.

The I^2C device address of the LM48100Q-Q1 is 1111 10X, where X is determined by the ADR line. ADR = 1 sets the device address to 1111 101. ADR = 0 sets the device address to 1111 100.

The device is programmed by sending a byte (MSB first) to the device address (MSB first) with the meanings given in Table 23.2.

In the event of one of the above error conditions, the /FAULT line is pulled from open-drain to ground and can thus be read out via a pin.

Specifically, the bits mean:

- **Power On**: 1 switches the amplifier on, 0 switches the amplifier off
- **INPUT_X**: 1 switches input path X on, 0 switches it off
- **DG_EN**: Switches on the diagnostics with 1 (The diagnostics are always run through once during a reset).
- **DG_CONT**: With 1 switches on a continuous diagnostic mode, i.e. output short-circuit on V_{DD} and GND, outputs short-circuited against each other and "no load", i.e. line break are measured continuously. With 0, diagnostics is only performed once after a start with DG_EN. Even without diagnostics, the protective function is maintained.
- **DG_RESET**: After an error, the error state is cleared with a 1 and the/FAULT line is set to open-drain again.

The output tests of the LM48100Q are individually controlled by the Fault Detection Control register. If one of the bits in the Fault Detection Control register is set to 1, the FAULT circuit ignores the associated test. For example, if B2 (RAIL_SHT) = 1 and the output is shorted to V_{DD}, the FAULT output will remain high. Although the FAULT circuit ignores the selected test, the LM48100Q's protection circuitry still remains active and shuts down the device. When DG_EN = 1 and a diagnostic sequence is initiated, all tests are performed regardless of the state of the fault detection control register. When DG_EN = 0, the RAIL_SHT, OUTPUT_OPEN, and OUTPUT_SHT tests are not performed, but the thermal overload and output overcurrent detection circuits remain active [5].

The bits mean:

- **OUTPUT_SHT**: Short circuit between the balanced speaker lines
- **OUTPUT_OPEN**: Line break
- **RAIL_SHT**: Short circuit to V_{DD} or GND
- **OCF**: Over current
- **TSD**: Thermal overload

Table 23.2 Messages for the LM481000Q

Adr	Name	B7	B6	B5	B4	B3	B2	B1	B0
0	MODE CONTROL	0	0	0	Power on	Input 2	Input 1	0	0
1	DIAGNOSTIC CONTROL	0	0	1	DG_EN	DG_Cont	DG_Reset	ILimit	0
2	FAULT DETECTION CONTROL	0	1	0	TSD	OCF	RAIL_SHT	OUTPUT_OPEN	OUTPUT_SHORT
3	VOLUME CONTROL 1	0	1	1	VOL_1_4	VOL_1_3	VOL_1_2	VOL_1_1	VOL_1_0
4	VOLUME CONTROL 2	1	0	0	VOL_2_4	VOL_2_3	VOL_2_2	VOL_2_1	VOL_2_0

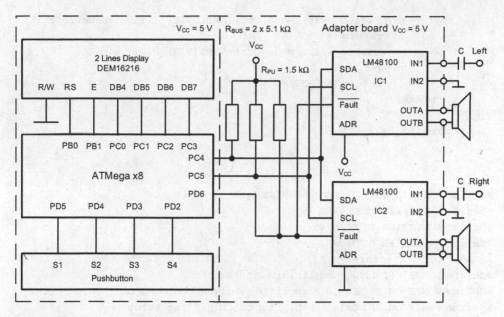

Fig. 23.6 Wiring of a stereo amplifier with LM48100

Finally, the VOLUME CONTROL bits are set in steps of $0 \ldots 31$ which corresponds to a gain of -80 dB $\ldots +18$ dB, according to a listening curve. This is tabulated in the data sheet [5].

Figure 23.6 shows the circuitry of a stereo amplifier with the LM48100 device. The two amplifiers are controlled by an ATmega88 microcontroller.

After initialization of the TWI interface, the two amplifiers are switched on and off with the LM48100_Set_Power function.

```
uint8_t LM48100_Set_Power(uint8_t ucampl_enable,
                          uint8_t ucpowerleft,
                          uint8_t ucpowerright)
{
    switch(ucampl_enable)
    {
    case AMPLIFIER_LEFT:
        if(LM48100_Write_Mode_Ctrl(AMPLIFIER_LEFT, ucpowerleft))
            return TWI_ERROR;
        break;
    case AMPLIFIER_RIGHT:
        if(LM48100_Write_Mode_Ctrl(AMPLIFIER_RIGHT,
                                   ucpowerright))
            return TWI_ERROR;
        break;
    case AMPLIFIER_BOTH:
```

```
            if(LM48100_Write_Mode_Ctrl(AMPLIFIER_LEFT, ucpowerleft))
                return TWI_ERROR;
            if(LM48100_Write_Mode_Ctrl(AMPLIFIER_RIGHT,
            ucpowerright))
                return TWI_ERROR;
        break;
    }
    return TWI_OK;
}

Whereas:
#define LM48100_DEVICE_TYPE_ADDRESS 0xF8
#define AMPLIFIER_LEFT 0x00
#define AMPLIFIER_RIGHT 0x01
#define AMPLIFIER_BOTH 0x02
//Mode Control Register
#define POWER_OFF 0x00 //amplifier switched off
#define POWER_ON_MUTE 0x10 //amplifier switched on, mute on
#define POWER_ON_IN1 0x14 //Amplifier on, input 1 selected
#define POWER_ON_IN2 0x18 //amplifier on, input 2 selected
#define POWER_ON_IN1_IN2 0x1C //amplifier on, input1 and 2 selected
```

With the call: LM48100_Set_Power(AMPLIFIER_BOTH, POWER_ON_IN1, POWER_ON_IN1); the two amplifiers are switched on and, according to the wiring shown in Fig. 23.6, input IN1 is configured as signal input for both.

The loudness can be set separately for each amplifier. When calling the function LM48100_Set_Loudness, the address of the device and the gain factors for each signal input are entered as parameters. The gain factors must be in the range 0 ... 31, otherwise they are set to 0 in the function after the check. This check is intended to prevent incorrect setting of the amplifier, since bits 5, 6 and 7 determine the internal address of the amplifier.

```
uint8_t LM48100_Set_Loudness(uint8_t ucdevice_address,
                             uint8_t ucgain1,
                             uint8_t ucgain2)
{
    uint8_t ucDeviceAddress, ucVolumeCtrlReg1, ucVolumeCtrlReg2;
    //address of the LM48100 audio amplifier form
    //(up to 2 audio amplifiers can be connected to one I2C bus)
    ucDeviceAddress = (ucdevice_address << 1) |
                    LM48100_DEVICE_TYPE_ADDRESS;
    ucDeviceAddress |= TWI_WRITE; //write mode
    if(ucgain1 > 31) ucgain1 = 0;
    if(ucgain2 > 31) ucgain2 = 0;
```

```
// VOLUME_REG_1_ ADDR = 0x60
ucVolumeCtrlReg1 = VOLUME_REG_1_ADDR | ucgain1;
// VOLUME_REG_2_ ADDR = 0x80
ucVolumeCtrlReg2 = VOLUME_REG_2_ADDR | ucgain2;
TWI_Master_Start();
if((TWI_STATUS_REGISTER) != TWI_START) return TWI_ERROR;
//send device address
TWI_Master_Transmit(ucDeviceAddress);
if((TWI_STATUS_REGISTER) != TWI_MT_SLA_ACK) return TWI_ERROR;
//the gain factor for input IN1 is transmitted
TWI_Master_Transmit(ucVolumeCtrlReg1);
if((TWI_STATUS_REGISTER) != TWI_MT_DATA_ACK) return TWI_ERROR;
//the gain factor for input IN2 is transmitted
TWI_Master_Transmit(ucVolumeCtrlReg2);
if((TWI_STATUS_REGISTER) != TWI_MT_DATA_ACK) return TWI_ERROR;
//stop TWI_Master_Stop();
return TWI_OK;
}
```

References

1. Werner, M. (2006). *Nachrichten-Übertragungstechnik*. Vieweg.
2. Silicon Labs. SI4840/44 – Broadcast Analog Tuning Digital Display AM/SW/FM Radio Receiver. (2015). www.silabs.com. Accessed 27 Feb 2021.
3. Silicon Labs. AN610 – SI48XX ATDD Programming Guide V3.0/2013. (2015). www.silabs.com. Accessed 27 Feb 2021.
4. Silicon Labs. AN602 – SI4822/26/27/40/44 Antenna, Schematic, Layout and Guidelines. (2015). www.silabs.com. Accessed 27 Feb 2021.
5. Texas Instruments LM48100Q Boomer™ 1.3-W Audio Power Amplifier. https://www.ti.com/product/LM48100Q-Q1. Accessed 27 Feb 2021.

Networkable Integrated Circuits

Abstract

In this chapter some networkable ICs are presented: A port expander, a digital variable resistor and a real-time clock.

This chapter describes a loose collection of further networkable devices that are practical in various contexts. Port expanders are used to significantly increase the number of digital inputs and outputs of a microcontroller, in the case of the PCF8574 presented here up to 128. Adjustable resistors decouple the analog from the digital world and a real-time clock can be found in networks with intelligent power management, in which nothing but the clock has to run in idle mode.

24.1 PCF8574: Port Expander

In some applications, the host microcontroller has too few I/O pins to serve all elements of the circuit. One way to solve this problem is to use a so-called port expander. The PCF8574 [3] device is such a port expander device that can add an 8-bit port to the peripherals of a microcontroller. It requires only the two lines of an I^2C bus for communication. The device is manufactured in two versions which are identical except for the device type address: 0x40 for PCF8574 and 0x70 for PCF8574A. Thanks to the three address ports, up to eight identical devices from each version can be connected to the same bus, which corresponds to a peripheral expansion of a microcontroller of up to 16 ports. The internal power-on reset circuit (see Fig. 24.1) ensures that all I/O pins are switched high after the device is powered up. This behavior is different from that of the microcontrollers from the ATmega family.

A. Meroth, P. Sora, *Sensor networks in theory and practice*,
https://doi.org/10.1007/978-3-658-39576-6_24

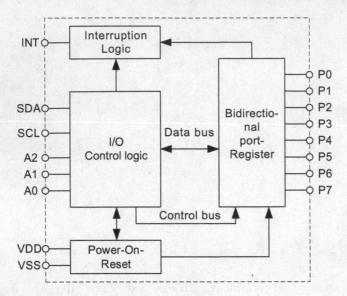

Fig. 24.1 PCF8574 – Block diagram

The microcontroller software must ensure that the I/O pins are controlled appropriately during the initialization phase. Communication with the device takes place via I²C in standard mode with up to 100 kBit/s.

24.1.1 Output Stage of an I/O Pin

When an I/O pin is switched low, transistor T2 in Fig. 24.2 can conduct a current of 25 mA to ground. When the pin is connected high, the current is determined by the constant current source (between 30 . . . 300 µA). This wiring allows each I/O pin to be used as an output or input independently of the other pins under the following conditions:

- the minimum output current in the high state must be sufficient to control an actuator;
- before a pin is used as an input, it must first be switched high. At maximum current of the current source, the level of the input voltage with a connected component at the input should still be in the range where it is detected as low.

24.1.2 Output Port Mode

Figure 24.3a shows an example of switching all I/O pins of a PCF8574 device from low to high. The device with the device type address 0x40 has the address input A0 switched to high and the other two to low. The host microcontroller, which is configured as the master,

Fig. 24.2 PCF8574 – Final
stage of an I/O pin

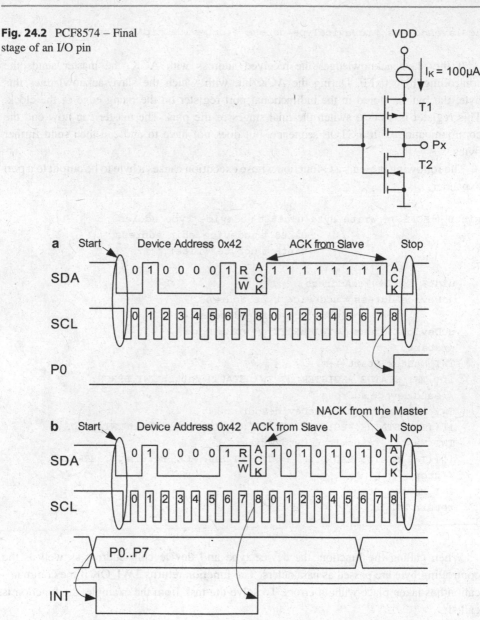

Fig. 24.3 PCF8574 – Input/output of a byte

initiates communication with the slave with a START sequence and checks whether the bus
is free. If so, it addresses the slave with its address and the R/W bit set to 0. The slave
address is composed of the OR operation of the device type address and the device chip
address shifted to the left by one digit:

```
ucSlaveAddress = ucDeviceTypeAddress | (ucDeviceChipAddress << 1);
```

If the slave acknowledges the received address with ACK, the master sends the controlling byte 0xFF. During the ACK bit, with which the slave acknowledges this byte, the byte is stored in the bidirectional port register on the rising edge of the clock. This register is used to switch the final stages of the pins. The master can now end the communication with a STOP sequence, but does not have to and can then send further bytes.

The following presents a C-function whose execution causes a byte to be output to a port expander.

```c
uint8_t PCF8574_Write_Byte(uint8_t ucdevice_type_address,
                           uint8_t ucdevice_chip_address,
                           uint8_t ucbyte2write)
{
    uint8_t ucDeviceAddress;
    ucDeviceAddress = ucdevice_type_address |
                    (ucdevice_chip_address << 1);
    ucDeviceAddress |= TWI_WRITE; //write mode
    //start
    TWI_Master_Start();
    if((TWI_STATUS_REGISTER) != TWI_START) return TWI_ERROR;
    //send device address
    TWI_Master_Transmit(ucDeviceAddress);
    if((TWI_STATUS_REGISTER) != TWI_MT_SLA_ACK) return TWI_ERROR;
    TWI_Master_Transmit(ucbyte2write);
    if((TWI_STATUS_REGISTER) != TWI_MT_SLA_ACK) return TWI_ERROR;
    //stop
    TWI_Master_Stop();
    return TWI_OK;
}
```

When calling the function, the device type and device chip address as well as the controlling byte are passed as parameters. The function returns TWI_OK if the communication has taken place without errors. To solve the task from the example, the function is called:

```c
PCF8574_Write_Byte(0x40, 0x01, 0xFF);
```

24.1.3 Input Port Mode

Figure 24.3b shows the time sequences that occur when the entire port of a PCF8574 device is read out once. This device is wired as described above and the logic levels 1010 1010 are applied to its pins P7 ... P0. The slave is addressed this time with the R/W bit set to 1. During the ACK bit, which the slave uses to acknowledge the received address, the logic input levels on the rising edge of the clock are stored in the bidirectional port register and output clocked on the serial data line. The master receives the byte, acknowledges it with NACK, and terminates the communication with a STOP sequence or may respond with ACK and receive another byte. When the communication is terminated, the contents of the data register are cleared. The following C-function illustrates the reading of an extension port, which is addressed with the addresses ucdevice_type_address and ucdevice_chip_address and stores the read byte at the address ucbyte2read.

```
uint8_t PCF8574_Read_Byte(uint8_t ucdevice_type_address,
                          uint8_t ucdevice_chip_address,
                          uint8_t* ucbyte2read)
{
    uint8_t ucDeviceAddress;
    ucDeviceAddress = ucdevice_type_address |
                      (ucdevice_chip_ address << 1);
    ucDeviceAddress |= TWI_READ; //write mode
    //start
    TWI_Master_Start();
    if((TWI_STATUS_REGISTER) != TWI_START) return TWI_ERROR;
    //send device address
    TWI_Master_Transmit(ucDeviceAddress);
    if((TWI_STATUS_REGISTER) != TWI_MT_SLA_ACK) return TWI_ERROR;
    *ucbyte2write = TWI_Master_Read_NAck();
    if((TWI_STATUS_REGISTER) != TWI_MR_DATA_NACK) return TWI_ERROR;
    //stop
    TWI_Master_Stop();
    return TWI_OK;
}
```

The bit duration at maximum clock frequency is:

$$t_{\text{Bit}} = \frac{1}{f_{\max}} = \frac{1}{100.000} = 10\mu s$$

which means that the time interval between two consecutive ACK bits is 90 μs. Level changes shorter than 90 μs that occur in the time between two ACK bits cannot be detected, even if the microcontroller continuously reads the port.

24.1.4 Interrupt Mode

In a microcontroller system, it is necessary to ensure the detection of changes in all input levels. The PCF8574 device has digital logic that allows an output to be activated whenever there is a logical level change on an I/O pin. This activation can signal to the microcontroller via an interrupt that the input constellation has changed and the microcontroller can specifically start a read operation. This principle is similar to the pin change interrupt of a microcontroller with the difference that with the PCF8574 it can only be determined which input has changed by comparing it with the old value. Figure 24.3b shows the change in inputs P7 ... P0 that caused the INT output to go low. The microcontroller responds and starts reading a new value. During the ACK bit, which the slave uses to acknowledge its own address, the input value is stored in the bidirectional port register on the rising edge of the clock and the interrupt is disabled. The new value must be present unchanged at the input of the device for at least 90 µs so that it can be detected by the microcontroller.

The INT output uses an open-drain transistor as the output stage, which requires an external drain resistor (pull-up). Multiple INT outputs connected in parallel use the same resistor and allow a microcontroller to monitor multiple input ports. Compound INT outputs multiply the time that the input values must remain unchanged in order to detect any change. To detect the trigger of an interrupt, the microcontroller must read all devices in sequence (1 address byte +1 data byte per device). A triggered interrupt can only be deactivated by accessing the triggering device. Short pulses can trigger interrupts, but are not always detected.

24.1.5 PCA9534

The PCA9534 [4] is a further development of the PCF8574 port expander with the same pin assignment. The device even has the same device type address 0x40, but meets the requirements of the I^2C Fast-mode and can transmit and receive data at up to 400 kBit/s. In this case, a duration of approx. 22.5 µs is calculated for the transmission of a byte. Internally, it has a similar structure, but also has four additional registers that control the data flow.

The configuration register is accessed via command code 0x03 and similar to the DDR register of the ATmegax8 microcontroller, it determines the data flow direction of each port pin. Resetting a bit in this register switches the corresponding port pin to output. When the device is powered up, the entire port is initialized to input. Using the output register (command code 0x01), setting each port pin to high with 1 and low with 0. Setting a bit to 1 in the Polarity Inversion register (command code 0x02) causes bitwise inversion when reading the corresponding input. The contents of these three registers can be changed with an I^2C write command as shown in Fig. 24.4, or read with a read command. The ACK bit stores the data byte into the selected register. Further data bytes can be transferred after the first one. Attention: reading the output register provides the state of the output flip-flops and

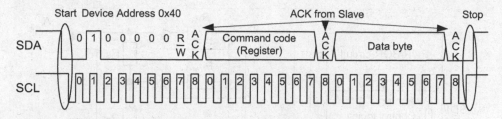

Fig. 24.4 PCA9534 – Writing registers

not the output levels! If the input port register was previously selected in a write cycle, reading in data proceeds exactly as with the PCF8574.

24.2   MCP41X1 Digital Variable Resistors

Digital variable resistors are integrated circuits containing a voltage divider consisting of 2^n equal resistors, where n is the resolution in bits. They are manufactured as potentiometers or as rheostats[1] (see Fig. 24.5). The electrical resistance between the end B and the wiper can only be changed in steps. The devices can be controlled via I^2C or SPI. The wiper position can be stored either in a RAM or in an EEPROM. In the latter case, the position is retained even in the de-energized state. Well-known manufacturers such as Microchip Technology Inc., Xicor, Analog Device, Maxim Integrated or Dallas Semiconductor produce digital resistors with different values and bit resolutions. In the following example, the MCP41X1 series of digital potentiometers is examined in more detail.

MCP41X1 [1] is a series of single digital potentiometers that have 7/8-bit resolution and are controlled via SPI. They can be supplied with voltages between 2.7 V and 5.5 V, while the other terminals can tolerate voltage levels of up to 12.5 V. The digital potentiometers can be used as voltage dividers with adjustable divider ratio to replace analog potentiometers in audio amplifiers, or for setting an offset voltage. They can also be used in filters with adjustable amplitude response. In this case, attention must be paid to the bandwidth of the devices. They have a $-$ 3-dB bandwidth of 2 MHz for the 5 kΩ- and of

Fig. 24.5 Adjustable resistors:
left potentiometer, right rheostat

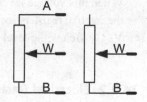

---

[1] Literally translated from the Greek as flow adjuster: an adjustable resistor.

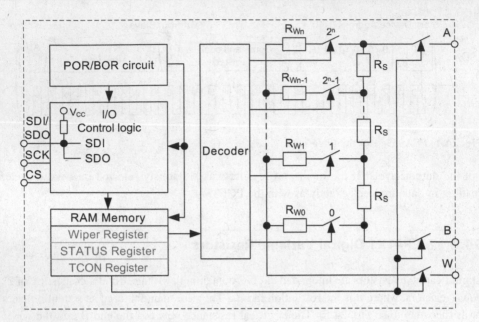

Fig. 24.6 MCP4151 – Block diagram

100 kHz for the 100 kΩ-potentiometers. A block diagram of an MCP4151 device is shown in Fig. 24.6.

24.2.1 Power On/Brown Out Reset Circuitry

The POR/BOR circuit in the MCP4151 ensures that the device assumes a defined state after power-up and when the supply voltage falls below a critical threshold (brown out). This is done in the following way:

1. the digital communication is enabled and the internal registers are loaded with the default values as soon as the supply voltage exceeds a level V_{BOR};
2. When this voltage level is undershot, digital communication is disabled. The contents of the internal registers are retained below this level until the supply voltage reaches the value V_{RAM}. Below the level V_{RAM} the contents of the individual registers can be damaged

24.2.2 Electrical Resistance

The electrical resistance of the MCP4151 device consists of a series connection of 2^n equal partial resistances R_S, where in this case n = 8.

Table 24.1 MCP4151 –
Registers

Register name	Register address	Default value
Wiper register	0x00	0x80
TCON register	0x04	0x1FF
Status register	0x05	0xE0

$$R_S = \frac{R_{AB}}{256} \qquad (24.1)$$

The two end connections of the potentiometer as well as the wiper are made at pins A, B and W respectively via analog switches. According to the manufacturer, the current flow at pins A and W can be bidirectional. An analog switch is connected at each junction of two partial resistors. The outputs of all switches are connected together and form the wiper of the potentiometer. In addition to these $2^n - 1$ places, the wiper should also be able to be connected directly to the end connections A and B. This results in $2^n + 1 = 257$ wiper positions. The analog switches are realized with FET[2] transistors and symbolized by the series connection of an electrical contact and a resistor in the block diagram. The partial resistances and the total resistance R_{AB} change their value only slightly with temperature and supply voltage. The wiper resistance depends on the position and strongly on the temperature and supply voltage. It has a greater influence on the linearity of the voltage with smaller R_{AB} resistors.

24.2.3 Potentiometer Registers

The module has a RAM memory in which the wiper positions and the configuration of the module are stored. After switching on, the desired settings must be transferred to the memory. It consists of three registers with a size of 9 bits to store the 257 wiper positions. The addresses of the registers and the default values, which are loaded immediately after switching on, can be found in Table 24.1.

The **Wiper register** stores and controls the wiper position via a decoder. After power-up, this register is initialized with the value 0x80 (decimal 128) for the 8-bit, or 0x40 (64) for the 7-bit potentiometers, which corresponds to the center of the total resistance. If the register is loaded with 0x00, the slider is switched to the B-end and for all values greater than 0xFF, the wiper is switched to the A-end of the potentiometer.

The status register can only be read.

- **Bit 8:5** – are reserved, permanently switched to 1;
- **Bit 4:2** – are reserved;

---

[2]FET – Field Effect Transistor.

Fig. 24.7 MCP41X1 in active
mode (**a**) and in shutdown mode
(**b**)

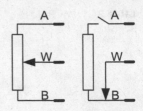

- **Bit 1** – SHDN; this bit indicates whether the device is in hardware shutdown mode when it is set to 1 (only for devices that have a SHDN pin). In this mode, pin A of the device is disconnected from the end terminal of the potentiometer and the wiper is directly connected to pin B (Fig. 24.7).
- **Bit 0** – is reserved;

The **Terminal Control Register** (TCON) controls the connections of the digital potentiometer. After power-up all bits are set to 1.

- **Bit 8** – is reserved;
- **Bit 7:4** – these bits are relevant for the devices with 2 potentiometers
- **Bit 3** – Mode control bit (1 – active mode, 0 – shutdown mode)
- **Bit 2** – switch control of the A pin (1 – the A pin is connected to the end terminal, 0 – the A pin is disconnected from the end terminal)
- **Bit 1** – switch control of the W pin (1 – the W pin is connected to the wiper, 0 – the W pin is disconnected from the wiper)
- **Bit 0** – switch control of the B pin (1 – the B pin is connected to the end terminal, 0 – the B pin is disconnected from the end terminal).

If bit 3 in register TCON is set to 1, the device is set to shutdown mode. However, this changes neither the bits of this register nor those of the other registers. After resetting this bit, the old state is restored.

The I/O control logic implements a 3-line SPI-compliant interface that allows data to be exchanged between an SPI master and the device's internal memory. With this interface, the internal SDI[3] and SDO[4] signals use a single external line, which is equivalent to a "wired and". For this wiring to work without problems, the SDI/SDO pin must be connected to the microcontroller's MOSI pin via a resistor (see Fig. 24.8) so that two outputs are not connected together during data readout. The internal pull-up resistor is activated by the SDO signal during data output and enables bidirectional data communication, but limits the clock frequency of the interface. This clock frequency can be increased by connecting an external pull-up resistor.

---

[3] SDI – Slave Data In.
[4] SDO – Slave Data Out.

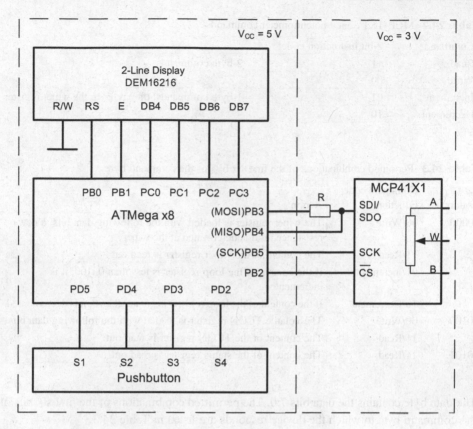

Fig. 24.8 MCP41X1 – Connection to a microcontroller

Thanks to the compatibility of voltage levels above the supply voltage on all SPI pins, the device can function without additional level switches in circuits in which the master and the slave are supplied with different voltages (split-rail applications).

24.2.4 Control Functions of the MCP4151 Device

In order to ensure the data flow to and from the internal memory, the device provides four general functions, which are summarized in Table 24.2.

The first byte is called the command byte, the second, if it exists, is called the data byte. The structure of the command byte is as follows:

- **Bit 7:4** – these bits form the address of the internal register (Table 24.1);
- **Bit 3:2** – the bits encode the command code (Table 24.2);
- **Bit 1** -is evaluated as data bit 9 and used as ERROR bit;
- **Bit 0** – is used as data bit 8 to control all switches.

Table 24.2 MCP41X1 digital potentiometer commands

Command	2-bit instruction code	
Read	11	2-byte commands
Write	00	
Increment	01	1-byte commands; they refer to the wiper register
Decrement	10	

Table 24.3 Permitted combinations of the first six bits of the command byte

Address	Command code/ Function	Action
0000	00/Write	The wiper register is loaded with the following data bits; a new resistance ratio is determined at the wiper
	11/Read	The content of the wiper register is read out.
	01/Increment	If the content of the loop register is less than 0x100, it is incremented.
	10/Decrement	If the content of the loop register is greater than 0, it is decremented.
0100	00/Write	The volatile TCON register is loaded with the following data bits
	11/Read	The content of the TCON register is read out.
0101	11/Read	The content of the status register is read out.

The data byte contains the data bits 7:0. The permitted combinations of the first six bits of the command byte to which the device responds are listed in Table 24.3.

24.2.5 SPI Communication

The MCP41X1 devices are controlled via an SPI interface in mode 0 or 3 with a clock frequency of up to 10 MHz (up to 250 kHz when executing a read command). The most significant bit of a byte is transmitted first. For the wiring shown in Fig. 24.8, the SPI data structure of the device is (see Chap. 14, para. 5):

```
MCP41X1_pins MCP41X1_1 = {{/*CS_DDR*/ &DDRB,
                          /*CS_PORT*/ &PORTB,
                          /*CS_pin*/ PB2,
                          /*CS_state*/ ON}}; //ON = 1
```

To ensure control compatibility with other families of digital potentiometers, a master can initiate communication either:

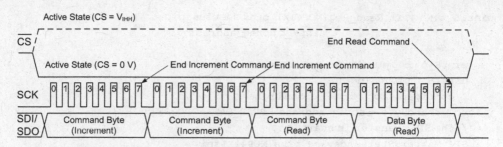

Fig. 24.9 MCP41X1 – Chained functions

- by switching the chip select line from high (+5 V) to low (0 V), or
- by switching the line from high (+5 V) to a higher voltage level V_{IHH} (+8.5 V ... +12 V) (see Fig. 24.9).

At the beginning of the communication the pin SDI/SDO is switched to input and receives the incoming bits. The state machine of the device evaluates the first six bits of the command byte and compares them with the possible combinations listed in Table 24.3. During these first six bits of the transmission, the signal SDO is internally switched to "1". At the end of the sixth clock period the pin SDI/SDO is switched to output and if no allowed combination was detected, the signal SDO is internally switched to low. This also switches the physical output low. The output retains this state until communication is terminated by switching the chip select line to the inactive state (+5 V). If the 6-bit combination of the command byte is evaluated as correct:

- the SDI/SDO pin remains on output if an incoming read command was detected, or
- the pin is switched to input to receive the data bits.

The individual instructions are executed while the chip select line is in the active state at the end of the eighth clock period for the 1-byte instructions or the 16th clock period for the 2-byte instructions and not – as one might think – after the chip select line has switched from the active to the inactive state. Thus, it is possible to chain the functions if necessary (Fig. 24.9). In this example, the Increment function is executed twice in succession and the Read function once.

24.2.6 Software Example

In the following the functions Read() and Increment() are presented as examples.

```
uint16_t MCP41X1_Read_Reg(MCP41X1_pins sdevice_pins,
                          uint8_t ucreg_address)
{
    uint8_t ucCommandByte, ucDataIn;
    uint16_t uiDataIn = 0;
    //configuration of the command byte
    ucCommandByte = ucreg_address | READ_DATA_OPCODE;
    //the SPI transfer is started
    SPI_Master_Start(sdevice_pins.MCP41X1spi);
    //the command byte is transmitted, the ninth data bit of the register
    //is read in
    ucDataIn = SPI_Master_Write(ucCommandByte);
    uiDataIn = ucDataIn << 8;
    //the data bits 7:0 are read in; the transfer of the dummy byte
    //0xFF enables the detection of a possible transmission error
    ucDataIn = SPI_Master_Write(0xFF);
    //the content of the read register is merged
    uiDataIn |= ucDataIn;
    //the SPI transmission is terminated
    SPI_Master_Stop(sdevice_pins.MCP41X1spi);
    return uiDataIn;
}
uint8_t MCP41X1_Increment_Reg(MCP41X1_pins sdevice_pins,
                              uint8_t ucreg_address)
{
    uint8_t ucCommandByte, ucDataIn;
    //configuration of the command byte
    ucCommandByte = ucreg_address | INCREMENT_OPCODE;
    //the SPI transfer is started
    SPI_Master_Start(sdevice_pins.MCP41X1spi);
    //the feedback is read in
    ucDataIn = SPI_Master_Write(ucCommandByte);
    //the SPI transmission is terminated
    SPI_Master_Stop(sdevice_pins.MCP41X1spi);
    return ucDataIn;
}
```

From these general functions, other specific functions can be created, such as a function that checks if the device is in shutdown mode. This function could look like this:

```
uint8_t MCP41X1_Get_Status(MCP41X1_pins sdevice_pins)
{
    uint8_t uiDataIn;
    uiDataIn = MCP41X1_Read_Reg(sdevice_pins, STATUS_REGISTER);
    if(uiDataIn & 0x0002) return SHUTDOWN_ON;
    else return SHUTDOWN_OFF;
}
```

where the constants STATUS_REGISTER, SHUTDOWN_ON and SHUTDOWN_OFF must of course be defined in the .h file as distinguishable integer constants.

24.3   MAX31629: Real Time Clock (RTC)

A real-time clock is an integrated component that implements a digital clock that measures time in human-readable units and provides the information for retrieval. The clock is set externally and runs as long as it is powered. When the main power source is turned off, the clock must continue to be powered by a battery, accumulator, or supercapacitor (supercap) because the time is stored in SRAM memory. For this reason, the power requirements of such devices must be low. Some devices have an additional power source integrated in the package. Manufacturers such as Maxim Integrated, NXP Semiconductors (formerly Philips), Texas Instruments and others produce RTC devices. A clock crystal with a frequency of 32,768 Hz is used as the clock source. This number is a power of two, so by repeatedly halving the basic frequency (15 times), a frequency of 1 Hz is achieved, which corresponds to a period of one second. This frequency is used to drive the clock. For a given period of time (usually a century), the leap years and the different lengths of the months are taken into account when calculating the date. As an example for such devices, a MAX31629 [2] is considered below, which implements the functions of a real-time clock with calendar up to the year 2100. In addition, it has a temperature sensor (see Fig. 24.10) with which the temperature can be measured from $-55\,°C$ to $+125\,°C$ with an accuracy of $\pm 2\,°C$ and a 32-byte SRAM memory that is freely available to the user.

An oscillator drives the real-time clock. Via an adjustable frequency divider the oscillator can drive an open-drain output, which can serve as a clock input for a microcontroller.[5] The frequency divider is set via the configuration register, as shown in Table 24.4.

When the clock output is not needed, it should be turned off to save power.

---

[5] For example, for the lowered clock in sleep mode, see Sect. 3.3.2.

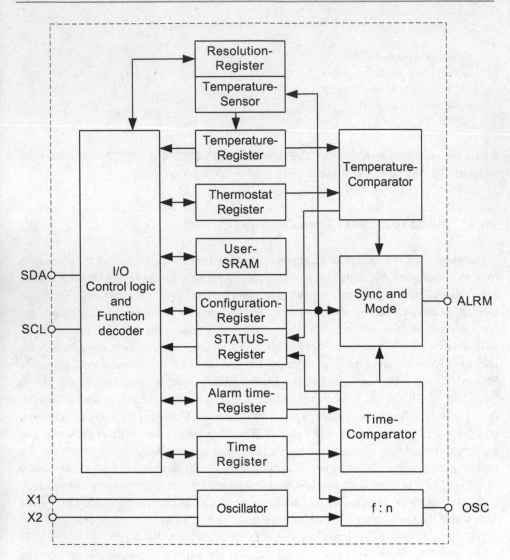

Fig. 24.10 MAX31629 block diagram

Table 24.4 Frequency divider clock

Configuration register		
Bit 7	Bit6	Clock output
0	0	Switched off
0	1	:8
1	0	:4
1	1	:1

Table 24.5 Memory organization of the MAX31629 device

Register-designation	Register elements	Logical address	Command code	Storage type	Access
Time register	Seconds	0x00	0xC0	SRAM	Write-read
	Minutes	0x01			
	Hours	0x02			
	Weekdays	0x03			
	Days	0x04			
	Months	0x05			
	Years	0x06			
Alarm time register	Seconds	0x00	0xC7	SRAM	Write-read
	Minutes	0x01			
	Hours	0x02			
	Weekdays	0x03			
Thermometer register	–	2 byte	0xAA	SRAM	Read only
Resolution register	–	1 byte	0xAD	EEPROM	Write-read
Thermostat-TH	–	2 byte	0xA1	EEPROM	Write-read
Thermostat-TL	–	2 byte	0xA2	EEPROM	Write-read
Configuration/STATUS register	–	1byte	0xAC	EEPROM	Write-read
	–	1byte		SRAM	Read only
User SRAM	–	32 bytes	0x17	SRAM	Write-read

24.3.1 Timing

The elements of the time: seconds, minutes, hours, weekdays, days, months and years occupy logical addresses byte by byte in a volatile memory area in ascending order starting with 0x00 (Table 24.5) and together form the memory area of the time registers. The values can be read or changed via the command code 0xC0. The seven values all have two digits because the years are counted from 0 to 99 (corresponding to the years 2000–2099) and are stored in BCD[6]-coded form. To binary encode a decimal digit requires four binary digits or a nibble, so a two-digit number requires eight bits or a byte. The least significant nibble stores the ones digit and the most significant nibble stores the tens digit of the decimal number. The following program code excerpt is intended to explain the BCD-decimal or the decimal-BCD conversion:

---

[6]BCD – Binary Coded Decimal.

```
//Decimal/BCD conversion
uint8_t ucDecimalNumber, ucBCDNumber;
//the ones digit of the decimal number is stored in the
//least significant nibble of the variable ucBCDNumber
ucBCDNumber = ucDecimalNumber % 10;
//the tens digit of the decimal number is calculated
ucDecimalNumber = ucDecimalNumber / 10;
//the decimal, binary coded number is completed
//with the higher value nibble
ucBCDNumber = ucBCDNumber + (ucDecimalnumber << 4);
//BCD/Decimal conversion
//the least significant nibble (the ones digit) is assigned
//to the decimal number
ucDecimalNumber = ucBCDNumber & 0x0F;
//the tens digit is calculated
ucBCDNumber = ucBCDNumber >> 4;
//the decimal number is completed with the tens digit
ucDecimalnumber = ucDecimalNumber + (ucdecimalNumber * 10);
```

The seconds byte counts the seconds from 0 to 59 (binary 0101 1001). These numbers in BCD format require only seven bits. The more significant bit of this byte, called the clock hold bit, turns the oscillator on at 0, or off at 1. The hours are counted in 12 h or 24 h format. When bit 6 from the hour byte is set to 1, the clock runs in 12 h format, the hour count takes values between 1 and 12, and bit 5 (AM/PM),[7] when 0, indicates that it is morning. If bit 6 of the hour byte is 0, then the clock runs in 24 h format and the number of hours takes values between 0 and 23. The days of the week are coded with numbers from 1 to 7, where 1 stands for Sunday.

24.3.2 Alarm Time

The MAX31629 device also offers an alarm time function. The alarm time, consisting of seconds, minutes, hours and day of the week, is stored in a volatile part of the memory starting at logical address 0x00. The default alarm time is Sunday, 12:00:00 AM. The hour formats of the time and the alarm time must be the same. The alarm can be activated or deactivated via the configuration register (see Table 24.6). An alarm is triggered when the reached time is equal to the set alarm time. This causes bit 7 (CAF = Clock Alarm Flag) of the status register to be set to 1 and if alarm mode 3 or 4 is selected in the configuration register, then the alarm output becomes active. The triggered alarm can be turned off as a result of an access (read or save) to the clock or alarm time registers. Bit 5 (CAL = Clock Alarm Latch) of the status register is set to 1 when the time alarm is triggered for the first time and is reset with the next power-up.

---

[7] AM – ante meridiem (morning), PM – post meridiem (afternoon).

Table 24.6 MAX31629 –
Alarm modes

Configuration register		
Bit5	Bit4	Alarm mode
0	0	1 – Deactivated
0	1	2 – Temperature only
1	0	3 – Time only
1	1	4 – Both

Table 24.7 MAX31629 – Temperature resolution of the thermometer

Bit resolution /bit	Temperature steps /°C	Max. Measuring duration /ms	Resolution register Bit 1	Bit 0
9	0.5	25	0	0
10	0.25	50	0	1
11	0.125	100	1	0
12	0.0625	200	1	1

Table 24.8 Temperature measuring mode

Configuration-register Bit 2	Bit 0	Measuring mode after switching on
0	0	Freewheel
0	1	One measurement; then single measurement
1	0	Standby; after the start of a temperature measurement, free-running mode
1	1	Single measurement

24.3.3 Temperature Measurement

The internal temperature sensor measures the temperature with a bit resolution of 9 to 12 bits, adjustable via the resolution register, as listed in Table 24.7.

Increasing the bit resolution by one doubles the measurement duration. This duration must be taken into account when reading out the temperature register, because the end of a temperature measurement is not signalled by the device. The measured temperature value is stored as an integer in a 2-byte register in two's complement with the $(16-n)$ least significant bits set to 0, where n is the set bit resolution. The high-order byte of the temperature register contains the measured temperature value in two's complement with a resolution of 1 °C, the low-order byte contains the decimal part. In this way, fixed-point numbers can be stored as integers.

The temperature measurement can be operated in free-running or single measurement mode. The measuring mode is determined via the configuration register (Table 24.8).

In free-running mode the last measured temperature value is available for reading in the temperature register, a read access to this register does not influence a running measurement. With a stop measurement function (command code 0x22) the free-running mode is temporarily stopped after the current conversion is completely finished and the measured value is stored. With a start measurement function (command code 0xEE), the stopped measurements are continued in free-running mode, or a new measurement is started in single measurement mode. With the following program excerpt a temperature measurement can be started:

```
char MAX31629_Start_TempMeasure(void)
{
    #define MAX31629_DEVICE_TYPE_ADDRESS 0x9E
    #define TEMP_MEASURE_START_OPCODE 0xEE
    uint8_t ucDeviceAddress;
    //write mode
    ucDeviceAddress = MAX31629_DEVICE_TYPE_ADDRESS | TWI_WRITE;
    //I²C-Master initiates communication
    TWI_Master_Start();
    if((TWI_STATUS_REGISTER) != TWI_START) return TWI_ERROR;
    //send device address
    TWI_Master_Transmit(ucDeviceAddress);
    if((TWI_STATUS_REGISTER) != TWI_MT_SLA_ACK) return TWI_ERROR;
    //the command code of the operation is sent
    TWI_Master_Transmit(TEMP_MEASURE_START_OPCODE);
    if((TWI_STATUS_REGISTER) != TWI_MT_DATA_ACK) return TWI_ERROR;
    //I²C-Master terminates the communication
    TWI_Master_Stop();
    return TWI_OK;
}
```

24.3.4 Thermostat with Alarm Function

In order to implement a thermostat function, the MAX 31629 has two 16-bit registers that store the upper and lower temperature thresholds. Two values are needed to ensure an adjustable hysteresis. The storage format of these two values is the same as the format of the thermometer. As soon as the measured temperature reaches or exceeds the set upper temperature threshold, the temperature comparator (Fig. 24.10) switches over and bit 6 (TAF – Thermal Alarm Flag) in the status register is set to 1. If alarm mode 1 or 3 (Table 24.6) is selected, the alarm output becomes active (Fig. 24.11). When the TAF bit is set for the first time, bit 4 (TAL – Thermal Alarm Latch) in the Status Register is also set and is not reset until the next power-up. This bit, when set to 1, indicates that the

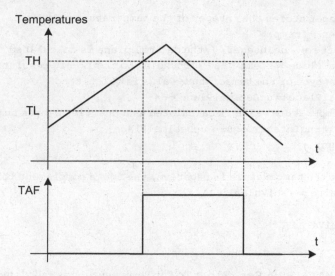

Fig. 24.11 MAX 31629 – Temperature alarm

Temperature Register																	
MSB								LSB									
±	2^6	2^5	2^4	2^3	2^2	2^1	2^0	,	2^{-1}	2^{-2}	2^{-3}	2^{-4}	2^{-5}	2^{-6}	2^{-7}	2^{-8}	°C

Fig. 24.12 MAX31629 – Temperature coding

temperature has reached or exceeded the upper temperature threshold at least once. The TAF bit is reset as soon as the measured temperature reaches or falls below the lower temperature threshold. The temperature resolution of the thermometer and the thermostat do not have to be the same.

The temperature threshold values must be coded according to Fig. 24.12 before transmission. The following program code can be used for this purpose.

```
uint16_t MAX31629_Set_FloatToIntTemp(float ftemp)
{
    uint8_t ucInteger, ucDecimalPlace, ucSign = 0;
    uint16_t uiValue;
    if(ftemp < 0)
    {
        //the variable ucSign stores the sign of the temperature value
        ucSign = 1;
        //the amount of the negative value is calculated
        ftemp = ftemp * (float)(-1.);
    }
```

```
//ucInteger stores the integer of the temperature value
ucInteger = ftemp;
ftemp = ftemp - ucInteger; //the decimal place is calculated
ucDecimalPlace = ftemp * 16; //rounding to the 12-bit resolution
//the low byte of the temperature value is calculated
ucDecimalPlace = ucDecimalPlace << 4;
//the temperature value is calculated as a positive 2-byte number
uiValue = (ucInteger << 8) + ucDecimalPlace;
if(ucSign)
{
    //if the temperature is negative, the two's complement formed
    uiValue = (~uiValue) + 1;
}
return uiValue;
}
```

When the `MAX31629_Set_FloatToIntTemp` function is called, the temperature value is passed as a fixed-point number as a parameter, the function converts this value and limits the result to a 12-bit resolution and then returns the coded value as a 2-byte number. The function can encode both positive and negative temperature values.

24.3.5 I²C Communication

Communication between a microcontroller configured as a master and the MAX31629 device as a slave takes place via I²C at a bit rate of max. 400 kBit/s. The device type address of the RTC device is 0x9E. The individual memory areas of the device (Table 24.5) can be addressed and read or changed via command codes. Different access procedures are shown in Fig. 24.13.

(a) Storing in a single 2-byte register, as with the thermometer or thermostat, among others, is shown in (a). After the master has initiated communication with a start sequence and determines that the bus is free, it sends the device address with the R/W bit set to 0 (write access). Further, it transmits the command code for the desired register and then transmits two bytes of data with the most significant byte first. With a stop sequence, the master terminates the communication and releases the bus again. The slave responds with ACK after each byte received. Saving to the resolution or configuration register is similar, except that the master terminates communication after the first data byte has been sent.

(b) Reading a 2-byte register such as the thermometer, thermostat or configuration and status registers is done after operation (b), which is the same as the operation described at (a) until the command code is sent. With a restart sequence after the command code is sent and the slave responds, the master sends the device address with the R/W bit set

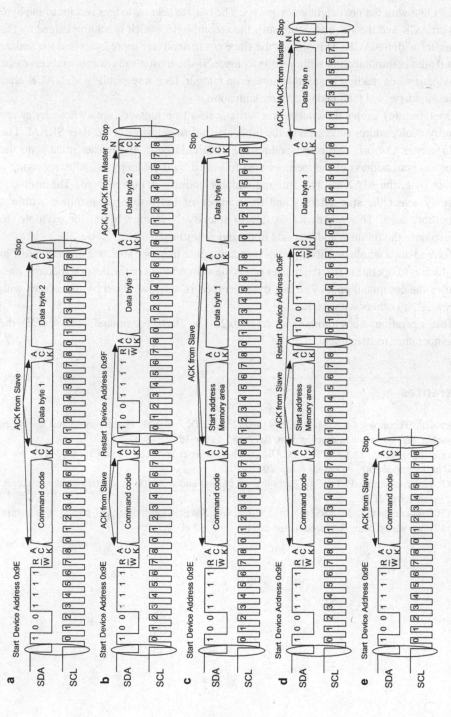

Fig. 24.13 MAX31629 – I²C communication

to 1. This signals the slave to output the higher order byte of the desired register on the data line with the next eight clock pulses. The master acknowledges receipt of the byte with ACK and the slave also transmits the second byte, which is acknowledged by the master with a NACK. This signals the slave not to send any more bytes and the master ends the communication with a stop sequence. If the master only wants to receive one byte, as when reading the temperature as an integer, then it responds with NACK after the first byte and then ends the communication.

(c) Operation (c) shows the course of a write access to a memory area whose bytes are individually addressable such as the time, alarm time registers or the user SRAM. The difference to (a) is that after sending the command code, the master must send the logical start address of the memory area followed by the data bytes. After receiving a data byte, the MAX31629's internal address counter is incremented. The user can freely select the start address and the number of bytes to be transmitted within a memory area. This makes it possible to selectively change bytes, for example, to zero only the minutes and seconds in the clock setting.

(d) To read out a whole or only a part of an addressable memory area, use operation (d). In addition to operation (b), the master sends the start address of the desired memory area after the command code. The last byte received is acknowledged by the master with NACK, all others with ACK.

(e) This operation stops (command code 0x22) or starts (command code 0xEE) the temperature measurement.

References

1. Microchip Technology Inc. MCP413X/415X/423X/425X – digitale Potentiometer mit flüchtigen Speicher. (2008). www.microchip.com. Accessed 4 Apr 2021.
2. Maxim Integrated. MAX31629 – I2C Digital Thermometer and Real-Time-Clock. (2015). www.maximintegrated.com. Accessed 4 Apr 2021.
3. NXP Semiconductors. PCF8574 – Remote 8-bit I/O expander for I^2C-bus. (2013). www.nxp.com. Accessed 4 Apr 2021.
4. NXP Semiconductors. PCA9534–8-bit I^2C-bus and SMBus low power I/O port with interrupt. (2017). www.nxp.com. Accessed 4 Apr 2021.

Displays

2

Abstract

The chapter gives an introduction to display technology and introduces a family of graphical and textual displays that can be easily integrated into the environment of an AVR controller.

The human-machine interface is not always visible or necessary in embedded systems. Nevertheless, there are always cases where the system state needs to be visualized. The chapter therefore gives an introduction to display technology and introduces a family of graphical and textual displays that can be integrated into the environment of an AVR controller without effort.

Displays provide an interface between a technical system and users of that system. They provide information about the state of the system, display measured values and provide feedback on user-controlled actions. The displays differ in the amount of information, information density, energy consumption, control complexity, readability and price.

25.1 Introduction

25.1.1 Display Layout

Alphanumeric displays are designed to display m characters using a fixed arrangement of n picture elements. The seven-segment and fourteen-segment displays are among the best-known arrangements and are used to display the digits and, to a certain extent, the letters.

The picture segments are directly controlled by a display controller. To reduce the number of control pins, the same segments of all characters are connected in parallel and the segments are controlled in multiplex mode.

In dot matrix displays, the picture elements are arranged as pixels in a matrix with m rows and n columns. By placing three picture elements with the colors red, green and blue close to each other at each matrix node, a fully color-capable display is created. With control in multiplex mode, the brightness of each of the $3 \times m \times n$ picture elements can be changed. By superimposing the three colors, the human eye perceives the pixel color. With a dot matrix display, information can be presented in alphanumeric form in the same way as images.

25.1.2 Emissive and Non-emissive Displays

A distinction is made between emissive and non-emissive displays. A non-emissive display requires backlighting, which is realized either with an artificial light source or by reflection of daylight. The LC[1] displays belong to the non-emissive displays and can be built as passive or active matrix.

25.1.2.1 Liquid Crystal Displays (LCD)

Crystals have the property of anisotropy, which means that certain properties of the material are direction-dependent. These include, among others, the speed of light and thus also the refractive index. These are related to the direction of polarization of light. The direction of polarization is the direction of the plane of oscillation of a transverse wave, in the case of light the direction of the electric field strength. Unpolarized light can be divided into a vertically and a horizontally polarized component. If it passes through the crystal, the two components are refracted to different extents, and two beams of polarized light emerge from the crystal. The superposition results in a polarization that is different from the incident polarization. This effect is called *birefringence* or *double refraction*. Optically anisotropic media can therefore influence the polarization direction of light. Normally, when crystals are heated, they pass from the solid phase, which is subject to long-range order, to the liquid phase, which is characterized by the absence of long-range order and isotropy. However, special organic crystals have the peculiar property that they are already liquid, but are subject to a (elastic) long-range order and remain anisotropic in this phase. This is called orientational remote order, while positional remote order is lost (liquid). These crystals were discovered in 1888 by the biologist Friedrich Reinitzer and are called liquid crystals (LC).

Since the end of the 60s, this effect has been used in display technology. In principle, the cavity (cell) between two glass plates is filled with liquid crystals which have this property

---

[1]LCD – Liquid Cristal Display.

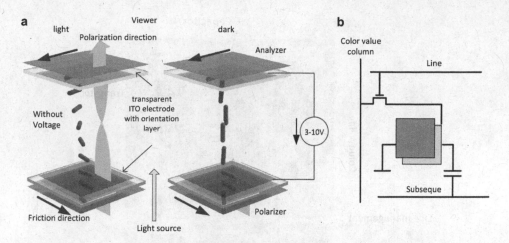

Fig. 25.1 Principle of a TN cell (**a** light, **b** dark), next to it the switching principle of an active TFT cell

at room temperature. Most displays in use today make use of the *Schadt-Helfrich* effect, which is shown schematically in Fig. 25.1a. By appropriate surface treatment (rubbing), it is ensured that the preferred directions (director) of the molecules directly on the two glass plates are rotated by 90° relative to each other. Due to the interactions between the molecules, the intervening layers twist up in a helical fashion, thereby rotating the polarization direction of passing light by 90°. This is referred to as a *twisted nematic (TN) cell*.

If the glass plates are coated with a transparent but conductive ITO (indium tin oxide) layer and a voltage is applied to it, the electric field in the cell causes the liquid crystals to straighten up and allow the light to pass through unaffected. Obviously, light can pass through two polarization foils whose polarization directions are perpendicular to each other unimpeded in the de-energized state because it is itself rotated in its polarization direction. When the voltage is switched on, the polarization direction is not rotated and the light is not transmitted by the polarizing filter lying beyond the cell. The cell is "actively dark." With parallel filters, the cell is "actively bright".

In the production of a (passive) LC display, the electrodes made of ITO are sputtered onto the glass and structured (etched) in a photochemical process. Care must be taken to ensure that opposing electrodes only cross at visible points. A polymer layer is then applied, which is conditioned by rubbing. Small beads (spacers) are filled between the discs to ensure a constant distance. The panes are glued together and the liquid crystal is filled in through an opening. Many LC displays already have permanently structured symbols or 7- or 13-segment digits.

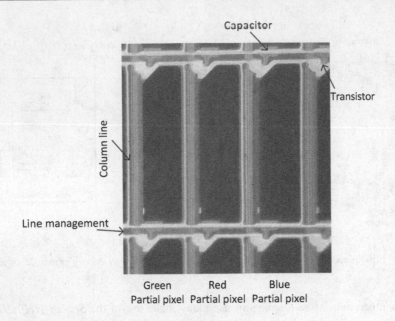

Fig. 25.2 Microscopic image of a pixel from a TFT monitor

These passive cells only allow a low refresh rate with a limited viewing angle. LC displays with IPS[2] and VA[3] cells improve contrast at increased viewing angles [1]. The advancement of TN cell to allow different background colors and higher multiplexing rate has led to Enhanced, Modulated, Super, Double Super and Enhanced Super Twisted Nematic cells [2]. High switching times and low storage times of a passive TN display matrix reduce the contrast as the display size increases, because the drive time per pixel and frame becomes smaller at a constant multiplex rate.

Active matrix LC displays overcome this disadvantage by using a capacitor for each picture element to maintain the electric field. This capacitor can be recharged via a thin film transistor (TFT). This TFT technology enables the construction of full-color displays. In this context, one also speaks of active matrix LC displays (AMLCD). Figure 25.1b shows the principle circuit diagram, and Fig. 25.2 shows a microscope image taken by the author with a classic light microscope in overhead light with a TFT monitor.

All LC displays require backlighting. This can consist of a simple cold cathode tube or a white LED line whose light is evenly distributed via a light guide. Some displays also use electron luminescent foils as backlighting, but these must be controlled with a higher voltage. Larger displays have LED matrices that are arranged behind the entire display surface and backlight it area by area, whereby darker parts of the image are illuminated

---

[2] IPS – In plane switching.

[3] VA – vertical alignment.

more weakly and brighter parts more strongly. This process increases the contrast of the display and significantly saves energy (full LED backlight). Backlighting is the largest energy consumer of an LCD display and requires more energy the brighter the ambient light. For displays that are to be read in bright sunlight, mirror foils are often used instead of backlighting; the sunlight then passes through the cell twice (so-called reflective displays). If semi-transparent mirrors and an active light source are also used, we speak of transflective displays, which can work in the dark and in the sun.

25.1.2.2 LED Indicators

As emissive displays only the displays built with LED[4]s are mentioned here. Nowadays, large-area displays can be manufactured from inorganic light-emitting diodes using semiconductor technology.

Organic LEDs (OLED) are based on the property of some organic materials to produce light when electric current is passed through them. Full-color displays made with OLEDs can be built as passive or active matrices. They can be extremely thin and flexible, offer a wide viewing angle, but currently have higher power consumption than LC displays. Active OLED matrix displays have a somewhat more complex structure than active LC displays because, in addition to the image information, the energy to generate the pixel must be delivered to each pixel. Figure 25.3 shows such a circuit schematically for a single pixel. However, no backlighting is necessary for this.

25.1.3 Image Composition

The driving controller of a dot matrix display receives the information necessary for image display from a microcontroller and refreshes the image regularly. The state or color of each pixel is mapped in a volatile memory (frame buffer). Direct access to the memory allows changes to be made to the image. The microcontroller transfers the coordinates (row and column) and the state or color of the pixel to be changed.

To build up an image, embedded displays often store elementary raster graphics in a non-volatile memory and then transfer them line by line to the corresponding position in the frame buffer. This technique is also called sprite technique and the graphic components are called sprites (goblins, ghosts).

To reduce the data flow between the microcontroller and the display and to scale the images easily, vector graphics is often used. In vector graphics, the character sets and graphics are defined vectorially by line arcs, circular arcs, polynomial curves and area elements. The microcontroller only transmits the vector elements that are used by the display controller to build up the image.

---

[4]LED – Light Emitting Diode.

Fig. 25.3 Controlling a pixel in an actively lit (LED or OLED) display

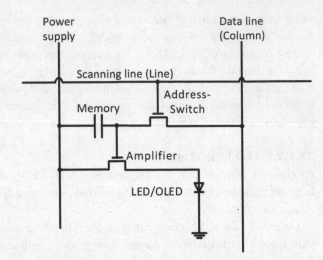

In this case, rasterization algorithms must be built directly into the graphics controller, which then control the respective affected pixels in the frame buffer after appropriate scaling.

25.1.4 Display Control

Some displays with a high information content provide a parallel bus for the transmission of the command codes and the data. This bus consists of 4, 8, 16 or more lines. Also older display controllers have a parallel bus because of the easy control.

Modern displays with low to medium information content for embedded systems use a fast serial interface such as SPI for the information flow. Displays with integrated vector graphics can also use a serial interface. High resolution displays are only partly controlled analog (RGB, VGA), most of them via LVDS[5] based bus systems like DisplayPort or TMDS[6] based connectors like DVI or HDMI. Here the color values and the pixel clock are transmitted partly separately and partly combined serially, mostly combined with audio signals and with signals for digital rights management. In this chapter, however, the display controls are to be presented which make do with the lowest possible processor resources and can therefore be directly controlled by an 8-bit processor.

---

[5] Low Voltage Differential Signaling.
[6] Transition Minimized Differential Signaling.

25.2   Dot Matrix LCD Display with Parallel Control

Dot matrix LC displays are often used in devices with ATmega class microcontrollers because of the easy parallel control. These displays usually use a controller compatible with the Hitachi HD44780. They are STN[7] displays, which are monochromatic LC displays with green or blue backgrounds and reduced power consumption. They may also have a backlight built in to improve readability in sunlight. They are manufactured as 1-, 2- or 4-line displays.

25.2.1 Structure of a Display with a KS0070B Controller

A rough block diagram of a display connected to a controller KS0070B, which is compatible with the HD44780 and manufactured by Samsung [3], is shown in Fig. 25.4. Data exchange between the host microcontroller and the display controller takes place via an 8-bit data bus (DB7 ... DB0) or, alternatively, via a 4-bit data bus consisting of pins DB7 ... DB4. In the following, the alternative control is considered in more detail, because it saves I/O pins of the microcontroller. The double time for the transmission of a byte is accepted in this case. In addition to the data pins, the controller also has three control pins to control the data flow:

- **RS** (Register Select) is an input that controls the incoming byte when high into one of the RAM memories and when low into the function decoder.

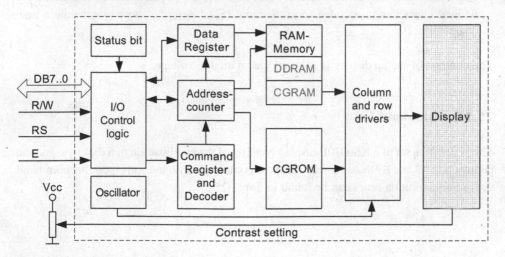

Fig. 25.4 LC display with controller KS0070B – block diagram

---

[7]STN – Super-Twisted Nematic.

- **E** (Enable) via a low-high-low at this input, the bits from the data bus are stored in a data buffer. The data buffer is selected with input RS.
- **R/W** (Read/Write) determines the direction of the data transfer. From the microcontroller's point of view, high means read data, low means write data.

The character set of the display, which consists of 192 5x7-dot and 32 5x10-dot characters, is stored in the CGROM.[8] This character set can be taken from the data sheet. The user can define 5x8 point characters himself and use up to eight of them at the same time.

The RAM memory is divided into two areas:

- DDRAM[9] is an 80 byte memory which maps the display positions. The corresponding DDRAM addresses in hexadecimal number format of a 4-line display such as DEM 16.481 [4] from Display Elektronik can be seen in Table 25.1. The ASCII values of those characters which are to be displayed are stored in the DDRAM. The following program excerpt:

```
Display_Set_DDRAMAddress(0x04);
Display_Write('A'); //transmission of the character A
```

causes the cursor to be set to the fifth position of the first line and the character A to be displayed at this position. When writing or reading a data byte, the address counter is incremented or decremented according to the cursor settings.

- CGRAM[10] is a 64-byte RAM memory that can store the bit patterns of up to eight newly defined characters in encoded form. Section 25.2.4 describes how to generate a new character.

The contrast of such a display is adjusted via an analog voltage.

25.2.2 Command Set

The instruction set of a KS0070B display consists of the combination of a data byte and the control bits RS and R/W. A summary of these commands with the corresponding command codes and execution times can be found in Table 25.2.

---

[8] CGROM – Character Generator ROM.
[9] DDRAM – Display Data RAM.
[10] CGRAM – Character Generator RAM.

Table 25.1 Display positions in DDRAM

Display position	1	2	3	4	5	6	7	8	9	10	11	12	13	14	15	16
1st line DDRAM-address	00	01	02	03	04	05	06	07	08	09	0 A	0B	0 C	0D	0E	0 F
2nd line	40	41	42	43	44	45	46	47	48	49	4 A	4B	4 C	4D	4E	4 F
3rd line	10	11	12	13	14	15	16	17	18	19	1 A	1B	1 C	1D	1E	1 F
4th line	50	51	52	53	54	55	56	57	58	59	5 A	5B	5 C	5D	5E	5 F

Table 25.2 Commands of the KS0070B (HD44780) controller

Command	RS	R/W	D7	D6	D5	D4	D3	D2	D1	D0	Execution time
Clear display	0	0	0	0	0	0	0	0	0	1	1.53 ms
Display basic setting	0	0	0	0	0	0	0	0	1	x	1.53 ms
Display input mode	0	0	0	0	0	0	0	1	I/D	SH	39 µs
Display/cursor on/off	0	0	0	0	0	0	1	D	C	B	39 µs
Shift cursor/ display	0	0	0	0	0	1	S/C	R/L	x	x	39 µs
Set function	0	0	0	0	1	DL	N	F	x	x	39 µs
Set CGRAM address	0	0	0	1	A5	A4	A3	A2	A1	A0	39 µs
Set DDRAM address	0	0	1	A6	A5	A4	A3	A2	A1	A0	39 µs
Read status bit/address	0	1	BF	A6	A5	A4	A3	A2	A1	A0	0
Write byte to RAM	1	0	D7	D6	D5	D4	D3	D2	D1	D0	43 µs
Read byte from RAM	1	1	D7	D6	D5	D4	D3	D2	D1	D0	43 µs

- **Clear display** fills the entire DDRAM memory with blanks (0x20), sets the DDRAM address to 0x00 (first position from the left of the first line of the display) and sets the input mode to increment; with each data byte the cursor is shifted to the right and the address counter is incremented.
- **Display basic setting** sets the DDRAM address to 0x00, the DDRAM content remains unchanged and any display shift is reversed;
- **Display input mode** determines the direction of movement of the cursor and the shift direction of the display when a data byte is transferred, depending on bits I/D (increment/decrement) and SH (shift); no display shift takes place when reading the RAM memory or writing to the CGRAM. Otherwise the effect of the two bits can be seen in Table 25.3.
- **Display/cursor on/off,** if $D = 1$ the display is switched on, if $C = 1$ the cursor is shown and $B = 1$ causes the cursor to flash;
- **Shift cursor/display** – the command affects the shifting of the cursor or display depending on the S/C (Displayshift/Cursor) and R/L (Right/Left) bits (Table 25.4) in the absence of a data byte transfer.

Table 25.3 Effect of the I/D and SH bits

I/D	SH	Description
0	0	Cursor is shifted to the left, the address counter is decremented, no display shifting
0	1	Cursor is shifted to the left, the address counter is decremented, display shift to the right
1	0	Cursor is shifted to the right, the address counter is incremented, no display shift
1	1	Cursor shifts to the right, address counter increments, display shifts to the left

Table 25.4 Effect of the S/C and R/L bits

S/C	R/L	Description
0	0	Cursor is shifted to the left, the address counter is decremented; no display shifting
0	1	Cursor is shifted to the right, the address counter is incremented; no display shifting
1	0	Entire display with cursor are shifted to the left
1	1	Entire display with cursor are shifted to the right

- **Set function** – The command determines whether communication with the display takes place via an 8-bit ($DL = 1$) or via a 4-bit ($DL = 0$) data bus, whether the display is initialized as a single-line ($N = 0$) or double-line ($N = 1$) display, and whether 5×7-dot ($F = 0$) or 5×10-dot ($F = 1$) characters are used.
- **Set CGRAM Address** – This command sets the cursor to the desired address in CGRAM;
- **Set DDRAM Address** – This command sets the display position where the next character is to be displayed.
- **Read Status Bit/Address** – When this command is executed, the microcontroller reads the contents of the address counter and the status bit which, when $BF = 1$, indicates that the display controller is busy.
- **Write byte in RAM** – The command stores a byte to the current RAM address (DDRAM or CGRAM).
- **Read byte from RAM** – With this command the microcontroller reads a byte from the current RAM address (DDRAM or CGRAM).

When the display controller has received a byte, an internally controlled sequence begins, the duration of which depends on the command. During this time, the status bit BF is set to signal to the microcontroller that the display controller cannot receive any further bytes. Via a blocking wait, the microcontroller can query the state of bit BF before sending another byte. Via a suitably set timer interrupt, which is started when a new byte is stored, the correct time for a new transmission can be determined. The execution time of the commands depends on the operating voltage, the controller used and the internal clock frequency of the controller and can be taken from the corresponding data sheet.

25.2.3 4-Bit Communication

To save I/O pins of the microcontroller, the communication with the display controller can be set to four bits. In this case, only the inputs DB7 ... 4 are used and a byte must be transferred and stored in two steps: first the high-order byte half D7 ... 4 and then the low-order byte half D3 ... 0, as shown in Fig. 25.5. After the controller has booted up, the entire contents of the DDRAM memory are filled with blanks, communication is set to eight bits, the display is switched off and the cursor is hidden.

The display can be initialized by software with the commands: Set Function, Display/ Cursor On/Off, Clear Display, and Display Input Mode initialization. The select bits of these commands are selected according to the specific task. The flowchart for display initialization in 4-bit mode ($DL = 0$) is shown in Fig. 25.6 and has the following settings:

- two-line display ($N = 1$),
- 5×7-dot characters ($F = 0$),
- Display on ($D = 1$),
- Cursor on ($C = 1$),
- Flashing switched off ($B = 0$),
- with a new data byte the address counter is incremented ($I/D = 1$) and
- the display should not be shifted ($SH = 0$)

Each transmitted byte half is loaded into the input register with the falling edge of the **E** signal.

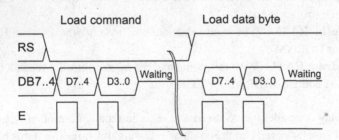

Fig. 25.5 Display communication via a 4-bit data bus

Fig. 25.6 Display initialization
in 4-bit mode

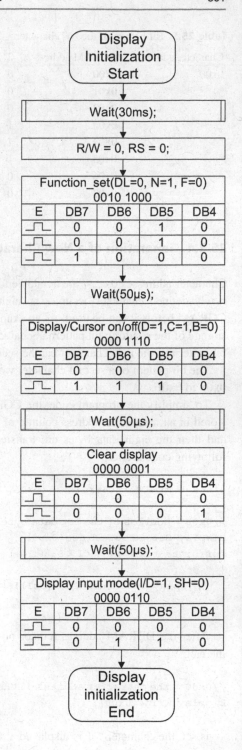

Table 25.5 Bit pattern of the "ö" character

Character address	CGRAM address	2^4	2^3	2^2	2^1	2^0	CGRAM value
0x00	0x00	0	1	0	1	0	$2^1 + 2^3 = 0x0A$
	0x01	0	0	0	0	0	0x00
	0x02	0	1	1	1	0	$2^3 + 2^2 + 2^1 = 0x0E$
	0x03	1	0	0	0	1	$2^4 + 2^0 = 0x11$
	0x04	1	0	0	0	1	0x11
	0x05	1	0	0	0	1	0x11
	0x06	0	1	1	1	0	0x0E
	0x07	0	0	0	0	0	0x00

25.2.4 Generation of a New Character

Character addresses 0 ... 7 are available to the user to define new 5x8 pixel characters. A newly defined character occupies an eight-byte memory area and must be loaded into CGRAM before it can be used. As an example, the generation of the character "ö" and the storage of the resulting bit pattern at character address 0 is explained in the following. The bit pattern of the new character can be seen in Table 25.5.

The bit pattern of a second character would have to be stored in CGRAM from address 0x08 to 0x0F.

To simplify the program code, the CGRAM values calculated for a new character are stored in an array. The address counter of the CGRAM memory is set to the first address and then the eight data bytes are transferred to store them in the CGRAM, as in the following code section:

```
uint8_t ucOeCharacters[8] = {0x0A, 0x00, 0x0E, 0x11, 0x11, 0x0E, 0x00};
...
//0x00 = address of the 1st data byte of the bit pattern in CGRAM
Display_Set_CGRAMAddress(0x00);
for(uint8_t ucI = 0; ucI < 8; ucI++)
{
    Display_Write(ucOeChar[ucI]);
}
```

After the DDRAM address (cursor position) has been set in the main program with the call:

```
//0x00 = the character address of the newly defined character
Display_Write(0x00);
```

is set, the character "ö" is displayed at the current cursor position.

25.2.5 Execution of the Display Commands Without Blocking Wait

The further example refers to a display whose R/W input is permanently connected to GND (=0 V) in order to save another I/O pin of the microcontroller. In this case, no read commands can be executed and the only way to determine the end of a command remains to measure the execution time. To accurately determine this time, a timer interrupt is used that is initialized to 50 μs. The execution of a display command means:

- the output of a byte (command code or data byte) on the 4-bit data bus. This is done in two steps according to Fig. 25.5,
- the corresponding control of the RS and E inputs of the display and
- the execution of the other tasks of the main program while the display is busy with the internal process.

To avoid a blocking wait of the microcontroller after a call of a display function and at the same time to ensure that no instruction is ineffective, the control of the display can be realized as a state machine. An n-digit ring buffer is defined as shown in Fig. 25.7, to which two pointers are directed, both initialized to 0. With the call of a display function:

- the byte to be transmitted (command code or data byte) is stored in the array ucDisplayCommandByte together with a second byte in the array ucDisplayDataByte that encodes the state of the RS input (0/1) and the corresponding wait time (as a multiple of 50 μs); the pointer ucDisplayFuncCalls points to the memory location of the ring buffer;
- the pointer ucDisplayFuncCalls is then incremented and if it reaches the value n, it is set to 0;

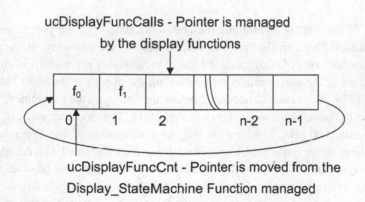

Fig. 25.7 Display function buffer

- the counter `ucDisplayFuncOpen` is incremented. This counts how many unexecuted functions are in the queue.

The second pointer `ucDisplayFuncCnt` is managed by a function `Display_StateMachine`, whose structure and call is illustrated in Fig. 25.8. This pointer points to the next function to be executed. With the call of a display function:

- the pointer `ucDisplayFuncCnt` is incremented (when it reaches the value n, it is set to 0),
- the counter `ucDisplayFuncOpen` is decremented. If this counter, which was initialized with 0 at the beginning, reaches the value 0, then no more functions are in the queue.
- the timer interrupt is started with the new waiting time. When the set waiting time is reached, the timer is stopped and the possible call of a further function is enabled.

The size n of the ring buffer should be chosen as small as possible, depending on the concrete application, in order to use as little memory as possible, but it must be large enough for the text display to function properly.

25.3   Serial Control of a Parallel LC Display

An LC display (see Sect. 25.2) connected to a port expander PCF8574 described in Sect. 24.1 can be controlled via I^2C as shown in Fig. 25.9. The microcontroller only requires the I^2C bus for control, to which other devices or displays can be connected.

25.3.1 Display Control Via I^2C

The timing of the display control and the communication with the port expander were described in Sect. 25.2.3 and in Sect. 24.1 respectively. In the following, the display at the current cursor position of the character A (ASCII value 0x41) is presented as an example. The output path of the port expander for controlling the display is shown in Table 25.6.

The PCF8574 device must be addressed before the five bytes are transmitted. The entire timing of the I^2C communication is shown in Fig. 25.10. The clock frequency of the bus in this case is max. 100 kHz, which means that the transmission of one byte takes at least 90 µs. For the example considered, a total of six bytes are transmitted with the address byte of the port expander, which means that the display of a first character takes about 540 µs. To subsequently display another character, only four bytes are transmitted in the same I^2C message because the address and setting of the RS bit are omitted. The transmission of a display command looks similar with the difference that the RS bit is set to 0. With this

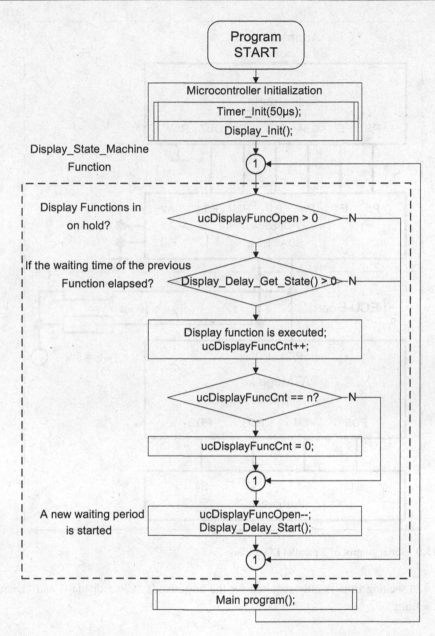

Fig. 25.8 State machine display functions

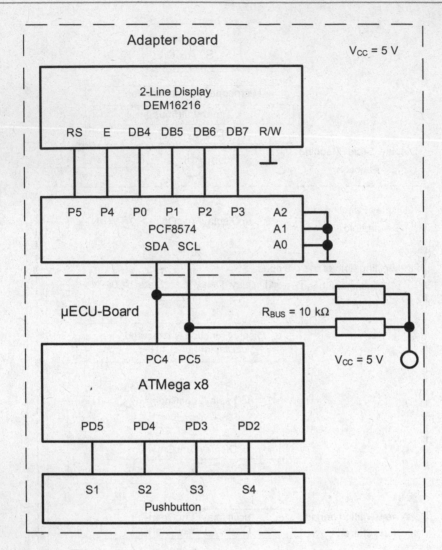

Fig. 25.9 Serial control of a parallel LC display

control, a waiting time is only required for the commands "Clear display" and "Display basic setting".

25.3.2 Software Example: Transmission of a Data Byte

For portability reasons of the software, the connections between the display and the port expander are defined as follows:

Table 25.6 PCF8574 – output history to display data byte 'A'

Description	P7	P6	P5 RS	P4 E	P3 DB7	P2 DB6	P1 DB5	P0 DB4
The RS bit is set to 1, followed by the transmission of a data byte.	0	0	1	0	0	0	0	0
The E-bit is set to 1, the higher-order byte half is transmitted.	0	0	1	1	0	1	0	0
The E-bit is set to 0 in order to store the higher-order byte half in the input register of the display.	0	0	1	0	0	1	0	0
The E-bit is set to 1, the low-order byte half is transmitted.	0	0	1	1	0	0	0	1
The least significant byte half is taken over into the input register of the display.	0	0	1	0	0	0	0	1

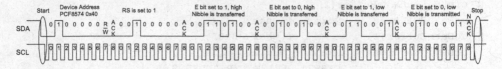

Fig. 25.10 I2C communication – indication of character 'A' on the display

```
#define DISP_TWI_RS_BIT 0x20
#define DISP_TWI_E_BIT 0x10
#define DISP_TWI_DATA_DB7_BIT 0x08
#define DISP_TWI_DATA_DB6_BIT 0x04
#define DISP_TWI_DATA_DB5_BIT 0x02
#define DISP_TWI_DATA_DB4_BIT 0x01
#define DISP_TWI_RS_SET_ON DISP_TWI_RS_BIT
#define DISP_TWI_RS_SET_OFF (~DISP_TWI_RS_SET_ON)
#define DISP_TWI_E_SET_ON DISP_TWI_E_BIT
#define DISP_TWI_E_SET_OFF (~DISP_TWI_E_SET_ON)
#define DISP_TWI_DATA_MASK (~(DISP_TWI_DATA_DB7_BIT |
                             DISP_TWI_DATA_DB6_BIT |
                             DISP_TWI_DATA_DB5_BIT |
                             DISP_TWI_DATA_DB4_BIT))
```

It was taken into account that the bit P7 has the highest and P0 the lowest value. This part may have to be changed depending on the concrete wiring in order to be able to use the code unchanged.

The function `DispTWI_Write_Data`, whose code is listed further, is used to control the display connected to the port expander in order to display a character. The device chip address of the PCF8574 device and the ASCII value of the character to be displayed are passed as parameters. The function returns the value TWI_OK in case of error-free I²C transmission, otherwise TWI_ERROR. The variable `ucDataBuffer` stores the two byte halves of the character. The bold part of the function provides a flexible pin connection between the display and the port expander. The bytes calculated in the function are transferred according to Table 25.6.

```
uint8_t DispTWI_Write_Data(uint8_t ucdevice_address,
                           uint8_t ucascii_of_char)
{
    uint8_t ucDispDataBit[4] = {DISP_TWI_DATA_DB7_BIT,
                                DISP_TWI_DATA_DB6_BIT,
                                DISP_TWI_DATA_DB5_BIT,
                                DISP_TWI_DATA_DB4_BIT};
    int8_t ucDeviceAddress, ucDispTWIData;
    uint8_t ucMask, ucI, ucJ;
    uint8_t ucDataBuffer[2];
    ucDataBuffer[0] = ucascii_of_char >> 4;
    ucDataBuffer[1] = ucascii_of_char & 0x0F;
    ucDeviceAddress = (ucdevice_address << 1) |
                      PCF8574_DEVICE_TYP_ADDRESS;
    ucDeviceAddress |= TWI_WRITE; //write mode
    TWI_Master_Start(); //start
    if((TWI_STATUS_REGISTER) != TWI_START) return TWI_ERROR;
    TWI_Master_Transmit(ucDeviceAddress); //send device address
    if((TWI_STATUS_REGISTER) != TWI_MT_SLA_ACK) return TWI_ERROR;
    ucDispTWIData = DISP_TWI_RS_SET_ON; //RS bit is set to 1
    TWI_Master_Transmit(ucDispTWIData); //RS line is set to high
    if((TWI_STATUS_REGISTER) != TWI_MT_DATA_ACK) return TWI_ERROR;
    for(ucJ = 0; ucJ < 2; ucJ++)
    {
        ucDispTWIData |= DISP_TWI_E_SET_ON; //EN is set to high
        ucMask = 0x08;
        ucDispTWIData &= DISP_TWI_DATA_MASK; //the data bits are set to 0
        for(ucI = 0; ucI < 4; ucI++)
        {
            if(ucDataBuffer[ucJ] & ucMask)
                ucDispTWIData |= ucDispDataBit[ucI];
            ucMask = ucMask >> 1;
        }
        //the data nibble is sent
        TWI_Master_Transmit(ucDispTWIData);
        if((TWI_STATUS_REGISTER) != TWI_MT_DATA_ACK)
            return TWI_ERROR;
        ucDispTWIData &= DISP_TWI_E_SET_OFF; //EN is set to Low
        TWI_Master_Transmit(ucDispTWIData);
```

```
        if((TWI_STATUS_REGISTER) != TWI_MT_DATA_ACK)
            return TWI_ERROR;
    }
    TWI_Master_Stop(); //stop
    return TWI_OK;
}
```

25.4   DOGS102-6: Graphic Display with Serial Control

The displays of the DOGS102-6 series are passive STN dot matrix LC displays [5]. The size of the displays, the low power consumption and the good readability even in sunlight enable their use in hand-held devices. They are driven by a controller of type UC1701x [6]. The controller has implemented an 8-bit parallel interface and a serial SPI interface for data communication with the driving microcontroller. It allows driving a display with up to 65x132 dots. An example circuit is shown in Fig. 25.11.

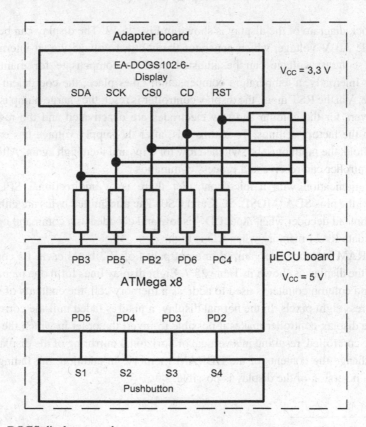

Fig. 25.11 DOGS display control

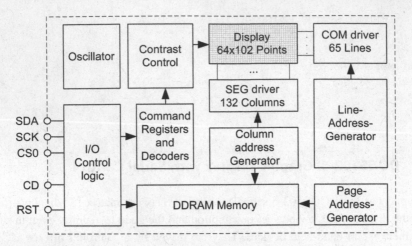

Fig. 25.12 Display – DOGS102-6 – block diagram

25.4.1 Structure of the Graphic Display DOGS 102-6

A rough block diagram of the display is shown in Fig. 25.12. The display can be supplied with a single 3.3 V voltage, which generates the contrast voltage via an internal charge pump. The contrast voltage can be adjusted once to compensate for manufacturing differences. Internally, a temperature compensation takes place, the coefficient of which is adjustable. Via the RST input, the display controller is reset, the charge pump is switched off, the drivers for the column and row electrodes are deactivated and the registers are loaded with the factory settings. To realize this, after the supply voltage has exceeded a preset threshold, the input must be switched low for 1 ms and then high again. After another 5 ms the controller can receive and process commands.

For communication with a microcontroller there is a unidirectional SPI interface consisting of the pins SDA (MOSI), SCK and CS0. The transmitted bytes are either loaded into the command decoder when input CD[11] is low and decoded as a command or stored in the display data RAM memory when CD is high.

The DDRAM memory is organized in eight pages of 102 bytes each. Its contents are mapped by the display as shown in Table 25.7. Eight display lines form one memory page. The page and column counter is used to address a memory cell, the contents of which are used to address eight pixels. In the normal display, a pixel is faded in if the corresponding bit is 1. The display controller makes it possible to invert the order in which the rows and columns are controlled, resulting in a vertical or horizontal mirroring of the display without having to change the contents of the DDRAM memory. In addition, the fading in of all pixels or an inversion of the display is possible.

---

[11] CD Control Data.

Table 25.7 Relationship between DDRAM content and pixel display

DDRAM		Display	
Page	Data byte m 0xD9	Column m	Line number
n	D0 = 1		8 * n
	D1 = 0		8 * n + 1
	D2 = 0		8 * n + 2
	D3 = 1		8 * n + 3
	D4 = 1		8 * n + 4
	D5 = 0		8 * n + 5
	D6 = 1		8 * n + 6
	D7 = 1		8 * n + 7

This relation is valid for normal display (no inversion) and start line equal 0

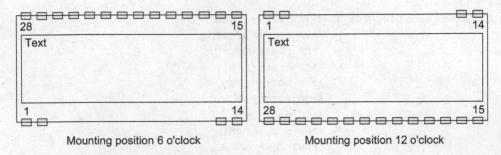

Mounting position 6 o'clock   Mounting position 12 o'clock

Fig. 25.13   DOGS – display mounting position

The control of the display makes both a 6 o'clock and a 12 o'clock mounting position possible, as shown in Fig. 25.13. The page numbering always starts from the top at 0. In the left picture the columns are numbered starting from the left from 0 to 101, in the right picture from 30 to 131.

25.4.2 SPI Communication

The display is preset as SPI slave and can receive data in SPI mode 3, most significant bit first, at a clock frequency of up to 33 MHz. A host microcontroller must be set as master accordingly. If the supply voltages of the display and microcontroller are different, the voltage level at the display input must be adjusted, as shown in Fig. 25.11.

The SPI data structure encodes the following control inputs: Chip-Select, Control-Data and Reset and looks like this for the example circuit:

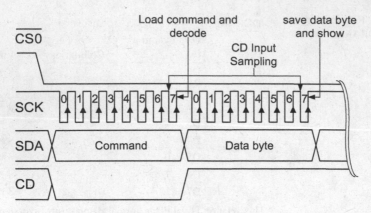

Fig. 25.14 DOGS – display SPI communication

```
DispDOGS_pins DispDOGS_1 ={/*CS_DDR*/ &DDRB,
                           /*CS_PORT*/ &PORTB,
                           /*CS_pin*/ PB2,
                           /*CS_state*/ ON},
                           /*CD_DDR*/ &DDRD,
                           /*CD_PORT*/ &PORTD,
                           /*CD_pin*/ PD6,
                           /*CD_state*/ ON,
                           /*RST_DDR*/ &DDRC,
                           /*RST_PORT*/ &PORTC,
                           /*RST_pin*/ PC4,
                           /*RST_state*/ ON};
```

Figure 25.14 shows the principle of SPI communication with the display, initiated by the microcontroller switching the chip select signal low. On the falling edge of every eighth clock of a byte, the state of the CD input is sampled internally and on the rising edge of that clock, the received byte is either loaded into the instruction decoder and decoded or stored into the DDRAM. The decoded command is executed immediately, the stored byte is represented immediately, there is no wait for the end of communication. No waiting time is required between the transmission of individual bytes.

An example of the implementation of the sequence shown in Fig. 25.14 is shown in the following code section. In this example, the entire display is cleared by storing a 0 at each address.

```
//the CS line is set to low
SPI_Master_Start(sdevice_pins.DispDOGSspi);
for(ucPage = 0; ucPage < 8; ucPage++)
{
    //with the CD line set to low, the following byte
```

```
//is interpreted as a command
DispDOGS_Set_CDLow(sdevice_pins);
//the page address is set
SPI_Master_Write(SET_PAGE_ADDRESS | ucPage);
//the initial column is set
DispDOGS_Set_Column(sdevice_pins, 0x00);
//the following bytes are data bytes
DispDOGS_Set_CDHigh(sdevice_pins);
for(ucColumn = 0; ucColumn < 102; ucColumn++)
{
    SPI_Master_Write(0x00);
}
}
SPI_Master_Stop(sdevice_pins.DispDOGSspi); //end of transmission
```

25.4.3 Command Set

The command codes (see Table 25.8) are transmitted via the SPI bus and the state of the CD input determines whether the received byte is stored in the DDRAM memory or in the command decoder. The DOGS102-6 series implements the following functions from the entire instruction set of the UC1701x display controller:

- **Select Page Address** (PA) The four bits $PA0:3$ are used to select the page (0 the uppermost, 7 the lowermost page) for storing the next byte;
- **Select Column Address** (CA – Column Address) is a 2-byte command to select the column for storing the next byte. In the bottom view, the column numbering starts from the left at 0 to 101, in the top view it starts from the left at 30 to 131;
- **Save Data Byte** (Write Data Byte) with the CD input set to 1, the byte is saved to the DDRAM memory in the currently addressed memory cell. The column address is incremented;
- **Reset display** (System Reset) the command sets the internal registers to the preset values;
- **Select display mode** (Set Display Enable, DC – Display Control) puts the display in active mode for $DC2 = 1$ or in sleep mode for $DC2 = 0$;
- **Control charge pump** (Set Power Control, PC – Power Control) switches the charge pump for generating an LCD high voltage on at $PC2 \ldots 0 = 111$ or off at $PC2 \ldots 0 = 000$;
- **Contrast Coarse Adjustment** (Set VLCD Resistor Ratio) coarsely adjusts the high voltage for contrast;
- **Contrast fine adjustment** (Set Electronic Volume) finely adjusts the high voltage for the contrast. To ensure good reliability of the display, the VLCD contrast voltage should not exceed 11.5 V (see Table 25.9);

Table 25.8 Command set of a display of type DOGS102-6

| Command | Command code | | | | | | | | |
	CD	D7	D6	D5	D4	D3	D2	D1	D0
Select page address	0	1	0	1	1	PA3	PA2	PA1	PA0
Select column address	0	0	0	0	0	CA3	CA2	CA1	CA0
	0	0	0	1	CA7	CA6	CA5	CA4
Save data byte	1	Data byte							
Reset display	0	1	1	1	0	0	0	1	0
Select display mode	0	1	0	1	0	1	1	1	DC2
Control charge pump	0	0	0	1	0	1	PC2	PC1	PC0
Contrast coarse adjustment	0	0	0	1	0	0	PC5	PC4	PC3
Contrast fine adjustment	0	1	0	0	0	0	0	0	1
	0	0	PM5	PM4	PM3	PM2	PM1	PM0
Select LCD bias ratio	0	1	0	1	0	0	0	1	BR
Select start line	0	0	1	SL5	SL4	SL3	SL2	SL1	SL0
Show all pixels	0	1	0	1	0	0	1	0	DC1
Display inversion	0	1	0	1	0	0	1	1	DC0
Vertical mirroring	0	1	1	0	0	MY	0	0	0
Horizontal mirroring	0	1	0	1	0	0	0	0	MX
Special settings	0	1	1	1	1	1	0	1	0
	TC	0	0	1	0	0	WC	WP

Table 25.9 Adjustable contrast voltage ranges

PC5 ... 3	PM5 ... 0	VLCD voltage range/V
000	0 ... 63	3.87 ... 6.33
001	0 ... 63	4.51 ... 7.38
010	0 ... 63	5.15 ... 8.43
011	0 ... 63	5.79 ... 9.48
100	0 ... 63	6.43 ... 10.53
101	0 ... 63	7.08 ... 11.51
110	0 ... 63	7.72 ... 11.46
111	0 ... 63	8.36 ... 11.48

- **Select LCD** Bias Ratio (Set LCD Bias Ratio, BR – Bias Ratio) sets the ratio between the LCD voltage and bias to 1/9 for BR = 0 or 1/7 for BR = 1;
- **Select Start Line** (Set Scroll Line, SL – Scroll Line) selects the start line of the display via bits SL5 ... 0. Executing this command causes the screen to scroll up by SL display lines;
- **Show All Pixels** (Set All Pixel On) shows all pixels for DC1 = 1, but does not change the contents of the DDRAM memory; for DC1 = 0 the display is driven to show the RAM contents;

- **Display inversion** (Set Inverse Display) fades in a pixel for $DC0 = 0$ if the corresponding bit from memory is 1, or for $DC0 = 1$ if the bit is 0;
- **Vertical irroring** (Set COM Direction, MY – Mirror Y-axle) determines the order of the lines; for $MY = 0$ the order is $0 \dots 63$ while for $MY = 1$ it is $63 \dots 0$, resulting in a vertical mirroring of the display;
- **Horizontal mirroring** (Set SEG Direction, MX – Mirror X-axle) sets the order of the columns; for $MX = 0$ the order is $0 \dots 131$ while for $MX = 1$ it is $131 \dots 0$, resulting in a horizontal mirroring of the display;
- **Special settings** (Set Advanced Program Control 0); the command sets the temperature compensation coefficient of the LCD bias to $-0.05\%/°C$ for $TC = 0$ or to $-0.11\%/°C$ for $TC = 1$; when the last column is exceeded, it jumps to the first one if $WP = 1$, similarly for the memory pages for $WA = 1$.

The display settings are stored in a RAM memory. Therefore, the display must be reinitialized after power-up. For a 6 o'clock mounting position, the initializations can be realized by the following steps:

- Set start line to 0 ($SL5 \dots 0 = 000000b$);
- Horizontal mirroring on ($MX = 1$);
- Vertical mirroring off ($MY = 0$);
- switch off the fading in of all pixels ($DC1 = 0$);
- Display inversion off ($DC0 = 0$);
- Select LCD bias voltage ($BR = 0$);
- Switch on charge pump ($PC2 \dots 0 = 111b$);
- Coarsely adjust contrast ($PC5 \dots 3 = 111b$);
- Fine-tune contrast ($PM5 \dots 0 = 010000b$);
- Special settings ($TC = 1$, $WC = WP = 0$);
- Select display active mode ($DC2 = 1$);

To implement these steps, the parameters suggested in the brackets must be inserted in the corresponding instruction codes and transmitted via SPI while the CD line is low. Combining the codes in an array results in a more compact program code as in the following example.

```
#define SET_SCROLL_LINE 0x40 //set start line 0x40
#define SET_SEG_DIRECTION_MIRROR 0xA1 //horizontal mirroring on
#define SET_COM_DIRECTION_NORMAL 0xC0 //vertical mirroring off
#define SET_ALL_PIXEL_ON_DISABLE 0xA4 //show all pixels off
#define SET_INVERSE_DISPLAY_OFF 0xA6 //display inversion off
#define SET_LCD_BIAS_RATIO_1_9 0xA2 //select LCD bias voltage
#define SET_POWER_CONTROL_ALL_ON 0x2F //switch on charge pump
#define SET_VLCD_RESISTOR_RATIO 0x27 //contrast coarse adjustment
```

```
//1st byte of the contrast fine adjustment
#define SET_ELECTRONIC_VOLUME_1 0x81
//2nd byte of the contrast fine adjustment
#define SET_ELECTRONIC_VOLUME_2 0x07
//temperature coefficient setting (1st byte)
#define SET_ADVANCED_PROGRAM_CONTROL_0 0xFA
//selection temperature coefficient (2nd byte)
#define TEMP_COMP_011 0x91
//set display to active mode
#define SET_DISPLAY_ON 0xAF
uint8_t ucDispDOGSInitValue[13] ={SET_SCROLL_LINE,
                                  SET_SEG_DIRECTION_MIRROR,
                                  SET_COM_DIRECTIONNORMAL,
                                  SET_ALL_PIXEL_ON_DISABLE,
                                  SET_INVERSE_DISPLAY_OFF,
                                  SET_LCD_BIAS_RATIO_1_9,
                                  SET_POWER_CONTROL_ALL_ON,
                                  SET_VLCD_RESISTOR_RATIO,
                                  SET_ELECTRONIC_VOLUME_1,
                                  SET_ELECTRONIC_VOLUME_2,
                                  SET_ADVANCED_PROGRAM_CONTROL_0,
                                  TEMP_COMP_011, SET_DISPLAY_ON};
void DispDOGS_Init(DispDOGS_pins sdevice_pins)
{
    //the Slave Select pin of the display is initialized
    //(output set to high)
    SPI_Master_SlaveSelectInit(sdevice_pins.DispDOGSspi);
    DispDOGS_Init_RST(sdevice_pins); //initialization of the RST pin
    DispDOGS_Init_CD(sdevice_pins); //initialization of the CD pin
    DispDOGS_Set_RSTHigh(sdevice_pins); //RST line set to high
    //20ms waiting time
    while(!Timer1_get_10msState());
    while(!Timer1_get_10msState());
    DispDOGS_Set_CDLow(sdevice_pins);
    //the transfer of the initialization values is started
    SPI_Master_Start(sdevice_pins.DispDOGSspi);
    //the initialization values are transferred
    for(uint8_t ucI = 0; ucI < 13; ucI++)
        SPI_Master_Write(ucDispDOGSInitValue[ucI]);
    SPI_Master_Stopp(sdevice_pins.DispDOGSspi);
    //the entire content of the memory is deleted
    DispDOGS_Clear_Memory(sdevice_pins);
}
```

Table 25.10 Bit pattern of the character "*A*"

Page		Spalte					
		m	m + 1	m + 2	m + 3	m + 4	m + 5
n	2^0	0	1	1	1	0	0
	2^1	1	0	0	0	1	0
	2^2	1	0	0	0	1	0
	2^3	1	0	0	0	1	0
	2^4	1	1	1	1	1	0
	2^5	1	0	0	0	1	0
	2^6	1	0	0	0	1	0
	2^7	0	0	0	0	0	0
		0x7E	0x11	0x11	0x11	0x7E	0x00

25.4.4 Generation of a Character

To display text on a DOGS display, all characters have to be generated by software, because the display itself has no character generator. There are no restrictions regarding the size of the character set. Considering the design of the display, a 5x7 pixel character set is convenient to implement. This character set size fits on the height of a memory page (1 byte) and one can use the bottom or top pixel to separate between two lines of text. As an example, the generation of the character "*A*" is further presented, whose bit pattern can be seen in Table 25.10. The lowest pixel of the character set is used for the separation between the text lines. The sixth generation byte is not strictly necessary, but it is used for the separation between two adjacent characters and offers the advantage that only the starting address of the representation has to be set for displaying a text. With these considerations, the character set is 6x8 pixels in size and 8x17 characters of this size can be displayed on the entire display.

The resulting generation bytes for the character "*A*" are combined into an array to make the program code more compact:

```
uint8_t ucchar_A[6] = {0x7E, 0x11, 0x11, 0x11, 0x7E, 0x00};
```

The following program section from the main function is used to display the character A on a display of type DOGS102-6 with the SPI data structure DispDOGS_1 in page *n*, starting with column *m*. The page number *n* must be a number between 0 and 7, the column number, depending on the orientation of the display, a number between 0 and 101 or between 30 and 131.

```
//Set Page Address
#define SET_PAGE_ADDRESS 0xB0
//Set Column Address
#define SET_COLUMN_ADDRESS_LSB 0x00
```

```
#define SET_COLUMN_ADDRESS_MSB 0x10
//SPI communication is started
SPI_Master_Start(DispDOGS_1.DispDOGSspi);
DispDOGS_Set_CDLow(DispDOGS_1); //the CD input is set to low
/*page selection*/
SPI_Master_Write(SET_PAGE_ADDRESS | n);
/*column selection - 2-byte command*/
SPI_Master_Write(SET_COLUMN_ADDRESS_LSB | (m % 16));
SPI_Master_Write(SET_COLUMN_ADDRESS_MSB | (m / 16));
DispDOGS_Set_CDHigh(DispDOGS_1); //the CD input is set to high
/*the six generation bytes of character A are transmitted*/
for(uint8_t ucI = 0; ucI < 6; ucI++)
{
    SPI_Master_Write(ucCharacter_A[ucI]);
}
SPI_Master_Stop(DispDOGS_1.DispDOGSspi); //SPI communication is stopped
```

According to this example code, nine bytes are transmitted serially to the display for the representation of a character of this size: three for the selection of the text line and the initial column and six for the visualization of the bit pattern. At a clock frequency of 4 MHz, the transmission time of the 9×8 bits is 18 µs. For the visualization of another character to the right of the previous one, only 12 µs are needed (transmission time of six data bytes), because the adresses of the line, resp. of the initial column do not have to be set explicitly. The easiest way is to address the single characters (letters, digits and displayable special characters) by their ASCII-value. A character set of all characters of the ASCII table with values between 32 and 127 contains 96 elements of six bytes each. The address of a character is calculated from the ASCII value of the character minus 32 (offset of the new table). Any character uccharacter from the character set described as above, which is stored in the table ucChar6x8[96] [6], can be represented with the sample code by swapping the command SPI_Master_Write(uccharacter_A[ucI]) with the SPI_Master_Write(ucChar6x8[uccharacter - 32] [ucI]).

References

1. Meroth, A., & Tolg, B. (2008) *Infotainmentsysteme im Kraftfahrzeug*. Friedr. Vieweg & Sohn Verlag | GWV Fachverlage GmbH.
2. Grünhaupt, G. (Hrsg.). (2006). *Handbuch der Mess- und Automatisierungstechnik im Automobil*. Springer-Verlag.
3. Samsung Electronics. (2015). *KS0070B – Driver & controller for dot matrix LCD*. www.datasheetcatalog.com.
4. Display Elektronik. (2015). *LCD Module DEM 16481 SYH-LY*. www.display-elektronik.de.
5. Electronic Assembly. (2015). *DOGS graphic series 102×64 dots*. www.lcd-module.de.
6. UltraChip Inc. (2015). *UC1701x 65x132 STN controller driver*. www.ultrachip.com.

Example Projects

Abstract

The chapter describes two example projects that build on the knowledge in the book.

In this chapter a simple data logger is presented, which can be used in model flight or for transport safety. In the version presented, the data logger writes the acceleration in two axes, the air pressure and the temperature, as well as the barometric altitude derived from these, to a flash memory and reads them out again via a mini-USB connection with a virtual serial port. In this project, it was used to measure the flight characteristics of model rockets. Another example project is a distributed measuring station with CAN networking, which is also presented here.

26.1 Data Logger

26.1.1 Structure of the Model Rocket

Model rockets are small and light missiles that are propelled by solid fuel. Below a certain size or weight and with approved engines, they may be flown without a permit if certain conditions are met. In Germany, the weight of the fuel may not exceed 20 g for a permit-free model. Operation is only permitted by persons over 14 years of age and with the consent of a parent or guardian, and engines may only be supplied to persons over 18 years of age. Larger missiles may require a permit and operator's license. The rocket was built by the Schoollab of the DLR Campus Lampoldshausen (German Aerospace Center). It weighs about 155 g including the motor, has a diameter of 4.1 cm and is 91.7 cm long. Its altitude

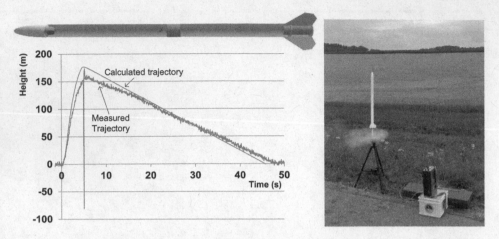

Fig. 26.1 Model rocket with flight path

of ascent was calculated in advance with approx. 168 m at a maximum acceleration of 124 m/s^2 and a maximum velocity of 69 m/s. The calculation as well as the drawing are taken from the software open rocket [1].

Figure 26.1 above shows a model of the rocket. Clearly visible in the sectional view are the motor and one of the two parachute assemblies. The two main tubes are held together by a sleeve. The motor used is a WECO D7-3 class propellant which is available from model shops. This develops a total impulse of 13.9 Ns and a maximum thrust of 20.4 N during a burn time of 1.48 s. The engine can be purchased from model shops. After burning the charge the rocket flies in free ballistic flight for another 3 s, then the tip with the payload is ejected by a propellant charge, rocket body and tip fall back to the ground on parachutes.

Since the climb altitude was determined from the air pressure and temperature using the barometric altitude formula, the pressure surge is perceived as an altitude pulse when the tip is ejected. Section 26.1.4 also discusses acceleration curves.

26.1.2 Description of the Project

The project was realized with an ATmega88 in a TQFP package. The MPL3115 described in Sect. 6.3 was used as pressure sensor. As acceleration sensor a MMA6525 was used and as external flash memory a SST25PF (Chap. 7). The conversion to the USB interface was carried out by the FT232RL module from FTDIchip (Sect. 5.2.4).

The whole circuit is clocked with 8 MHz and powered by a lithium battery, which is soldered on the board because of the high accelerations. To supply the circuit via the USB interface during readout, a TS5204 voltage regulator reduces the 5 V voltage of the USB interface to the 3.6 V supply voltage of the board. The size of the components played a decisive role, the third acceleration axis was not necessary.

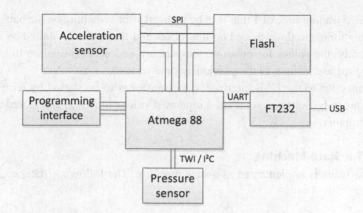

Fig. 26.2 Block diagram of the entire project.

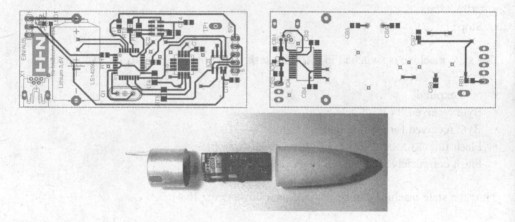

Fig. 26.3 Board layout and finished board with rocket tip

Figure 26.2 shows the block diagram with the various bus systems used. A detailed circuit diagram is included in the supplementary material of the book.

The board was designed to be 30 mm × 71 mm to fit in most rocket tips. Fig. 26.3 shows the layout of the double-sided board, which was assembled by hand on both sides using SMD technology, as well as the finished board with the rocket tip and the housing. The openings for the USB connection in the housing and the hole for the pressure compensation in the tip are clearly visible.

26.1.3 Description of the Software

The software of the board has two tasks:

- First of all, a waiting time of 1 min is to be realized after switching on the battery supply. During this time, the tip is placed on the rocket and the rocket is started by an igniter. Subsequently, the values for pressure, temperature and acceleration are to be read out every 100 ms and written to the previously deleted flash.
- After connection to the USB interface, data transfer is to be started by terminal input. The data are to be output as semicolon separated values in ASCII and stored as .csv file on the computer.

26.1.3.1 The State Machine
The entire software is implemented as a state machine. The following states exist:

- Waiting
- Delete Flash
- log data
- extract data
- Stop

The state machine is switched in a 100 ms task. The following events are evaluated:

- 60 s expired
- Byte received for flash delete
- Byte received for reading data
- Flash full (FLASH_FULL) or byte received to stop
- Flash completely read out (SEND_COMPLETE)

Thus the state machine can be realized as follows (Fig. 26.4).

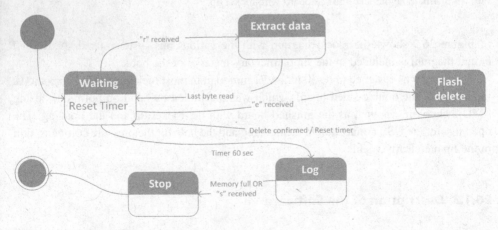

Fig. 26.4 State machine of the rocket logger

26.1.3.2 Logging on the Flash

Logging on the flash happens regularly in a task whose frequency can be set in the timer interrupt. This can be 100 ms, for example.

First, the data is read from the MPL3115 and written to the flash memory. Pressure and temperature are stored in a data set of 5 bytes, of which 3 bytes are pressure and 2 bytes temperature:

```
//setting the OST bit in the Control1 Reg. to update the measured value
MPL3115_Write_ByteReg(CTRL1_REG, maske_TD);
//read out the first 5 bytes (data relevant)
//of the OUT_P_MSB_REG in drucktempdata
MPL3115_Read_DataReg(OUT_P_MSB_REG,drucktempdata,5);
for (int i = 0; i < 5; i++)
{
    SST25PFXX_Write_Byte(SST25PFxx_1, OP_CODE_WRITE_BYTE,
                        ulAddress_pointer, drucktempdata[i]);
    ulAddress++;
}
```

26.1.3.3 Read Out Data

When reading out the data, a counter must globally count the current flash address. At the same time, a total of nine bytes are read from the flash:

- 3 bytes for printing
- 2 bytes for the temperature
- 2 bytes for acceleration in x-direction
- 2 bytes for acceleration in y-direction

```
CntFF = 0;
for (i = 0;i < 9;i++)
{
    fbuf[i] = SST25PFXX_Read_Byte(SST25PFxx_1, OP_CODE_READ_BYTE,
                                ulAddress_pointer);
    if (buf[i] == 0xFF) CntFF++;
    ulAddress++;
}
```

ulAddress_pointer points to the address in the flash ulAddress. If 0xFF is detected more than eight times in succession in the flash, this is a sign that the blank flash area has been reached and the transmission can be aborted. The variable CntFF is responsible for this, which triggers the event SEND_COMPLETE at the end.

The data is read out in such a way that the data is transferred semicolon separated as ASCII values via the serial interface. This allows an existing terminal program to write a .csv file directly, which can be read with Excel.

The pressure is first divided by 64 (shifted to the right by 6 digits), after which it is available in 0.01 mBar.

```
pres = (long)fbuf[0] << 16;
pres |= (long)fbuf[1] << 8;
pres |= (long)fbuf[2];
pres = pres >> 6;
len = Convert2ASCII(pres,tbuf);
TransmitNumber(tbuf,len);
uartTransmitChar(';');
```

Here a separate routine Convert2ASCII(pres,tbuf) was implemented as described in Sect. 2.9.1, instead of itoa(). The same is done with the temperature, which is available in °C * 256, and the accelerations, which are available in 2 m/s².

26.1.4 Evaluation

The flown altitude results from the well-known barometric altitude formula, which can be looked up in any physics book, and which was derived under the assumption of an isothermal atmosphere. With the expected flight altitudes this assumption is still permissible and proves to be valid also with the temperature measurement. With

- M as the mean molar mass of atmospheric gases (0.02896 kg mol^{-1}),
- g as gravitational acceleration (9.807 m s^{-2}),
- R as the universal gas constant (8.314 J K^{-1} mol^{-1}) and
- T as the absolute temperature

is the flown altitude $\Delta h = h_1 \text{-} h_0$.

$$\Delta h = - \frac{R \cdot T}{M \cdot g} \ln\left(\frac{p(h1)}{p(h_0)}\right) \qquad (26.1)$$

This can be easily calculated in Excel from the measured values t (temperature in °C) and p or p (h_0) on the ground, whereby for p only the ratio is important. By applying the barometric altitude formula, flight paths are calculated as shown in Fig. 26.1.

Figure 26.5 shows the acceleration curve in flight direction. The fuel thrust build-up does not reach the full calculated altitude, but in real flight the burn closure occurs about 0.5 s after the calculated time at 2 s. The rocket is then subjected to the acceleration due to

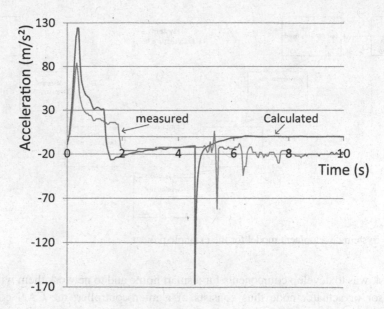

Fig. 26.5 Acceleration curve in flight direction

gravity. In the next moment the rocket is subjected to the acceleration due to gravity, continuing to fly upwards. At 5 s (measured) the parachute is ejected, resulting in a negative acceleration and, in the altitude plot Fig. 26.1, a pressure surge, which is of course misinterpreted as altitude. From here on the rocket falls downwards at the parachute, whereby the acceleration values are no longer plausible because of the position of the rocket tip.

26.2   Smart Home with CAN

In teaching, examples of CAN networks are always sought as examples. Especially in the case of automotive components, it is often difficult to get hold of suitable documentation with which the component can be integrated into a network. However, the devices described in this book can be used to build a wide variety of CAN networks that have both practical and didactic benefits. The following example outlines an application that was successfully tested with students. The students were given a circuit proposal consisting of an ATmega88 controller, a CAN controller of type MCP2515 and a CAN transceiver TJA1050 according to Chap. 10.

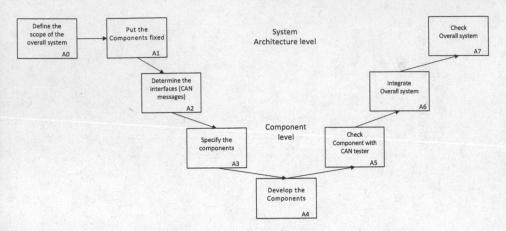

Fig. 26.6 System development model for the example project

The task was to develop components for a smart home and to network them with CAN. Each sensor or actuator node thus consists of a microcontroller, the CAN controller/ transceiver and the respective sensor, actuator or the micro switches mentioned below for configuration, according to the descriptions in the previous chapters. The entire development flow is outlined in Fig. 26.6.

26.2.1 Structure

The Smart Home example project consists of a code lock with a 12-key keypad, a fine dust sensor, a radio transmitter and receiver, a radiator thermostat, a motion detector, a smoke detector, a roller shutter drive with PWM, a thermometer and a central unit with display. In total, there are 13 basic CAN messages Table 26.1.

These are:

- SetTemp: The set temperature is sent from the central unit to the radiator. The message consists of the setpoint temperature between 0 °C and 40 °C in steps of 0.5 K and the identifier of the heating thermostat. If the identification of the heating thermostat is 0xFF, all thermostats are addressed, if it is 0xFE, the thermostat can be taught and its new identification is in byte 3.
- HeatingStatus: The actual temperature (resolution as setpoint) and the identifier are transmitted by the radiator.

Table 26.1 CAN matrix of the Smart Home

Message	Name	CycleTime	Code lock	Fine dust	Radio alarm	Radiator	Move	Smoke	Roller shutter	Thermo/Hygro	Central
0x100	SetTemp	1 s				Rx					Tx
0x101	Heating status	1 s				Tx					Rx
0x200	CodeStatus	100 ms	Tx					Rx			Rx
0x201	CodeDoorLock	Event	Rx								Tx
0x300	DustSensor	30 s		Tx							Rx
0x310	Radio bridge	100 ms			Tx			Rx			Rx
0x400	Move	1 s					Tx	Rx			Rx
0x401	MotionSharp	100 ms					Rx				Tx
0xF01 n	SmokeStatus	100 ms						Tx			Rx
0x500	GlobalTime	1 s							Rx		Tx
0x501	Roller shutter Prog	100 ms							Rx		Tx
0x502	Roller shutter status	1 s							Tx		Rx
0x220	TempHum	1 s				Rx				Tx	Rx

- CodeStatus: The code lock sends a message when three incorrect entries have been made, as well as when the door has been opened. An identifier can be supplied for each lock (analogous to the radiator).
- CodeDoorLock: A message from the control panel to open or close the lock (identifier of the lock plus status). If the identifier is 0xFE, a new identifier must be transmitted in a further byte.
- DustSensor: The fine dust sensor sends its two fine dust values and an identifier (see Chap. 15)
- Wireless bridge: An 868 MHz receiver serves as a wireless bridge to forward messages via CAN. In transmitter mode it forwards a CAN message.
- Motion: A PID infrared sensor is connected to the input of a microcontroller. An identifier is encoded via a bit pattern (DIP switch).
- MotionArmed: The control panel sends one byte and the identifier of the motion detector. In another byte the time until arming in seconds is indicated, in a fourth byte the delay time until the alarm in seconds.
- SmokeStatus: Each smoke detector sends its own CAN message (via 4-bit DIP switch)
- GlobalTime: The time per 1 byte hour, minute, second, day, month, year, weekday is distributed by the control panel.
- ShutterProg: The central sends the time in hour and minute, as well as the weekday and an identifier of the shutter control. If the identifier is 0xFE, a new identifier can be taught via a further byte.
- ShutterStatus: The shutter control sends the current status of the shutter as well as its identifier.
- TempHum: A temperature/humidity sensor sends its data and an identifier (8 bit DIP switch)

The respective messages must then be specified in the project. In the following, only the keypad block for the code lock is shown, as the other circuits result in each case from the previous chapters. It should be noted that a system implemented in this way can only be used for demonstration purposes; in the area of roller shutter controls and heating valves in particular, installations must be implemented in accordance with the relevant standards and there is a risk to life in the event of incorrect handling. Smoke detectors from such laboratory set-ups are also not recognised under insurance law. However, the project is very well suited for training purposes and generated a great deal of commitment among the students. In particular, the project can be used to demonstrate the procedure of a system development. The students learn to first describe the overall system scope, then to define a system architecture in which the interfaces are described in the form of CAN messages. Subsequently, the components are developed and tested independently of each other and finally integrated into a complete system.

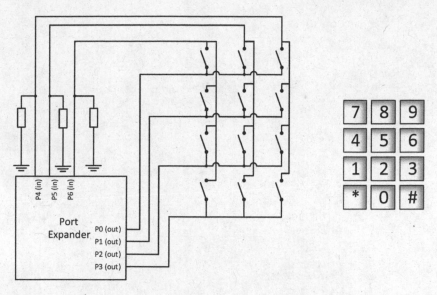

Fig. 26.7 Connecting a keypad block

26.2.1.1 Keyboard Block

The keyboard block is implemented as a matrix. For various reasons, it was decided to use a PCA9534 port expander (Sect. 24.1) for control. In the code lock the correct code (three digits) is preset with 000. The entry is to be made according to the following criteria:

- Entering the correct code plus * opens the door (example: 000*)
- Reprogramming is done by < old code > # < new code > #, example 123#456#.
- After three false entries, an alarm is triggered by sending a CAN message which is received by the control panel and the smoke detectors, causing them to activate their alarm sound.

A red and a green LED are connected to the microcontroller, as well as a driver that switches a door buzzer. The door is opened when the correct code is entered or by a CAN message. The green LED lights up when the door is opened, the red one when a wrong code has been entered.

The reading of the keypad is done scanning as sketched in Fig. 26.7. Three pushbuttons each are combined into one row, the four rows are on a port output, which is set to 0 by default. Now one row each is set to 1 in succession and read out in the columns. If a button is pressed, then a 1 is located at the corresponding column (Fig. 26.7).

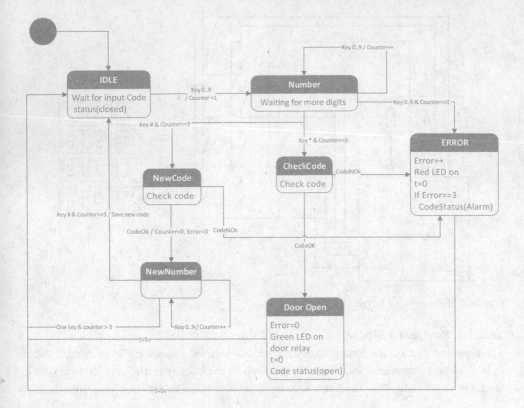

Fig. 26.8 State machine of the code lock

A state machine (Sect. 5.2) must be designed for the sequence.

The code lock sends a cyclic CAN message indicating whether the door is closed, has been opened, or if a wrong entry has been made three times, which also results in an alarm (Fig. 26.8).

Reference

1. Open rocket. http://openrocket.info/. Accessed Feb 2021.

rinted in the United States
y Baker & Taylor Publisher Services